P9-AEZ-334

Glow Discharge Spectroscopies

MODERN ANALYTICAL CHEMISTRY

Series Editor: David M. Hercules
University of Pittsburgh

ADVANCES IN COAL SPECTROSCOPY
Edited by Henk L. C. Meuzelaar

ANALYTICAL ATOMIC SPECTROSCOPY
William G. Schrenk

CHEMICAL DERIVATIZATION IN ANALYTICAL CHEMISTRY
Edited by R. W. Frei and J. F. Lawrence
Volume 1: Chromatography
Volume 2: Separation and Continuous Flow Techniques

COMPUTER-ENHANCED ANALYTICAL SPECTROSCOPY
Volume 1: Edited by Henk L. C. Meuzelaar and Thomas L. Isenhour
Volume 2: Edited by Henk L. C. Meuzelaar
Volume 3: Edited by Peter C. Jurs

GLOW DISCHARGE SPECTROSCOPIES
Edited by R. Kenneth Marcus

ION CHROMATOGRAPHY
Hamish Small

ION-SELECTIVE ELECTRODES IN ANALYTICAL CHEMISTRY
Volumes 1 and 2
Edited by Henry Freiser

LIQUID CHROMATOGRAPHY/MASS SPECTROMETRY
Techniques and Applications
Alfred L. Yergey, Charles G. Edmonds, Ivor A. S. Lewis, and Marvin L. Vestal

MASS SPECTROMETRY: CLINICAL AND BIOMEDICAL APPLICATIONS
Volume 1
Edited by Dominic M. Desiderio

PRINCIPLES OF CHEMICAL SENSORS
Jiri Janata

TRANSFORM TECHNIQUES IN CHEMISTRY
Edited by Peter R. Griffiths

A Continuation Order Plan is available for this series. A continuation order will bring delivery of each new volume immediately upon publication. Volumes are billed only upon actual shipment. For further information please contact the publisher.

Glow Discharge Spectroscopies

Edited by

R. Kenneth Marcus

Clemson University
Clemson, South Carolina

Plenum Press • New York and London

Library of Congress Cataloging-in-Publication Data

Glow discharge spectroscopies / edited by R. Kenneth Marcus.
p. cm. -- (Modern analytical chemistry)
Includes bibliographical references and index.
ISBN 0-306-44396-1
1. Gas discharges. 2. Spectrum analysis. 3. Solid state physics.
4. Chemistry, Analytical. I. Marcus, R. Kenneth. II. Series.
QC711.8.G5G58 1993
543'.0858--dc20 93-10300
CIP

ISBN 0-306-44396-1

A Division of Plenum Publishing Corporation
233 Spring Street, New York, N.Y. 10013

Printed in the United States of America

To the true joys in my life:

Melanie, Drew, Courtney, and Kendalee

Contributors

J. A. C. Broekaert • Department of Chemistry, University of Dortmund, D-4600 Dortmund 50, Germany

Sergio Caroli • Department of Applied Toxicology, Istituto Superiore di Sanità, 00161 Rome, Italy

Douglas C. Duckworth • Analytical Chemistry Division, Oak Ridge National Laboratory, Oak Ridge, Tennessee 37831

Duencheng Fang • Materials Research Corporation, Orangeburg, New York 10962

James M. Harnly • United States Department of Agriculture, Agricultural Research Service, Beltsville Human Nutrition Research Center, Nutrient Composition Laboratory, Beltsville, Maryland 20705

W. W. Harrison • Department of Chemistry, University of Florida, Gainesville, Florida 32611

Kenneth R. Hess • Department of Chemistry, Franklin and Marshall College, Lancaster, Pennsylvania 17604

Hubert Hocquaux • IRSID UNIEUX, BP 50, 42702 Firminy, Cedex, France

F. L. King • Department of Chemistry, West Virginia University, Morgantown, West Virginia 26506

R. Kenneth Marcus • Department of Chemistry, Howard L. Hunter Chemical Laboratories, Clemson University, Clemson, South Carolina 29634

Terry A. Miller • Laser Spectroscopy Facility, Department of Chemistry, The Ohio State University, Columbus, Ohio 43210

Edward H. Piepmeier • Department of Chemistry, Oregon State University, Corvallis, Oregon 97331

Bryan L. Preppernau • Laser Spectroscopy Facility, Department of Chemistry, The Ohio State University, Columbus, Ohio 43210

Philip G. Rigby • School of Biological and Chemical Sciences, University of Greenwich, Woolwich, United Kingdom

Oreste Senofonte • Department of Applied Toxicology, Istituto Superiore di Sanità, 00161 Rome, Italy

David L. Styris • Battelle, Pacific Northwest Laboratory, Richland, Washington 99352

Michael R. Winchester • Inorganic Chemistry Research Division, National Institute for Standards and Technology, Gaithersburg, Maryland 20899

Preface

One of the greatest challenges remaining in the area of analytical atomic spectrometry is the development of more universal methods for the direct analysis of solid materials. While the success of flame and furnace atomic absorption spectrophotometries and inductively-coupled plasma optical and mass spectrometries is undeniable, there still remain great demand and effort directed toward improving the techniques of solid sample dissolution. It is the dissolution step that often limits the ultimate analytical utility of such methods. Dissolution procedures, despite their payback in terms of presenting a homogeneous, easily manipulated (flow injection, chromatography, etc.) sample to the spectrochemical source, can fall short in terms of time constraints or loss of analytical quality due to sample dilution or contamination. Irregardless of such possible problems, aqueous solution nebulization will continue to be the predominant method of sample introduction.

Different from the requirements of just a decade or two ago, contemporary solid materials analysis requires greater powers of detection, speed, and precision than presently available with standard arc and spark technologies. In addition, spatially (depth) resolved elemental profiles are required to assess systems such as galvanized coatings or multilayered automotive glass. It is for these reasons that glow discharge (GD) devices are receiving increased interest within the analytical community. These reduced pressure, inert atmosphere plasmas rely on a cathodic sputtering step to atomize solid samples directly. Subsequent atomic excitation and ionization processes in the adjacent plasma (negative glow) render the devices useful for atomic absorption (AA), fluorescence (AF), emission (AE), and mass spectrometries (MS).

Glow Discharge Spectroscopies is a multiauthored volume having a goal to assess the state of the art in this diverse and continuously expanding area of atomic spectroscopy. Given the wide variety of source geometries, powering schemes, and analytical applications, no single-authored volume could do just service to the field. Each of the authors is acknowledged as a leader in the particular area described in the respective chapters. While each author has written in his own individual style, most of the chapters begin with some historical perspective of the general area and then deal extensively with fundamental aspects relevant to the application at hand. Analytical characteristics and figures of merit are detailed, with some visions of future developments and applications concluding each chapter. It is hoped that this volume will be useful as both a survey of the use of glow discharge devices in atomic spectroscopy and as a technical reference for practitioners in the field.

As editor, I would like to express my appreciation to the authors (each of whom was my first choice for the respective topics) for their thorough and thoughtful contributions. As this was my first time undertaking such a challenge, I greatly appreciate the staff at Plenum Publishing under the direction of Amelia McNamara who were most helpful and patient.

R. Kenneth Marcus

Clemson, South Carolina

Contents

1. Introduction

R. Kenneth Marcus

2. Fundamental Plasma Processes

Duencheng Fang and R. Kenneth Marcus

3. *Atomic Absorption and Fluorescence Spectroscopies*

Edward H. Piepmeier

4. *Atomic Emission Spectrometry*

J. A. C. Broekaert

5. *Glow Discharge Mass Spectrometry*

F. L. King and W. W. Harrison

6. *Hollow Cathode Discharges*

Sergio Caroli and Oreste Senofonte

7. *Analysis of Nonconducting Sample Types*

Michael R. Winchester, Douglas C. Duckworth, and R. Kenneth Marcus

8. *Thin Film Analysis*

Hubert Hocquaux

9. Discharges within Graphite Furnace Atomizers

James M. Harnly, David L. Styris, and Philip G. Rigby

10. Laser-Based Methods

Kenneth R. Hess

11. Laser-Based Diagnostics of Reactive Plasmas

Bryan L. Preppernau and Terry A. Miller

1

Introduction

R. Kenneth Marcus

1.1. Rationale

Driven by advances in materials research, government regulations, and interdisciplinary collaborations, analytical chemistry has been one of the most active areas of chemical research over the last three decades. Fundamental advances in the fields of environmental modeling, medicine, and materials science are evidence that the discipline has met many of the challenges that have been posed. In developing methods, analytical chemists have become more adept at interpreting and implementing the findings of biologists and applied physicists. The role that computers have had in revolutionizing chemical instrumentation and techniques cannot be underestimated. Advances in analytical chemistry have shown up in the vernacular of the trade; units of quantitation have switched from weight percent to parts per billion (and trillion). Simple functional-group analysis has been replaced by methods of extraordinary selectivity and specificity. Each step toward better sensitivity or specificity in analytical measurements can, in many cases, cause more difficulties than provide answers. For example, an analytical method can be too sensitive for many uses and environments. This volume addresses some of the advances that have taken place in the area of elemental spectrochemical analysis.

Glow discharge devices were some of the earliest spectrochemical sources, first used in fundamental spectroscopic studies of atomic structure in the 1910s.[(1-3)] From that time until the 1960s, their application in analytical

R. Kenneth Marcus • Department of Chemistry, Howard L. Hunter Chemical Laboratories, Clemson University, Clemson, South Carolina 29634.

Glow Discharge Spectroscopies, edited by R. Kenneth Marcus. Plenum Press, New York, 1993.

chemistry was very limited. Over the last 30 years, the applications of these devices have grown slowly but steadily with commercial instrumentation now being readily available. Their primary uses have been as line sources in atomic absorption spectrometry and in bulk solids elemental analysis by atomic absorption, emission, and mass spectrometries. The forte of the devices is their ability to allow the direct elemental analysis of materials in the solid state. It is the purpose of this book to outline the developments in analytical applications of glow discharge devices over the last two decades and to highlight future trends as the techniques continue to evolve. Experts in the various applications of glow discharge devices have contributed to this volume and put their own applications into perspective with competing and complementary methods. It is hoped that the reader will begin to gain an appreciation for the fundamental processes occurring in glow discharges and also the scope of the current and expected analytical applications.

1.1.1. Scope of Elemental Analysis

The challenges facing all of analytical chemistry can at times seem most complicated in the area of elemental analysis. Obviously, the gross elemental composition of a given system (material) defines its chemical and physical characteristics and in the end, its utility. Beyond elemental quantitation, additional information such as ionic speciation, isotopic composition, and spatial distribution may be of importance. These other considerations take elemental analysis far beyond the determination of "X-in-Y." In many instances, it is the presence and concentration of an alien element that is of interest, making minor and trace analysis of utmost importance in many applications.

Often, generation of the appropriate elemental information is complicated by the variety of sample types that the atomic spectrometrist is presented. Table 1-1 is a short compilation of the analytical systems (sample types) that often require elemental analysis. It is when the atomic spectrometrist is faced with this wide array of sample matrices that one can see why

Table 1-1. Sample Types Requiring Elemental Analysis

Metals and alloys
Geological specimens
Potable, waste, and ground waters
Glasses and ceramics
Biological specimens
Semi- and superconducting materials
Catalyst materials
Composites and polymers
Thin films

sample preparation and introduction continue to be at the forefront of spectrochemical research. In fact, one of the most important considerations in deciding between the various atomic spectrometry techniques is their respective susceptibility to matrix effects. In many cases, the sample form will dictate the ultimate choice in methodology.

1.1.2. Aqueous Sample Elemental Analysis

Going through Table 1-1, it is first evident that there are a wide variety of sample matrices that must be covered, making the possibility of using a single analytical method or instrument for all of these systems extremely unlikely. Municipal, industrial, and environmental regulations are such that the third entry (waters) is by far the most prevalent of the sample types. Human body fluids (e.g., serum, urine) must also be among the most common of elemental analysis matrices. Beyond basic elemental analysis, these essentially aqueous samples may require that elemental concentrations be broken down into specific oxidation states (i.e., speciation). Sample preparation generally involves nothing more complex than an organic extraction step or possibly analyte preconcentration with an ion exchange polymer or the like. Advantages in direct solution analyses include ease in calibration, sample homogeneity, and ease of sample delivery (automation and throughput). Even so, the introduction of aqueous solutions continues to be the "Achilles' heel" in atomic spectroscopy.[4] The relatively new approaches incorporating flow injection[5] or ion chromatography[6] have become powerful tools for sample introduction particularly in dealing with matrix effects and allowing on-line speciation. It is these advantages that have led to the near-total dominance of solution-based analytical methods in the atomic spectrometry market.

A number of excellent review articles have been written that are devoted almost exclusively to those techniques that are best suited for solution-phase samples.[7–9] The most widely applied of these methods are flame and furnace atomic absorption spectrophotometry, and inductively coupled plasma atomic emission and mass spectrometries. As is the case of the sample forms, techniques that rely on aqueous sample nebulization have dominated the field since the early 1960s (as opposed to solids techniques). Flame atomic absorption spectrophotometry (AAS) was the first commercially available of this group and continues to be the most widely applied. Samples are introduced by pneumatic nebulization under continuous solution flow or by discrete injection methods. Furnace AAS was first introduced as a supplementary atomization source to the conventional flame, sold for those specific applications where sample volumes are limited (< 0.1 ml). Samples originally in the liquid state are dried in the furnace and then atomized as residues. Currently, the technique is acknowledged as being the most sensitive in terms of absolute analyte mass. Inductively coupled plasmas (ICP) are generated by coupling radio frequency (multiples of 13.56 MHz) energy (1–2.5 kW) to a flow of argon discharge gas. The resulting plasma is characterized by gas temperatures that reach upwards of 5000 K. As such, the ICP is very efficient at breaking down the most refractory species to free atoms. The atomic emission mode has the distinct advantage of providing truly simultaneous multielement analysis when direct reading spectrometers are employed. Since its first commercial availability in the early 1980s, ICP mass spectrometry has become the class of the atomic spectrometry field in terms of "bulk solution" sensitivity. It is this group of time-tested solution analysis methods that make aqueous sample introduction the most popular means of elemental analysis, regardless of the initial sample form. It must be emphasized, in discussions of the competing direct solids methods, that figures of merit for these techniques are quoted in terms of mass per unit of solution *after* ~100–1000× dilution [i.e., 1 ppm (μg/ml) in solution is 0.01–0.1% by weight in the solid].

Beyond the liquid matrix samples presented in Table 1-1, there are a very wide variety of solid sample types that require elemental analyses. These matrices can be electrically conducting or nonconducting (to one degree or another), organic or inorganic, thermally stabile or heat sensitive, and either homogeneous or have a well-defined structure. From an atomic spectroscopy standpoint, the possible mechanisms or steps required to generate a free atom for analysis are greatly complicated by the physical properties of the sample. Indeed, if one is solely equipped for the analysis of aqueous samples, by what method can the sample type be placed into solution? Certainly, the number of methods and range of difficulties in sample dissolution are as wide as the number of possible matrices.[(10)] Sample preparation can range from simple nitric acid dissolutions to sodium or lithium fusions followed by treatment with HF. The diversity in procedures is particularly apparent in the analysis of refractory materials, glasses, and ceramics where acidic or basic media may be employed depending on the composition/mineralogy. In fact, the dissolution procedures for specialty alloys such as those used in the aerospace industry (e.g., Al, Ti) can be very complex in their own right. Fortunately, there is extensive documentation provided by instrument manufacturers and in the analytical literature which covers almost any case of sample dissolution. In addition, the advent of microwave technology has added a great deal of power and speed to the sample preparation process, often reducing digestion times from days to hours and minutes.[(11)]

Despite any complexities found in solid sample dissolution procedures, the analyst gains all of the benefits of solution sample analysis including ease in quantitation (e.g., calibration curves, standard additions) and the use of chromatographic or flow injection introduction schemes. Possibly of great importance is the inherent matrix homogenization of aqueous samples.

Many of the complicating factors with regard to obtaining representative sampling from solids are circumvented by placing the solid samples into solution. Finally, dissolution of a solid sample leaves the analyst with a matrix with which he or she is well accustomed.

1.1.3. Direct Solids Elemental Analysis

Given the many attributes of solution-based sample analysis methods, there are a number of factors that the analyst must consider in determining the ultimate procedure for the analysis of a particular solid sample type. This decision goes far beyond the relative precision, accuracy, and sensitivities of the respective methods. In spite of the wealth of sample preparation methods, the bottom line in many production laboratories is sample turnaround time. Therefore, ease of sample preparation can be a determining factor for those laboratories contemplating solution phase analyses of solid materials. The classic example of the necessity of speed is in metallurgical foundries, where analytical time frames are on the order of minutes rather than hours. In high-purity material applications, time may not be as important as absolute sensitivity. Sample dissolution procedures dilute the apparent analytical concentrations of analytes and, in some instances, result in the introduction of contaminant species at levels exceeding the trace impurities. The final consideration in the choice of methodology in solids analysis is the desire to obtain regio-specific information, either laterally (x/y) or depth resolved (z). If such information is desired, then aqueous sample preparation is practically obviated. It is these special considerations that have led to the development and continued use of direct solids elemental analysis techniques.

1.1.3.1. The Ideal Technique for Direct Solids Elemental Analysis

As is the case for spectrochemical sources employed for solution-based analyses, each of the solids techniques has its unique analytical characteristics. Clearly, the range of physical and chemical properties of solid samples places much more stringent demands on the applicable spectrochemical methods than for liquid matrix samples. The number of possible solid sample matrix types, each of which requires special sample preparation and analytical standards, makes solid-sampling techniques much more specialized than liquid-sampling methods. Borrowing in part from other atomic spectrometry reviews,[12,13] Table 1-2 lists a series of characteristics that the ideal spectrochemical source for solids elemental analysis would possess. In choosing the most appropriate solids source for a given application/laboratory, each method should be assessed relative to the ideal case.

In going through the characteristics of the ideal source, the first six entries refer to the atomization properties of the technique. As is the case

Table 1-2. The Ideal Spectrochemical Source for Solids Elemental Analysis

1. Applicability to all possible sample matrices (bulk and particulate, electrically conductive and nonconductive)
2. Easily controlled atomization/excitation/ionization rates
3. Stoichiometric atomization for all elements within a sample
4. Nondestructive
5. Rapid, high-precision depth profiling
6. High degree of lateral resolution
7. No sample preparation
8. Easy to operate
9. Useful for all elements under the same operating conditions
10. Low capital and maintenance costs
11. Total elemental coverage
12. Infinite linear dynamic range (1 atom to 100%)
13. Easy to quantitate (no matrix effects)
14. Multimode operation (e.g., AA, AE, MS)
15. Ability to analyze liquid and gaseous samples

with any sample introduction scheme for atomic spectrometry, the success or failure of the method can be determined solely by the atomization efficiency of the device. In direct solids analysis, this is of course a tall order because of the wide variety of possible matrices. Depending on the actual atomization mechanism, the sample may be an integral part of the method of signal generation; as such, the electrical and mechanical characteristics of the sample may determine the applicability of a particular method, or at least the rate at which sample constituents are evolved. In the best of all worlds, the technique would be nondestructive. Unfortunately, because of the limited sampling depth of charged particle/photon probing techniques, surface erosion is required to generate depth-resolved analyses.

Characteristics 7 through 10 refer to the user friendliness of the method. Given the availability of solution-based spectrochemical methods, any solids technique must be more convenient to use than the corresponding sample dissolution scheme. This includes sample preparation, optimization of instrument response characteristics, and the actual purchase and maintenance costs of the instrument. Items 11–14 deal most specifically with the basic analytical characteristics of the techniques. While the desire for total elemental coverage is a noble one, it is almost an impossibility given the fact that most atomic spectrochemical techniques employ support gas atoms/ions to generate the analytical signal. The qualities of infinite dynamic range and lack of matrix effects are the most daunting of the characteristics, but are at the same time possibly the most desired for any analytical system. Techniques of direct solids analysis do not have the luxuries of sample dilution or preconcentration inherent to solution analysis methods, nor is outright matrix removal possible. The complete elemental makeup is continuously present during the analysis, with the matrix element(s) dictating the energy absorption and atomization characteristics of all the constituents, inflicting spectral and excitation/ionization interferences as well. In short, the minor and trace species would ideally be separated from the matrix by some mechanisms before the actual analysis.

The final analytical quality would be the capability to operate in more than one sampling mode. For example, in some instances only bulk, single-element analysis might be required, making atomic absorption (AAS) a good option. Of course, not all elements are well suited for study by AAS. Atomic emission spectroscopy (AES) is very useful for multielement analysis, providing the capabilities for depth-resolved analysis when polychromators are employed. As a final example, mass spectrometry (MS) is the most practical option for those instances where isotopic information of constituent elements is needed. Thus, the flexibility of the device and its ability to generate specific types of information are favourable characteristics.

In the best of all worlds, the source could be operated as an auxiliary "solids" source directly compatible with commercial "solution" instrumentation. The final characteristic of the ideal solids source is the ability to analyze solutions (residues) and gases in addition to solid materials. Admittedly, this is turning the table on the solids device, but in the ideal world the capability to accept all sample types is desirable.

1.1.3.2. Comparison of Solids Techniques

In the real world, the analyst is often forced to make a number of compromises from the ideals of Table 1-2. The analyst must be willing to admit that no single technique can meet all of his or her analytical objectives. The realistic points of comparison are far more traditional and must take into account the current available technologies. Table 1-3 is a compilation

Table 1-3. Ten Points of Comparison for Solids Elemental Analysis Techniques

1. Accuracy
2. Precision
3. Limits of detection
4. Dynamic range
5. Matrix effects
6. Elemental uniformity
7. Sample pretreatment
8. Analysis time
9. Depth profiling
10. Cost and complexity

of ten practical points of comparison for direct solids elemental analysis techniques. In essence, this list is the same as comparable compilations for any set of analytical techniques. The only glaring difference is the preferred ability to perform depth-resolved analyses.

Accuracy, precision, and limits of detection are as easily defined for solids as they are in solution analyses. Entries 4–6, on the other hand, take on new meaning in direct solids analysis. In the presence of matrix elements ranging from 10's to 100%, dynamic range and matrix effects are virtually out of the analyst's control. As mentioned previously, separation of analytes from the sample matrix along with its concomitant spectral and spectrochemical influences is not generally an option. In all of the techniques to be discussed presently, the sample matrix has a direct influence on the absolute analyte sensitivity (signal per unit concentration). In the majority of instances, the sample matrix affects sample introduction (atomization) rates both between different samples (matrices) and individual elements within a given sample. Alternatively, in x-ray fluorescence, electron spectroscopies, and secondary ion mass spectrometry, the matrix controls the yield of secondary particles that reach the detector. The overabundance of matrix atoms in a spectrochemical plasma will most probably affect the excitation/ionization characteristics of the device. It is this propensity for matrix effects of one form or another that has limited the greater application of most direct solids techniques.

If there are any straightforward arguments for direct solids analysis it is the possible freedom from sample pretreatment and subsequent reduction in total analysis time. The bottom line is often "time is money." This point can go a long way for solids methods as long as there are no large compromises to be made in the quality of the analytical data. In many laboratory settings, the fact that a sample need not be dissolved is insignificant in comparing the different solids techniques. In some instances, the actual analysis time (from sample in to numbers out) can be a determining factor.

The ninth item in Table 1-3 is one of increasing importance. In many applications, simple bulk analyses are not as important as the spatial distribution of sample constituents. Such requirements are no longer limited to the semiconductor industry, but are spreading into fields such as the automotive industry. In that field, depth-resolved elemental analyses are required of plated bumpers, painted sheet metal, and tinted safety glass (note the wide range of matrices and coatings). In the vast majority of such applications, the materials are constructed as multilayered systems that require depth profiling of "matrix" species along with the detection of trace species at the same time. For example, analysis of a galvanized coating on a steel base material would require both the determination of the thickness of that layer (Zn) and also the presence of contaminant species such as oxygen that may be located at the Zn/steel interface. Therefore, the requirements are the ability to make proper depth assignments and concentration determinations simultaneously over the full range of trace to matrix concentrations. A final set of depth profiling considerations is the resolution of the profiling and the speed at which the profiles can be generated. These two quantities are for the most part mutually exclusive with the criteria varying widely with applications.

The final points of comparison for solids elemental analysis methods are the cost and complexity of the instrumentation. In general, these two attributes are directly related to one another. The range of costs is from 10s to 100s of thousands of dollars, with the corresponding level of operator expertise ranging from technicians of moderate skill to Ph.D.-level scientists.

Table 1-4 summarizes the most common commercially available types of direct solids elemental analysis techniques along with their relative analytical figures of merit and characteristics. As described in the previous section, the salient points of comparison for the analytical chemist are sample turn-around time, freedom from matrix interferences, matrix compatibility, sample size, instrumentation complexity, ease of quantitation, and depth-resolving capabilities. The reader is encouraged to refer to the scientific literature and spectroscopy monographs for more detailed descriptions of the respective methods.

A cursory comparison of the characteristics of the techniques in Table 1-4 indicates that there are a number of methods that are ideally applied in

Table 1-4. Techniques for Direct Solids Elemental Analysis[a]

Technique	Accuracy	Precision	LOD	Dynamic range	Matrix effects
High-current arc AES	++	+	++	+	+
High-voltage spark AES	++	++	+	+	+
X-ray fluorescence	+++	+++	+	+	+
Secondary ion MS	++	++	+++	+++	+
Laser ablation MS	+	+	++	++	+
Laser ablation ICP-MS	++	++	++	++	+
Glow discharge AES/MS	+++/++	+++/++	++/+++	++/+++	++/++
	Elemental uniformity	**Sample pretreatment**	**Analysis time**	**Depth profiling**	**Cost and complexity**
High-current arc AES	++	+++	+++	NA	++
High-voltage spark AES	++	+++	+++	NA	++
X-ray fluorescence	+	++	+++	+	+++
Secondary ion MS	+	+	+	+++	+
Laser ablation MS	+	+++	++	+	++
Laser ablation ICP-MS	++	+++	++	+	+
Glow discharge AES/MS	++/+++	++/+	++/+	++/++	++/+

[a]+, unfavorable; ++, moderate; +++, favorable; NA, not applicable.

certain analytical situations. For example, spark emission analyses are well suited for rapid survey of metallurgical samples with limits of detection on the order of 1 ppm. They are limited by fairly severe matrix matching requirements and a lack of depth-profiling capabilities. These capabilities are nearly ideal for foundry applications. Secondary ion mass spectrometry (SIMS) is characterized as having excellent depth-resolving capabilities along with high absolute sensitivities. Unfortunately, SIMS is a rather slow technique that requires a high level of operator skill. Thus, the technique finds its major use in the analysis of materials associated with the production of integrated circuits.

Of the techniques listed in Table 1-4, glow discharge spectroscopies seem to have the most advantageous characteristics across-the-board. As such, there is growing interest in the devices and their applications across a broad range of analytical problems. Of current interest are the analysis of bulk metals and alloys, the depth-resolved analysis of layered metallic systems, and the development of new methods for the analysis of nonconductive materials such as geological materials, glasses, and ceramics. Throughout this volume, the operating principles and applications of glow discharge devices will be described in detail and compared with the competitive methods listed here.

1.2. Glow Discharge Devices: Basic Operating Principles

Glow discharge devices would seem to offer a number of advantages over other methods in the area of direct solids elemental analysis. In order to familiarize the reader with the scope of this text, it is instructive to discuss briefly the basic operational aspects of the devices. A much more detailed description of glow discharge processes is presented in Chapter 2.

Glow discharge devices are reduced-pressure, inert-atmosphere, gaseous conductors.[14,15] A depiction of their most simple geometry and structure is presented in Fig. 1-1. A glow discharge is initiated by the application of a

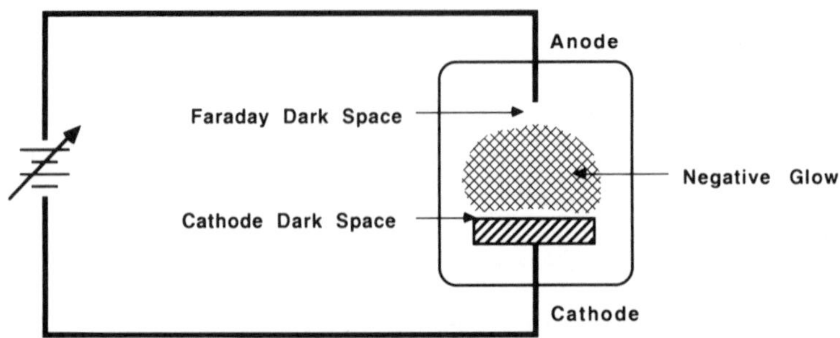

Figure 1-1. Schematic representation of a simple glow discharge device.

sufficiently high voltage between two electrodes in contact with the discharge gas (most typically Ar). The potential difference (250–2000 V) causes the breakdown of the discharge gas to form positively charged ions and free electrons. The relative potentials on the cathode (−) and anode (+) result in the establishment of electric field gradients such that positively charged ions are accelerated to the cathode surface. The impinging ions transfer their momentum to the surface (and lattice), setting off a cathodic sputtering event. The products of the sputtering process are ejected atoms and small clusters of cathode material, ionic species, and secondary electrons. The process of cathodic sputtering is the means of sample atomization and the basis for depth-resolved analyses. The secondary electrons sustain the discharge through gas-phase ionization of sputtered material and discharge gas atoms. Electrons are also efficient at producing excited state atoms of the sputtered and discharge gas atoms. Evidence of these electron impact collisions is seen in the characteristic luminous "negative glow" region. Important as well in bulk plasma ionization are Penning-type collisions between highly excited, metastable discharge gas atoms and neutral atoms of the sputtered material. The result of these collisions is the formation of ions of the sputtered atoms.

In summary, glow discharge devices are inherently capable of generating a representative atomic population of a solid sample and producing both excited state and ionic populations of those atoms as well. Thus, as shown in Fig. 1-2, the devices are directly applicable for bulk and depth-resolved analyses by the traditional techniques of atomic absorption, emission, and mass spectrometries. In addition, interaction of external light sources with the sputtered populations allows for the use of the devices in atomic fluorescence, optogalvanic effect, and resonance ionization spectroscopies. As can be seen, glow discharge devices are indeed some of the most versatile of the elemental analysis methods.[16]

1.3. Volume Outline

In an effort to cover the most relevant applications of glow discharge spectroscopies in the most informative way, the chapters of this volume have been written by acknowledged research and application leaders in the respective areas. It is they who are the most up-to-date regarding literature coverage and can furnish the most insight into the nuts and bolts of source operation in their particular fields. The chapters have been arranged so as to build first on the fundamentals of glow discharge operation, discuss the general application of the devices in the various atomic spectrometric modes, and then look at specific fields of application. The authors have been encouraged to provide historical background in these areas as appropriate and also

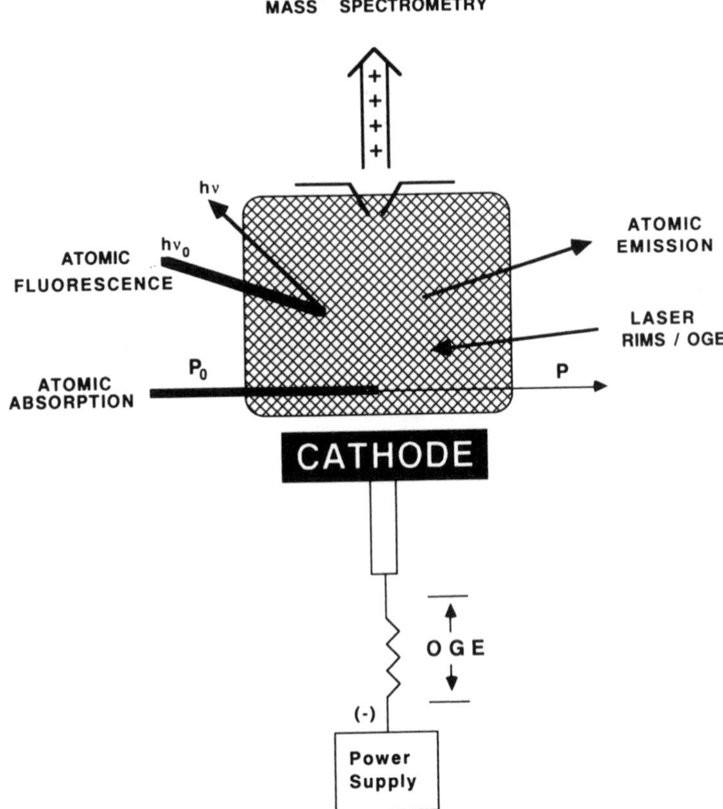

Figure 1-2. Analytical sampling modes of glow discharge sources.

to speculate on future directions in which they see advances being made. In an effort to give completeness to the volume, spectroscopic methods of interrogating molecular plasmas such as those employed in plasma etching and deposition systems are also included.

The fundamental processes occurring within glow discharge devices are covered in detail in Chapter 2. The authors, Duencheng Fang (Materials Research Corporation) and R. Kenneth Marcus (Clemson University), discuss in detail the fundamental aspects of cathodic sputtering in terms of the role that the choice of discharge gas and source operating parameters have on sample atomization. Methods for characterizing electron populations in the plasma negative glow are described. Gas-phase collisional processes including electron impact and Penning-type interactions are discussed with reference to creating excited state and ionic populations of the sample atoms.

It is hoped that this chapter will provide for a better understanding of the plasma processes that are manifested in the analytical applications.

Professor Edward Piepmeier (Oregon State University) describes in Chapter 3 the application of glow discharge devices in atomic absorption and fluorescence spectroscopies. In these applications, the devices are employed as atomization sources for bulk solid and solution residue analysis. The reduced-pressure, inert plasma is highlighted as being advantageous in providing a nearly ideal environment for such studies.

The most widespread commercial application of glow discharge devices is in the area of atomic emission spectroscopy. In Chapter 4, Professor J. A. C. Broekaert (University of Dortmund) describes the fundamental aspects of emission spectroscopy and how these plasmas generate useful emission spectra. Also described is the evolution of the common Grimm-type cell geometry and its many applications in bulk solid analysis. Methods of modifying the basic design in order to optimize the source characteristics are also discussed.

Chapter 5 concerns the design considerations and applications of glow discharge devices in the area of mass spectrometry. Professors Fred L. King (West Virginia University) and Willard W. Harrison (University of Florida) review the pertinent plasma processes responsible for the ionization of sputtered atoms and the roles of plasma parameters, operating modes, and mass analyzers in producing quantitatively useful mass spectra. Applications in the analysis of bulk metals, geological materials, and thin-film systems are highlighted. Finally, the use of the devices in the area of molecular (organic) mass spectrometry of gaseous or liquid samples is described.

Hollow cathode discharge geometries are the oldest and most widely studied of the glow discharge sources. Drs. Sergio Caroli and Oreste Senofonte (Istituto Superiore de Sanità) outline, in detail, the historical development of hollow cathode atomic emission devices in Chapter 6. Applied primarily as line sources in atomic absorption spectrophotometry, the devices have in fact been applied extensively in elemental analysis of bulk solids, solution residues, and gases. The hollow cathode geometry is also well suited for the analysis of metal filings and compacted nonconducting samples.

One of the limiting features of glow discharge devices is the requirement that the cathode (sample) be electrically conductive in nature. Drs. Michael R. Winchester (National Institute for Standards and Technology) and Douglas C. Duckworth (Oak Ridge National Laboratory), and Professor R. Kenneth Marcus (Clemson University) describe in Chapter 7 two basic approaches to the analysis of nonconducting sample types. First, methods of compacting oxide powders in metal powder matrices are outlined, with emphasis placed on sample preparation considerations. Second, the operating principles and preliminary applications of radio-frequency-powered glow discharge devices are presented. The radio frequency discharges allow for direct analysis of both conducting and nonconducting samples.

The ability of glow discharge devices to remove successive layers of cathode material makes them applicable for depth-resolved elemental analyses of metals and semiconductors. Dr. Hubert Hocquaux (IRSID UNIEUX) describes this specialized area in Chapter 8. In particular, those source design and operation considerations for atomic emission monitoring of such systems are detailed. The special requirements of optical spectrometers in these applications are also described.

In Chapter 9, Drs. James Harnly (USDA–NCL), David Styris (Pacific Northwest Laboratory), and Philip Rigby (University of Greenwich) present the applications of glow discharge atomic emission devices for the analysis of solution residues, a growing area of interest. The techniques of furnace atomization nonthermal excitation spectroscopy (FANES) and furnace atomization plasma emission spectroscopy (FAPES) rely on sample volatilization in a graphite furnace (tube) with subsequent excitation within the hollow cathode discharge. As such, the techniques offer high-sensitivity, multielement analysis of volume-limited samples.

The use of high-intensity, monochromatic laser light sources has opened up a number of analytical and diagnostic opportunities in the field of glow discharge spectrometries. Professor Kenneth R. Hess (Franklin and Marshall College) describes in Chapter 10 the wide variety of laser/plasma interaction schemes and the types of information they generate. The techniques available include resonance ionization, optogalvanic effect spectroscopy, and laser ablation of cathode material.

In Chapter 11, Bryan L. Preppernau and Professor Terry A. Miller (The Ohio State University) review the optical techniques applied to those types of glow discharges used in the semiconductor industry. These molecular plasmas require extensive characterization of particle densities and energies. The laser-based methods applied in these systems include single- and multiphoton fluorescence (LIF), optogalvanic spectroscopy, and absorption spectroscopy.

In the development of this volume, it was the editor's goal to put together a comprehensive overview of the quickly expanding area of glow discharge spectroscopies. It is hoped that the volume will be a useful reference source for those entering the field and practitioners alike.

References

1. F. Paschen, *Ann. Phys.* 50 (1916) 901.
2. H. Schuler, *Z. Phys.* 59 (1929) 149.
3. H. Schuler, and J. E. Keystone, *Z. Phys.* 72 (1931) 423.
4. R. F. Browner and A. W. Boorn, *Anal. Chem.* 56 (1984) 786A.
5. J. F. Tyson, *Spectrochim. Acta Rev.* 14 (1991) 169.

6. A. Siriraks, H. M. Kingston, and J. M. Riviello, *Anal. Chem.* 62 (1990) 1185.
7. M. L. Parsons, S. Major, and A. L. Forster, *Appl. Spectrosc.* 37 (1983) 411.
8. G. Tölg, *Analyst* 112 (1987) 365.
9. W. Slavin, *Spectroscopy* 6(8) (1991) 16.
10. R. Bock, *A Handbook of Decomposition Methods in Analytical Chemistry*, Wiley, New York, 1979.
11. H. M. Kingston and L. B. Jassie, *Introduction to Microwave Sample Preparation: Theory and Practice*, American Chemical Society, Washington, D.C., 1988.
12. G. M. Hieftje, *Fresenius Z. Anal. Chem.* 337 (1990) 528.
13. J. D. Ingle and S. R. Crouch, *Spectrochemical Analysis*, p. 227, Prentice–Hall, Englewood Cliffs, N.J., 1988.
14. A. M. Howatson, *An Introduction to Gas Discharges*, Pergamon Press, Elmsford, N.Y. 1976.
15. B. Chapman, *Glow Discharge Processes*, Wiley–Interscience, New York, 1980.
16. *Analytical Atomic Spectrometry: Use and Exploration of Low Pressure Discharges* (P. W. J. M. Boumans, J. A. C. Broekaert, and R. K. Marcus, eds.), *Spectrochim. Acta* 46B, Nos. 2 and 4 (1991).

2

Fundamental Plasma Processes

Duencheng Fang and R. Kenneth Marcus

2.1. Gaseous Discharges

When a sufficiently high voltage is applied across two electrodes immersed in a gaseous medium, atoms and molecules of that medium will break down electrically, forming electron–ion pairs and permitting current to flow. The phenomenon of current flowing through a gaseous medium is termed a "discharge," also known as a plasma. The breakdown of the medium is characterized by the transition of the gas from a poor electrical conductor with resistivity (resistance × area/separation) of some 10^{14} ohms · m to a good conductor with resistivity (dependent on the particular conditions) of $\sim 10^3$ ohms · m. The potential difference between electrodes at which this transition occurs is called the breakdown potential, V_b, which depends on the identity and density of the gas, the electrode material and interelectrode separation, and the degree of preexisting ionization.

Breakdown of this gas results in the formation of positively and negatively charged ions and electrons. Electric fields developed between the electrodes dictate that the positively and negatively charged species be accelerated toward the cathode (−) and the anode (+), respectively. After the initial breakdown of the support gas, collisional processes within the discharge serve to produce more charged species. At this point the discharge is said to be "self-sustaining" provided that a suitable voltage, V_n, which is usually lower than the breakdown voltage, is continually applied.

Duencheng Fang and R. Kenneth Marcus • Department of Chemistry, Howard L. Hunter Chemical Laboratories, Clemson University, Clemson, South Carolina 29634. *Present address of D.F.*: Materials Research Corporation, Orangeburg, New York 10962.

Glow Discharge Spectroscopies, edited by R. Kenneth Marcus. Plenum Press, New York, 1993.

Discharges may be classified as a function of their voltage–current characteristics.[1] Figure 2-1 illustrates the most common types of discharges and their operating regimes. Of the three major classifications, the Townsend discharge, the glow discharge, and the arc discharge, only the latter two have been applied extensively in analytical chemistry. In Fig. 2-1, V_b is the breakdown voltage, V_n is the normal operating voltage, and V_d is the operating voltage of arc discharge.

The electrical characteristics of a gas discharge can be best understood by beginning with the Townsend discharge regime. This discharge is generally operated in the submillitorr pressure regime and is characterized by having only a small degree of ion and free electron production. Following the Townsend discharge is a transition region, resulting from the increased energy exchange through collisions (owing to higher gas pressures), wherein the electrical current increases while actually decreasing the required discharge maintenance voltage.

After the transition region, a luminous glow forms between the electrodes and is accordingly named a "glow discharge." At the onset of the glow discharge regime, increases in the current do not change the current density because the cathode surface is only partially covered by the discharge; as such, no increase in voltage is required. This is classified as the "normal" glow discharge regime. As shown in Fig. 2-2, as the current is

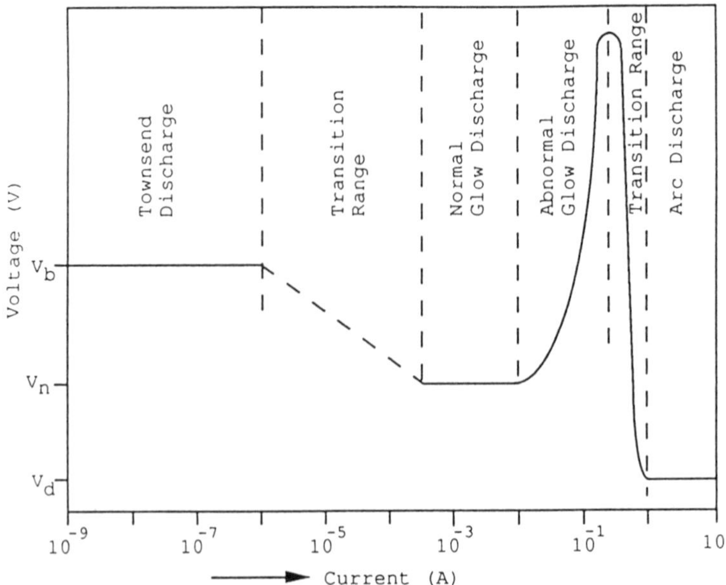

Figure 2-1. Operating regimes of various gas discharges. Adapted from Howatson.[1]

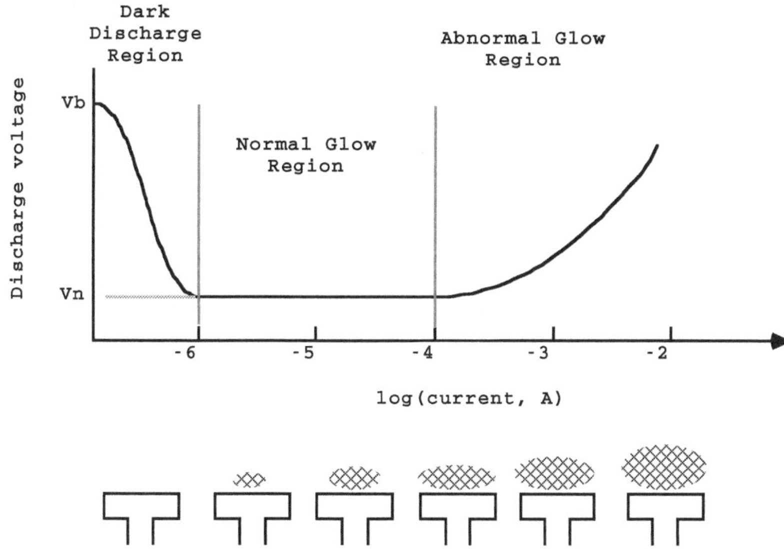

Figure 2-2. Transition between normal and abnormal glow discharge operation.

further increased, the discharge glow will eventually cover the entire cathode surface. At this point, any increases in discharge current will result in an increase in current density, requiring an increase in the discharge voltage. Discharges that display this type of increasing *i–V* relationship are termed "abnormal" glow discharges. It is the abnormal glow discharge mode that is used most often in atomic spectroscopy. Analytical glow discharge devices generally operate in reduced pressure (0.1–10 Torr), inert gas atmospheres and at powers of less than 100 W.

As the current is increased further in the glow discharge, the current density becomes so high that intensive heating of the cathode through bombardment by filler gas ion species causes thermal vaporization of the cathode. Under these conditions, the availability of high number densities of analyte perturbs the potential fields and the *i–V* characteristics of discharge become "normal," i.e., the current then increases while decreasing the required discharge voltage, as is the situation for a dc arc.[(2)]

Usually operating at atmospheric pressure, the dc arc is characterized by its large currents and bright discharge plasma. At typical operating currents, 10–1000 A,[(3)] the cathode surface is heated to the point that thermionic electron emission becomes a prominent current-carrying mechanism. If the cathode is made of some type of metal alloy, cathode heating will result in vaporization of large amounts of material. Gas temperature in arc discharge can be up to 2×10^4 K with charged particle densities up to 10^{16} cm^{-3}. The combination of high vaporization rates and collisionally energetic plasma has made dc arcs a mainstay in analytical spectrochemical analysis of metallurgical samples.[4–6]

2.2. Glow Discharges

Glow discharge devices have a rich history of use dating back to 1912.[7] Since their introduction, experimenters have taken advantage of the glow discharge's low operating powers and rich collisional environment for such studies as atomic structure.[8] Another important early application of glow discharges was in the first-generation mass spectrometers.[9] As will be discussed in detail in subsequent chapters, glow discharge devices have found wide application in a number of fields of analytical spectroscopy including atomic absorption, emission, fluorescence, and mass spectrometries. More recent applications include laser-enhanced ionization and resonance ionization mass spectrometry. In addition, the devices are now widely used in the production of electronic devices and components in metal vapor lasers. The following sections will give the reader insight into the fundamental processes occurring in glow discharge devices and how they are utilized in analytical spectroscopy.

2.2.1. Glow Discharge Processes

A glow discharge is initiated in a reduced-pressure environment when the voltage applied between two electrodes exceeds the necessary energy to cause breakdown of the rare gas, leading to the creation of electron–ion pairs. The supplied voltage is usually 500 to 2000 V in a rare gas atmosphere at a pressure of approximately 0.1 to 10 Torr. The resultant operating current, which is dependent on the gas pressure and the impedance in the power supply, is usually in the range of 5 to 100 mA. The maintenance voltage is dependent on the discharge current, fill gas identity and its pressure in the source, cathode identity, and the particular electrode configuration. Figure 2-3 is a schematic diagram of a simple diode glow discharge source. Three prominent regions are observed in most analytical glow discharge devices: the cathode dark space, the negative glow, and the Faraday dark space. While the two electrodes are specifically designated as the cathode and the anode, in order to simplify instrumentation design the grounded source housing is employed as the anode in most glow discharge devices.

In the vicinity of the cathode, electrons are repelled, resulting in the creation of a positive space charge near the cathode. Consequently, the majority of the potential difference between the two electrodes is dropped

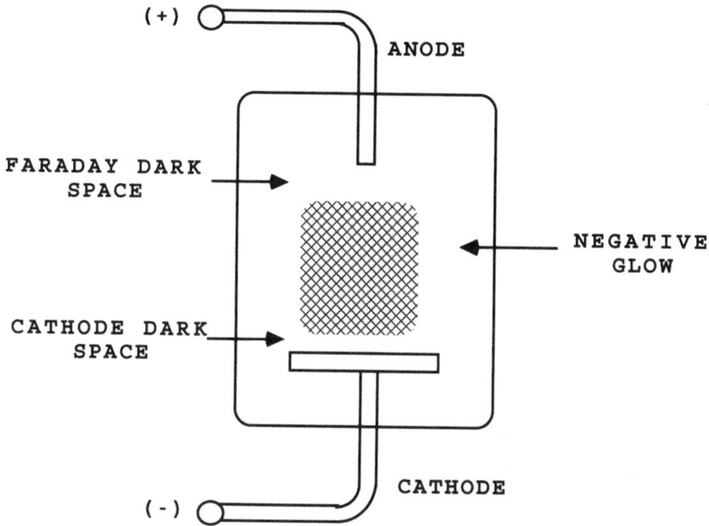

Figure 2-3. Schematic diagram of a simple diode glow discharge configuration.

across a narrow region surrounding the cathode, known as the cathode dark space (CDS) because of its noticeably low luminosity. Adjacent to the CDS is the bright, collision-rich negative glow (NG) region. The visible emission is the result of gas-phase excitation and ionization collisions and, therefore, is where most analytical information is acquired. In addition, a glow discharge plasma may exhibit Faraday dark space, positive column, anode dark space, and anode glow regions. However, most analytical glow discharge sources are designed in such a way that only the CDS and NG regions are observed between the two electrodes because the other portions of the glow discharge plasma offer little useful analytical information.

The eight distinct regions of a normal glow discharge plasma are shown in Fig. 2-4 along with the potential and charged particle distributions.[10] In a dc discharge, the negative potential of the cathode creates fields that accelerate positively charged gas ions to its surface. Ions colliding with the cathode surface cause secondary electron emission, resulting in a net negative space charge defining the Aston dark space. In the voltage range in which most glow discharges operate, the secondary electron (γ) emission coefficient, γ_i, is on the order of 0.1 electron/ion for argon ion bombardment.[11] The electrons, which have not yet been accelerated by the cathode potential, can undergo inelastic collisions with gaseous species. Emission from these species characterizes the cathode layer. Electrons that pass through the cathode layer without experiencing any collisions may

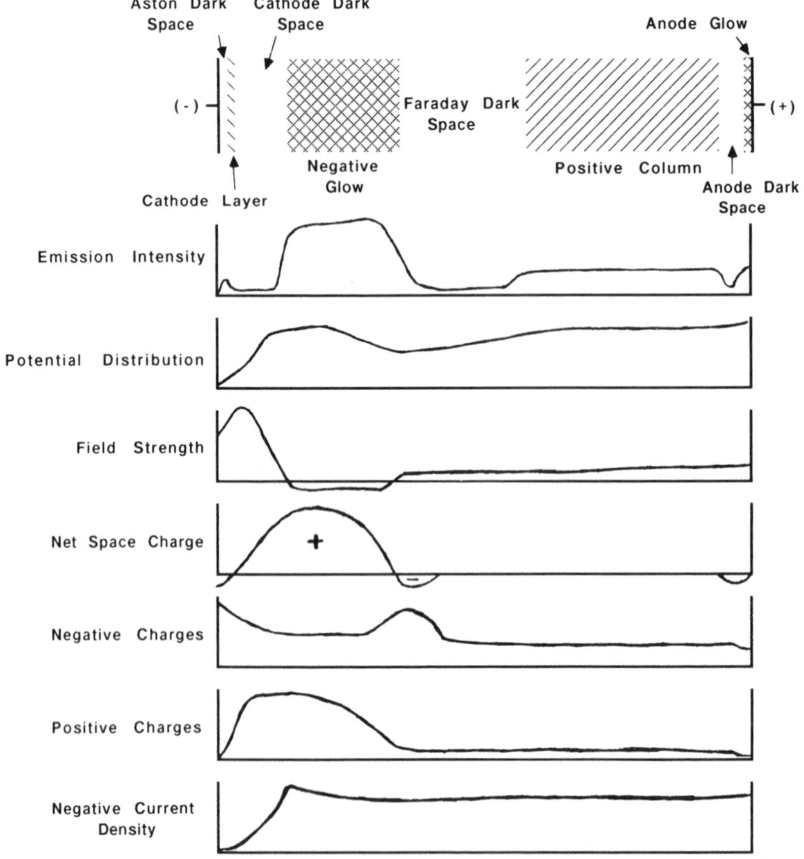

Figure 2-4. Properties of the different regions of a dc glow discharge device.

acquire energies up to that of the cathode fall[12] and as such have a low probability (low cross section) for excitation or ionization collisions. This region of low luminosity is the CDS. The CDS is characterized by its relatively high positive charge density, ρ, such that if V is the potential fall across the dark space and D its length, $\rho = V/D^2$ for a planar diode glow discharge device.[13] The vast majority of the total discharge potential drops, fairly linearly,[13] in this cathode fall region. The actual thickness of the CDS is an integral consideration in source design. At the anode edge of the CDS, a large number of electrons exist because of multiplication via ionization processes; therefore, the highest density of ions is present at this, the beginning of the NG.[14]

The onset of the NG region corresponds to the position where the fast cathode electrons have lost a large fraction of their energy through elastic collisions with other particles and, therefore, a large population of electrons with energies from ~25 eV to near thermal values is found.[15] The actual distance over which this energy loss occurs, the CDS thickness, can be approximated as being three times the Debye length (λ_D)

$$\lambda_D = (kT_e\varepsilon_0/n_e e^2)^{1/2} \tag{2-1}$$

where k is the Boltzmann constant, T_e the electron temperature, ε_0 the permittivity of free space, n_e the electron density, and e the electron charge. The resultant range of electron energies is low enough to allow electron impact excitation and ionization collisions with the various atomic species present.

The NG is the region of the discharge with which most people are familiar and is of most importance in analytical chemistry applications. Two general groups of electrons enter the NG region: fast secondary electrons, which have not undergone collisional energy losses, and slow electrons, which may be collisionally cooled secondary electrons or those created as a result of an ionization reaction. Fast electrons are only capable of ionizing collisions while the slow electrons may either excite atomic (or molecular) species or ionize excited state species. The NG is characterized by its high luminosity, which is the result of collisions of slow electrons having densities of 10^9–10^{11} cm^{-3}. The electron density is matched by an equal (order of magnitude) number of positive ions making the NG an essentially field-free region. The electric field strength in the NG region is negligible relative to that in the dark space as reported by Aston.[13,16] The intrinsic electric field-free characteristic in the NG region has been further confirmed for analytical devices by electrical probe experiments of Fang and Marcus.[14] They used a single cylindrical Langmuir probe system to characterize a planar diode analytical glow discharge device (similar to Fig. 2-3). Their findings showed that the plasma potential in the NG region is almost invariable, less than 1 V difference, in comparison with changes in discharge voltage between 800 and 1000 V.

Electrons diffusing from the anode end of the NG have experienced enough excitation/ionization collisions to deplete their energies. This region of low luminosity is called the Faraday dark space. In the Faraday dark space, electrons begin to feel the effects of the positive anode potential and are accelerated toward it, albeit with a relatively low field strength. As the electrons gain energy, they are again capable of excitation collisions in the positive column. Depending on the source size and discharge pressure, the positive column may be the largest region of the discharge. Eventually, electrons may be accelerated to the point where they are only capable of ionization. This region, analogous to the CDS, is called the anode dark space. Likewise, as in the cathode region, close to the anode surface there exists an anode glow, the last region in the dc glow discharge.

The number and size of the regions in a glow discharge plasma depend on the size of the vessel and pressure of the fill gas. In general, glow discharges employed in analytical chemistry applications are made up only of the three regions shown in Fig. 2-3. In actuality, a glow discharge can exist without many of the discrete regions but never without a CDS.[10]

2.2.2. Alternate Operating Modes

To this point, discussions of the fundamental glow discharge processes and electrical characteristics have been limited to relatively simple, diode-type source geometries, i.e., a single metal cathode and its adjacent NG in contact with a single counter electrode (generally the grounded vacuum chamber). The use of dc voltage sources has also been assumed. There are two variations of this model that should be mentioned here. The first, the hollow cathode effect, deals with operation of the source with the NG region trapped within a cathode cavity. The second, radio frequency-powered glow discharges, entails maintaining the discharge with high-frequency potentials rather than simple direct current. Both topics will be discussed in greater detail in Chapters 6 and 7, respectively. It must be stressed that these modes of operation still rely on the fundamental plasma processes that characterize glow discharge devices in terms of analyte atomization, excitation, and ionization.

2.2.2.1. Hollow Cathode Effect

By placing two planar cathodes (with a common anode) some distance d apart, they may act as two discrete glow discharges. As d is decreased so that the respective NGs coalesce, the operating voltage will decrease if in constant-current operation or the current will increase if in constant-voltage regulation. The ability to operate the resultant discharge at a voltage lower than the sum of the distinct discharge voltages, for a constant current, is the simplest case of what is generally termed the "hollow cathode effect." The effect occurs when operating pressures allow a single NG to form between two planar cathodes or inside of a cup-shaped cathode.[17]

The drop in maintenance voltage in hollow cathode geometries can be attributed to enhanced electron production. In conventional (simple diode) electrode configurations, the required electrons are generated by gas-phase ionization near the boundary of CDS and NG or as a by-product of sputtering. In a hollow cathode arrangement, secondary electrons ejected from the cathode surface are accelerated toward the CDS/NG boundary where they may ionize atoms or molecules, producing electrons. These electrons may diffuse through the NG and eventually be drawn toward the opposite CDS/NG boundary by its positive space charge, or possibly enter some small distance into the opposite CDS. If the electron travels close enough to the opposing cathode wall, it will eventually be repelled backward with a net gain in kinetic energy. In this way, electrons are trapped within the cathode body and attain higher energies than in a diode electrode geometry, resulting in enhanced electron multiplication through ionizing collisions. In this way, the required discharge maintenance voltage is lowered relative to two independent discharges. It is the enhanced production of electrons, which can excite and ionize atomic species in the plasma, that leads to the widespread use of hollow cathode discharges in atomic spectroscopy.

2.2.2.2. Radio-Frequency-Powered Discharges

As seen to this point, the flow of charged particles (current) is implicit in glow discharge plasma operation. Positive ions are attracted to the cathode, followed by the release of secondary electrons fed through the power supply from the anode. As such, both electrodes must conduct electricity (electrons). Unfortunately, solid analytical samples (cathodes) are not only metals and alloys, but also glasses, ceramics, and the like, which are electrically nonconductive. As will be discussed in Chapter 7, analysis of these types of samples employing dc voltage sources requires modification of the sample matrix to become conductive in nature. This involves grinding and mixing the sample with a metal powder and pressing a composite conducting sample.

To address the inability to atomize nonconducting matrices in glow discharge sputter deposition systems, Wehner and co-workers[(18)] proposed the use of high (radio)-frequency potentials to power the plasmas. Very briefly, the placement of a high voltage on the surface of a nonconductor induces a capacitor-like response where the surface acquires the applied potential only to be neutralized by charge compensation by (depending on the polarity) ions or electrons. The result is no net current flow and an unsustained discharge. Rapid polarity reversals of voltage pulses allow for rapid charge compensation and reapplication of the desired high voltage, overcoming the inherent decay time constant. To achieve a "continuous" discharge, pulse frequencies on the order of 1 MHz are required. A necessary by-product of the capacitor-like response is the self (dc) biasing of the electrodes such that the smaller of the two electrodes acquires an average negative bias potential sufficient to maintain the discharge processes, establishing it as the cathode. Duckworth and Marcus[(19)] have demonstrated the utility of this technology for mass spectrometric analysis of such materials as alloys, oxide powders, and glass samples.

2.3. Atomization via Cathodic Sputtering

As an atomic spectrochemical source, atomization of the sample is the first step in analysis. The atomization in a glow discharge device is accomplished by "cathodic sputtering." Cathodic sputtering makes glow discharge devices useful in analytical spectrometry as atomization sources, for it provides a means of obtaining directly from a solid sample an atomic population for subsequent excitation and ionization.

2.3.1. Cathodic Sputtering

The first observation of metal deposits sputtered from the cathode of a glow discharge device was reported in 1852.[20] At that time, cathodic sputtering was considered a nuisance because it caused erosion of the electrodes and led to undesired deposits that blackened glass walls and observation windows of discharge tubes. This attitude changed greatly as interest in glow discharge sputtering has been nourished over the years by experimentalists in the fields of electrical engineering, material analysis, physics, and analytical chemistry.[21] All of these disciplines are attempting to clarify, enhance, and apply various aspects of glow discharge sputtering phenomena.

When a metal is bombarded by high-velocity (>30 eV) ions, atoms can be ejected from the surface. The phenomenon is referred to as "sputtering." In a glow discharge, the positive gas ions in the NG region, having energies corresponding to the neutral gas temperature in random motion ($\sim kT$), may be accelerated across the cathode fall region if their motion brings them close to the CDS/NG boundary. An energetic ion impinging on the solid surface is either backscattered from a surface atom (a low-probability event, estimated at 10^{-3}) or penetrates the solid and transfers its energy to surface atoms. The potential energy of the bombarding ion goes into effecting secondary electron emission, while its kinetic energy and momentum are transferred to lattice atoms through a number of elastic and inelastic collisions. This dissipation results in a collision cascade in the lattice as shown in Fig. 2-5, propagating in random directions in the vicinity of the collision site and lasting about 10^{-12} s[22] for a given ion impact. (A simple mental picture of the resultant processes is a three-dimensional billiards break.) Recoil atoms acquiring velocity vectors near the surface normal can escape the solid and enter the gas phase if their energies are greater than the surface binding energy.[23]

This bombardment results in the emission of atoms, secondary electrons, ions (positive and negative), photons, and atom clusters[23] from the cathode surface. However, neutral atoms make up the vast majority of the ejected particle flux. In early glow discharge sputtering studies, Von Hippel[24,25] found in spectrograms of light taken just at the cathode surfaces

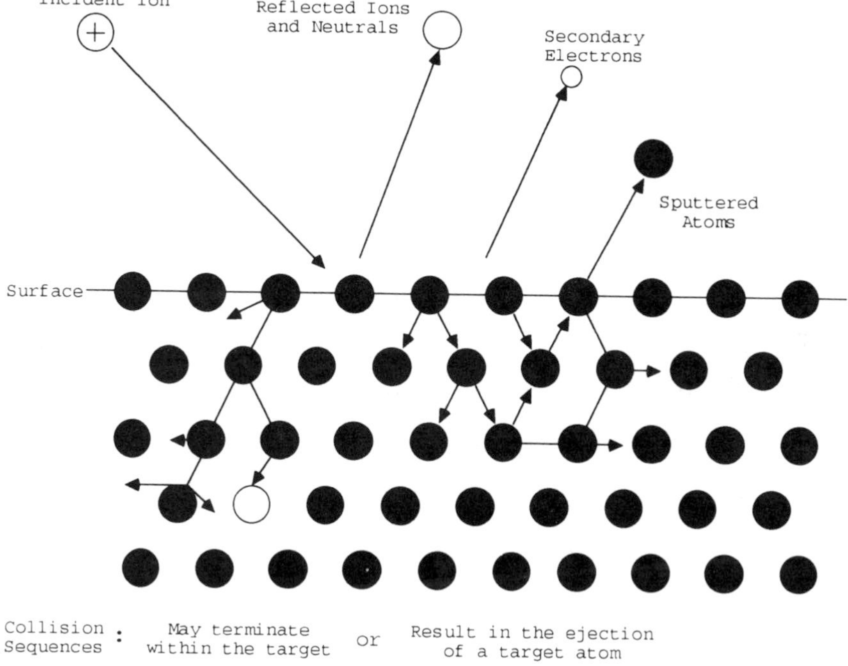

Figure 2-5. The process of cathodic sputtering.

that the only spectral lines of the cathode material that appeared were the resonance lines of the atom. From this he concluded that the material is given off in an atomic state and arrives in the NG by a diffusion process. These studies laid the groundwork for the use of glow discharges for bulk solids, elemental analysis.

It is the process of cathodic sputtering that is exploited in most glow discharge applications as the means of sample atomization. It is relevant, therefore, to discuss the fundamental aspects of sputtering that have been determined in high-vacuum ion beam experiments. The discussions that follow are intended to illustrate basic phenomenology, and not to assess the state of the art in high-vacuum sputtering theory.

2.3.2. Sputter Yield

The most basic measure of sputter efficiency is the sputter yield, the ratio of the number of atoms sputtered from the surface to the number of incident sputtering particles, usually expressed as the number of atoms per ion. Sigmund[(26)] has presented a general theory of sputter yields deriving the following equation, semiquantitatively evaluating the physical parameters affecting them. The sputter yield (S) is given by

$$S = (3/4)\pi^2(\alpha)[4M_1M_2/(M_1 + M_2)^2](E/U_0) \qquad (2\text{-}2)$$

where α is a function of the relative masses and the angle of incidence of the incoming ion,[27] M_1 and M_2 are the respective masses of the ion and sputtered atom, E is the incident ion energy, and U_0 is the surface binding energy that must be overcome for sputtering to occur. The third term is called the mass transfer term. Those parameters used in Eq. (2-2) will be discussed in detail in the following subsections.

The practical (experimental) expression for sputter yield is defined as the number of sputtered atoms per primary ion colliding with the solid surface:[28]

$$S = 9.6 \times 10^4\ (W/M \cdot i^+ \cdot t) \qquad (2\text{-}3)$$

where W is the weight loss (in grams), M is the relative atomic mass of the sputtered species (in grams), i^+ is the ion current (in amperes), and t is the sputtering time (in seconds). The bombarding ion current is related to the total discharge current, i, by

$$i^+ = i/(1 + \gamma_i) \qquad (2\text{-}4)$$

where γ_i is the number of secondary electrons released, on the average, by one ion. For argon, the gas most often used in glow discharge devices, this value is approximately 0.1.

2.3.2.1. Mass of Sputtering Ion

In glow discharge devices, as well as most other techniques utilizing sputtering, ions of noble gas atoms are used as the primary sputtering species. These ions are quite advantageously applied in these applications because of their chemical inertness, ease of ion formation, and availability of the gas in high purity. However, for example, cesium ions have found widespread use in organic mass spectrometry.[29]

The mass transfer term of Eq. (2-2), $4M_1M_2/(M_1 + M_2)^2$, maximizes as M_1/M_2 approaches unity. Therefore, Ar^+ would be the best choice for sputtering the first row transition elements, Kr^+ for the second, etc. The other mass factor in Sigmund's theory, α, increases from 0.17 to 1.5 over a target atom-to-sputtering ion mass ratio of 0.1 to 10, so that lower ion masses are favorable. For this reason, $^{40}Ar^+$ shows better overall sputtering characteristics than $^{84}Kr^+$ or $^{131}Xe^+$ even though the latter two would deliver more kinetic energy, and momentum, at the cathode surface.[30,31]

2.3.2.2. *Angle of Incidence*

The angle of incidence between the incoming ion and the cathode surface has a large effect on the subsequent sputter yield.[32] Figure 2-6 illustrates the effect of bombarding angle of rare gas ion beams on a polycrystalline copper target. The enhancement at angles away from normal incidence comes from the increased probability of the collisional cascade propagating back to the cathode surface. At severe angles (>80°) the incoming ion is more likely to reflect off the surface without any penetration or momentum transfer. The high angular dependency explains, in part, the general enhancement in glow discharge sputtering rates as the cathode surface is sputter-roughened. The angle of incidence also affects the relative elemental sputtering yields in high-vacuum sputtering systems.[33]

2.3.2.3. *Incident Ion Energy*

Equation (2-2) states that the sputter yield of a given target material is proportional to the energy of the bombarding ion. In practice this is not the

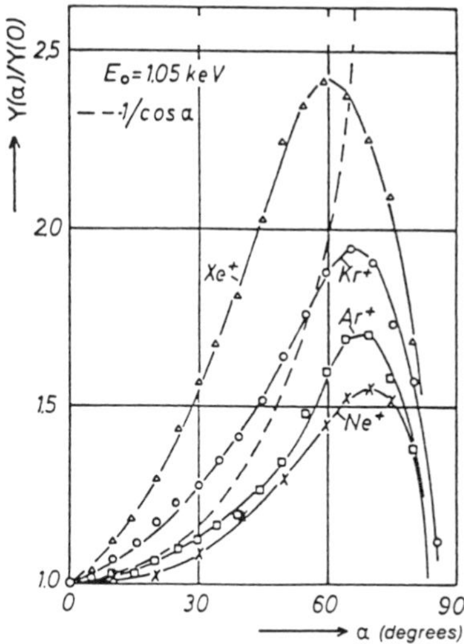

Figure 2-6. Effect of angle of incidence in ion beam sputtering.[32]

case. As illustrated in Fig. 2-7, at relatively low ion energies, sputter yields increase rapidly (nonlinearly) with energy up to about 100 eV.[34] From this point, the sputter yield is seen to increase fairly linearly with energy, until a plateau is reached at energies on the order of 1 keV. The leveling off of sputter yields for ion energies larger than 1000 eV is the result of ion implantation phenomena, where the ions begin to become embedded within the lattice. The penetration depth for a 1-keV Ar^+ ion is roughly 10 Å in Cu. The exact ion energy at which these transitions occur are dependent on the specific sputter ion–target atom pair.

2.3.2.4. Target Material

The largest variable in sputter yields, and most important in multielement analyses, is the identity of the target material. Equation (2-2) indicates that sputter yields are dependent on the ion–atom mass ratio. Wehner[31] has shown, though, that there is a more complex relationship than explained

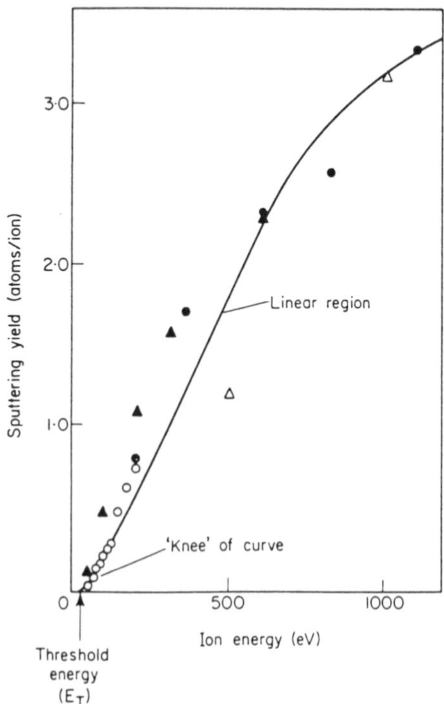

Figure 2-7. Effect of incident ion energy on sputter yield; Cu target, Ar ions.[34]

by simple mass considerations. Figure 2-8 is a compilation of the sputter yields obtained for various elements under sputtering conditions close to those found in glow discharge devices, 400-eV Ar^+. The author found that the sputter yields seem to vary according to the *d*-shell filling. For example, copper, which has nine *d*-shell electrons, has a higher yield than iron (six *d* electrons) even though it has a larger mass and thus should sputter less readily. The *d*-shell filling dependence is due to the fact that as the *d* shells are filled the respective atomic radii decrease, increasing the atomic density in the matrix. High atomic densities decrease the ion penetration depth, resulting in more efficient energy transfer to the surface and more sputtering.

In the sputtering of multicomponent targets, the binding energy for each element in the alloy is determined not only by the bulk composition but also by the surface concentration, which differs from the bulk composition because of differential sputtering (see Section 2.3.4). The ejection probability of the atoms sputtered by momentum transfer in the collision cascade is inversely proportional to its surface binding energy,(26) thus

$$c_A/c_B = c_{0A}/c_{0B} \cdot E_{bA}/E_{bB} \tag{2-5}$$

where c and c_0 are the respective surface and bulk composition of each component A and B, and E_b is the surface binding energy at that combination of bulk and surface compositions.

2.3.2.5. *Target Temperature*

At low bombardment energies (<1 keV), the sputter yields tend to decrease with increased temperature because of the annealing of more loosely bound atoms on the surface (such as created in a previous ion impact) to

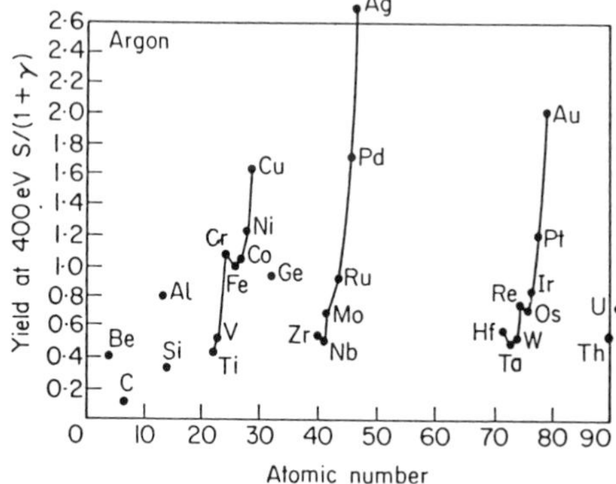

Figure 2-8. Sputter yields for various elements under bombardment of 400-eV Ar ions.(34)

positions of stronger binding. In addition, at surface temperatures approaching the target melting point the "relaxation" of the metal lattice lowers the binary momentum transfer efficiencies. At high bombarding ion energies (>10 keV) when metals are heated to within 250°C of their melting point, "thermal sputtering" becomes superimposed on binary collisional sputtering so that the apparent sputter yield increases with temperature.[(35)] Cathodic sputtering in a glow discharge, with relatively high power densities, is prone to suffer from thermal effects. For metals and some oxides, the sputter yield is reported to be independent of temperature up to about 250°C below the melting point.[(35)] For some oxides, the sputter yields have been reported to be temperature dependent with "thermal" sputtering (i.e., a higher sputter yield than that predicted by Sigmund's collision theory of sputtering) reported to occur.[(36)] While the effects can be severe, most glow discharges employ sample cooling schemes that nearly eliminate thermal complications.

2.3.3. Sputter Rate

The analytically pertinent quantity describing the sputtering phenomenon is the sputter rate. The sputter rate of a system describes the amount of cathode (sample) material removed per unit time. This of course is a direct result of the sputter yield under a given set of conditions. The sputter rate of a sample cannot be thought of as a sum of the rates of the sample's constituents; the chemical (metallurgical) form of the sample may also play an important role. For sputter-based techniques, the sputter rate may loosely define the capabilities of the source in that it defines the rate of sample introduction.

Stocker[(37)] has studied the atomization properties of a planar dc glow discharge employing a separate collector electrode to measure weight losses in a neon–argon discharge gas mixture. He found that the sputter rate varied as $(i/p)^{2.5}$ (where i is the current and p is the discharge pressure) under a variety of discharge conditions. Musha[(38)] obtained similar sputtering characteristics using molybdenum electrodes in a parallel planar configuration (approximating a hollow cathode discharge). By passing resonant radiation from a separate molybdenum discharge between the plates, absorbance measurements were made. Measurements were also made using a single cathode acting as a conventional glow discharge. Musha found that atomization with the hollow cathode configuration varied with $i^{2.5}$, while that of the planar discharge had an $i^{2.2–2.4}$ dependence. These results seem reasonable in light of the fact that the bombarding ion energies should be similar at the cathode surface even though the single-cathode discharge operates at a much higher potential. A much greater difference was observed when only the molybdenum atomic emission was monitored for the hollow and planar cathode geometries. In the hollow cathode configuration, the emission intensity varied with $i^{3.8}$, while the corresponding dependence in the planar cathode case was $i^{1.2-1.7}$. The enhancement in the hollow cathode emission can be attributed to the increased electron activity in the NG region.

Fang and Marcus[39] studied in detail the sputter rate dependencies on the discharge current and gas pressure in planar, diode glow discharge. As shown in Fig. 2-9, the studies revealed a definite i^2 dependence over a range of discharge pressures. The relative slopes of the response functions show much stronger dependence on the current-squared for the lower argon pressures. This is most likely the result of the significantly higher voltages of the low-pressure plasmas. Lower amounts of redeposition of sputtered atoms are also a likely contributor. The authors also found that the sputter rate is proportional to the discharge power $(i \cdot V)$ and the reduced power $[i(V - V_0)]$. The reduced power is based on the discharge operation voltage beyond that required to cause breakdown (V_0) of the gas, which is a function of the source pressure. Though the sputter rates in Fig. 2-9 exhibit distinct pressure dependencies, the data in Fig. 2-10 show that the use of reduced power units removes the influence of pressure on the sputter rates. Plots of the simple discharge power dependence, though linear, show definite discharge pressure effects.

As illustrated in Fig. 2-8, sputter rates (yields) will certainly vary according to the identity of the cathode (target) material. Furthermore, as will be alluded to in the next section, sputter rates of multicomponent alloys of different lattice structure (phase) can be different. In general, however, the sputter rates across a wide variety of matrices are still within the range of elemental values shown in Fig. 2-8.[14,28,40]

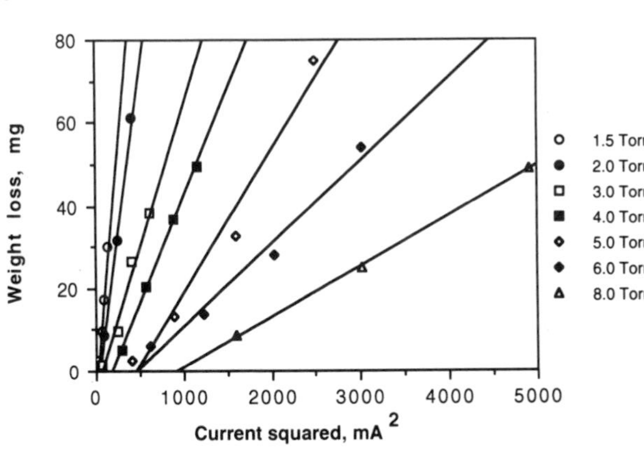

Figure 2-9. Effect of discharge current on sputter weight loss for OFHC (Cu) at various discharge pressures; Ar discharge gas.[39]

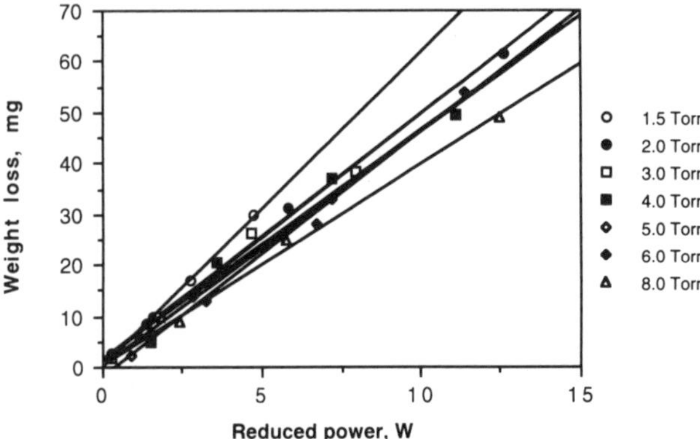

Figure 2-10. Effect of reduced power on sputter weight loss for OFHC at various discharge pressures; Ar discharge gas.[39]

2.3.4. Differential Sputtering

The sputter atomization step in glow discharges has certain advantages over thermal volatilization methods such as arcs, sparks, and lasers. In particular, elemental sputter yields over the range of bombarding ion energies likely to be found in the devices vary by only a factor of 3–5,[45] whereas elemental volatilities differ by many orders of magnitude.[41] Ideally, the sputter yields (sample introduction rates) would be uniform across the periodic table. As is found in the sputtering of solids in high-vacuum environments [such as in secondary ion mass spectrometry (SIMS)], the sputtering characteristics of the elements may be affected significantly when they are incorporated in multicomponent alloys. Betz[42] and Betz and Wehner[43] have reviewed extensively the work performed in this field. Complications arise from the fact that atoms of different elements may sputter preferentially from the cathode surface. This "preferential sputtering" is not as significant as the disparities in atomization rates found in thermal vaporization processes. The effects that alloy constituents have on the overall sputter characteristics of a sample are even more complex.

The general phenomenon illustrated in Fig. 2-8 has important consequences for techniques utilizing cathodic sputtering for multielement analysis, such as glow discharge methods and SIMS.[44] In multielement analysis, the rate at which each given element comes off the surface during the analysis time frame depends on its relative sputter yield. Since these rates differ from element to element, the phenomenon of "differential sputtering" may occur. Differential sputtering occurs on a cathode surface as elements with high sputter yields are sputtered away first, leaving those of lower yields enriched on the surface. A simplistic view of how differential sputtering might occur in a simple binary (say Cu:Al) alloy matrix is shown in Fig. 2-11. Starting with a 1:1 (atom:atom) mixture on the surface, copper atoms will be sputtered away preferentially, leaving the surface enriched with aluminum atoms. As sputtering continues, aluminum atoms will be sputtered in larger numbers than the copper until a new "layer" of copper atoms is exposed, at which point copper makes up the majority of the flux. In actuality, this process occurs at a very rapid rate, with equilibrium (steady state) sputtering conditions reached after sputtering through a few monolayers. In fact, the

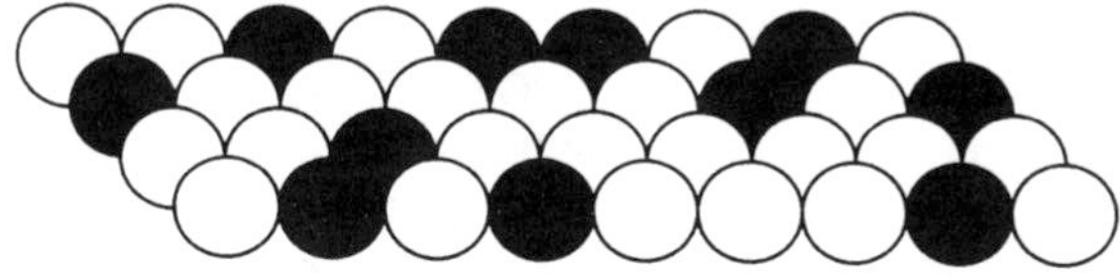

Surface; t = steady state

Surface; t = x

Bulk Solid; t = 0

Figure 2-11. Depiction of differential sputtering in a Cu–Al binary alloy (1:1). Solid circles are Cu.

conservation of mass dictates that a steady-state flux composition must be established. In the steady state, the relative surface composition of a two-component system (C_A/C_B) can be calculated from the simple relationship

$$C_A/C_B = (P_B/P_A)(C_{0A}/C_{0B}) \tag{2-6}$$

where P is the ejection probability of the respective atoms.

As mentioned previously the absolute, and differential, sputtering rates for each element in an alloy matrix may not be simply proportional to the pure elemental sputtering rates of the elements. The target material composition and structure also play important roles when quantitative information is of interest. Hammer and Shemenski[(46)] used argon and xenon ions at energies between 1 and 4 keV to bombard a series of one- and two-phase brass alloys with concentrations of copper ranging from 18 to 48% to study differential sputtering phenomena. The authors found that all the specimen surfaces studied were depleted of zinc (the higher-sputter-yield element) for all sputtering conditions. However, significant differences in the relative Cu:Zn ratios were found between the two phases. The steady-state surface composition was dependent on ion energy for the α (fcc) phase, and energy and mass for the $\alpha + \beta$ (bcc) phase, with the relative zinc concentrations decreasing with increasing ion energy and mass. Greater deviations from the original bulk composition were found in the mixed phase alloys, along with a more rapid change in composition as a function of depth immediately beneath the surface. Therefore, for analytical applications, differential sputtering effects must be considered and characterized for a given sample type.

An important result concerning the differential sputtering effects in the analytical use of glow discharges has been obtained by Fang and Marcus.[(47)] The authors used Auger electron spectroscopy to study the surface compositions of a brass alloy before and after sputtering along with a deposited thin film that was collected from the NG region during sputtering. In this way, the Auger spectrum of the thin film would be representative of the steady-state flux into the NG region. The results of these analyses are shown in Fig. 2-12. As can be seen, the surface composition after sputtering indicates an enrichment of the low-sputter-yield element, copper, relative to the bulk sample. The composition of the deposited thin film shows a nearly identical Cu:Zn peak height ratio to that of the bulk sample. This indicates and ensures that once the discharge has reached steady-state sputtering conditions, the plasma composition does reflect the bulk sample composition. While the absolute rates at which different samples sputter will vary, the conservation of mass dictates that within a given sample *sputtering is stoichiometric*. Of course, the fact that the sampled volume does reflect the actual sample composition is a desired quality for any analytical technique. This work shows, as well as a previous work by Boumans,[(28)] that the sputter

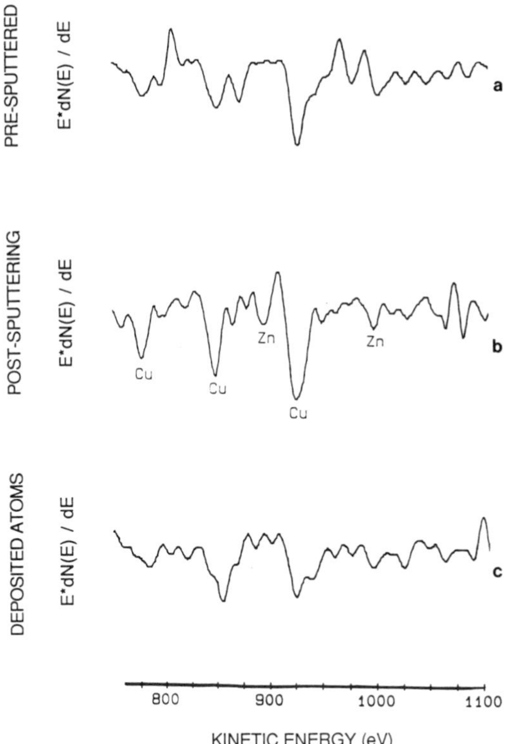

Figure 2-12. Auger electron spectra of a brass alloy (72% Cu:28% Zn) taken (a) before and (b) after sputtering, and (c) of a deposited thin film.[47]

yields for simple binary alloys can be approximated from the relative elemental sputter yields.

2.4. Physical Characteristics of Glow Discharge Devices

Glow discharge devices are characterized by a more complex set of processes than the easily controlled high vacuum sputtering systems. Figure 2-13 illustrates the cumulative collisional processes occurring in a glow discharge device. We concern ourselves here only with the factors related to cathodic sputtering. In the glow discharge, electric field gradients and ion–neutral collisions control ion trajectories and energies. The majority of the potential between the electrodes (ζ) is dropped across the CDS, which accelerates positively charged discharge gas (argon) ions to the cathode surface with secondary electrons accelerated away from the surface. [The

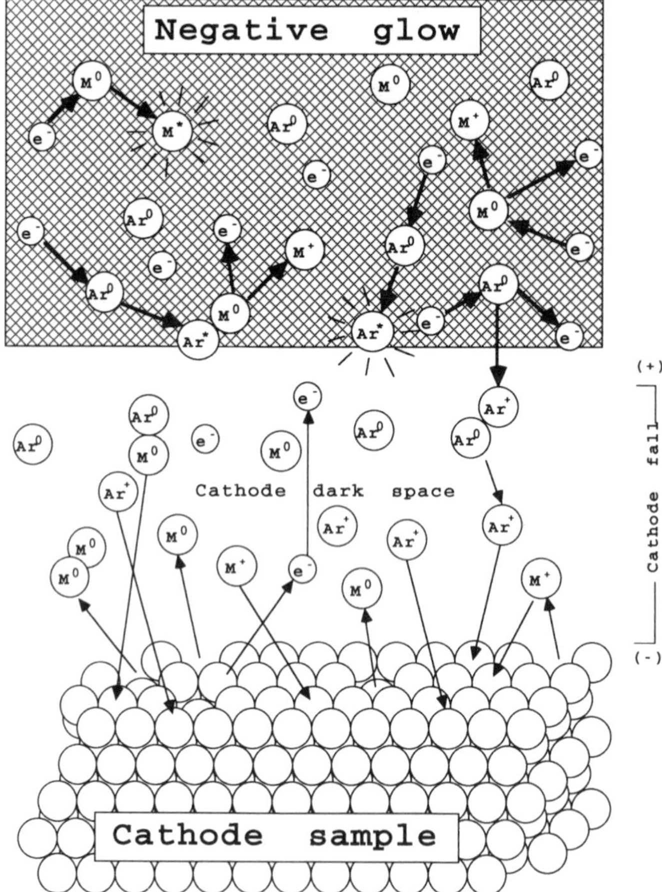

Figure 2-13. Fundamental collision processes in glow discharge devices.

current flow in the electric field-free region (NG) is reduced by coulombic collisions of electrons and ions.[23,48)] In actuality, sputtering ions generally have only a fraction of ζ kinetic energy (~40%) because of charge exchange reactions with neutral gas atoms[49] and the proximity of ion formation to the cathode surface. The ion bombardment will sputter-release atoms, atom clusters,[23] and ions of the cathode material as well as secondary electrons (γ)[11] and UV photons.[50] The negative bias of the cathode dictates that positively charged ions (M^+), which make up about 1% of the total particle flux,[51] will be returned immediately to the cathode surface. Conversely, negative ions (M^-) will be accelerated away. The vast majority of the particle flux is made up of neutral atoms (M^0), generally leaving the surface in their electronic ground state with kinetic energies up to 5–15 eV.[52–54] In the pressure regime of 0.1–10 Torr, atomic mean free paths are less than 0.1 mm. As a result of these short mean free paths, the sputtered atoms quickly lose their momentum through elastic collisions with other discharge species (atoms and ions). In this way, sputtered neutral atoms may be knocked back to, or redeposited, onto the cathode surface. Sputter weight loss studies by Harrison and Bruhn[55] indicate that up to 95% of sputtered atoms in a glow discharge environment may be returned to the cathode surface. The use of directed discharge gas flow to sweep atoms away from the cathode surface will be discussed in Chapter 3.

While the analytical spectra (e.g., absorbance, emission) obtained from glow discharge sources can be characterized as being "atomic" in nature, it must be stated that a number of studies indicate that the majority of atoms in the NG region seem to originate as "molecular" species.[56–58] Spatially resolved atomic absorption profiles of sputtered species indicate that atom densities are at a maximum directly above the cathode surface (as close as possible for sampling), with a sharp decrease through the dark space region, and finally a second maximum in the region of the CDS/NG interface. From the second maximum, atom densities drop almost exponentially as would be predicted from diffusional losses. A typical spatial profile of this phenomenon is illustrated in Fig. 2-14. The initial maximum (not seen here) seems to be related to released free atoms with the subsequent loss indicating a high degree of redeposition. The appearance of the second maximum indicates an "injection" of atoms within the plasma. The most probable mechanism for such an increase is the dissociation of sputtered clusters, most likely by energetic electrons and/or metastable atoms, which are concentrated in this region of the plasma.

2.4.1. Positive Ion Energies

The energy of the positive ions impinging on the cathode surface is an important quantity because this will determine the sputtered atom and secondary electron yields in the plasma. The energy of a bombarding ion depends on the number of collisions it experiences between its point of formation (e.g., NG/CDS interface) and the cathode. Positive ions carry a substantial fraction of the total discharge current in the cathode fall. The motion of the positive ions is limited by symmetric charge exchange

$$X^{+}(\text{fast}) + X^{0}(\text{slow}) \rightarrow X^{0}(\text{fast}) + X^{+}(\text{slow}) \tag{2-7}$$

Each collision produces a fast neutral and an ion with only thermal energy.[34,59] At typical discharge voltages, collisions between ions and atoms will tend to be elastic, so that no energy will be lost through excitation. The

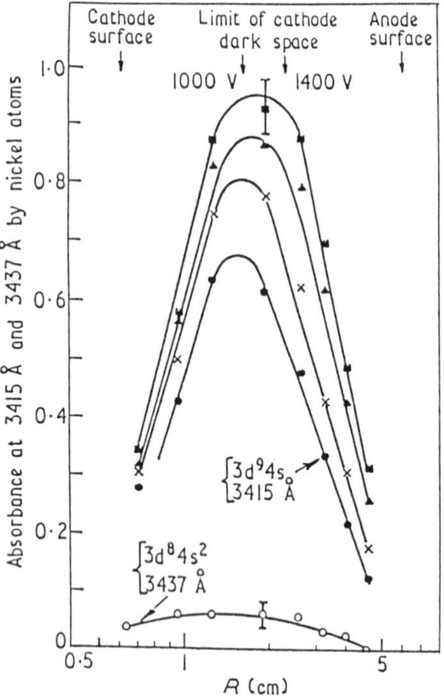

Figure 2-14. Spatially resolved absorption profiles of "atomic" species from a Ni target.[56]

major source of energy loss will be through symmetric charge exchange.[60] The produced thermal ion then is accelerated by the fraction of the fall potential existing between that point and the cathode. The frequency of these collisions will be determined by the source pressure, the fall potential, and the dark space thickness.

It is now generally accepted that positively charged particles passing through a gas do not necessarily retain their ionic identity, but by capturing and losing electrons may lose and regain their charge a number of times. It appears probable that a certain percentage of positive ions formed in the NG actually do reach the cathode with the full velocity corresponding to a free fall. However, a much larger proportion reach the cathode with velocities far less than the fall potential because of successive collisions.

The energy of bombarding ions is generally determined by mass spectrometric sampling through a small hole in the wall of the electrode of interest. Bodarenko[61] has studied the energy distribution of argon discharge gas ions in both abnormal, planar, and hollow cathode glow discharge configurations. As shown in Fig. 2-15, the singly charged argon ions from the

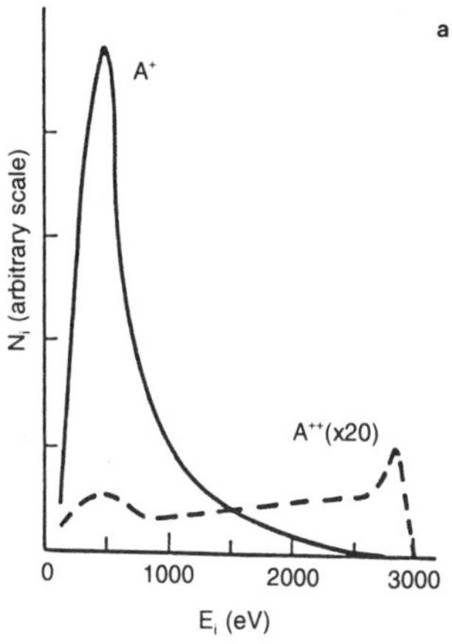

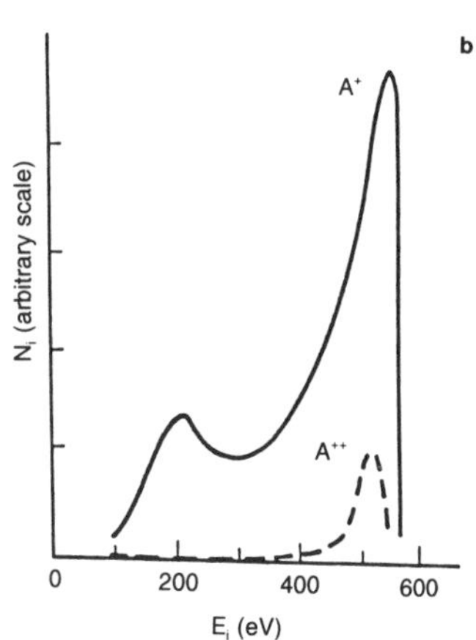

Figure 2-15. Kinetic energy distributions for argon ion species in (a) planar and (b) hollow cathode configurations.[61]

planar discharge (panel a) are centered at an energy of about one-sixth of the cathode fall potential while those sampled from the base of a hollow cathode (panel b) have energies closer to the applied potential. This is likely due to the reduced dark space distance in the hollow cathode discharge.

Howorka *et al.*[62] used a nonconducting orifice plate mounted at the end of a cylindrical hollow cathode to map radially the argon ion density of the NG. As was the case with electron density distribution measurements (see Section 2.4.3.1),[63] they found that the ion densities concentrated toward the center of the cathode at increasing pressures, although the maximum for each pressure was not on the cathode axis as was the case with the electrons, but more toward the respective NG/CDS interfaces.

2.4.2. Electrical Characteristics

As discussed in the previous section, the kinetic energy of sputtering ions is intimately related to the potential at which a plasma is maintained. This will, of course, affect the sputter yield for a particular target. The second component affecting sputtering rates is the discharge current. At each discharge pressure there is a unique current–voltage (i–V) relationship that controls the sputtering rate. Therefore, for a given discharge geometry one must be concerned with the i–V characteristics and the effects that they have on absolute sputtering rates. While sputtering rate is the analytical variable in atomic absorption and fluorescence applications, it must be kept in mind that for atomic emission and mass spectrometry applications, discharge conditions will also affect the extent of analyte excitation and ionization.

2.4.2.1. Planar, Diode Devices

The relationship between discharge current and voltage at a given source pressure in a glow discharge device, along with the effects of redeposition, determine the rate of cathodic sputtering. High-vacuum sputtering studies of copper with argon ions over the range of energies most likely occurring in glow discharges indicate that the sputter yields (No. of sputtered atoms/incident ion) increase nearly linearly with ion energy.[34] Likewise, one would expect the overall sputter rate of a material to increase proportionally to the sputtering ion current.

For glow discharges operating in an abnormal mode (as do most sputtering plasmas) at a constant pressure, increases in discharge current must be accompanied by increases in operating voltage. Fang and Marcus[39] have reported the relationships between discharge current, voltage, and gas pressure in a planar, diode glow discharge device. Figure 2-16a is a plot of the operating voltage of the glow discharge source as a function of current for a range of argon pressures from 1.5 to 8 Torr. These curves closely resemble those determined by Tong and Harrison[64] who employed a similar source design to sputter chromium and niobium disks in a lower-pressure (0.3–1.0/Torr) regime. Dogan *et al.*[65] and Boumans[66] observed these sorts of relationships in Grimm-type discharges. It is clear that at lower source pressures the dependence of operating voltage on current is greater. The response of voltage to increasing current reflects increases in current density at the cathode surface. The pressure effects are attributable to the more efficient (higher frequency) collisional processes occurring at high pressures, requiring lower initial secondary electron energies for plasma ionization to maintain the discharge.[11]

It must be emphasized that the emission of secondary electrons from the cathode surface, with their subsequent acceleration by the cathode fall potential, is the major means of maintaining plasma ionization. The electron energy requirement also leads to the dependence of CDS thickness on source pressure. Under high-pressure (collision frequencies) conditions, electrons lose their energy efficiently, terminating the dark space at shorter distances. The inverse relationship between source pressure and discharge voltage and dark space thickness is an important factor in glow discharge source design.

Figure 2-16b illustrates a more specific relationship between discharge voltage and current. A plot of voltage versus the discharge current squared (mA^2) shows linear relationships for all examined discharge pressures. Similar plots have not been explicitly depicted in the literature, although the comparable i–V curves of others suggest that they may also follow an i^2 term. Fang and Marcus's explanation for this relationship involves an Ohm's law analogy where the resistance of the plasma is controlled by the discharge current ($V = iR\alpha i^2$).[39] The physical quantity, which does in fact change

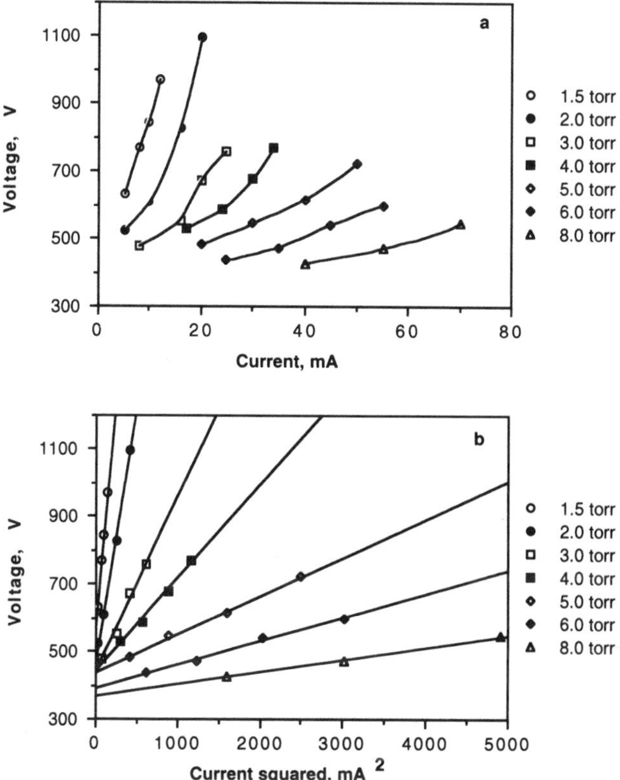

Figure 2-16. Current–voltage characteristics for simple, diode glow discharge device at various pressures (Cu target). (a) Voltage versus current; (b) voltage versus current squared.[39]

proportionally with the discharge current, is current density. As the current density at the cathode surface increases, more charge carriers are being forced to the same conductive area, effectively causing a bottleneck. The response to the increase in current density is an increase in resistance and therefore a higher operating voltage.

2.4.2.2. Hollow Cathode Devices

Most studies dealing with the comparison of the *i–V* characteristics between planar and hollow cathode configurations are approximated using a parallel plane configuration. This approach allows experimenters to vary the distance between the plates, analogous to varying the cathode diameter, and source pressure to determine the degree of the hollow cathode effect. Guntherschulze[67] took this approach using two planar cathodes (50 cm^2)

spaced 1 cm apart and wired such that he could operate either a single cathode or both at the same time. These studies showed that the CDS length of the single cathode in a 450-V hydrogen discharge was 14 electron free paths (efp's) while the comparable two-electrode discharge had a pair of CDSs of only 5 efp's each. The lengths of the two dark spaces varied less with pressure than those of the single-cathode case, which varies inversely with pressure. The shortened dark space in the hollow cathode reflects the maintenance of higher electron energies and densities in that geometry relative to the planar cathode case.

Little and Von Engel[(68)] determined that the discharge voltage dropped linearly across the CDS of a hollow cathode going from near zero at the NG/CDS interface to the applied voltage at the cathode surface. (In this particular case, the cathode was at a negative potential and the anode grounded.) Kirichenko *et al.*[(69)] found that while the CDS length is independent of pressure from 0.1 to 1 Torr, there is a decrease as the pressure is increased to 2 Torr. They also determined that a decrease occurred as the current was raised from 30 to 200 mA; the change being from 4 to less than 1 efp. Those studies indicate that an electron emitted from the cathode surface will reach the NG after few or no collisions, which would lower the voltage requirements for a given discharge current as compared with the planar cathode case. In fact, most hollow cathode discharges operate in a near-normal discharge mode with very little or no increase in operating voltage as current is increased. We are not aware of any specific mathematical functions that predict current–voltage relationships in hollow cathode discharges.

2.4.3. Energetic Particle Densities and Energies

The preceding discussions have dealt with those factors that determine the sputtering (introduction) rate and identity of analyte species sputtered from the cathode sample. Once in the gas phase, the atoms (and small clusters) are subject to elastic and inelastic collisions causing excitation, ionization, and possibly association/dissociation. These reactions are depicted schematically in Fig. 2-13. In terms of performing emission or mass spectrometries, the principal collisions of interest are those between analyte atoms and electrons or with metastable (excited state) discharge gas atoms. The following discussions deal with the densities and energies of these important gas-phase collision partners.

2.4.3.1. Electrons

Three electron groups are believed to exist in the NG region of a glow discharge:[(70)] group I are the secondary electrons emitted from the cathode surface (so-called fast or primary electrons), which gain kinetic energy through the CDS and attain electron temperatures (T_e) of ~20–25 eV and number of densities on the order of 10^6 cm^{-3}; group II are the secondary electrons of gas-phase ionization collisions with electron temperatures of ~2–10 eV and number densities in the range of 10^7–10^8 cm^{-3}; and group III (ultimate or slow electrons) are the electrons from either group I or group II that have experienced several elastic and inelastic collisions with particles in the plasma and thus have electron temperatures of only 0.05–0.6 eV, having densities in the range of 10^9–10^{11} cm^{-3}.

Electron energy (temperature) and density measurements in a thermal equilibrium environment can be carried out, in principle, by optical methods that include the slope or two-line methods for excitation temperature measurements and H_β line broadening for electron number density determinations.[71] However, none of these methods are applicable in diagnosing a glow discharge plasma. The reasons are that (1) the glow discharge plasma is not at thermal equilibrium or even local thermal equilibrium (LTE). This excludes the validity of excitation temperature measurements. (2) H_β line broadening is only useful/practical for plasmas that have electron number densities larger than 10^{13} cm^{-3}, which is much larger than those determined in glow discharges by other methods.

Given the properties and chemical environment of the glow discharge, electrostatic (Langmuir) probe techniques may be the best way to measure localized electron temperature (T_e), electron number density (n_e), average electron energy ($\langle \varepsilon \rangle$), and electron energy distribution functions (EEDF).[14,72,73] Langmuir probes have been used widely in diagnosing nuclear plasmas. The electron temperatures and number densities found for different glow discharge plasma types are listed in Table 2-1 along with the particular measurement technique employed. As can be seen, the n_e values measured by H_β line broadening are much higher than those determined by Langmuir probes because of the fact that the H_β line broadening method can only generate electron number densities larger than ~10^{13} cm^{-3}.

A typical experimental configuration for a Langmuir probe system is shown in Fig. 2-17.[14] Basically, a Langmuir probe is a small metal wire or disk submerged in a plasma (or flame) to which a dc bias is imposed. According to the polarity and magnitude of the applied voltage, either positively or negatively charged species will be attracted to its surface, causing a resulting current to flow through the probe circuit. The magnitude of the current is a function of the number of charged species incident on the probe, the applied probe potential, and its surface area. Scanning the applied voltage across a range of values, say from −12 to +12 V, generates a current–voltage response curve of the form shown in Fig. 2-18. As the scan progresses, positive ions and then electrons are attracted to the probe, with the magnitude of the current as a function of potential being indicative of the particles'

Table 2-1. Electron Temperatures and Electron Number Density in Low-Pressure Discharges[75]

Plasma	Fill gas pressure	Power	T_e (eV)	n_e (10^{12} cm^{-3})	Ref.
MIP[a]	3.0 Torr Ar	25 W	1.87	15.0	43
MIP[a]	3.0 Torr He	25 W	4.48	10.5	43
RF[a]	2.0 Torr Ar	50 W	3.71	1.00	10
RF[a]	2.0 Torr Ne	50 W	4.74	0.85	10
HCD (Ni)[a]	2.3 Torr Ar	10 mA × 325 V	1.21	—	28
HCD (Ni)[a]	1.9 Torr He	10 mA × 325 V	1.38	—	28
MIP[a]	3.0 Torr Ar + Hg	25 W	2.73	19.5	43
RF[a]	2.0 Torr Ar + Tl	50 W	3.45	2.00	44
RF[a]	2.0 Torr Ar + Ti	50 W	3.19	0.15	44
RF[a]	2.0 Torr Ar	50 W	2.41	0.80	44
HCD (Cu)[a]	1.0 Torr Ar	25 mA × 350 V	2.24	—	63
HCD (Ni)[a]	1.0 Torr Ar	25 mA × 350 V	1.98	—	63
HCD (steel)[a]	1.0 Torr Ar	25 mA × 350 V	0.83	—	63
GD (Al)[b,c]	4.0 Torr He + Hg	8 mA[h]	0.05[d]	0.82[d]	69
GD (Al)[b,c]	12.0 Torr He + Hg	8 mA[h]	0.07[d]	0.22[d]	69
GD (Cu)[b]	2.0 Torr Ar	10 mA × 620 V	0.27	0.16	75
GD (Cu)[b]	3.0 Torr Ar	10 mA × 530 V	0.27	0.16	75
GD[e]	15.3 cm^3/min Ar	40mA × 800 V	0.08[f]	203[g]	44
GD[e]	29.1 cm^3/min Ar	60 mA × 800 V	0.10[f]	287[g]	44
GD[e]	35.5 cm^3/min Ar	80 mA × 800 V	0.11[f]	319[g]	44
GD[e]	17.5 cm^3/min Ar	40 mA × 1000 V	0.07[f]	201[g]	44

[a]Measured by double cylindrical probe.
[b]Measured by single cylindrical probe.
[c]Discharge tube.
[d]Ultimate electron.
[e]Grimm-type glow discharge lamp with unspecified cathodes.
[f]Doppler temperatures.
[g]Hβ method.
[h]No operating voltage has been given.

kinetic energies and densities. Detailed theory and pertinent equations dealing with Langmuir probe operation and data evaluation are described in Refs. 14, 72, and 73.

Fang and Marcus have developed a low-cost computer-controlled single Langmuir probe system.[74,75] This system has been employed to determine the axial dependences (away from the cathode surface) of electron temperature, electron number density, average electron energy, and electron energy distribution function under various discharge conditions for a planar, diode glow discharge device. In an argon glow discharge, they found that the electron number density is proportional to the discharge current, while the axial density distribution in the NG region is fairly constant as shown in Figs. 2-19 and 2-20 respectively. The electron number densities are found to be of the magnitude of 10^{10} cm^{-3}.[14] The electron energy distribution functions in the NG region are shown in Figs. 2-21 and 2-22 as a function

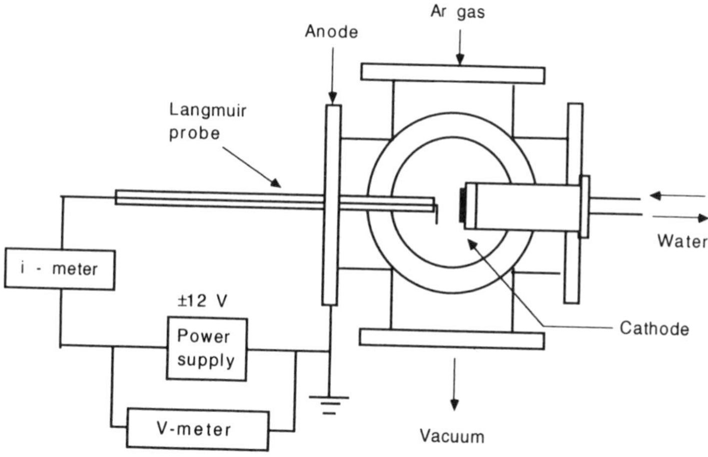

Figure 2-17. Typical experimental configuration for Langmuir probe measurements.[14]

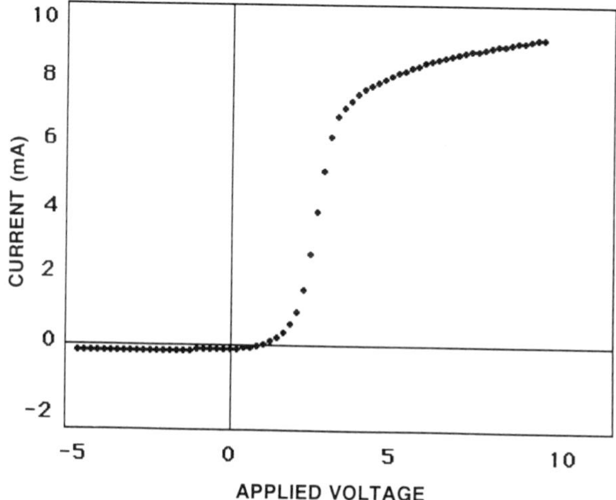

Figure 2-18. Typical current versus voltage plot obtained from a Langmuir probe experiment.

of the axial position and discharge current, respectively. The electron populations are centered at energies of less than 2 eV, showing a depletion of higher-energy electron populations relative to those predicted from a Maxwell–Boltzmann distribution (which would be characteristic of a system at thermodynamic equilibrium). Deviations from Maxwellian behavior are also evident from differences observed in the measured electron temperature values and those of the average electron energy. In the dc glow discharge

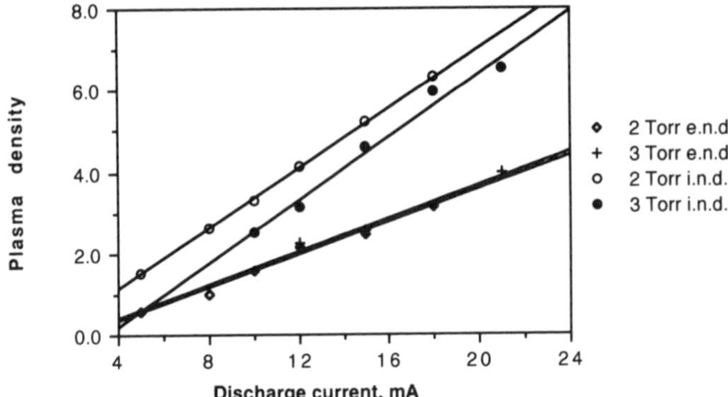

Figure 2-19. Effect of discharge current and pressure on electron (e.n.d.) and ion (i.n.d.) number densities for a simple, diode discharge configuration (Cu target).[14] (Units are $10^{10}\ cm^{-3}$.)

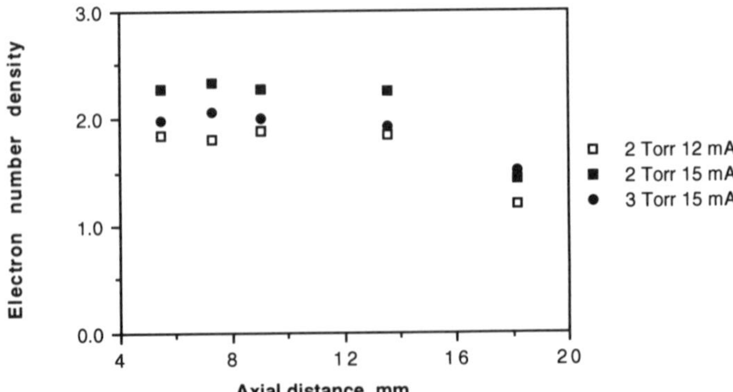

Figure 2-20. Axial profile of electron number density at various discharge conditions (Cu target).[14] (Units are $10^{10}\ cm^{-3}$.)

studied here,[14] the average electron energies were consistently higher than those of the electron temperatures (~0.9 versus 0.3 eV). Based on the way these quantities are derived, this relationship again points to a relative depletion of higher-energy electrons. Because the average electron energy is derived from the complete distribution function, the authors tend to believe that it is a more realistic characteristic of the plasma excitation conditions.

The observed depletion of the high-energy electrons is likely related to the first excitation threshold of the gaseous atomic species. Figure 2-23 depicts the electron energy distribution functions for the case of sputtering a copper cathode at discharge conditions of 2 Torr argon pressure and 8 mA

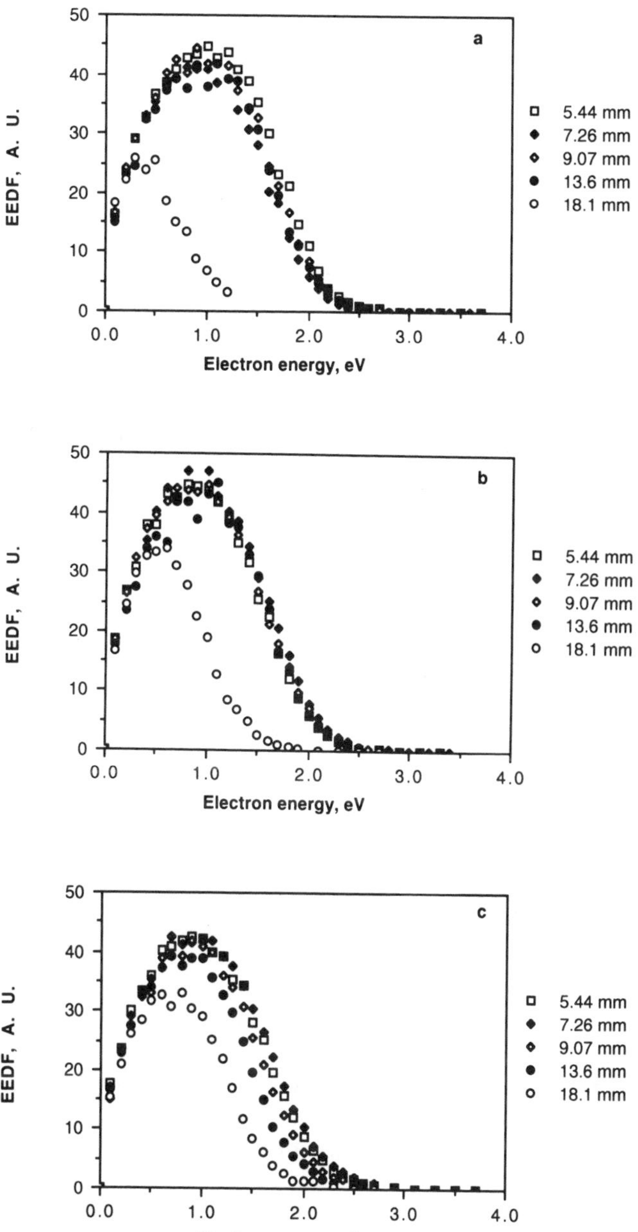

Figure 2-21. Axial dependence of electron energy distribution function (EEDF) (Cu target). (a) 2 Torr, 12 mA; (b) 2 Torr, 15 mA; (c) 3 Torr, 15 mA.[14]

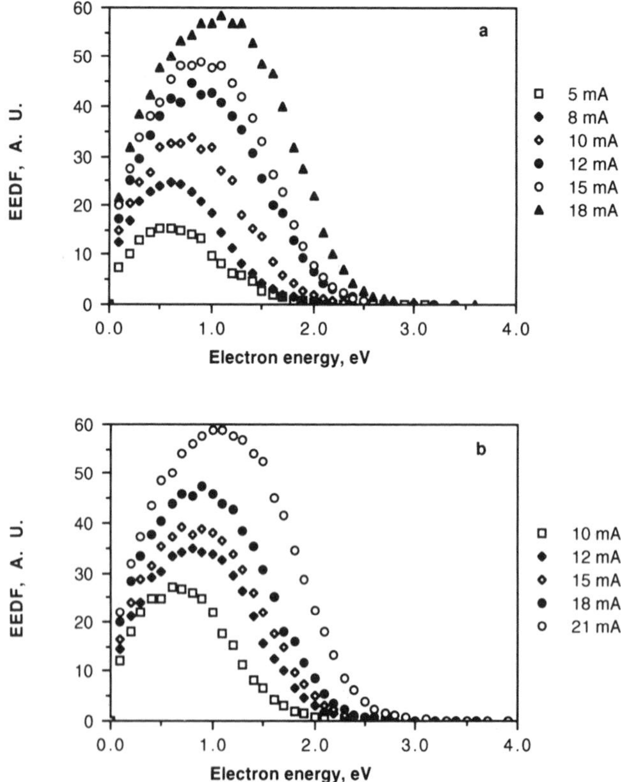

Figure 2-22. Effect of discharge current and pressure on EEDF (Cu target). (a) 2 Torr; (b) 3 Torr.[14]

discharge current.[75] The respective plots represent the Maxwell–Boltzmann distribution based on (A) 0.286 eV electron temperature (calculated), (B) average electron energy of 0.832 eV (calculated), and (C) the actual EEDF. According to the two- and three-electron group models for low-pressure gas discharges developed by Vriens,[73] the discrepancy between the three plots indicates non-Maxwellian behavior of group III electrons. The intersection point of the curves indicates a depletion of electrons above the first inelastic threshold (required for atomic excitation), which in this case is ~1.2 eV. This value is fairly reasonable considering the electronic excitation of the atomic species (Cu and Ar) in this plasma. The plots indicate that the electron temperature best describes the electrons of energies >1.2 eV, while the average electron energy represents the bulk of the electrons in the plasma.

In detailed studies by Fang and Marcus,[75] the role of cathode identity on the charged particle characteristics has been studied in order to ascertain

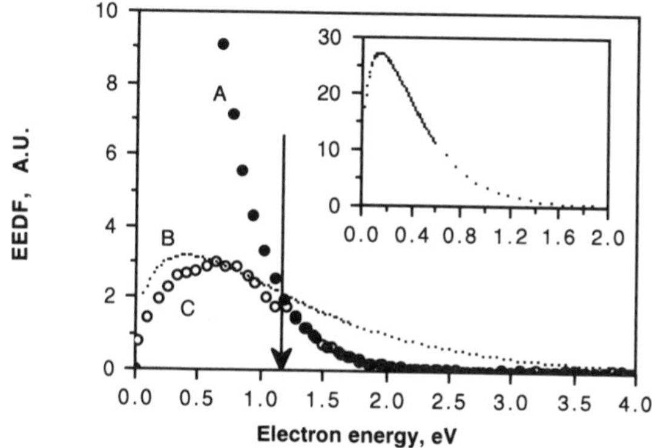

Figure 2-23. Maxwell–Boltzmann distribution functions determined by (A) electron temperature (0.286 eV), (B) average electron energy (0.832 eV), and (C) experimentally determined EEDF. Insert shows the full distribution of (A). Cu target; 2 Torr, 8 mA.[75]

the origins, or lack, of spectrochemical matrix effects. The studies employed cathodes of copper, various brass alloys, stainless steel, titanium, molybdenum, and a compacted oxide sample. In considering the average electron energies and densities as the most important analytical characteristics, it was found that there are no appreciable matrix effects. The electron energies and densities were found to vary by less than 15% when the effects of discharge voltage and cathode surface area were taken into account (i.e., constant power densities). It must be mentioned that the densities of gas-phase ions (discharge gas and sputtered species) are of the same order of magnitude as the electron populations, but generally two- to fivefold larger.

In a hollow cathode discharge, Borodin and Kagan[76] used a Langmuir probe to study electron energies and densities in the NG and positive column of a parallel plane discharge using helium as the fill gas. They determined that the majority of electrons in the NG exist at low energies (~2 eV) although there is a substantial population of fast electrons of 19–25 eV kinetic energy. The higher-energy electrons were not as prominent in the positive column. The observed energies, however, do not account for the high degree of emission observed from high-lying helium atom and ion states.

Subsequent studies of hollow cathode discharges were performed by Kagan and co-workers[77] to determine the extent to which electrons existed at energies above 25 eV by employing an electrostatic energy analyzer to sort the electrons. The authors found that electrons of higher energy do indeed exist with a large number centered around the cathode fall potential. These would be electrons formed at the cathode surface that had not experienced inelastic collisions as they traversed the NG from side to side.

Howorka and Pahl[63] found similar electron energy distributions in discharges operating at lower pressures (0.1–0.7 Torr). As an extension of these studies, a Langmuir-type probe was used to map the electric fields and electron densities. They found that over the length of the hollow cathode (7.4 cm long; 2 cm diameter), the field changes were on the order of 0.1 V cm^{-1} indicating that the electron and ion densities were equivalent, as would have been expected. The authors also studied the radial electron distribution in an effort to explain observed spatial variations in hollow cathode emission studies. Figure 2-24 illustrates the radial electron densities as a function of discharge pressure. The density maximizes toward the cathode walls at high pressures and toward the middle at lower ones, implying that there may be CDS thickness dependency.

2.4.3.2. Metastable Atoms

In addition to their chemical inertness, ease of handling, and ready availability, rare gas atoms have the fortuitous characteristic of high-lying, long-lived metastable states. Metastable (triplet) electronic states have long radiative lifetimes because radiative transitions to lower-energy levels are spin-forbidden. Table 2-2 lists the energy levels and spectroscopic notation for the metastable levels of the rare gas elements. As can be seen, the energy levels for these states are quite high relative to the excitation and ionization potentials of most other elements. In all cases, the lifetimes are longer than the collisional lifetimes of gaseous atoms. Therefore, given the extended lifetimes of the states, metastable atoms are effectively sources of potential

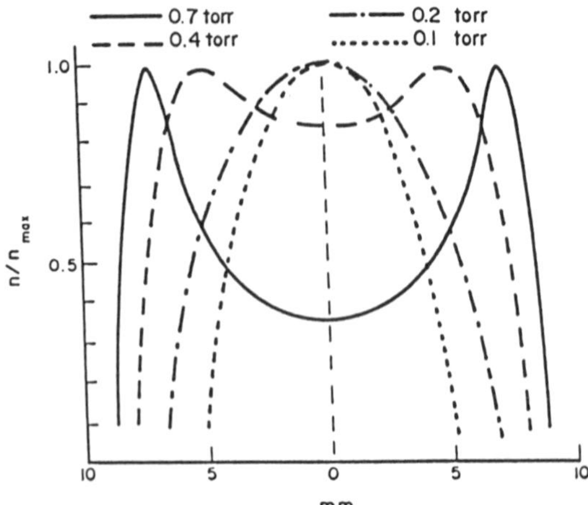

Figure 2-24. Radial electron densities as a function of discharge pressure for a hollow cathode configuration.[63]

Table 2-2. Metastable Energy Levels for Rare Gas Atoms

Atom	Spectroscopic notation (L-S)	Energy		Ionization potential (eV)
		cm^{-1}	eV	
He	2^1S	166 272	20.6	24.5
	2^3S	159 850	19.8	
Ne	$^1P_0$	134 820	16.7	21.6
	$^3P_2$	134 043	16.6	
Ar	$^3P_0$	94 554	11.7	15.8
	$^3P_2$	93 144	11.5	
Kr	$^3P_0$	85 192	10.5	14.0
	$^3P_2$	79 973	9.9	
Xe	$^3P_0$	76 197	9.4	12.1
	$^3P_2$	67 086	8.3	

energy within the plasma, with number densities on the order of 10^{11}–10^{12} cm^{-3}.

The actual role of metastable atoms in glow discharge devices is still of considerable debate.(11,78–88) Their roles in analyte excitation and ionization will be discussed in Section 2.5. Hess and co-workers(78–80) have studied extensively the discharge parameters that affect the production and quenching of argon and neon metastable atoms in simple, diode geometry devices. The role of metastables in discharge maintenance has been established through optogalvanic effect (OGE) spectroscopy in a hollow cathode wherein the discharge currents are seen to decrease (in a constant-voltage mode) as the metastable atom states are depopulated by a laser (see Chapter 10).(78) Thus, the relative ease of ionization of metastable atoms, relative to direct ionization from the ground state, is important for maintaining the total discharge ionization.(81–83) The density of metastable atoms (as determined by atomic absorption spectrometry) was shown to be reduced by use of quenching species introduced into the plasmas.(79,80,84,88) Gases such as methane, nitrogen, and mixtures of rare gases have clearly shown that collisional energy transfer from a metastable atom to a target of lower ionization potential is a very efficient process. These sorts of findings point out the importance of vacuum integrity in glow discharge devices, as residual atmospheric gas can greatly perturb the source operation characteristics. Larkins(89) has recently demonstrated that use of argon discharge gas with as little as 140 ppm of water vapor can reduce sputtering rates as much as 77%, with the actual percentage being a function of discharge conditions and the cathode material.

As would be expected, the production of metastable atoms is a function of source pressure (collision frequency) and discharge voltage/current (electron density and energy). Hardy and Sheldon(90) have presented a theoretical model for the production of helium, neon, and argon metastable atoms in low-pressure (0.01–1 Torr) glow discharge environments. Most important for the discussions here is the breakdown of the metastable densities as a cumulative set of loss mechanisms. Figure 2-25 illustrates the calculated cross sections for metastable neon atom depopulation via diffusion to the chamber walls (D), electron impact ionization (E), and quenching by collisions with ground state atoms (TB) as a function of cell pressure. Comparisons between calculated density response functions and actual densities determined by atomic absorption show much more extensive depopulation at higher source pressures than predicted, with the density values peaking at ~0.5 Torr. The authors suggest that the discrepancies seen for the higher source pressures are likely the result of the nonideal spectroscopic sampling of the plasma, rather than a flaw in their model. The observed response to discharge pressure at constant discharge current (a maximum density occurring over a limited pressure range) has in fact been observed by other researchers as illustrated in Fig. 2-26a, which is typical for glow discharge devices.[70] Likewise, the response to increases in discharge current at constant pressure (Fig. 2-26b) reflects the increases in electron density and energy as the voltage is increased. These effects are general to all glow discharge sources and, therefore, must be elucidated for particular source

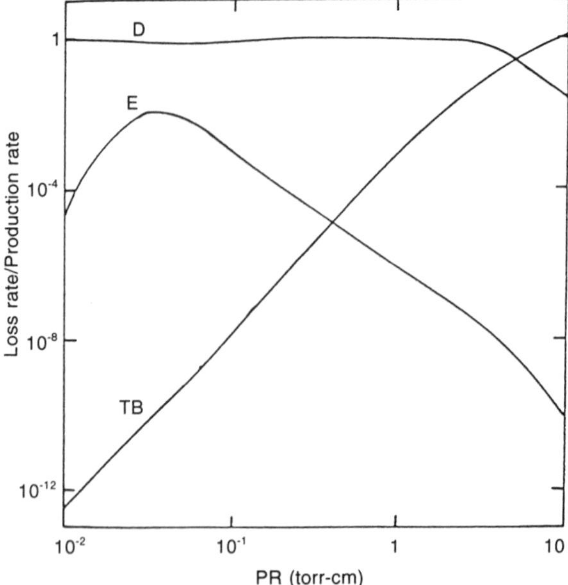

Figure 2-25. Calculated loss cross sections for metastable neon atoms lost by diffusion to the chamber walls (D), electron impact ionization (E), and quenching collisions with ground state atoms (TB).[90]

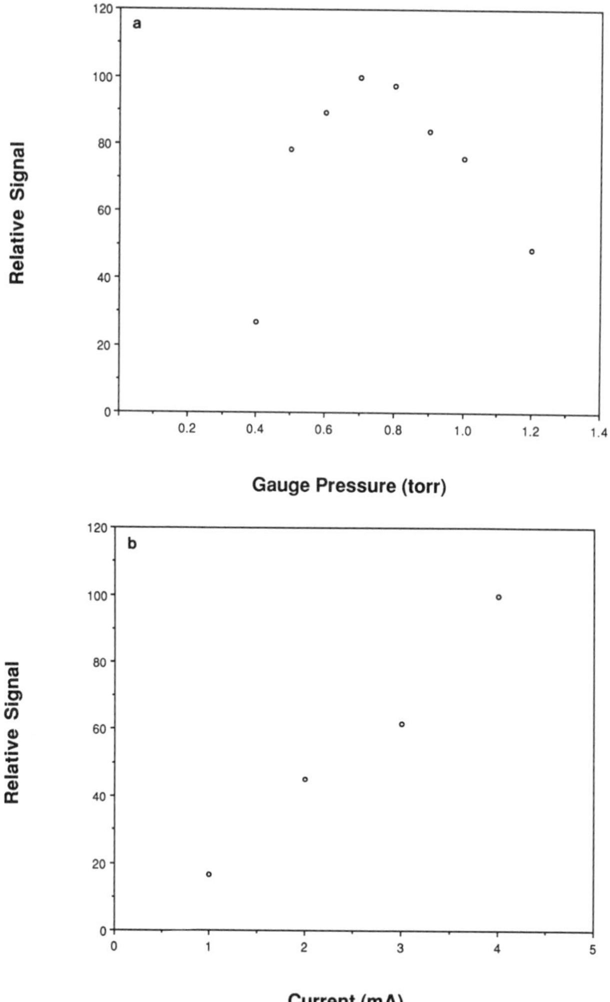

Figure 2-26. Effect of (a) discharge pressure (current = 3 mA) and (b) current (pressure = 0.8 Torr) on argon metastable atom absorption (811.5 nm) for a coaxial pin configuration; brass target. Adapted from Smith *et al.*[79]

geometries to gain a more thorough understanding of source operating characteristics.

If metastable discharge gas atoms play an important role in spectrochemical determinations, then the spatial distribution of the species within the glow discharge source is also an important consideration. The distribution of metastable argon atoms in the modified Grimm-type discharge has been studied by Ferreira and co-workers.[87] The populations were monitored by the measurement of the absorbance of argon metastable atoms of the Ar(I) 696.5-nm transition. A typical spatial profile is shown in Fig. 2-27 for a range of discharge conditions when sputtering a Cr:Cu (1:99) cathode. The plots reveal that the maximum metastable population exists at the cathode surface, with a steep decline throughout the CDS. In all cases, a second maximum is seen at a distance just beyond the CDS/NG interface. Similar types of profiles were presented by Doughty and Lawler[83] who measured the metastable atom densities and electric field strengths in a He glow discharge by optogalvanic spectroscopy. They found that the concentration of metastable atoms had a maximum value at the end of the electric field, i.e., the boundary of dark space and NG. Nearly identical plots to those in the Grimm source have been presented for a diode geometry device by Winchester and Marcus,[58] who offered the following explanation. The

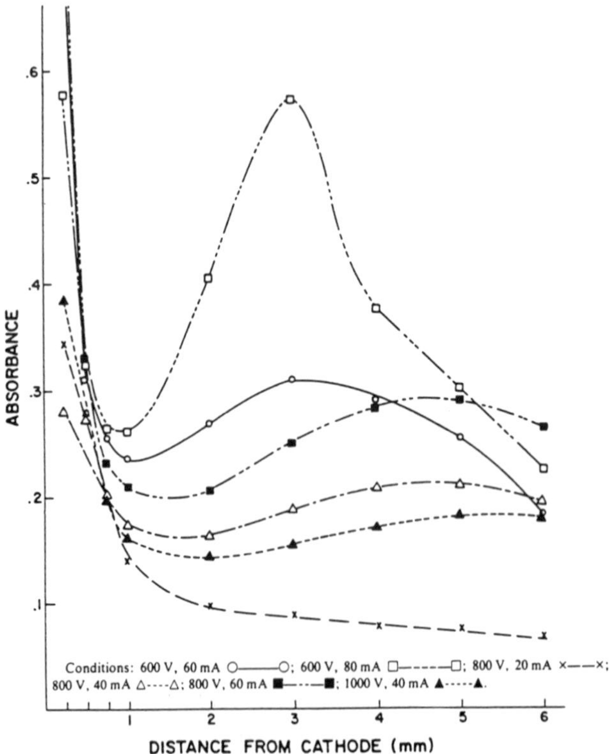

Figure 2-27. Spatial profile of argon metastable atom absorbance (696.5 nm) in a Grimm-type configuration (Cr–Cu target).[87]

high density of metastable atoms at the surface is likely the result of neutralizing collisions of sputtering argon ions at the cathode surface. The metastable atoms produced may then diffuse into the bulk of the plasma, or more likely be redeposited onto the cathode surface, thus the minimum within the CDS. The correlation between the second maximum and the CDS/NG interface corresponds to the region of the discharge where secondary electrons have lost sufficient energy to efficiently populate metastable levels. The decrease in metastable atom densities at farther distances reflects the lower-energy electron distributions in the NG as shown previously. The differences in the profiles as a function of discharge conditions can most likely be explained on the basis of electron energies and densities (production), as well as the density of sputtered and discharge gas atoms (loss).

2.5. Analyte Excitation and Ionization

At the pressures at which glow discharges operate, collisional processes determine the observed excited and ionized states. In order for sputtered target atoms to become electronically excited or ionized, there must be some sort of inelastic collision between the atom and a particle with either kinetic or potential energy. Inelastic collisions with plasma electrons of various kinetic energies are termed collisions of the "first kind." Collisions with massive particles, followed by subsequent potential energy transfer, are grouped together as collisions of the "second kind." Listed in Table 2-3 are the most likely excitation and ionization processes occurring in the NG region. Of course the relative roles of these processes will be determined by the actual discharge conditions and source geometry, which also dictate number densities and energies of the respective collision partners, and the physical cross sections of the fundamental processes.

2.5.1. Excitation

Atoms that complete the fortuitous journey to the NG are subject to inelastic collisions with electrons, metastable atoms, and ions. Based simply on number density considerations, these mechanisms for electronic excitation should be roughly equivalent. Geometrically speaking, the latter two processes should be dominant. Electronic excitations due to electron impact are more generally termed "thermal excitation." The key factor in assessing the relative roles of these processes is the resonant nature of electronic excitation. On this basis, electron impact excitation is clearly most favorable. Electrons in the NG, a virtually field-free region, are usually of low energy as shown in Figs. 2-21 and 2-22. The resonant character reflects the fact that those energy levels existing at ≤ 3 eV above the ground state will be

Table 2-3 Excitation and Ionization Processes in the Glow Discharge

I. Primary excitation/ionization processes

A. Electron impact

$$M^0 + e^- \text{ (fast)} \rightarrow M^* + e^- \text{ (slow)}/M^+ + 2e^-$$

B. Penning collisions

$$M^0 + Ar_m^* \rightarrow M^* + Ar^0/M^+ + Ar^0 + e^-$$

II. Secondary processes

A. Charge transfer

1. Nonsymmetric

$$Ar^+ + M^0 \rightarrow M^+ \ (M^{+*}) + Ar^0$$

2. Symmetric (resonance)

$$X^+ \text{ (fast)} + X^0\text{(slow)} \rightarrow X^0\text{(fast)} + X^+ \text{ (slow)}$$

3. Dissociative

$$Ar^+ + MX \rightarrow M^+ + X + Ar^0$$

B. Associative ionization

$$Ar_m^* + M^0 \rightarrow ArM^+ + e^-$$

C. Photon-induced excitation/ionization

$$M^0 + h\nu \rightarrow M^*/M^+ + e^-$$

D. Cumulative ionization

$$M^0 + e^- \rightarrow M^* + e^- \rightarrow M^+ + 2e^-$$

[a] M^0, sputtered neutral; Ar_m^*, metastable Ar atom (for example); X, any gas-phase atom.

preferentially populated. For this reason, *atomic* transitions in high UV and visible regions of the spectrum dominate the emission from the NG.

In addition to the production of excited state analyte atoms, electron impact is also responsible for the populating of the metastable levels of the discharge gas atoms. A second pathway would be the recombination of discharge gas ions leaving the atom in the metastable level,[(87)] though at typical glow discharge pressures the required three-body collisions are improbable. Collisions between the metastable gas atoms and sputtered neutral atoms may lead to an internal energy transfer from the metastable to the atom resulting in the excitation or ionization of the latter. Hess and coworkers,[(79)] in studying the quenching of metastable discharge gas atoms, have suggested that Penning-type collisions do contribute to the formation of excited state *ionic* species. This process will likely have a resonance dependence in that the excited state ion energy should be close to that of the metastable energy value or that of an excited state atom plus the metastable energy.

In discussing the production of excited state ions, asymmetric charge exchange has been postulated to be a prominent mechanism. Steers and coworkers[91,92] have studied the population of Cu(II) states with hollow cathode lamps, conventional Grimm lamps, and microwave-boosted Grimm lamps using neon and argon as discharge gases. The authors argue that, on the basis of state selectivity, direct production of excited state ions via charge exchange from argon ions is a prominent mechanism. Analogous production of higher excited states by neon was not investigated. Populating the states via a two-step mechanism of Penning ionization followed by electron impact excitation of the ion was said to be improbable. The excited state–Penning ionization mechanism was not addressed. In a similar study, Wagatsuma and Hirokawa[93] have used He-doped Ar and Ne discharge gases in Grimm-type discharges to study charge exchange reactions. They also concluded this to be a mechanism for the production of excited state ions. Again, the extent of this process will be highly dependent on the identity of the discharge gas and the sputtered atom in question. It is interesting to note that while the high-current-density hollow cathode and Grimm geometries exhibit asymmetric charge exchange characteristics, the low-current diode geometry employed by Hess[79] and most mass spectrometry applications does not seem to have asymmetric charge exchange as a primary excitation mechanism. While these differences are quite interesting, it must be kept in mind that ionic emission is a very minor component of glow discharge optical spectra.

2.5.2. *Ionization*

In principle, each of the collisional processes listed in Table 2-3 is a viable candidate for analyte ionization in glow discharge mass spectrometry so long as the ionization potential can be met. In many respects, knowledge of the dominant ionization processes is more important than for analyte excitation. In mass spectrometry, there is no analogy to degree of excitation where simply different states are populated; in glow discharge environment ionization is simply a question of “ion or atom.” Once in an ionic state, the analyte can be detected mass spectrometrically. Therefore, the primary concern is the production of a maximum number of ions with as much elemental uniformity as possible.

When looking at the source of kinetic and potential energies for glow discharge ionization in the NG (ions cannot be extracted from the dark space), the mechanisms of electron impact, Penning ionization, and charge exchange are again the most likely candidates. Referring to the electron energy distributions shown in Fig. 2-21 and 2-22, it is fairly clear that there is not a sufficient electron population in the 5–10 eV range to produce extensive ionization. Given that electron and ion densities are both on the order of

10^{11} cm^{-3}, the populations of these higher-energy electrons are not likely greater than $\sim 10^7$–10^8 cm^{-3}. Vieth and Huneke[94] have used glow discharge mass spectrometry to measure the relative populations of singly and multiply charged argon, matrix, and argide ions in an Ar glow discharge for different sample matrices. They found that the measured relative densities of multiply charged argon ions (from +3 to +6) are in good agreement with the Lovett rate model[95] for electron impact ionization/three-body recombination with an electron temperature of about 7 eV. The 7-eV electrons were said to be the result of a double-Penning collision of the form

$$Ar_m^* + Ar_m^* \rightarrow Ar^+ + Ar + e^-(7.1\text{–}7.6\ \text{eV}) \tag{2-8}$$

However, there is no experimental evidence for the existence of electron populations of these energies. Regardless of the exact mechanism, such highly charged ions ($>M^{2+}$) are extremely rare for analyte species.

Penning-type collisions have generally been assigned as the dominant ionization mechanism of sputtered atoms in most types of glow discharge devices.[78,79,82,85,86] Coburn and Kay[85] demonstrated the importance of Penning ionization in the rf sputtering of a europium monoxide (EuO) cathode (containing 6% Fe) in both argon and neon discharges. Eu^+ and Fe^+ signals were observed in both discharges, but an O^+ peak was observed only in the neon discharge. The authors explained this by the difference in the metastable energies of argon and neon. The energy of metastable argon atoms is sufficient to Penning-ionize Eu and Fe (IP = 5.68 and 7.83 eV, respectively) but unable to Penning-ionize oxygen atoms (IP = 13.55 eV), which can be Penning-ionized by the metastable neon atoms. Their experiments also implied that the contribution of electron impact ionization and dissociative charge-transfer ionization is less significant relative to the Penning ionization.

Coburn and co-workers[86] extended the previously mentioned studies employing simultaneous mass spectrometric and atomic absorption measurements to relate sputtered copper atom and Ne metastable densities to Cu^+ signals. Figure 2-28 illustrates the relationship between the product of the atom × metastable density and ion signal over a range of operating powers and pressures in an rf glow discharge. The direct relationship is strong evidence for a Penning ionization mechanism since that process would be first-order in the respective collisional partner concentrations. Indirect evidence of Penning ionization has been obtained by Hess and co-workers[79] who related the depopulation of neon and argon metastable atoms through methane quenching and reductions in analyte ion signals measured optically.

Laser sources have been useful tools for elucidating the possible role of metastable atoms in glow discharge ionization. Harrison and co-workers[82] used a combination of optogalvanic spectroscopy and mass spectrometry to

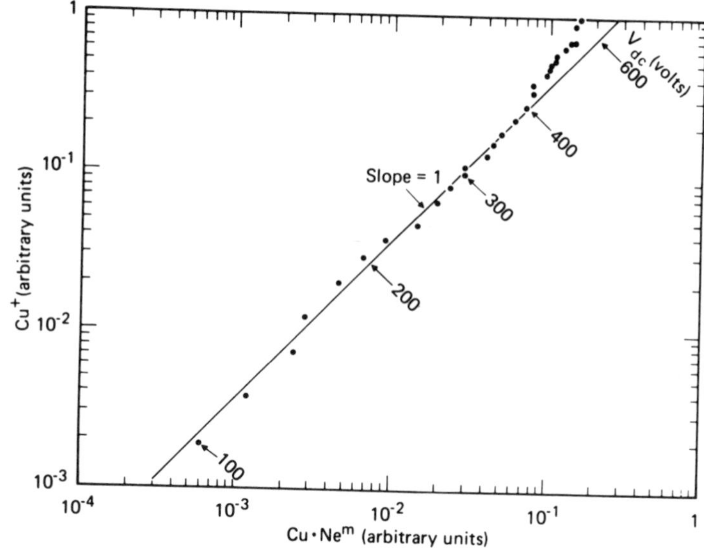

Figure 2-28. Cu^+ versus Cu.Ne^m as a function of rf power for a diode configuration at a pressure of 20 mTorr.[86]

illustrate that depopulation of neon metastable levels in a hollow cathode discharge affected discharge voltage characteristics and reduced the monitored copper ion signal. These findings support a Penning mechanism. Hess and Harrison[78] also used a tunable laser to depopulate metastable atom levels in a coaxial pin discharge geometry and monitored the effects by optogalvanic spectroscopy and mass spectrometry. In the neon discharge, they found that depopulation of the metastables caused a general reduction in all measured ion signals, while similar studies with argon showed that many species with ionization potentials greater than metastable energies of argon were not affected by laser depopulation. Figure 2-29 depicts the use of gated detection to monitor the decrease of analyte ion signal in the time frame of the metastable state depopulation.

In terms of quantitating the extent of Penning ionization, no definitive experiments have been performed. Although most researchers believe that Penning ionization is the dominant mechanism in the NG region, electron impact ionization is still likely to play a role in analyte ionization. It is difficult to conclude whether electron impact ionization will compete with the Penning ionization, since secondary electron populations exist at two or three orders of magnitude less than those of thermal electrons and the lifetimes of metastables are long enough to dominate the ionization events. Vieth and Huneke[96] have deduced through elemental quantitation studies that Penning ionization accounts for between 50 and 95% of the overall

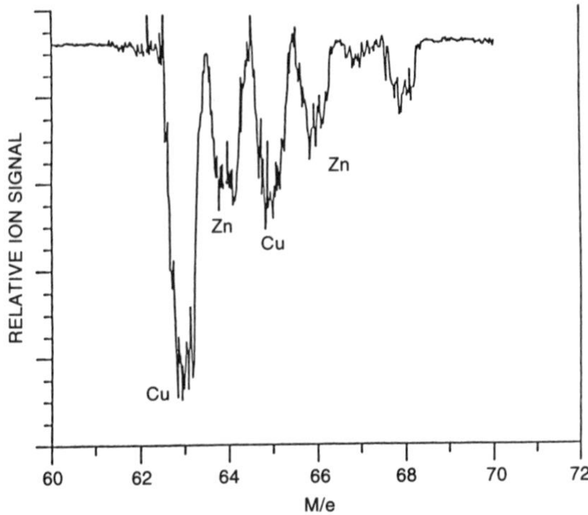

Figure 2-29. Mass spectrum produced by the subtraction of the steady-state glow discharge ion signal from the ion signal generated during laser metastable depopulation [Ne(I) 594.48 nm]; Ne discharge, 0.4 Torr, 3 mA.[78]

analyte ionization. The given range reflects the relative cross sections of elements for electron impact ionization.

The final mechanism that might be a meaningful contributor to NG ionization is asymmetric charge exchange. Given the resonant nature of the charge exchange process, as discussed in the previous section, it seems quite unlikely that charge exchange from (presumably) discharge gas ions is a major contributor to the singly ionized analytes present in glow discharge mass spectra. Hecq and Hecq[97] have, however, implicated charge exchange from doubly charged argon discharge gas ions to the formation of doubly charged ions of sputtered atoms.

2.6. Conclusions

In this chapter, the fundamental plasma properties of glow discharge devices have been presented in an effort to give current and future practitioners insight into those parameters that affect analytical source performance. While glow discharges are easily controlled and operate under rather mild conditions (in terms of power and gas temperature), the spectrochemical behavior is the result of a very complex set of collisional processes. The cathodic sputtering (atomization) step presents real advantages over thermal vaporization methods such as those found in arcs and laser/solid interactions. Elemental sputtering yields were shown to vary by only a factor of 3 to 5. For this reason, nominal differences are found between the sputter rates of vastly different sample matrices. Even so, a truly outstanding feature of the sputtering process is the fact that once steady–state sputtering conditions are met (1 to 5 monolayers), sputter rates for all sample components are the same and the sample introduction is stoichiometric.

The gas-phase collisions most prominent in glow discharge devices are those of electron impact, Penning, and charge transfer reactions. Based on number density and energy considerations, electron impact is clearly the dominant atomic excitation mechanism. Under discharge conditions that involve large current densities (Grimm and hollow cathode sources), asymmetric charge exchange from discharge gas ions is a likely mechanism for production of excited state ions. In lower-density sources, such as those employed in mass spectrometry, excited state ions seem to be produced via Penning ionization of excited state atoms.

Ionization in glow discharge devices can be classified according to the region of the plasma in which the ions are formed. Discharge gas ions, which are the incident particles in the sputtering process, are formed by direct electron impact ionization in the vicinity of the CDS/NG interface. In this region, secondary electrons emitted from the cathode surface have been cooled by collisions with heavy particles. In the NG the electron populations at atomic ionization energies are negligible; therefore, collisions of the second kind are dominant. (Ionization of discharge gas ions is most likely the result of electron impact ionization out of atomic metastable levels.) Most experimental evidence points to the Penning ionization mechanism as the dominant reaction in the production of analyte ions.

Throughout the remainder of this volume, the respective chapters will emphasize the fundamental processes that underlie the specific spectrochemical application.

Acknowledgment

Many of the results presented here from Clemson University are based on work supported by the National Science Foundation under Grant CHE-8901788, whose support is gratefully acknowledged.

References

1. A. M. Howatson, *An Introduction to Gas Discharges*, Pergamon Press, Elmsford, N.Y., 1976.
2. J. A. C. Broekaert, *J. Anal. At. Spectrom.* 2 (1987) 537.
3. A. von Engel, *Electric Plasmas: Their Nature and Applications*, Taylor & Francis, New York, 1983.

4. P. W. J. M. Boumans, *Theory of Spectrochemical Excitation*, Hilger & Watts, London, 1966.
5. L. Paksy and I. J. Lakatos, *Spectrochim. Acta* 38B (1983) 1099.
6. L. Radermacher and H. E. Beske, *Spectrochim. Acta* 37B (1982) 769.
7. F. Paschen, *Ann. Phys.* 50 (1916) 901.
8. H. Schuler, *Z. Phys.* 59 (1929) 149.
9. F. W. Aston, *Mass Spectra and Isotopes*, 2nd ed., Longmans, Green, New York, 1942.
10. E. Nasser, *Fundamentals of Gaseous Ionization and Plasma Electronics*, Wiley–Interscience, New York, 1971.
11. B. Chapman, *Glow Discharge Processes*, Wiley–Interscience, New York, 1980.
12. K. Wiesemann, *Phys. Lett.* 29A (1969) 691.
13. F. W. Aston, *Proc. R. Soc. London Ser. A* 84 (1911) 526.
14. D. Fang and R. K. Marcus, *Spectrochim. Acta* 45B (1990) 1053.
15. P. F. Knewstubb and A. W. Tickner, *J. Chem. Phys.* 36 (1962) 684.
16. F. W. Aston, *Philos. Mag.* 46 (1923) 211.
17. M. E. Pillow, *Spectrochim. Acta* 36B (1981) 821.
18. G. S. Anderson, W. N. Meyer, and G. K. Wehner, *J. Appl. Phys.* 33 (1962) 2991.
19. D. C. Duckworth and R. K. Marcus, *Anal. Chem.* 61 (1989) 1879.
20. W. R. Grove, *Philos. Trans. R. Soc. London* 142 (1852) 87.
21. W. D. Westwood, *Prog. Surf. Sci.* 7 (1976) 71.
22. A. Benninghoven, *Surf. Sci.* 53 (1975) 596.
23. R. F. K. Herzog, W. P. Poschenrieder, and F. K. Satidiewics, *Radiat. Eff.* 18 (1973) 199.
24. V.A. Von Hippel, *Ann. Phys. (Leipzig)* 80 (1926) 672.
25. V.A. Von Hippel, *Ann. Phys. (Leipzig)* 81 (1926) 1043.
26. P. Sigmund, *Phys. Rev.* 184 (1969) 383.
27. H. F. Winters, in: *Topics in Current Chemistry No. 94, Plasma Chemistry III* (S. Veprek and M. Vervgopalen, eds.), Springer-Verlag, Berlin, 1980.
28. P. W. J. M. Boumans, *Anal. Chem.* 44 (1972) 1219.
29. M. A. Rudat and C. N. McEwen, *Int. J. Mass Spectrom. Ion Phys.* 46 (1983) 351.
30. G. K. Wehner, *Phys. Rev.* 108 (1957) 35.
31. G. K. Wehner, *Phys. Rev.* 102 (1956) 690.
32. H. A. Oeschner, *Physics (N.Y.)* 261 (1973) 37.
33. J. Roth, J. Bohdansky, and W. Eckstein, *Nucl. Instrum. Methods Phys. Res.* 218 (1983) 751.
34. G. Carter and J. S. Colligan, *Ion Bombardment of Solids*, Chapter 7, Heinemann, London, 1968.
35. R. S. Nelson, *Philos. Mag.* 11 (110) (1965) 291.
36. R. Kelly and N. Q. Lam, *Radiat. Eff.* 19 (1973) 39.
37. B. J. Stocker, *Br. J. Appl. Phys.* 12 (1961) 465.
38. J. Musha, *Phys. Soc. Jpn.* 17 (1962) 1440.
39. D. Fang and R. K. Marcus, *Spectrochim. Acta* 43B (1988) 1451.
40. M. Dogan, K. Laqua, and H. Massmann, *Spectrochim. Acta* 26B (1971) 631.
41. G. F. Weston, *Cold Cathode Discharge Tubes*, Iliffe, London, 1968.
42. G. Betz, *Surf. Sci.* 92 (1980) 283.
43. G. Betz and G. K. Wehner, in: *Sputtering by Particle Bombardment II*, (R. Behrisch, ed.), Springer-Verlag, Berlin, 1983.
44. K. Wittmack, *Nucl. Instrum. Methods* 168 (1980) 343.
45. G. K. Wehner, in: *Methods and Phenomena: Their Applications in Science and Technology* (S. P. Wolsky and A. W. Czanderna, eds.), Vol. 1, Elsevier Scientific, New York, 1975.
46. G. E. Hammer and R. M. Shemenski, *Surf. Interface Anal.* 10 (1987) 355.
47. D. Fang and R. K. Marcus, *J. Anal. At. Spectrom.* 3 (1988) 873.

48. V. L. Ginsburg, *J. Phys.* (*USSR*) 8 (1944) 253.
49. G. F. Weston, *Cold Cathode Discharge Tubes*, Iliffe, London, 1968.
50. N. H. Tolk, I. S. T. Tsong, and C. W. White, *Anal. Chem.* 49 (1977) 16A.
51. J. M. Schroeer, R. N. Rhodin, and R. C. Bradley, *Surf. Sci.* 37 (1973) 571.
52. R. V. Stuart and G. K. Wehner, *J. Appl. Phys.* 35 (1964) 1819.
53. V. I. Veksler, *Sov. Phys. JETP* 11 (1960) 235.
54. S. Komiya, K. Yoshikawa, and S. Ono, *J. Vac. Sci. Technol.* 14 (1977) 1161.
55. W. W. Harrison and C. G. Bruhn, *Anal. Chem.* 50 (1978) 16.
56. A. J. Sterling and W. D. Westwood, *J. Phys. D* 4 (1971) 246.
57. N. P. Ferreira and H. G. C. Human, *Spectrochim. Acta* 36B (1981) 215.
58. M. R. Winchester and R. K. Marcus, *16th Annu. Meet. Fed. Anal. Chem. Spectrosc. Soc.* 1989, Paper No. 327.
59. D. K. Doughty, E. A. Den Hartog, and J. E. Lawler, *Appl. Phys. Lett.* 46 (1985) 352.
60. F. W. Aston, *Proc. R. Soc. London Ser. A* 104 (1923) 565.
61. A. V. Bodarenko, *Sov. Phys. Tech. Phys.* (*Engl. Transl.*) 18 (1973) 515.
62. F. Howorka, W. Lindinger, and M. Pahl, *Int. J. Mass Spectrom. Ion Phys.* 12 (1973) 67.
63. F. Howorka and M. Pahl, *Z. Naturforsch. Teil A* 27 (1972) 1425.
64. S.-L. Tong and W. W. Harrison, *Anal. Chem.* 56 (1984) 2028.
65. M. Dogan, K. Laqua, and H. Massmann, *Spectrochim. Acta* 26B (1971) 631.
66. P. W. J. M. Boumans, *Anal. Chem.* 44 (1972) 1219.
67. A. Guntherschulze, *Z. Tech. Phys.* 11 (1930) 49.
68. P. F. Little and A. Von Engel, *Proc. R. Soc. London Ser. A* 224 (1954) 209.
69. V. I. Kirichenko, V. M. Tkachenko, and V. B. Tyutyunnik, *Sov. Phys. Tech. Phys.* (*Engl. Transl.*) 21 (1976) 1080.
70. W. Stern, *Beitr. Plasmaphys.* 9 (1969) 59.
71. C. T. J. Alkemade, T. Hollander, W. Snelleman, and P. J. T. Zeegers, *Metal Vapours in Flames*, Pergamon Press, Oxford, 1982.
72. F. F. Chen, in: *Plasma Diagnostic Techniques* (R. H. Huddlestone and S. L. Leonard, eds.), Academic Press, New York, 1965.
73. L. Vriens, *J. Appl. Phys.* 45 (1974) 1191.
74. D. Fang and R. K. Marcus, *J. Anal. At. Spectrom.* 5 (1990) 569.
75. D. Fang and R. K. Marcus, *Spectrochim. Acta* 46B (1991) 983.
76. V. S. Borodin, and Y. M. Kagan, *Sov. Phys. Tech. Phys.* (*Engl. Transl.*) 11 (1966) 131.
77. V. S. Borodin, Y. M. Kagan, and R. I. Lyaguschenko, *Sov. Phys. Tech. Phys.* (*Engl. Transl.*) 11 (1967) 887.
78. K. R. Hess and W. W. Harrison, *Anal. Chem.* 60 (1988) 691.
79. R. L. Smith, D. Serxner, and K. R. Hess, *Anal. Chem.* 61 (1989) 1103.
80. M. K. Levy, D. Serxner, A. D. Angstadt, R. L. Smith, and K. R. Hess, *Spectrochim. Acta* 46B (1991) 253.
81. R. B. Green, R. A. Keller, G. G. Luther, P. K. Schenck, and J. C. Travis, *Appl. Phys. Lett.* 29 (1976) 727.
82. K. C. Smyth, B. L. Bentz, C. G. Bruhn, and W. W. Harrison, *J. Am. Chem. Soc.* 101 (1979) 797.
83. D. A. Doughty and J. E. Lawler, *Plasma Processing, Mater. Res. Soc. Symp. Proc.* 68 (1986) 141.
84. N. P. Ferreira, J. A. Strauss, and H. G. C. Human, *Spectrochim. Acta* 37B (1982) 273.
85. J. W. Coburn and E. Kay, *Appl. Phys. Lett.* 18(10) (1971) 435.
86. E. W. Eckstein, J. W. Coburn, and E. Kay, *Int. J. Mass Spectrom. Ion Phys.* 17 (1975) 129.
87. J. A. Strauss, N. P. Ferreira, and H. G. C. Human, *Spectrochim. Acta* 37B (1982) 947.
88. L. G. Piper, J. E. Velazco, and D. W. Stetzer, *J. Chem. Phys.* 59 (1973) 3323.

89. P. L. Larkins, *Spectrochim. Acta* 46B (1991) 291.
90. K. A. Hardy and J. W. Sheldon, *J. Appl. Phys.* 53 (1982) 8532.
91. E. B. M. Steers and R. J. Fielding, *J. Anal. At. Spectrom.* 2 (1987) 239.
92. E. B. M. Steers and F. Leis, *J. Anal. At. Spectrom.* 4 (1989) 199.
93. K. Wagatsuma and K. Hirokawa, *Spectrochim. Acta* 46B (1991) 269.
94. W. Vieth and J. C. Huneke, *Spectrochim. Acta* 45B (1990) 941.
95. R. J. Lovett, *Spectrochim. Acta* 37B (1982) 985.
96. W. Vieth and J. C. Huneke, *Spectrochim. Acta* 46B (1991) 137.
97. M. Hecq and A. Hecq, *J. Appl. Phys.* 56 (1984) 872.

3

Atomic Absorption and Fluorescence Spectroscopies

Edward H. Piepmeier

3.1. Introduction

The analysis of alloys, powders, and dried solutions by atomic absorption and fluorescence spectrometry (AAS and AFS) using sputtering cells for atomization is receiving wider notice in the literature as the advantages of these methods come to be appreciated. For example, compared with emission spectrometry, these methods have higher spectral resolution and fewer spectral interferences. Rapid multielement AAS analyses are now practical because of recent technological advances. With recent developments, sensitivities approach and in some cases exceed flame and graphite furnace methods, and improvements are continuing.

AAS for determining chemical elements in solution was introduced as a viable analytical method in 1955 by independent and simultaneous papers by Alkemade and Milatz[1,2] who demonstrated the determination of Na in solutions introduced into a flame and discussed practical considerations, and by Walsh[3] who discussed in detail the theoretical and experimental considerations of the method. The direct analysis of metals using a sputtering chamber as a convenient atomic absorption cell was first proposed by Russell and Walsh in 1959[4] in a paper describing perceptive observations of the differences in the spectra emitted in different directions from a hollow

Edward H. Piepmeier • Department of Chemistry, Oregon State University, Corvallis, Oregon 97331.

Glow Discharge Spectroscopies, edited by R. Kenneth Marcus. Plenum Press, New York, 1993.

cathode lamp. Their observation that atomic fluorescence was occurring in the lamp eventually led to the use of AFS to observe the atomic vapor in a sputtering cell.[5] Their prediction that the sputtering chamber method would be useful "particularly for analyses requiring very high sensitivity" is continuing to materialize as a better understanding is being obtained of ways to improve sensitivity.

Figure 3-1 shows a block diagram of the main components of atomic absorption and fluorescence instruments. Modulated or pulsed hollow cathode lamps are commonly used as the primary radiation source for atomic absorption measurements although tunable lasers are used in special cases. A modulated Grimm-type glow discharge has been used as the primary radiation source for multielement simultaneous reading AAS because of the ease of changing the cathode to select the elements of interest.[6] Although hollow cathode lamps are used to excite atomic fluorescence, tunable-laser-excited fluorescence provides the best detection limits.[7]

The wavelength isolation system may be a tunable monochromator, an interference filter, or a resonance atomic fluorescence monochromator (RAFM; consisting of a cloud of atoms of the element of interest, which absorbs and then fluoresces radiation of that element[8]). Fluorescence measurements have been made without a wavelength isolation system,[5] but background interferences were apparent. After the appropriate wavelength has been isolated, a photomultiplier converts the radiation into an electrical signal for processing and eventual readout.

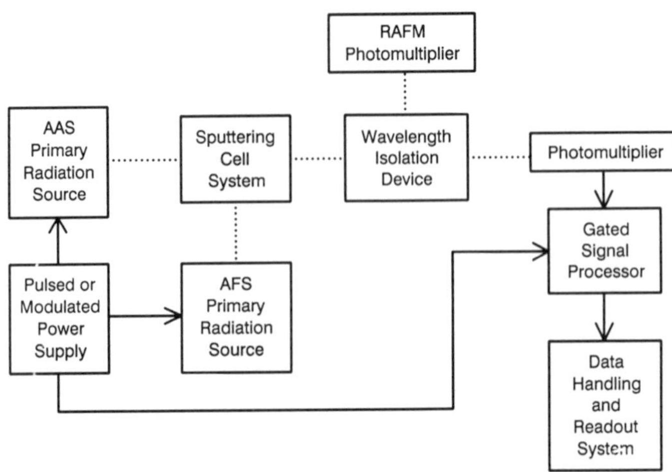

Figure 3-1. Block diagram of the main components of basic sputtering cell atomic absorption (AAS) and atomic fluorescence (AFS) spectrometers. The RAFM photomultiplier is used when the wavelength isolation device is a resonance atomic fluorescence monochromator.

Several types of sputtering cells are used to convert a solid sample into an atomic vapor for the atomic absorption or fluorescence measurements. In 1960, Gatehouse and Walsh[(9)] used a stagnant-gas sputtering chamber containing a sample in the form of a hollow cathode. Although the time required for an analysis was much longer than for other routine analytical atomic spectrometry methods, Goleb[(10)] found the narrow absorption lines of a low-pressure hollow cathode sputtering cell useful for the isotopic analysis of uranium samples. Goleb and Brody[(11)] applied this method to solutions by drying them on the inside of the hollow cathode. In 1971, Gandrud and Skogerboe[(12)] used flat spectroscopic-grade graphite and aluminum platrodes as replaceable cathodes on which sample solutions were dried and inserted into the sputtering cell, which had a flowing gas stream. An analysis rate of 15–20 determinations/h is possible with their system, and detection limits range from 1 to 10 ng.

In 1973, Gough *et al.*[(5)] used a sputtering cell for atomic fluorescence determinations. Flat samples were now easily mounted on the side of the cell by pressing the sample against an O-ring surrounding a replaceable silica annulus with a 1-cm hole. The O-ring provides a vacuum seal and the 1-cm hole defines the location of the discharge on the sample surface. Surrounding the hole is a ledge on the annulus that provides a 0.02-cm gap between the sample and the annulus to prevent deposits that might otherwise electrically short the sample to the anode. A flow-through gas control system aids the rapid interchange of samples. In 1976, Gough[(13)] reported an atomic absorption cell (Fig. 3-2) similar to the fluorescence cell, but the gas flow was specifically designed to improve the absorption signal by increasing the transport of the sputtered atoms into the observation zone, which was 1–2 cm from the flat sample.

In 1987, Bernhard[(14)] reported the discovery that gas jets that strike the sputtering surface significantly increase the sampling rate as well as the absorption signal in a sputtering chamber. This discovery led to a renewed interest in sputtering cells for the analysis of alloys, powders, and dried solutions by AAS and AFS. A commercial atomic absorption cell (Fig. 3-3), based on this principle and named the Atomsource[(14)] (Analyte Corp., Medford, OR), has further stimulated wider applications of gas-jet-enhanced sputtering because it can be mounted on many of the atomic absorption spectrometers that have appeared in analytical laboratories since the pioneering work of Walsh and his co-workers over three decades ago. For the 15 elements studied, the characteristic concentrations that produce an absorbance of 0.0044 for the Atomsource are comparable to, or better than, those for a flame, assuming for the flame that 1 g of alloy is dissolved in 100 ml of solution.[(15)]

The increased sampling rate of a gas-jet-enhanced sputtering cell has recently been applied by Chakrabarti *et al.*[(16)] to the analysis of dried solutions deposited on flat cathodes pressed against the sample port of the

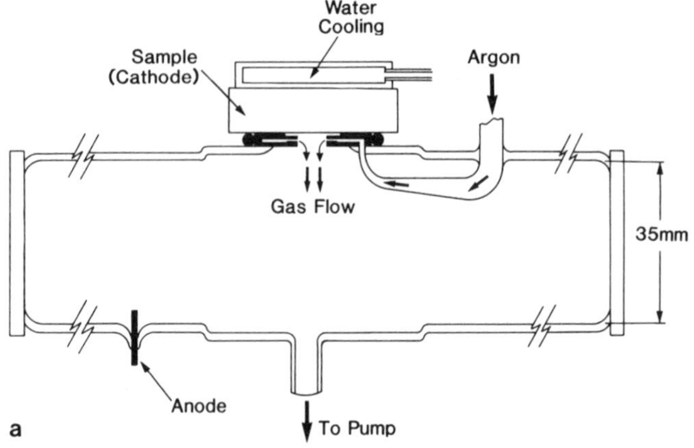

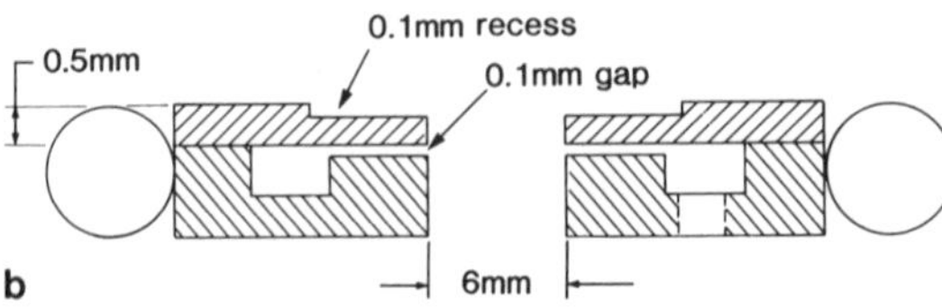

Figure 3-2. Schematic diagram of (a) Gough's sputtering chamber showing the replaceable hollow silica annulus (b) that admits high-velocity gas into the chamber. Reprinted with permission from D. S. Gough, *Anal. Chem.* 48 (1976) 1926, copyright 1976, American Chemical Society.

Atomsource. They used a high-electrical-power pulsed mode of operation for high atomization rates and obtained results that are comparable to those of the highly sensitive graphite furnace atomic absorption spectrometry (GFAAS) method for some elements. Absolute detection limits range from 0.2 to 600 pg. They predict that "the pulsed mode of atomization can be made at least as sensitive as GFAAS for most elements."

Although lasers are still relatively very expensive radiation sources, the high irradiances that are possible and their narrow spectral bandwidths make them attractive for special applications. Winefordner and co-workers have calculated limiting noise levels that should eventually allow the detection of only several thousand atoms (around 10^{-6} pg) in the observation region in a sputtering cell.(7,17,18) Even so, detection limits of only 0.5 pg for Pb and 20 pg for Ir in dried aqueous solutions have been demonstrated with their system, which used a hollow cathode sputtering cell at 30 mA, 500 V, and 0.7 Torr.(17) The difference between the detection limits for atoms in the

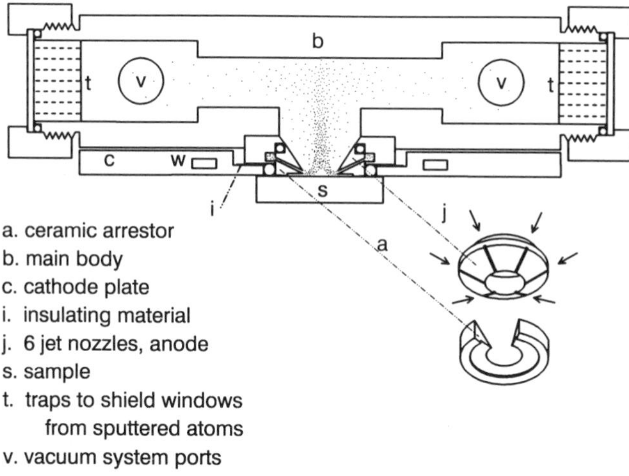

Figure 3-3. Schematic diagram of the Atomsource showing the six nozzles that direct high-velocity gas jets onto the sputtering surface of the sample. The isometric views of the nozzles and arrestor are adapted by permission from D. S. Gough, P. Hannaford, and R. Martin Lowe, *Anal. Chem.* 61 (1989) 1652, copyright 1989, American Chemical Society.

observation region and the amounts in the dried solution samples can be reduced by more efficient atomization and more efficient temporal and spatial probing of the sputtered atoms by the pulsed laser; ideally all of the sample atoms should be in the observation region when the atoms are observed. The increased sampling rate of a gas-jet-enhanced sputtering cell and high-current pulses of the type used by Chakrabarti and co-workers should help to improve detection limits.

The rest of this chapter considers some of the fundamental aspects of glow discharge AAS and AFS, followed by practical considerations, applications, and a look to the future.

3.2. Fundamental Considerations

3.2.1. Characteristics of Low-Pressure Atomic Spectra

The spectral profiles of the atomic absorption lines in low-pressure sputtering cells have a significant influence on spectral interferences and on the slopes and shapes of atomic absorption and fluorescence analytical working curves. Detailed information about spectral line profiles can help determine the shapes of analytical working curves so that appropriate mathematical models can be used to fit these curves to determine concentrations. Detailed information also helps predict how changes in operating conditions of the primary radiation source and the atom cell influence these curves, so that we can obtain better control of the operating conditions, and more efficiently update working curves as operating conditions drift with time. Factors that contribute to the spectral profile of an atomic absorption line will now be reviewed followed by the influences that spectral profiles have on analytical working curves and spectral resolution.

3.2.1.1. Shapes of Spectral Line Profiles

The shapes of atomic absorption line profiles vary significantly from one spectral line to another, even for different lines of the same element. Unfortunately, specific experimental and theoretical information about spectral profiles is available for relatively few analytically useful atomic lines, but more information continues to become available as the need for it is recognized. The theory of atomic spectral line profiles has been covered in more detail in several books[19–25] and papers.[26,27]

a. Hyperfine Components and Isotope Shifts. In general, the atomic absorption line profile may have several hyperfine components caused by different isotopes having slightly different energy levels, and by splitting of otherwise degenerate electron energy levels by nuclei that have a net nuclear spin (i.e., nuclei with an odd number of protons and/or neutrons). The overall absorption profile for a line is the sum of the spectral profiles of each of the hyperfine components.

The isotope shift for a handful of light and very heavy elements is large enough to form the basis for the determination of different isotopes using sputtering cells. For the lighter elements the relatively large isotope effect is due to the large relative differences in nuclear mass, and for the very heavy elements the isotope effect is mainly due to the nonuniform way in which nuclear charge is distributed over the volume of a nucleus.[28] That part of the isotope shift caused by mass difference can be estimated by

$$\Delta\lambda = \frac{-q\lambda}{1836}\,\frac{M_a - M_b}{M_a M_b} \tag{3-1}$$

where $\Delta\lambda$ is the wavelength shift (isotope shift) for a line at wavelength λ for two isotopes with masses M_a and M_b in daltons, and q is a constant that depends on the type of mass effect and the type of electrons involved in the energy level transition. For the *normal* mass effect, $q = 1$, and for the *specific* mass effect involving d electrons, q varies from -20 to $+16$.[26] For a mass difference of 1 at 33 daltons, these isotope wavelength shifts are on the order of −0.004 to +0.003 nm. The isotope shift caused by the nonuniform distribution of nuclear charge is comparable in size to the mass effect for intermediate elements and dominates for heavy elements (>140 daltons).[29] About two-thirds of the stable atomic nuclei have a net nuclear spin that causes hyperfine splitting, which typically amounts to 10^{-3} nm.

The relative intensities of these hyperfine components vary, depending on the relative abundances of the different isotopes and on Russell–Saunders coupling. Although the relative intensities can often be predicted, the positions of the hyperfine components along the wavelength axis usually must be determined experimentally.

b. Broadening. Each of the hyperfine components is individually broadened by natural lifetime, Doppler, and collisional (or pressure) broadening processes. For example, the vertical bars in Fig. 3-4 show the hyperfine components for the 403.3-nm Mo(I) line while the profile with several peaks corresponds to the low-pressure or Doppler-broadened profile. The broad profile corresponds to the atmospheric pressure, high-temperature Doppler- and pressure-broadened profile. All lines are influenced by natural lifetime broadening, which has a spectral profile in the shape of a Lorentzian function with a width at half-height of

$$\Delta\lambda_{\mathrm{L}} = \frac{\lambda_0^{\,2}}{2\pi c}\left[\frac{1}{\tau_{\mathrm{i}}} + \frac{1}{\tau_{\mathrm{j}}}\right] \tag{3-2}$$

where τ_{i} and τ_{j} respectively represent the lifetimes of the upper and lower energy states for the transition, c is the velocity of light (in nm s^{-1}), and λ is in nanometers. For a resonance line, where the ground state lifetime is very long, the term in brackets becomes the Einstein probability for spontaneous

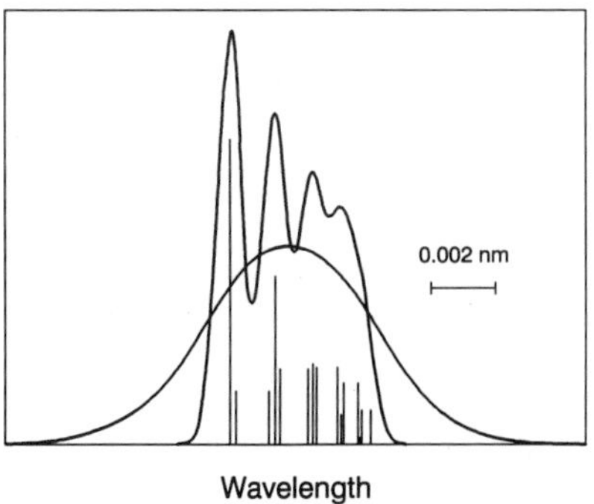

Figure 3-4. Shapes of the Mn(I) 403.3-nm spectral line in a low-pressure glow discharge (four-peaked curve) and an atmospheric-pressure flame (single-peaked curve). The unbroadened hyperfine components are indicated by vertical lines.

emission, A_{ij} in photons per second. A strong line, with a 1-ns lifetime ($A_{ij} = 10^9\ s^{-1}$), at 300 nm has a natural lifetime half-width of only 5×10^{-5} nm. Although this is narrower than half-widths caused by other broadening processes, the Lorentzian function has long tails, which become significant in extreme cases as we will see.

The main broadening process in a low-pressure sputtering cell is the Doppler effect. The Doppler effect causes the wavelength at which an atom absorbs to be shifted to the blue (shorter wavelength) for atoms traveling with a velocity component v (cm s^{-1}) toward the detector, while a redshift occurs for atoms traveling in the opposite direction. To a first-order approximation the blueshift is given by

$$(\lambda_0 - \lambda) = \lambda_0 v/c \tag{3-3}$$

where λ_0 is the wavelength at which a stationary atom absorbs, λ is the wavelength absorbed by an atom moving with a velocity v (in cm) relative to the detector, and c is the velocity of light (in cm s^{-1}). *The absorption profile and the emission profile are shifted by the same amounts and in the same direction.*

Net Doppler shifts in emission and absorption have been observed in pulsed hollow cathode lamps[(30,31)] and they may be present in other pulsed atomizers where a rapid gas expansion occurs. They may also be present in atomizers where the gas is rapidly flowing toward or away from the detector.

When the translational energy of a population of atoms is in thermal equilibrium so that the velocities of the atoms are given by the Maxwell–Boltzmann distribution, the result is a Doppler-broadened spectral profile that is described by a Gaussian function. The width of the Gaussian peak at half-height is given by

$$\Delta\lambda_D = 7.16 \times 10^{-7}\ \lambda_0(T/M)^{1/2} \tag{3-4}$$

where T is the temperature (in K), M is the atomic mass (in amu), and $\Delta\lambda_D$ has the same units as λ_0 (e.g., nm). A line at 300 nm for an atom with a mass of 60 at a temperature of 1000 K has a Doppler width of 9×10^{-4} nm, more than an order of magnitude greater than its natural lifetime width even if the line is relatively strong.

Although collisional broadening is usually negligible in sputtering cells, it is comparable in width to Doppler broadening in atmospheric-pressure flames, furnaces, and electrical plasmas that may be used as light sources for atomic fluorescence. Equation (3-2) may be used to describe collisional broadening. In this case the term in brackets is increased by adding the collision rate (in s^{-1}) for collisions that are close enough to interrupt the coherence of the interaction between the emitted or absorbed radiation field and the atom. These collisions may be *adiabatic* (no energy transfer) or *nonadiabatic* (quenching or exciting).

When collisional broadening is comparable in width to Doppler broadening, the overall profile for each hyperfine component is often described by the Voigt function.[24] This function assumes a Lorentzian profile for each set of atoms with a common Doppler shift, and adds these profiles together as though the population of atoms with common Doppler shifts were distributed in a Gaussian manner. The Voigt function is only a first-order approximation[25] but is adequate to fit experimentally determined spectral profiles.[32] The Lorentzian and Voigt functions have long tailing wings relative to the Gaussian function, which drops off rapidly on either side of its center. Therefore, there will tend to be more spectral overlap in high-temperature atmospheric-pressure cells than in low-pressure sputtering cells where pressure broadening is negligible and the profile is dominantly Gaussian in shape (Fig. 3-4). In extreme cases, long Lorentzian-type tails caused by the minor amount of collisional and natural line broadening that occurs in a low-pressure sputtering cell may become significant, such as in the emission spectral profile of self-reversed hollow cathode lines used for background correction,[30,31,33] and in the absorption spectrum when a laser beam is used to saturated a spectral transition to maximize atomic fluorescence signals.[19,34]

3.2.1.2. Influence on Working Curves

Compared with the relatively broad spectral profiles in atmospheric flames and furnaces, the spectral profiles in low-pressure sputtering cells are narrower and closer in shape to the spectral profiles of hollow cathode lamps that are commonly used as primary light sources to determine absorbance. Because of the greater spectral overlap between the spectral profiles of the lamp and the absorption cell, more photons from the source lamp are absorbed per atom in a low-pressure sputtering cell than in an atmospheric flame or furnace. This tends to improve the initial slope of an absorbance or fluorescence versus concentration working curve.

On the other hand, because the spectral profiles are narrower, the atomic absorption analytical working curve at higher absorbances tends to bend more in the direction of the concentration axis. The bending occurs because the absorptivity (or absorption coefficient) is not the same for all of the wavelengths of the photons that can pass through the monochromator to the detector. (This is sometimes known as the polychromatic effect, although different absorptivities as well as photons of different wavelengths must also be involved.)

This author has studied these effects on atomic absorption working curves for 17 lines belonging to eight elements.[35] The spectral profiles of the primary light source and the absorption cell were assumed to have the same shape. In general, the initial slope is higher but the bending is greater for lines that appear to have a single Doppler-broadened hyperfine component (e.g., Ca 422.7 nm) compared with lines that are wider because they consist of several noticeable but heavily overlapping Doppler-broadened peaks (e.g., Mn 403.3 nm, Fig. 3-4). Analytical lines of many elements tend to have the latter type of shape.

Svoboda *et al.*[(36)] have studied the theoretical shapes of fluorescence analytical working curves. This study was intended primarily for atmospheric pressure flames and so the authors do not consider the influence of hyperfine structure or cases where Doppler broadening is dominant. However, they discuss the considerations for several atomizer cell geometries and cell illumination geometries.

While studying isotopes of lithium, Goleb and Yokoyama[(37)] found large variation in the isotopic composition of some reagent-grade lithium salts. Since absorption spectral profiles may change if the isotopic composition of the element varies from sample to sample, such variations may produce errors and appear as matrix effects.

3.2.1.3. Influence of Hollow Cathode Lamp Current

When the absorption cell and primary radiation source both have the same spectral profile, the atomic absorption analytical working curves are more sensitive to changes in the spectral profiles than if the absorption cell had a broader spectral profile than the primary radiation source. Bernhard[(14)] and Chakrabarti *et al.*[(16)] have found that the lowest possible hollow cathode lamp currents give the best slope to the atomic absorption working curve, but not necessarily the lowest detection limits, because of the additional noise present at the lower lamp intensities. Attempts to increase intensity in order to decrease shot noise[(38)] by increasing the current of a hollow cathode lamp usually result in spectral broadening of the emission line and a decrease in slope of the absorbance working curve.[(13,14,16)] The increase in broadening occurs because more atoms are sputtered from the surface of the hollow cathode and diffuse out of the hollow cathode into the optical path where they are no longer excited, but absorb light that was emitted from inside the hollow cathode. Since the center of the spectral profile absorbs photons more readily than the wings, the intensity in the center of the profile does not increase as rapidly as the intensity in the wings, and the profile broadens. This *self-absorption* may become so bad at sufficiently high currents that *self-reversal* occurs, where the intensity at the center of the spectral profile drops below the wing intensities.

In an extreme—but useful—case of self-reversal, the intensity in the center of the spectral profile drops to near zero in a region comparable in width to the absorption cell spectral profile. The intensity in the far wings of the spectral profile of the radiation source passes on through the absorption cell to be used to determine background absorption.[39]

Self-absorption in a high-current hollow cathode lamp can be reduced by exciting the atoms that have diffused out of the excitation region inside the hollow cathode. This additional excitation outside the hollow cathode can be done with an auxiliary discharge such as a dc,[40] rf,[41] or microwave[42] discharge. The Analyte 16 (Analyte Corp., Medford, Oreg.) is a commercially available jet-enhanced sputtering cell atomic absorption spectrometer that uses an rf pulse to improve the intensity and spectral profile of commercial hollow cathode lamps operating in a pulsed mode.

Chakrabarti *et al.*[16] found that atomic absorption sensitivity depends on the design and manufacture of the lamp and its age, as well as on current. Wagenaar *et al.*[43] measured the line profiles of several different lamps under different operating conditions, and their results can help to explain the dependency of sensitivity observed by Chakrabarti and co-workers. As far as age is concerned, it may be that an older lamp has a lower atomization rate and therefore less self-absorption for the same operating current.

For an atomic fluorescence working curve, the initial slope will increase as the primary light source intensity increases, until self-reversal in the primary light source becomes so obvious that the intensity in the center of the spectral profile begins to decrease. As self-reversal increases, eventually there will be less light available in the center of the absorption (fluorescence excitation) spectrum and the slope will decrease as lamp current is increased. If scattering is a problem, the increase in self-absorption broadening of the spectral profile of the primary source will increase the scattering contributed by photons in the wings of the profile more than the increase in fluorescence signal, and the detection limits will eventually deteriorate.

3.2.1.4. *Fluorescence Spectral Resolution and Resonance Monochromators*

In some cases, atomic fluorescence will have better spectral resolution or selectivity than atomic absorption when both use a hollow cathode lamp as the primary radiation source. Consider a sufficiently low-pressure, low-temperature cell, in which an excited atom undergoes no perturbing collisions that change its Doppler velocity before it fluoresces (emits). In such cells, better spectral resolution will occur for spectral lines where Doppler broadening dominates over broadening caused by hyperfine structure. Under these conditions, the fluorescence will have a spectral profile given by the profile of the absorbed radiation, which is narrower than the profile of the absorption line because preferential absorption occurs for those photons that have wavelengths near the peak of the absorption profile.[5] This improvement in spectral resolution will only become apparent when the fluorescence detection system has high enough spectral resolution to observe it. Such a system would be based on an echelle monochromator, an interferometer, or a resonance atomic fluorescence monochromator.

In cases where broadening of the line caused by hyperfine structure is comparable to the Doppler broadening, the fluorescence peaks for each hyperfine component may be narrower, but their overlap will still cause the overall width of the line profile to be about the same. Then there will be relatively little improvement in spectral resolution.

When a laser is used to excite fluorescence, an irradiance that is high enough to begin to saturate the energy level transition is usually used.[(7)] Although this increases the fluorescence, the absorption spectral profile is broadened and the spectral resolution is reduced.[(19)] Although a potentially interfering line that did not interfere at low irradiances may interfere at high irradiances, such interferences are expected to be few if direct line fluorescence and an appropriate wavelength isolation system are used. Direct line fluorescence occurs from the same upper level as the absorption/excitation line, but to a different lower energy level, so that the fluorescence has a different wavelength than the excitation. It is rare that an interfering element would have both nearly the same excitation wavelength and nearly the same fluorescence wavelength.

3.2.2. Sampling and Transport to the Observation Region

For atomic absorption and fluorescence measurements, it is desirable to have the maximum number of ground-state atoms in the optical path, and to minimize emission because of its contribution to noise. Therefore, modern cells are designed to transport the sputtered atoms away from the cathode so they can be observed at some distance from the glowing discharge region that appears near the cathode. (As we will see, further reduction in emission interference is obtained by pulsing the sputtering current and making the measurement shortly after the current has been turned off and the emission subsides, but while a significant fraction of the atoms are still present.)

3.2.2.1. Experimental Studies

McDonald[(44)] studied diffusion of beryllium and copper in a sputtering cell to show the differences that occur when atoms differ by a factor of two in their diffusion coefficients (because of differences in atomic mass and radius). Absorbances were spatially resolved and made at different times up to 20 ms after the sputtering pulse was shut off. Their results show that Be atoms diffuse more rapidly from the light path than Cu atoms. Measurements during steady-state sputtering show that along the center of the discharge the Be/Cu absorbance ratio changes from 1.7 at a distance of 0.5 cm from the sample surface to 0.83 at a distance of 3.0 cm.

Winefordner and co-workers[(45)] used laser-excited fluorescence to study diffusion of Na atoms sputtered off a NaCl pellet. An axial scan at 2.9 Torr shows a shallow maximum 0.2 cm from the surface of the sample, indicating the presence of a minimum nearer the surface. Others have observed similar behavior during absorption measurements, and explain this behavior by the presence of clusters of the analyte atoms.[(46–48)] These authors propose an additional explanation that the minimum in Na atom population is caused by a higher degree of ionization in the dark space near the cathode, caused by energetic electrons in the cathode fall, compared with other locations. After the sputtering pulse was switched off, the temporal decay of the sodium density produced a decay in the fluorescence signal that dropped to e^{-1} of its initial value in 1.84 ms. When this value was used in a diffusion model, a diffusion coefficient of 160 ± 50 cm^2 s^{-1} was found for Na atoms in Ar at 1.5 Torr.

Gough[(13)] was the first to use a flowing gas cell instead of a static cell to move sputtered atoms into an observation region some distance from the cathode. The gas enters the cell with high velocity through a hollow annulus located 0.05 cm above the surface of the sample (Fig. 3-2). The sputtered atoms that diffuse into the rapidly moving gas are swept into the observation zone about 1 to 2 cm from the sample. Absorbances were improved by over an order of magnitude.

A decade later, Bernhard[(14)] reported the discovery that gas jets pointed directly at the surface of the sample cause craters to form where the gas jets impinge on the surface. Sputtered atoms at the surface are immediately entrained in the rapidly moving gas and diffusion of the atoms back onto the sample surface is significantly reduced. In the Atomsource, six gas jets are pointed toward the center of a circle and also at an angle to the sample so that they impinge on its surface. This design forms a column of gas containing entrained sputtered atoms that is directed away from the surface and into the observation zone, producing a significant increase in sample erosion and in the absorption signal.

In the cell used by Gough, the gas flow is perpendicular to the optical axis of the absorption path. This provides an absorption path length of 1–2 cm. He recognized that the path length and therefore the absorption could be increased by measuring absorption "along the axis of the plume of atoms swept down the center of the sputtering cell by the fast gas flow . . . ." In the Atomsource, the column of rapidly moving gas is split into a "T" so that the entrained atoms are bent in the direction of the optical path to increase the absorption path length.

Gough *et al.*[49] studied the Atomsource and atomizer cells that have observation regions in the shape of a straight tube, a tube with a 90° bend, and a T-shaped tube. The T-shaped tube and the tube with the 90° bend both show a higher absorbance measured along the length of the tube than when measured across the tube. A comparison of absorption measurements made through the sides of the straight tube and the 90° glass tube at various points along the tubes shows that the sputtered atoms are efficiently swept around the 90° bend. Their results show that for a given gas mass flowrate, there is an optimum diameter for the tube when a 90° bend in used. Atoms are more efficiently swept along the tube and around a 90° bend by the higher gas velocity of a smaller-diameter tube, but there is also a faster loss of atoms by diffusion to the tube wall compared with a larger-diameter tube. (Comparison of the 90° bend with the T-shape configuration is difficult from this report because different discharge voltages were used, and tubes of different diameters were not studied to find the optimum tube diameter for the T shape.)

The T-shaped glass tube that introduces gas through an annulus parallel to the sample surface, and the T-shaped Atomsource that uses gas jets that impinge on the sample surface show comparable absorbances for a relatively low cell current of 25 mA, cell voltage of 430 V, and a gas flowrate of 0.3 liter min^{-1}. The authors conclude that the enhanced absorption sensitivity of the Atomsource over the cell design of Gough[13] mainly results from the increased absorption path length under these operating conditions. Further studies are needed to determine whether this is also true for the higher currents normally used with the Atomsource, since the gas jets of the Atomsource and the gas annulus of Gough produce distinctly different sample erosion characteristics.

3.2.2.2. Transport Theory

The theory of gas transport helps us understand the influences of diffusion and entrainment so that optimum operating conditions may be chosen, and, eventually, improvements in cell design may be made, especially for special applications.

Ferreira and Human[46] used a differential equation model to study diffusion and transport in the Grimm-type glow discharge that uses a flowing gas in a cylindrical cell where the sputtered atoms enter the cell uniformly across one end of the 0.8-cm-diameter cylinder. Experimental determinations of spatially resolved number densities of Cu and Cr atoms were used to fit this mathematical model. Diffusion coefficients ranged from 240 to 320 $cm^2\ s^{-1}$ and carrier gas velocities from 400 to 730 cm s^{-1}. Their results show that most of the sputtered atoms are removed from the plasma within 0.5 cm of the cathode. In our laboratory this same differential equation has been used with boundary conditions that were modified so that the sputtered sample atoms enter through a sampling disk (i.e., at the surface of the sample) that may have a smaller diameter than the cylinder. This model more closely approximates the geometry of flow cells like the Atomsource, although it does not include any bends in the direction of the gas flow. The solution to the differential equation with these boundary conditions is given in Boumans.[50] It was assumed as a boundary condition that the concentration of sputtered atoms is zero at the wall of the cylinder where rapid deposition occurs. (The boundary condition for zero concentration at the wall was effected by letting $k/D = 100$ in Eq. 9.27 in Boumans.[50])

a. Model Limitations. It is worthwhile to briefly consider the influence of the assumptions made in this model before considering the results. The assumption that the concentration of sputtered atoms is zero at the wall is a conservative, worst-case condition, because it leads to the fastest loss rate and lowest concentration throughout the cell.

The model considers the flow in the cylinder to be uniform across the diameter of the cell rather than laminar, which would produce a parabolic velocity profile that is highest in the center and lower near the wall. Turbulence tends to make the average flowrate more uniform across the cross section of the cylinder.

Most of the calculations were done with a diffusion coefficient $D = 120\ \text{cm}^2\ \text{s}^{-1}$, corresponding to 5 Torr[50]; others with $D = 30$, corresponding to 20 Torr. These values produce slower diffusion than in Ref. 46, but are closer to the values determined in Ref. 45. The pressures are typical of those in the Atomsource. The cell pressure is assumed to be independent of flow velocity, which may not be true if the vacuum system is operating at full capacity. A velocity of $1000\ \text{cm s}^{-1}$ in a cylinder of radius 0.381 cm (radius of a typical annulus) corresponds to a flowrate of $0.8\ \text{liter min}^{-1}$ measured at atmospheric pressure.

The model assumes the analyte atoms enter the cell in a uniform manner through the sampling disk and immediately attain the velocity of the carrier gas. The mass flow of the analyte into the cell is assumed to be constant so that the transport characteristics of the gas flow can be studied independently of any influence that the gas flow has on sample erosion. Since the mass flow is constant and the carrier gas immediately dilutes the analyte, the concentration of the analyte just off the surface of the sampling disk is inversely proportional to the velocity of the carrier gas.

Figure 3-5 shows radial concentration profiles 0.3 cm from the surface, corresponding to the mouth of the arrestor region, for two values of the diffusion coefficient and two gas velocities. The diameter of the cylinder equals the diameter of the sampling disk, which corresponds to the geometry in this part of a typical cell containing an annulus. The lower two curves are

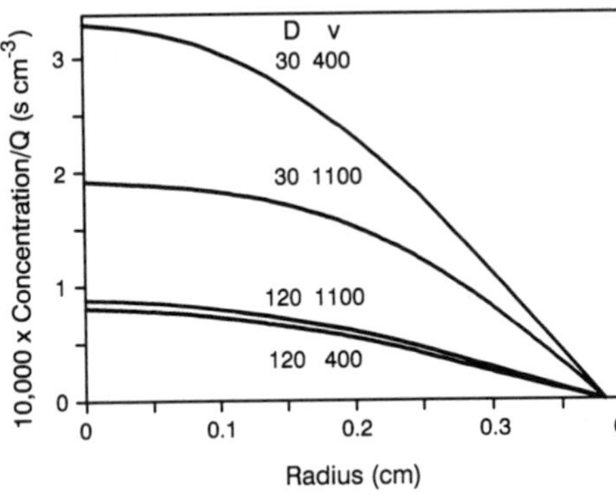

Figure 3-5. Calculated radial profiles for the concentration of sputtered atoms 0.3 cm downstream from a flat sample surface for diffusion coefficients $D = 30$ and 120 cm^2 s^{-1} and carrier gas velocities v of 400 and 1100 cm s^{-1}. The sputtered atoms are assumed to enter uniformly across the open end of a cylindrical cell that has a radius of 0.381 cm, where the concentration drops to zero. Concentration is in number density per cm^3 and Q is the number of atoms per second that enter the cell.

for the higher diffusion coefficient $D = 120$ cm^2 s^{-1} (5 Torr), and show that the two velocities 400 and 1100 cm s^{-1} produce about the same profiles at this distance from the surface, with the lower-velocity curve showing the lower concentration. Since the concentration just off the surface of the disk is inversely proportional to velocity, the lower-velocity curve would be expected to have the higher concentration. The reason it does not is that the higher initial concentration has been offset by diffusion losses to the walls that occur during the longer time that the gas spends in the annulus region at the low flowrate.

The concentration in the center of the 1100 cm s^{-1} profile has dropped to 43.7% of its initial value at the sample surface. Integration over the circular cross section at 0.3 cm from the surface shows that 80% of the analyte is deposited on the walls before the gas leaves the mouth of the annulus. This loss increases to 93% for the lower velocity.

The two upper curves in Fig. 3-5 for a smaller diffusion coefficient (at 20 Torr instead of 5 Torr) show a greater difference. There is much less loss of analyte to the walls than for the higher diffusion coefficient. The lower velocity (top curve) shows a higher concentration because of less dilution as expected; the loss by diffusion has not yet overwhelmed the increase in concentration produced by less dilution at the lower velocity.

These results indicate that there may be an optimum velocity that balances both (1) the increase in concentration caused by less diffusion loss to the wall at higher velocities and (2) the decrease in concentration that occurs because of the dilution effect with higher velocities. This optimum will now be considered.

b. Optimum Velocity. The five lower curves in Fig. 3-6 show how the concentration varies with gas velocity at five different points along the axis of the cell. The cylinder radius was increased to 1.0 cm to more accurately represent the geometry of flow cells such as the Atomsource at larger distances from the sample surface. Each curve has a maximum caused by the

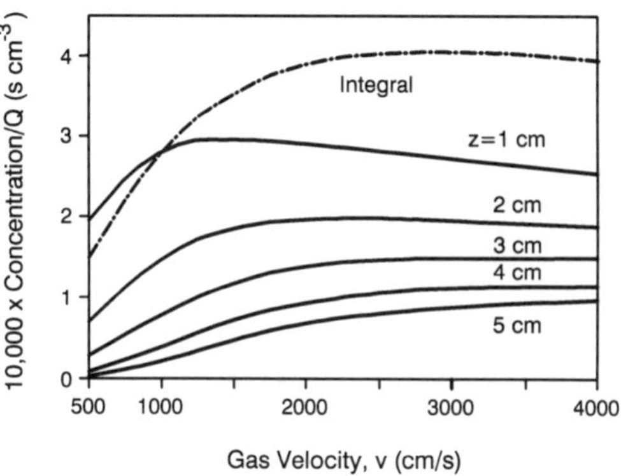

Figure 3-6. Calculated sputtered atom number density versus carrier gas velocity at five distances z from the sample surface. The cylinder radius was increased to 1 cm while the entry of atoms is still limited to a disk with radius 0.381 cm. The integral of the number density from 1 to 5 cm along the cylinder is shown as a dashed line whose values have been scaled arbitrarily so that it can be plotted on the same graph. Q is the number of atoms per second that enter the cell.

balance between more dilution and less diffusion loss as flow velocity increases. The flow required for the maximum concentration increases as distance from the surface increases.

The analytical absorbance signal is proportional to the integral of concentration along the optical path. In a flow cell where the carrier gas bends 90°, the absorbance is measured along a path that begins at some distance from the sample surface, say 1 cm, and ends at a window. To simulate this, the top curve in Fig. 3-6 is proportional to the absorbance that would be measured along the axis of the cell from 1 cm to 5 cm. A maximum in absorbance appears for a gas velocity of about 2700 cm s^{-1}. Although the geometry is somewhat different in a real cell, a similar maximum would be expected in a real cell where losses by diffusion at low velocity and dilution at high velocity also exist.

c. Radial Profiles for a Large Cylinder. Figure 3-7 shows how diffusion rapidly spreads the analyte so that a large fraction of it is outside the radius

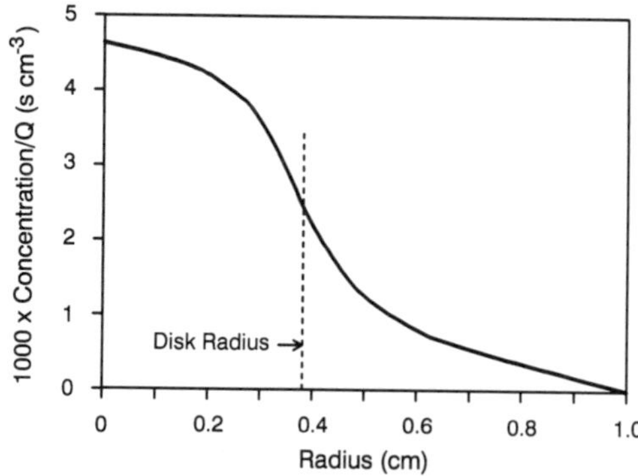

Figure 3-7. Calculated radial profile for the concentration of sputtered atoms 0.01 cm downstream from a flat sample surface disk of radius 0.381 cm located at one end of a cylindrical cell with a radius of 1.0 cm. The diffusion coefficient is 120 $cm^2\ s^{-1}$ and the carrier gas velocity is 160 cm s^{-1}. Q is the number of atoms per second that enter the cell.

of the sampling disk even at a downstream distance of only 0.01 cm when the diameter of the cylinder is larger than the diameter of the sampling disk. The radial profiles beyond 2 cm downstream have the general shape of the profiles in Fig. 3-5. The radial concentration profiles for small distances from the disk surface do not have the parabolic shape of Fig. 3-5, but have an inflection point near the radius of the disk. Since the concentration at the disk radius is not zero, as it would be if the cylinder were the same size as the disk, the concentration gradient from the center to the wall is lower, and there is correspondingly less diffusion of the analyte to the wall. Consequently, there is less loss of analyte whenever a larger-diameter cylinder is used.

d. Axial Concentration Gradients. The log-linear plots in Fig. 3-8 show that the concentration along the axis of the cylinder decreases exponentially with downstream distance for three different cylinder radii from 0.381 cm to 1.0 cm. The mass flow of the argon gas is the same (0.8 liter min^{-1} at 760 Torr) for these cases and the gas velocity decreases with the square of the radius. For the smallest-diameter cylinder (curve A), the concentration drops about three orders of magnitude from a downstream distance of 0.3 cm to 2.4 cm because of diffusion losses to the walls. For the 1-cm cylinder (curve C), the axial concentration is generally about an order of magnitude higher and drops off less than two orders of magnitude in the same distance. The larger-diameter cylinder decreases diffusion losses, and in this case also increases concentration because of reduced dilution caused by the lower carrier gas velocity.

The influence of velocity can be eliminated for the sake of discussion by adding makeup gas to increase the velocity for the 1-cm cylinder from 160 cm s^{-1} to 1100 cm s^{-1}. The added makeup gas causes the curve to change from C to D, which shows a considerable improvement in downstream

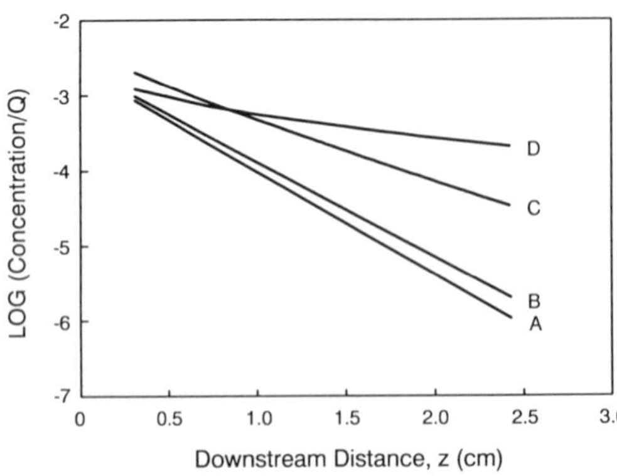

Figure 3-8. Influence of cell radius on axial concentration. Line A, cell radius = 0.381 cm, carrier gas velocity = 1100 cm s^{-1}; line B, radius = 1.0 cm, velocity = 160 cm s^{-1}; line C, radius = 1.0 cm, velocity = 1100 cm s^{-1}; line D, radius = 0.45 cm, velocity = 788 cm s^{-1}. Diffusion coefficient = 120 cm^2 s^{-1}.

concentrations. Clearly the change in cylinder radius from 0.381 cm (curve A) to 1.0 cm while maintaining the same gas velocity (curve D) causes a dramatic change in concentration, and emphasizes the significance of losses caused by diffusion to the walls when rapid deposition occurs there.

These studies of the influence of gas transport can be summarized as follows:

1. Most of the analyte is deposited on the walls around the sampling disk before it has a chance to enter the observation region.
2. Larger-diameter tubes help to reduce analyte losses caused by deposition on the wall.
3. The concentration decreases exponentially with distance from the sample.
4. An increase in pressure causes an increase in concentration, because the diffusion losses are reduced due to the lower diffusion coefficient.
5. Higher flowrates increase the absorption signal, but eventually a maximum is reached.

It should be kept in mind that gas flowrates and pressure may influence the sputtering process as well as the transport processes. For example, Kim and Piepmeier[51] found a maximum in absorbance when the carrier gas flowrate was changed, but the cell voltage and sample erosion rates also changed with flowrate. In any case, these maxima are important because they provide a relatively stable operating region where absorbance does not change with minor changes in flowrate.

3.2.2.3. Transient Samples

When the mass of the sample (such as a dried solution) is limited so that a transient absorbance peak occurs, additional considerations must be taken into account. Ideally the largest analytical signal would be obtained by quickly atomizing the entire sample and transporting all of the atoms to the observation region. Regardless of whether or not the ideal case can be approached, peak duration is important because a high, short (narrow) peak has a larger signal-to-noise ratio than a low long (wide) peak of the same area when background noise is the main source of noise, because more background noise is present during the long peak. This suggests that atomization should occur as quickly as possible, and that the atoms should be transported by the carrier gas *to* the observation region as fast as possible to minimize diffusion losses, but that the carrier gas should not transport the atoms out of the observation region. Although the fundamental theory of this transient transport problem remains to be studied, it seems clear, in the face of diffusion losses, that there will be an optimum carrier gas flowrate, and that it may be helpful to shut off the carrier gas flow at some optimum point in time near the maximum of the analytical signal.

Chakrabarti and co-workers[(16,52–54)] have been able to fit an equation derived from kinetic theory to transient absorbance peaks obtained using dried solutions deposited on the cathode of the Atomsource. The model assumes that the rate of loss of atoms from the observation region is proportional to the concentration of atoms $[B]_t$ in the observation region at any time t, and similarly that the rate of entry of atoms into the observation region is proportional to the concentration of atoms $[A]_t$ that can enter the observation region at time t. The proportionality constants are first-order rate constants k_1 and k_2 and the resulting equation is

$$[B]_t = [A]_0 \frac{k_1}{k_1 - k_2} [\exp(-k_2 t) - \exp(-k_1 t)] \tag{3-5}$$

where $[A]_0$ is the initial concentration of atoms that can enter the observation region at $t = 0$. This equation describes a peak that quickly rises in an exponential manner with a rate corresponding to the larger of the two rate constants, and that exponentially falls more slowly at a rate corresponding to the smaller of the rate constants. Values can be obtained for the two rate constants by fitting this equation to the shape of an experimentally determined absorbance peak. However, since $[A]_0$ is not known and since the same peak shape is produced simply by interchanging the values for the two rate constants, the shape of the peak alone does not reveal whether the rate constant for entry of atoms into the observation zone is larger or smaller than the rate constant for the loss of atoms. Which rate constant is smaller is determined from other measurements, such as the rate of loss when the power was suddenly turned off during a pulse.

It was concluded that for low-power sputtering conditions, the rate of entry of atoms into the observation region was the limiting step, whereas for higher-power sputtering conditions, the rate of loss was slowest. Even so, since the ratio of the entry rate constant to the loss rate constant is ideally large, but was experimentally found to be only in the range of 1.3 to 7.0 for 14 elements, they concluded that there is considerable room for improving the sensitivity of the transient method, for example, by increasing the entry rate by increasing the sputtering rate.

3.3. Practical Considerations

3.3.1. Methods Development

Although instrument manufacturers usually have recommended sets of operating conditions for a variety of samples, an analytical laboratory sometimes encounters other types of samples or requirements. Additional considerations that are helpful in choosing a set of operating conditions are therefore worthwhile discussing.

3.3.1.1. Choice of Spectral Line

Experience with the Atomsource has shown that a spectral line that is commonly used for flame or graphite furnace AAS may not be the best line for use with a low-pressure sputtering cell.[(55)] One reason for this may be differences in background emission interferences. Another reason may be the increased influence that the spectral profile of the emission line from the hollow cathode lamp has on the slope and shape of the analytical working curve in low-pressure cells as we have seen.

When pulsed operation is used, lines may vary in their pulsed mode behavior. Demers *et al.*[(56)] show the redistribution in ground-state and nearly ground-state lines of U that occurs in the few milliseconds after a sputtering pulse is turned off in a hollow cathode absorption cell.

The best atomic absorption lines for a sputtering cell are usually among the lines that have a lower energy level at or within a few hundred kaysers (wave numbers, cm^{-1}) of the ground state of the element and have the highest gf values, typically above 1.0, where f is Ladenburg's oscillator strength and g is the statistical weight of the upper energy level (also known as the degeneracy, the number of electrons that can occupy the upper energy level). Tables such as those in Ref. 57 often list the product of g and f, rather than individual values. The absorption coefficient is proportional to the gf value times the square of the wavelength.[(19)] Therefore, for two spectral lines with widely different wavelengths, the gf values found in a table should be multiplied by the square of the wavelengths before a comparison of the strengths of absorption (absorption coefficients) is made. On the other hand, the error in comparing the gf values alone directly from a table is only a factor of 4 when comparing a line at 200 nm with one at 400 nm, and it would not be uncommon to find the gf values to be in error by that amount anyway.

Additional care is necessary when tables of gA values are used to estimate the absorption coefficient, where A is the Einstein transition probability per second. Although gA values in tables are appropriate when estimating emission intensities (particularly for atom cells in thermal equilibrium), the absorption coefficient is proportional to the fourth power of the wavelength times the gA value. Therefore, estimating relative absorption coefficients by comparing gA values alone, without first multiplying by the fourth power of the wavelengths, will produce an error of a factor of 16 when comparing a line at 200 nm with one at 400 nm. The values of gf and gA are related by the equation

$$gf = 0.667 \times 10^{16} \lambda^2 gA \tag{3-6}$$

where the wavelength λ is in angstroms.

A more accurate comparison also requires dividing each gf or gA value by the statistical weight g_l of the lower energy state. Unfortunately, these values are not always readily available. Information about the hyperfine structure of the lines being compared is also needed for an accurate comparison. Again this information is usually not available. Therefore, it may be worthwhile to choose more than one line from a table and check the experimental performance of each one.

Fluorescence is a two-step process involving absorption followed by emission. Therefore, some function of the product of the absorption coefficient times the Einstein probability for emission is compared when choosing spectral lines. When the same wavelength is used for the absorption and fluorescence steps, it is appropriate to compare the square of the gf values (divided by g_l squared when known). When the lines are different, a comparison is made of the product of $\lambda^2 gf$ for the absorption (excitation) wavelength times gA for the fluorescence wavelength. These comparisons for fluorescence do not take into account quenching of the excited state. However, true quenching is uncommon in sputtering cells that use noble gases.

The choice of lines for fluorescence also depends on the relative intensities of the lines emitted by the primary light source. Fortunately, the strongest absorption line for the analyte often corresponds to one of the stronger emission lines from a hollow cathode lamp. However, when a laser is used to excite fluorescence, the choice of lines may be limited by the relative intensities of the lasing wavelengths available for that particular laser. When background correction is required, the choice of line may be influenced by the performance of the background correction system, which may vary from line to line. Although background has not been a problem for metal alloy analysis with the Atomsource, this is not true for cells that do not direct gas jets directly at the sample surface. Even with the Atomsource, background may be found when samples, such as organic materials (made into pellets with a conducting matrix), are used that contain major concentrations of elements that form gaseous molecules, for example C_2 and CN, that absorb in the visible and ultraviolet.

3.3.1.2. Plasma Gas

The quality of the plasma gas is important. The effect of 17 ppm water vapor in argon plasma gas on the determination of Cr in low-alloy steel is to decrease the absorbance by 7% and double the noise level[(49)] compared with runs using argon containing 3 ppm water. It is best to keep atmospheric gases out of the sputtering cell and its gas and vacuum tubing so that moisture does not adsorb on the walls and then outgas during an analysis. When in storage the cell should be evacuated. During sample changing, a positive pressure of argon in the cell along with an accompanying flow of gas out of the sample port helps to keep the atmospheric gases out. Larkins[(58)] describes a gas control unit for sputtering cells and presents important considerations in its design.

3.3.1.3. Discharge Conditions

High sputtering currents increase the atomic population and the analytical signal. For a given current, decreasing the pressure increases sputtering because the sputtering ions attain a higher kinetic energy due to the increase in their mean free path and the accompanying increase in cell potential that is necessary to keep the same current. Fang and Marcus[(59)] have found that the sample erosion rate for a flat cathode is proportional to the electrical power (current times voltage) applied to a cell above a particular threshold power that varies with pressure (Fig. 3-9.) Chakrabarti *et al.*[(16)] have found that controlling the power to the Atomsource instead of controlling current produces more reproducible results.

However, care must be used when lowering the pressure to increase voltage and power in order to increase sputtering, because reduced pressure increase losses by diffusion. Also, if the vacuum system is running at maximum capacity, then the pressure is reduced by reducing the flow of carrier gas into the cell chamber. The reduction in mass flow of the carrier gas reduces the transport of sputtered atoms from the surface of the sample into the observation zone. Therefore, there may be an optimum pressure that will provide the best analytical signal.

If detection limits are a consideration, then high power is desirable. Although a high sputtering power increases the atomic population and the analytical signal, it also increases sample heating (because most of the power is dissipated near the cathode) and tends to increase background emission. Although most instruments are able to compensate electronically for background emission, the *noise* in the background emission signal cannot be eliminated and still contributes to the total noise in the analytical signal. Therefore, when the background emission signal approaches the transmitted emission signal of the primary source in atomic absorption, it may be worthwhile to determine if an increase in sputtering current still increases the signal-to-noise ratio. A similar study of signal-to-noise ratio is worthwhile when the background emission signal approaches the scattering signal of the primary light source in atomic fluorescence.

3.3.1.4. Sample Temperature

High temperatures produced by sample heating may cause a change in the sputtering rate and preferential diffusion of atoms from the bulk to the

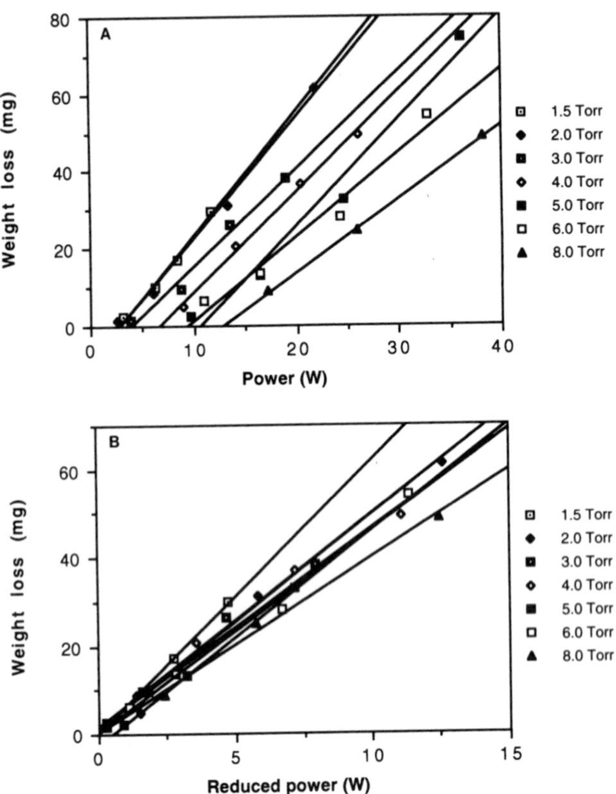

Figure 3-9. Effect of sputtering discharge power on sample erosion of a flat cathode at various pressures. (A) Weight loss versus applied power; (B) weight loss versus reduced power, $i(V - V_0)$, where V_0 is the threshold voltage at which a plot of weight loss versus discharge voltage intercepts the voltage axis. Reprinted with permission from *Spectrochim. Acta* 43B, D. Fang and R. K. Marcus, Parametric evaluation of sputtering in a planar glow discharge—I. Sputtering of oxygen-free hard copper (OFHC), copyright 1988, Pergamon Press.

surface of the sample.[60] Although it might be possible to compensate for sputtering rate drift, this adds an additional variable to the technique, and the thermal characteristics of samples may contribute to matrix effects if compensation is not complete. Preferential diffusion for a sample that is hot near its surface will cause the bulk material to become depleted in some elements, and the sputtered atomic vapor will then no longer be representative of the sample. Preferential diffusion could be a cause of matrix effects for samples that have different diffusion characteristics.

Sample temperature can be reduced by cooling the sample. Although cooling has been done using dry ice and liquid nitrogen for special applications,[61] water cooling is most commonly used. The Atomsource cools the sample by pressing the front surface of the sample against a water-cooled copper plate surrounding the sputtering region. With this configuration only one sample surface needs to be flat. Other methods include a water-cooled plate pressed against the flat back of a sample.[5] The latter method may be helpful particularly for samples with poor thermal conductivity if the samples are thin relative to the diameter of the sputtering region (e.g., some samples of superconductor materials). The cooling effect of gas jets in a gas-jet-enhanced sputtering cell remains to be determined.

The rise in temperature during sputtering of a water-cooled sample was studied experimentally by Gough *et al.*[5] Although the temperature continued to rise for several minutes after the discharge was turned on, stable fluorescence signals were obtained in 1–2 min.

Pulsing the sputtering discharge while maintaining the same average power, increases the instantaneous sputtering rate during the pulses without changing the average sample temperature. However, further studies are needed to determine the influence of the transient temperature rise during the pulses.

3.3.1.5. Preconditioning the Surface (Preburn)

The best accuracy and reproducibility are usually obtained for an analysis of the bulk of a sample when the sample surface is preconditioned by sputtering before the analytical measurements are made. Preconditioning removes surface contaminants and roughness, and allows the surface composition to adjust to an equilibrium composition that compensates for the differences in sputtering rates of different elements. When the equilibrium composition is reached, the surface is proportionally richer in elements with lower sputtering rates, and the composition of the sputtered material is therefore the same as that of the original sample.[5,62–64]

All surfaces pick up contamination from the atmosphere and other things with which they are in contact. Machining or polishing methods used to prepare a flat surface may cause contamination. Preconditioning the surface by sputtering helps to remove contamination so that the analysis is representative of the bulk material.

Sometimes the surface may be rough enough to influence the gas flow patterns when gas jets strike the surface as they do in the Atomsource. Therefore, preconditioning will usually be done at a low gas mass-flowrate so that the gas jets do not cause preferential erosion of the surface where the jets strike the surface. In a gas-jet-enhanced sputtering system where the vacuum system is removing gas at its maximum capacity, a low gas mass-flowrate is produced by restricting the entry of gas into the cell. The decrease in gas flow is accompanied by a reduction in cell pressure, which produces an increase in cell potential and in sputtering rate for a given current. Because the influence of the gas jets is minimal at the low flowrate, this high sputtering rate is uniform across the sample surface and is desirable for preconditioning the surface. Although the sample erosion rate may decrease because of increased redeposition at low gas flows, the high sputtering rates will cause the surface to quickly smooth out and come to equilibrium and lose unwanted adsorbed gases. On the other hand, surface contamination by solids may require a longer removal time because of the lower erosion rates. After preconditioning, the gas mass-flowrate is increased to increase the rate of sample erosion under each gas jet and to increase the transport of the sputtered atoms away from the surface to the observation zone for the analytical measurements. These are general guidelines, and systematic studies may be needed to obtain the very best results.

3.3.2. Relative Precision for Atomic Absorption Determinations

Because of the exponential relationship between concentration and the fraction of light transmitted in an absorption measurement, there is an optimum transmittance or absorbance at which the best relative precision in concentration is obtained. Although this optimum is commonly presented in textbooks when discussing spectrophotometric measurements with solutions, it is rarely considered in flame or furnace AAS, probably because most of these determinations are near the detection limit or do not require the highest obtainable precision. However, under proper conditions the precision of emission and absorption measurements using sputtering cells can be better than 0.1%.[14] Since sputtering cells are commonly used for the determination of major and minor elements in a sample as well as trace elements, it is worthwhile to briefly review what is required to obtain the best relative precision for an atomic absorption measurement.

Figure 3-10 shows a plot of the relative uncertainty, or relative standard deviation (RSD), in concentration versus absorbance for the Mn(I) 403.3-nm line in an atmospheric flame (lower curve) and in a low-pressure sputtering cell (upper curve). These curves were calculated from a model that assumes that shot noise is the limiting source of noise. The noise in the blank is arbitrarily assumed to be 0.01%; a different value would simply shift the curves to a different vertical location and their shapes would not change. At low absorbances, decreasing the concentration causes the RSD to become worse because the analytical signal decreases while the background noise in the analytical signal remains the same. Background noise for an absorption measurement corresponds to noise in the beam from the primary radiation source, almost all of which is being transmitted to the detector at low absorbances (high transmittances). At high absorbances, the fraction of the beam from the primary radiation source that reaches the

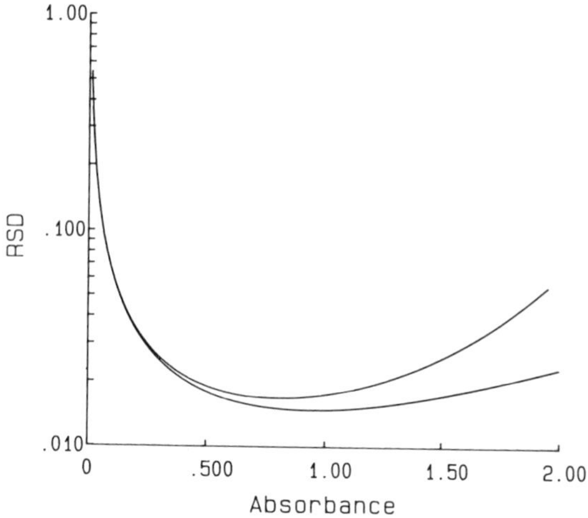

Figure 3-10. Relative standard deviation (RSD) in concentration versus absorbance for the Mn(I) 403.3-nm line in an atmospheric-pressure flame (lower curve) and a low-pressure sputtering cell (upper curve), for measurements where the RSD in the blank is 0.01 and the uncertainty in the spectrophotometer readings is limited by shot noise. Reprinted with permission from *Spectrochim. Acta* 44B, E. H. Piepmeier, The influence of the spectral profile of the source lamp for low pressure atomic absorption cells, copyright 1989, Pergamon Press.

detector is small, and low intensities are difficult to measure because of shot noise. Increasing the concentration reduces in an exponentially fast manner the fraction of primary source radiation reaching the detector, and the RSD continues to become worse as concentration increases.

Between these two extremes, at the bottom of the "U"-shaped curve lies an optimum absorbance where the relative uncertainty in absorbance or concentration is lowest. The minimum varies with the source of noise and the shape of the analytical working curve.[(38)] These curves have been studied in detail in our laboratory[(35)] for 17 lines that have widely different hyperfine structures, since hyperfine structure controls the shape of the analytical working curve for a given primary radiation source and absorption cell. The shapes of the spectral profiles of the primary radiation source and the absorption cell were assumed to be the same. The differences in the curves in Fig. 3-10 are caused by the greater curvature of the analytical working curve in a low-pressure sputtering cell compared with the curvature of the working curve in an atmospheric-pressure cell. For a given RSD in the blank, the location of the minimum along the absorbance axis is generally in the range of 0.40–0.45 for a low-pressure cell if the noise is background or dark current limited and is in the range of 0.62–0.80 if the noise is only limited by shot noise as in Fig. 3-10 (the best situation). The best RSD in concentration is three times worse than the RSD of the blank for the background- or dark current-limited case and for the shot noise-limited case is two times worse.

It is best to choose an analytical line and absorption cell operating conditions to obtain an absorbance in these ranges whenever the highest precision is desired. However, the differences in blank noise for different analytical lines must also be considered since the RSD of the sample is directly proportional to the RSD of the blank. In some cases it may be better to choose a line that gives an absorbance on one side of the optimum range if that line has a better blank RSD than a noisier line that causes the absorbance to fall with the optimum range.

3.3.3. Background Correction

Background emission from the anode glow and positive column of a glow discharge are usually eliminated by positioning the anode at the edge of the negative glow. Unfortunately, this does not eliminate all emission, which interferes with atomic absorption and fluorescence measurements. The background emission signal from the sample cell can be compensated by modulating the primary radiation source and synchronizing the detector with the modulation frequency. The detection system measures the background emission signal when the primary radiation source is off and subtracts this signal from the total signal when the primary light source is on.

Although the background emission signal is eliminated, its noise is not.[38] If background emission noise is limiting the ability to detect small concentrations of analyte, then background emission that is near but not at the wavelength of the analysis line may be reduced by narrowing the spectral bandpass of the detection system (e.g., by narrowing the slit width of the monochromator). This will improve the detection limit until the photon rate of the analysis line is reduced to the point where shot noise in the analytical signal is the dominant source of noise.

Background emission at the wavelength of the analysis line (e.g., analyte emission) may be reduced by changing the operating conditions of the discharge. For example, emission may be reduced by increasing pressure, if the accompanying loss in sputtering rate can be tolerated. An alternative is to pulse the discharge and make the analytical measurement after the pulse when the emission dies down, but while the atomic vapor is still present as was first done by Yokoyama and Ikeda[65] for atomic absorption measurements. This is also commonly done with fluorescence measurements.[18,45,66] The Analyte 16 atomic absorption spectrophotometer uses a similar pulsed mode of operation to eliminate emission with the Atomsource.[14,67] If background emission is still present in the pulsed mode, care must be exercised in measuring the background emission at a different time than the absorption signal because both signals are varying with time and an over- or undercompensated correction may occur.[67]

Molecules and small particles cause background absorption, fluorescence, and scattering.[13,44,68] McDonald[69] discusses Mie scattering and measured the beam attenuation from 300 to 450 nm for a silver sample in a glow discharge. He estimates that clusters of atoms may contain 30% of the sputtered atoms. In other papers he reports background absorbances up to 0.002 near 234.8 nm and 222.6 nm for a Be–Cu alloy,[44] and from 0.004 to 0.006 for silver samples, 0.001–0.002 for copper samples, and less for other alloys.[70] Gough[13] reports broadband beam attenuations of 0.25–1% for iron, copper, and aluminum samples and 2% and more for silver. He attributes the attenuation with silver samples to be caused by the aggregation of metal atoms in the discharge to form particles with diameters around 5 nm.

Patel and Winefordner[71] observed laser-excited fluorescence from CuO and diatomic molecules of copper and lead in a simple planar geometry glow discharge at 4.5 Torr and 25 mA. Broadband scattering attributed to large metal clusters was also observed. No evidence for background absorption has been reported for metal alloys run on a gas-jet-enhanced sputtering cell. Since aggregate formation is expected to be highest near the surface where the concentration of atoms is highest, perhaps the flow characteristics of the gas jets striking the surface reduce aggregate formation or direct the heavier aggregates out of the column of carrier gas that transports the atoms to the observation zone. However, until more is known about this, the presence of molecules and small particles may be anticipated when nonmetallic samples are analyzed, particularly those with an oxide or organic matrix.

The same methods that are used for other atomic absorption cells may also be used to correct for background absorption in a low-pressure sputtering cell. These include the use of a continuum source, lines near the analysis line, self-reversed hollow cathode lines (e.g., the Smith–Hieftje method), and lines shifted by the Zeeman effect for atoms in a magnetic field. These methods are discussed in more detail in textbooks.[e.g., 72,73] The Smith–Hieftje method[39] has been shown to cause rollover (bending back down toward the concentration axis) of the analytical working curve at high absorbances for flame atomic absorption when stray light or other nonabsorbable spectral lines reach the detector.[33,74,75] Similar problems are expected in low-pressure cells. Nonabsorbed lines and their influence can sometimes be reduced by reducing the spectral bandpass of the detection system, which then also changes the shape of the working curve and may increase the noise level.

Zeeman splitting methods will work if the magnetic field is placed around the hollow cathode lamp. However, a Zeeman method that places the sputtering cell in the magnetic field will distort the plasma. This may be beneficial if the magnetic field is stationary, but a *changing* magnetic field at the cell, a technique often used for background correction, may produce an unacceptably changing background absorption. Tie-Zheng and Stephens[76] found that placing the magnetic field around a hollow cathode lamp produced a worse signal-to-noise ratio than placing the magnetic field around the atom cell (a flame) when high lamp currents were used for atomic fluorescence. This was attributed to the emitting and the self-absorbing regions in the lamp not being subjected to the same magnetic field strengths. Clearly, Zeeman background correction methods need to be studied experimentally for low-pressure sputtering cells.

Because spectral lines with minimal hyperfine structure are narrower in a low-pressure sputtering cell than in atmospheric flames and furnaces, there will be less overlap of the background correction radiation with the absorbing analytical line. Therefore, for these lines all of the correction methods may be able to compensate over a larger range of background absorption than when they are applied to higher-pressure cells.

On the other hand, all of these background correction methods assume that the background absorption is the same for the analytical line as for the background correction wavelengths. Although this essentially may be true for broadband absorption caused by small particles and large molecules, the absorption spectra for small molecules may be more highly resolved at low pressures than at atmospheric pressure, and the background will be more structured than smooth. This increases the possibility that the background produced by small molecules will not be completely or properly compensated, and care in studying background correction must be exercised just as it must be for atmospheric cells.[73]

In atomic fluorescence, small particles and molecules produce background scattering, which ultimately limits detection limits. Scattering may also occur off of walls and other parts of the instrument that are within the optical aperture of the detector. Such scattering is usually eliminated, when possible, by using direct-line fluorescence, where the excitation and fluorescence wavelengths are different, but where they share a common upper energy state.

Background fluorescence may also occur from windows, lenses, and other parts of the instrument. This can be minimized by designing the system so that these parts are not excited by the primary light source or its reflections, and by the use of materials that are manufactured specifically to minimize fluorescence.

3.3.4. Pulsed Sputtering Cells

The analytical signal can be increased by increasing the sputtering rate. However, the highest electrical power that can be used in a sputtering cell to increase the sputtering rate is ultimately limited by overheating of the sample. Overheating tends to prevent surface equilibration, and will eventually convert the discharge into an arc. Overheating may also crack fragile samples.

To avoid overheating, short, high-power sputtering pulses may be used to produce higher instantaneous concentrations of analyte atoms than is possible with a dc current.[14,16] Although the temperature rise during a pulse may be high, if the pulses are spaced far enough apart, cooling between pulses occurs, especially in the presence of gas jets, and the average temperature rise can be limited to tolerable levels.

Capturing the analytical signal at its highest value, rather than observing its average value, requires a pulsed detection system that is synchronized with the sputtering pulses. Since a shorter signal integration time or response time tends to raise the noise level, pulsed primary radiation sources (e.g., pulsed hollow cathode lamps) that produce higher spectral radiances than dc sources are commonly used to minimize shot noise.

A pulsed mode of operation not only helps to improve detection limits by increasing the analytical signal, but also allows emission noise to be reduced. Emission noise can be reduced by measuring the analytical signal after the discharge pulse is turned off and the emission dies down, but before the atoms dissipate. Although the pulsed mode of operation is somewhat more complex than the dc mode, the lowest detection limits reported so far have been achieved by pulsing the sputtering cell.[16]

3.3.4.1. Pulse Duration

There are generally three different cases to consider with a pulsed sputtering cell: the case where the sample lasts for many pulses,[14,18] the case where the sample lasts for a limited number of pulses, and the case where the sample is consumed during one pulse.[16] A metal alloy or pressed pellet will repeatedly generate atom clouds for many pulses. This situation allows preconditioning sputtering of the sample to improve the accuracy and precision of the analysis. It also provides the greatest flexibility in making measurements that compensate for drift in the intensity of the primary light source, background emission, and background absorption. The time between pulses can be lengthened as necessary to ensure that these measurements can be made independently for each pulse.

When the sample is a thin film or a dried solution deposited on the cathode, the analytical signal may last for only a limited number of pulses, and the envelope of the sequence of pulses may appear in the shape of a peak rather than a steady plateau. In this case, any time difference between the time that the analytical signal is observed and the time that a background signal is observed will require proper interpolation of the results to produce accurate background correction.

When the sample is consumed during one current pulse, the response time of the detection system must be fast enough to keep up with the changing signal if the most sensitive results are to be obtained. Emission background correction may be possible by switching rapidly back and forth between the total signal and the background signals (emission and absorption). Piepmeier and de Galan[77] have studied the effect of a slow detection system that uses a modulated light source and detection system to observe a transient absorption peak. They modified the general expression for the atomic absorption signal to allow for possible time-dependence of the source emission and sample absorption spectral line profiles. A table in their paper summarizes the results for several cases of practical interest. They found that deviations from Beer's law may occur even with a perfectly monochromatic source if the sample absorption is time-dependent.

3.3.4.2. Peak Height or Peak Area?

When a peak is observed, there is the question of whether to use the peak height or the peak area as the analytical signal. This question has been actively discussed for years in graphite furnace AAS. Absorbance peak area is recommended if the width of the peak changes from sample to sample or run to run because of matrix effects, since the area of the peak often remains proportional to concentration. However, this assumes that the response time of the detection system is fast enough to give an accurate representation of absorbance *before* the integrations occur, that all of the analyte in the sample is eventually observed, and that each atom of analyte, when it is observed, contributes in the same proportion to the analytical signal. Confusion about whether area or peak height is better will undoubtedly occur when all of these requirements are not met.

Chakrabarti *et al.*[16] have shown that dried solutions of different salts sputter at different rates. Ohls[78] has shown that peak width depends on the amount of the same element present in a dried solution. These results indicate that peak width is expected to depend on the sample matrix for dried complex multielement solutions.

The use of peak area should have a signal-averaging advantage similar to increasing the time interval over which the analytical signal is observed. However, if the integration is carried too far into the wings of a peak, then the noise in the baseline begins to cause deterioration in the signal-to-noise ratio. If the area is to be used and the shape of the peak does not have a mathematical model to which it could be fit, then the locations of the beginning and end of the peak may not be clear in the presence of noise. The influence of baseline noise on the choice of how much of a transient absorbance peak to include in the integral to obtain the best signal-to-noise ratio has been considered in detail by Piepmeier.[79]

3.3.5. Multielement Determinations

The determination of more than one element in a sample is often desired. Various approaches to multielement atomic absorption determinations have been reviewed by Salin and Ingle.[80] There are three basic approaches to multielement determinations, depending on the available instrumentation: simultaneous measurements of all elements of interest, rapid sequential measurements of all elements while the sample is being sputtered, and the determination of one element at a time for all samples, then the next element for all samples, and so on. The last approach is used with most atomic absorption spectrophotometers because they cannot easily measure the absorbance at more than one or two wavelengths during sample atomization. A rapid sequential method to determine 4 elements is described by Salin and Ingle.[81] Light from four hollow cathode lamps is combined into one beam with semitransparent mirrors, passed through the absorption cell and into a polychromator. Light from each of the exit slits is reflected to the same photomultiplier. The lamps are pulsed in rapid sequence so that the signals for each lamp can be distinguished. Rapid sequential determinations of up to 24 elements are possible during the sputtering of one sample with the Analyte 16 atomic absorption spectrophotometer because of its ability to change reproducibly the hollow cathode lamps and the wavelength settings of the monochromator within a few seconds.[15]

Simultaneous measurements are needed when the sample is limited and atomizes in a transient manner. Simultaneous atomic absorption measurements are discussed by Mavrodineanu and Hughes.[82] Basically, a grating is used to combine the beams from several hollow cathode lamps, pass the combined beam through the absorption cell, and then another grating is used to disperse the different wavelengths to separate photomultiplier tubes. A modification of this approach is used in a Perkin–Elmer portable graphite furnace atomic absorption spectrophotometer for wear metal analysis.[83] This instrument uses a semitransparent flat mirror instead of a grating to combine the beams from two multielement hollow cathode lamps. Instead of several hollow cathode lamps, Ohls *et al.*[6] used a Grimm-type glow discharge as the primary radiation source for multielement atomic absorption because of the ease of selecting and changing the elements by choosing an alloy of those elements as the flat cathode. As an alternative to spatially dispersing different wavelengths with a grating, separate wavelengths in a combined beam can be detected with resonance detectors in series with each other, each consisting of a resonance atomic fluorescence monochromator to isolate a wavelength along with a photomultiplier tube to detect the fluorescence.[84] Although a continuum light source, such as a xenon lamp, might be used with a resonance atomic fluorescence monochromator, the low spectral radiance of such sources and the high potential for scattered light from all of the unused wavelengths would make this an unlikely choice for an instrument that would have good detection limits.

Multielement atomic fluorescence is somewhat less complicated because the beams from each hollow cathode lamp do not have to be combined, but can be focused from separate directions onto an appropriately designed atom cell. A polychromator with a photomultiplier for each wavelength can be used to detect the fluorescence if background emission is a problem and simultaneous determinations are needed. Alternatively, separate photomultipliers with interference filters in front of them could be positioned around the atom cell to observe the fluorescence simultaneously, as was done on an inductively coupled plasma AFS instrument manufactured by Baird Corporation (Bedford, Mass.). Background emission can be eliminated by pulsing the atom cell and observing the fluorescence after cell emission has subsided. Rapid sequential observations can be made with a single photomultiplier that observes the rapid sequence of pulses of fluorescence that occur when each of the hollow cathode lamps is pulsed in rapid succession.

3.3.6. Compensation for Sputtering Rate Variations

Because sputtering rates do not vary from one type of sample to the next nearly as much as vaporization rates vary for thermal methods of atomization, it is possible to use the same analytical working curves for samples that vary more widely in composition. For example, the same working curves can be used with the Atomsource for some high-alloy and low-alloy steel samples.[85] On the other hand, Fang and Marcus[64] have shown that the weight loss for a series of brass samples varies with the concentration of copper. In such cases, improved analytical results can be obtained by compensating for differences in sample erosion rates.[70,86]

One method of compensation is to weigh the sample before and after the analysis as suggested by Gough *et al.*[5] However, the loss of material during a typical analysis time is only in the microgram range. Even with a microgram balance, such a small weight loss may be difficult to determine because of the possibility of contamination during handling of the sample. To avoid these problems, Chakrabarti *et al.*[16] continued sputtering the sample long after the analysis time in order to obtain a weight loss that could be accurately measured.

Winchester *et al.*[67] used an internal reference (or internal standard) of $BaCO_3$ in pressed copper pellets to determine Pt and Rh in α-alumina catalysts. McDonald[70] used an internal reference of Ni in a pressed powder disk containing 1% Ni, 30% copper slag, and 69% silver for the analysis of copper slags. Ni was used because the silver lines were almost totally absorbed. Erosion rates varied by a factor of five for these samples. A series of Ni compounds was analyzed for Ni by mixing with copper powder and pressing the mixture into disks. Copper was used as the internal reference. The absorbances were kept below 0.15 to ensure linearity of the working curves. The excellent analytical results obtained by these studies show the usefulness of an internal standard in compensating for differences in sputtering rates.

When the concentration of none of the elements is known, an individual element cannot be used as an internal reference. In this case it may be possible to use the entire sample as an internal reference; that is, measure the absorbances of all of the elements in the sample and use for an internal reference the sum of the concentrations of all of the elements, which of course has a known value of 100%. This method was used by Jäger[87] for the Grimm discharge emission analysis of gold samples, but is applicable to absorption and fluorescence methods if the analytical working curves are linear and pass through the origin of the graph. For one standard, h, usually selected because it has relatively high concentrations of all of the elements, values of

$$k_{\text{elt,h}} = C_{\text{elt,h}}/A_{\text{elt,h}} \tag{3-7}$$

are calculated for each element, elt, where $C_{\text{elt,h}}$ is the concentration of element elt in standard h, and $A_{\text{elt,h}}$ is the absorbance of element elt in standard h. Then for each standard, the sum S_{std} over all elements of the absorbances A for each element times these $k_{\text{elt,h}}$ values is calculated as a measure of total concentration

$$S_{\text{std}} = \sum_{\text{elt}} k_{\text{elt,h}} A_{\text{elt,std}} \tag{3-8}$$

Then the corrected absorbance A' for each element in each standard is calculated as

$$A'_{\text{elt,std}} = \frac{k_{\text{elt,h}} A_{\text{elt,std}}}{S_{\text{std}}} \tag{3-9}$$

The analytical working curve for each element is a plot of the corrected absorbance for that element in each of the standards versus concentration. To determine the concentration of an element in a sample sam, values of S_{sam} and $A'_{\text{elt,sam}}$ are calculated for that sample and used with the working curve for that element to determine the concentration. This method has been used by Bernhard and co-workers[86] to analyze tool steels, which vary widely in composition and have different sputtering erosion rates.

When the absorbances are high enough that the working curves are not linear, then using this method that assumes linear working curves will still improve the analytical results, but even better results will be achieved if account is taken of the nonlinear shape of the working curve for each element. This can be done by using an appropriate mathematical model (e.g., a polynomial equation) for each of the nonlinear working curves, and using the absorbances obtained with the standards to determine the unknown parameters in these mathematical models. This produces a set of simultaneous nonlinear equations that must be solved. For example, we have used a second-order polynomial equation for the shape of each of the working curves to obtain a set of equations (one equation for each element in each standard) of the following form:

$$A_{\text{elt,std}} = b_{1,\text{elt}} R_{\text{std}} C_{\text{elt,std}} + b_{2,\text{elt}} (R_{\text{std}} C_{\text{elt,std}})^2 \tag{3-10}$$

where $b_{1,\text{elt}}$ and $b_{2,\text{elt}}$ are the first and second-order coefficients of the polynomial for the working curve for element elt, and R_{std} represents the sample erosion rate for standard std as well as the rate at which the sputtered material for that standard is transported to the observation zone. The values of $b_{1,\text{elt}}$, $b_{2,\text{elt}}$, and R_{std} are the unknown parameters for which this set of equations must be solved in order to construct the analytical working curves.

To analyze a sample sam, the absorbances $A_{\text{elt, sam}}$ for all of the elements in the sample are determined. These values and the known values of $b_{1,\text{elt}}$ and $b_{2,\text{elt}}$ [determined by solving Eq. (3-10) for the standards] are used to solve the set of simultaneous equations (one equation for each element) represented by Eq. (3-10) where the subscript std is replaced by sam to find the values of R_{sam} and the concentrations of all of the elements in the sample.

We have solved this set of equations by using a Simplex search routine to minimize the sum of the squares of the deviations of the experimental absorbances from the working curves. By using a quadratic model for each of the working curves, better results were achieved for the same set of experimental data for tool steels that was analyzed by assuming linear working curves.[86] The sum of the squares of the deviations from the working curves of the measured absorbances for all ten elements determined in all six standards for the linear method was 0.00257. For the global fit to the quadratic working curves, this was reduced to 0.00019. An example of the improvement in working curves using the nonlinear global fit to all of the data is shown in Fig. 3-11. Although this method gives better results, the Simplex search finds different minima when the initial starting conditions for the search are different. The minima all give results that are better than the linear fit, but we are searching for a better way to solve the equations along with a better mathematical model for the working curves.

A modification of this procedure involves determining the shape of the analytical working curves by some independent means such as another set of standards all of which have the same sample erosion rate. Values of $b_{1,\text{elt}}$

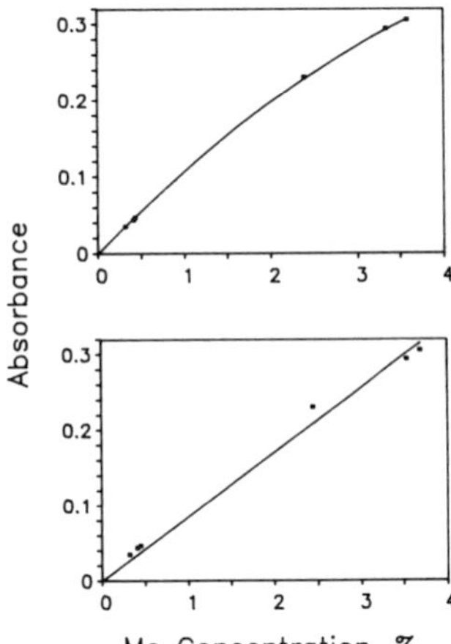

Figure 3-11. Curves for Mn in tool steel resulting from global fits to all of the data in a total analysis of the standards used to compensate for variations in sputtering rates. The upper curve shows the better fit that results when it is assumed that the working curves for all of the elements are quadratic polynomials. The lower curve results from assuming linear working curves.

and $b_{2,\mathrm{elt}}$ would then be known and only values of R_{std} would have to be determined from the set of equations represented by Eq. (3-10). This method assumes that the shape of the working curve does not change. For example, it assumes that the shape is controlled by the spectral profiles of the absorption line in the sputtering cell and the emission line of the primary radiation source, and that these spectral profiles do not change. However, since the emission profile of the primary source changes with lamp current and with age, the parameters describing the shape of the working curve may have to be updated from time to time.

3.3.7. Determining Element Ratios with Only One Standard

McDonald[70] has shown that only one standard containing the elements of interest can be used to determine the ratio of two elements in a variety of samples that have widely different sputtering erosion rates, provided the working curves are linear. Since ratios are being determined within the same sample, the sputtering erosion rates of the standard or sample need not be known. The standard is used to determine the proportionality factor F_{ij} between the ratio of the absorbances A and the ratio of the concentrations C of two elements i and j:

$$\frac{A_{\mathrm{i}}}{A_{\mathrm{j}}} = F_{\mathrm{ij}} \frac{C_{\mathrm{i}}}{C_{\mathrm{j}}} \qquad (3\text{-}11)$$

Once determined, the proportionality factors F_{ij} are valid for a wide variety of samples. This method is based on the idea that the absorbance of an element is related to its concentration by proportionality factors that are constants or that can be readily held constant in a sputtering chamber.[70] These factors include the oscillator strength of the line, the spectral profiles of the absorption line in the sputtering cell and the emission line of the primary radiation source, the fraction of atoms in the ground state, wavelength, the atomic weight, the diffusion coefficient, and in a flowing cell, a carrier gas transport factor that helps determine the fraction of atoms that reach the observation zone. The factor F_{ij} is the ratio of these proportionality factors for the two elements.

Also assumed is the absence of preferential effects, such as preferential sputtering, preferential ionization, or preferential agglomeration of the elements. As we saw earlier, although preferential sputtering occurs initially, its effects are eliminated (if the surface temperature is not too high) after an equilibrium surface layer has been established during preconditioning sputtering. However, ionization differences may be significant and agglomeration may be different for different elements.

If the concentration of only one element in a sample is known, then the concentrations of the other elements for which F_{ij} is known can be determined with this method. This comes close to the proposal by Walsh in his seminal paper in 1955[3] that atomic absorption could be used for the determination of concentrations without standards.

3.4. Applications

This section includes a brief summary of the types of analyses that have been done with sputter cells using AAS and AFS. The list continues to grow as the advantages of these methods continue to be recognized.

3.4.1. Alloys

In 1960 Gatehouse and Walsh[9] used a hollow cathode sputtering cell to determine Ag in copper. Gough[13] lists atomic absorption detection limits ranging from 1 μg/g to 800 μg/g for 16 elements in alloys of Fe, Al, Cu, and Zn. McDonald[70] used an internal standard to determine Cr in a series of low-alloy steels, high-speed steels, and product steel. He found that one sample was not homogeneous and that the sputtering cell correctly determined the concentration in the center of the sample, which was confirmed by another analytical method to be different from the certified bulk value. Ni was determined in steel samples, brass, bronze, and an aluminum alloy by using Cu as the internal standard, since its concentration was known in all of the samples. Only one reference sample of steel was used to determine the calibration factor for the Ni-to-Cu ratio. The calibration factor then remains the same for all of these different samples because surface equilibration is readily achieved during the preburn.

Ohls[78] was the first to publish the results of an atomic absorption analysis of alloys using the Atomsource. Elements included Cr, Cu, Mn, Nb, and Zr in steels and La in zinc. Bernhard and Kahn[15] determined Al in seven varied nickel-based alloys using only one standard with the Atomsource. Detection limits for 15 elements in common alloys range from 1 to 5 $\mu g/g$ for the Atomsource.[88]

Gough *et al.*[5] used nondispersive atomic fluorescence to determine Si, Mn, Ni, Cr, and Cu in low-alloy steels. Gough and Meldrum[66] used nondispersive atomic fluorescence to determine Cr, Mn, Ni, and Mo in low-alloy steels. Detection limits for 12 elements ranged from 1 to 100 $\mu g/g$ for various metals in low-alloy steel, aluminum, brass, and copper samples.

Winefordner and co-workers[18] used laser-excited nonresonance atomic fluorescence to determine Pb in copper with a 0.1 $\mu g/g$ detection limit.

3.4.2. Powders

A method of preparing pressed copper pellets for the analysis of powder samples has been studied in detail by Winchester and Marcus.[89] Winchester *et al.*[67] used the rapid sequential multielement capability of the Analyte 16 with an Atomsource to determine Pt and Rh in α-alumina catalysts by AAS with an internal reference. The catalyst was mixed with copper powder and $BaCO_3$ as an internal reference, and pressed into an electrically conducting pellet. The internal reference reduced the deviations from the analytical working curve by an order of magnitude.

McDonald[70] used Cu as an internal reference in pellets of pressed copper powder to determine Ni in powdered Ni compounds. Cu was determined in powdered copper slags pressed into silver powder pellets with 1% Ni added as an internal reference. Slags are well known for the lack of good standards because of their wide variation in composition, but the internal reference method alleviates this problem.

3.4.3. Dried Solutions

Before drying a solution on the surface of a cathode, the cathode is often preconditioned by sputtering. However, because of the small amount of dried material that is usually used, there is not enough sputtering time prior to the analysis to precondition the dried material. As we saw earlier, preconditioning by sputtering brings the surface to an equilibrium condition that compensates for preferential sputtering. Therefore, the lack of preconditioning may cause some elements to atomize more quickly than others, producing peaks of different durations. Even so, detection limits for some elements are better than those for graphite furnace AAS.

Goleb and Yokoyama[37] found that uniform deposits of lithium hydroxide solutions could be prepared in a hollow cylindrical copper electrode by placing 0.1 ml of solution in the copper electrode and adding about 0.1 ml of acetone to it while the electrode was rotating at 60 rpm under a heat lamp. Goleb and Brody[11] determined Na by atomic absorption after drying solutions of NaF on the surface of a hollow cathode. Li and Mg produced severe matrix effects. McCamey and Niemczyk[90] studied eight elements in a hollow cathode absorption cell. Matrix effects were generally not observed for 1:1 ratios of potential interferants to analyte, but became noticeable for ratios of 100:1 in most cases. Additional information about hollow cathode cell applications appears in Chapters 6 and 7.

Gandrud and Skogerboe[12] used flat spectroscopic-grade graphite and aluminum platrodes as cathodes on which sample solutions were evaporated and inserted into the sputtering cell. Cell current was 50 mA and cell pressure was 3 Torr. Peak widths at half-height were 12 s but the peaks had tails lasting several minutes. Detection limits ranged from 1 to 10 ng for eight elements.

Chakrabarti *et al.*[16] analyzed dried solutions deposited on flat cathodes pressed against the sample port of the Atomsource. Power regulation was found to give more reproducible results than current or voltage regulation. They used a high-electrical-power pulsed mode of operation for high atomization rates and found that an instrumental time constant of 2 ms on one of the faster commercial atomic absorption spectrophotometers is too slow to respond to the 5-ms peak of one of the elements. They obtained results that are comparable to those of the highly sensitive GFAAS method for some elements. Absolute detection limits for 14 elements range from 0.2 to 600 pg.

Ohls[78] showed that elements like Zr could be determined although they are not suitable for determinations by graphite furnace atomization because they form refractory carbides. The detection limit for Zr using 5 μl of solution was 100 ng using the Atomsource. Although peak height was linearly related to concentration up to 500 ng, a 5000-ng sample produced a much broader peak than the lower quantities of Zr.

Chakrabarti *et al.*[16] found that matrix effects for Fe caused by the presence of a $Ni(NO_3)_2$ matrix with a matrix-to-analyte ratio up to 100 to 1 could be eliminated by using peak area rather than peak height. In other cases this did not work. The severe matrix effects caused by alkali and alkali-earth chlorides on the determination of metals using dried solutions could not be compensated by using peak area. These differences were attributed to the differences in electrical conductivity of the dried deposits; nonconducting deposits produce large matrix effects because they allow a positive charge to build up that repels sputtering ions.

The problem of charge buildup on insulating samples has been studied by Drobyshec and Turkin.[61] They showed that pulses could be used to sputter insulators by studying the emission of Ga, Gd, Fe, and Y when an insulating monocrystalline film of $Y_3Fe_5O_{12}$ on a $Gd_3Ga_5O_{12}$ substrate was subjected to successive sputtering pulses. They also developed a model to predict how long sputtering will continue before significant positive charge builds up. In an example for a 0.5-mm particle, their model predicts that sputtering will continue for 0.3 s at 10 mA, or for 1.7 ms at 2 A before sputtering is stopped by charge buildup. The latter time is comparable to the shortest pulses observed for dried solutions by Chakrabarti *et al.*[16] and indicates that atomic absorption pulses could be used for the analysis of insulating materials, including dried solution residues that are insulating.

Winefordner and co-workers[18] used nonresonance atomic fluorescence to determine Pb in aqueous solutions. They used a frequency-doubled nitrogen pumped dye laser with a spectral bandwidth of 0.015 nm and a 5-ns pulse having a typical energy of 5 μJ. The linear dynamic range is six orders of magnitude. The detection limit is 20 pg for 5 μl of aqueous solution deposited on a graphite cathode.

3.4.4. Isotopes

In 1955 Walsh[3] suggested that AAS might be suitable for the isotopic analysis of certain elements. As mentioned earlier, the isotope shift for a handful of light and very heavy elements is large enough for isotopic analysis using sputtering cells. Larkins[28] determined the major isotopes of Pb using flames as well as the flowing sputtering cell of Gough.[13] Hannaford and Walsh[91] determined the isotopes of boron. Goleb[10,92] used water-cooled hollow cathode lamps and a water-cooled hollow cathode sputtering cell for the isotopic analysis of uranium. He confirmed that heavier sputtering ions produced a larger absorption signal for a given sputtering current. Krypton was used in the hollow cathode lamps and xenon in the absorption cell. Goleb and Yokoyama[37] used a similar water-cooled system for the isotopic analysis of Li. They found large variation in the isotopic composition of some reagent-grade lithium salts.

Doppler-free saturated absorption spectrometry[93] extends isotopic analysis to other elements and has been used for the isotopic analysis of Ta,[94] V,[95] Y,[94,96,97] and Zr.[91,94]. With this technique, the beam from a highly monochromatic tunable laser is split into two beams and the resulting strong and weak beams are passed through the sputtering cell in opposite directions. The strong laser beam is modulated and saturates the line while the weak beam detects the modulated decrease in absorption produced by the saturation. Since both laser beams have the same wavelength, only those atoms that have a zero velocity component along the direction of propagation of the beams produce a signal. Therefore, a Doppler-free spectrum results when the wavelength of the laser is scanned. Further experimental details are discussed by Hannaford and Walsh.(91)

Doppler-free saturated fluorescence methods are briefly reviewed by Hannaford(98) who used Doppler-free intermodulated fluorescence to study the isotopes of molybdenum. Two counterpropagating laser beams are used to excite the atoms, and their fluorescence is detected. Gough and Hannaford(94) studied the hyperfine structure of the 578.1-nm line of ^{181}Ta, and Gough *et al.*(95) studied the hyperfine structure of ^{51}V.

3.5. The Future

One of the most rapidly advancing areas of study is the atomic absorption analysis of dried solution residues using gas-jet-enhanced sputtering with high-power sputtering pulses. Chakrabarti *et al.*(16) have achieved detection limits that rival those of graphite furnace AAS (GFAAS) for some elements, particularly those that are difficult to study with GFAAS, and further improvements are continuing to be made. Although matrix effects appear to be less severe than those of GFAAS, they may be worse than for the sputtering cell analysis of alloys where a preburn allows the surface to come to equilibrium before analytical measurements are made. In any case a better understanding of matrix effects and sputtering processes in the analysis of dried solution residues is needed.

Although gas jets help to reduce redeposition of sputtered atoms and help to transport atoms into the observation region, the continued flow of gas also moves atoms out of the observation region. A higher population of atoms might be obtained by synchronizing gas-jet pulses with high-power sputtering pulses so as to transport a large number of atoms into the observation zone and then to stop the gas flow when the atom population has reached its maximum. Synchronized gas-jet pulses might have another advantage when dried solutions are used. A low-current preburn with the jets off could bring the surface to equilibrium; sample loss would be minimal because of redeposition of the sputtered atoms back onto the surface. Surface equilibration would help to reduce those matrix effects that are caused by the differences in sputtering rates of different elements and forms of elements.

The direct sputtering of electrically nonconducting (insulating) materials, discussed in more detail in Chapter 9, would have many applications. Although insulating powders are now analyzed by incorporating them into pressed pellets of conducting powders, it would be much easier to simply dust a powder onto a cathode surface and atomize the material with short pulses. As more is learned about surfaces, it should become easier to choose the type of surface on which to retain a particular type of powder.

Duckworth and Marcus[(99)] have shown that an rf discharge sputtering cell can produce mass spectrometer ion signals for a glass sample that are comparable to those produced by metals in dc sputtering cells. Although their results do not directly extrapolate to the generation of a population of ground-state atoms, the idea looks very promising for the direct sampling of insulating materials for AAS and AFS.

Sputtering of insulating materials using high-current pulses has been discussed and demonstrated by Drobyshev and Turkin.[(61)] Direct-current pulses with a controlled duty cycle would produce efficient sputtering during each current pulse, while each pulse would be followed by sufficient time to dissipate the electrical charge that has built up on the insulating sample. Perhaps a reversed current pulse or negatively charged gas-jet pulse could be used to help dissipate the accumulated charge. Some important advances in the rapid sputtering of insulating materials for AAS and AFS may be near.

Laser-excited fluorescence has the potential of reaching detection limits that are many decades lower than those that have been demonstrated so far by either AAS or AFS. To date only stagnant-gas cells have been used as atomizers. A combination of improvements of the type suggested above and those being used for atomic absorption measurements (e.g., gas jets and higher-power sputtering pulses) should help to improve detection limits and may lead to wider applications. However, improvements in laser reliability are needed to move the tunable laser methods out of the special applications laboratory and into the routine workplace.

Because sputtering cells use a carrier gas that is relatively transparent at vacuum ultraviolet wavelengths, they are good atomizers for the determination of elements such as C, P, and the halides whose strongest spectral lines are below 200 nm. Although emission has been used for such determinations, atomic absorption or fluorescence would be encouraged by the availability of hollow cathode lamps that are designed with windows and excitation conditions that emphasize the vacuum ultraviolet lines.

A properly designed gas-jet flow clearly helps to improve atomic absorption signals. Although transport processes are one reason for the improvement, little is known about how jets that strike the surface of the sample influence the spatial distribution of the flow of sputtering current. More information about the influences of the jets may lead to further improvements in sample erosion.

As we have seen, the spatial resolution that is inherent in the size of the sample ports of sputtering cells of Gough and the Atomsource has already been used to demonstrate differences in the bulk concentration and the concentration in the center of a large sample. Analyses with this spatial resolution may eventually lead to improved or more homogeneous products. Even higher spatial resolution is possible by reducing the diameter of the jets and/or bringing the jet nozzle closer to the surface. In our laboratory a jet has been used to drill a 0.05-cm hole through a copper strip. As more is understood about gas-jet-enhanced sputtering, a microscopic atomizer may be possible with spatial resolution and control comparable to those of a laser microprobe.

The low detection limits that can be achieved with sputtering cell atomic absorption and fluorescence methods will make them complementary to graphite furnace methods for the analysis of solutions and small samples. Because collisional broadening is so low, spectral resolution is so high, and sample containment is relatively easy, sputtering cell atomizers will continue to find use for isotopic analyses, especially using Doppler-free laser methods. Because of their high precision, multielement capability, and ease of use, atomic absorption and fluorescence with sputtering cell atomizers will become important analytical methods for a wide variety of samples, including superalloys and other new "high-technology" non-, semi-, and superconducting materials.

References

1. C. T. J. Alkemade and J. M. W. Milatz, *J. Opt. Soc. Am.* 45 (1955) 583.
2. C. T. J. Alkemade and J. M. W. Milatz, *Appl. Sci. Res.* B4 (1955) 289.
3. A. Walsh, *Spectrochim. Acta* 7 (1955) 108.
4. B. J. Russell and A. Walsh, *Spectrochim. Acta* 15 (1959) 883.
5. D. S. Gough, P. Hannaford, and A. Walsh, *Spectrochim. Acta* 28B (1973) 197.
6. K. Ohls, J. Flock, and H. Loepp, *Fresenius Z. Anal. Chem.* 332 (1988) 456.
7. B. W. Smith, J. B. Womack, N. Omenetto, and J. D. Winefordner, *Appl. Spectrosc.* 43 (1989) 873.
8. J. V. Sullivan and A. Walsh, *Spectrochim. Acta* 21 (1965) 727.
9. B. M. Gatehouse and A. Walsh, *Spectrochim. Acta* 16 (1960) 602.
10. J. A. Goleb, *Anal. Chem.* 35 (1963) 1978.
11. A. Goleb and J. K. Brody, *Anal. Chim. Acta* 28 (1963) 457.
12. B. W. Gandrud and R. K. Skogerboe, *Appl. Spectrosc.* 25 (1971) 243.
13. D. S. Gough, *Anal. Chem.* 48 (1976) 1926.
14. A. E. Bernhard, *Spectroscopy* 2 (1987) 24.
15. A. E. Bernhard and H. L. Kahn, *Am. Lab.* 20 (1988) 126.
16. C. L. Chakrabarti, K. L. Headrick, J. C. Hutton, Z. Bicheng, P. C. Bertels, and M. H. Back, *Anal. Chem.* 62 (1990) 574.
17. M. Glick, B. W. Smith, and J. D. Winefordner, *Anal. Chem.* 62 (1990). 157.
18. B. M. Smith, N. Omenetto, and J. D. Winefordner, *Spectrochim. Acta* 39B (1984) 1389.
19. E. H. Piepmeier, in: *Analytical Laser Spectroscopy* (N. Omenetto, ed.), Wiley, New York, 1979.
20. I. I. Sobel'man, *Introduction to the Theory of Atomic Spectra*, Pergamon Press, Elmsford, N.Y., 1972.
21. H. G. Kuhn, *Atomic Spectra*, 2nd ed., Academic Press, New York, 1969.

22. C. Candler, *Atomic Spectra and the Vector Model*, Hilger & Watts, London, 1964.
23. R. G. Breene, Jr., *The Shift and Shape of Spectral Lines*, Pergamon Press, Elmsford, N.Y., 1961.
24. A. C. G. Mitchell and M. W. Zemansky, *Resonance Radiation and Excited Atoms*, Cambridge University Press, London, 1961.
25. C. T. J. Alkemade and R. Herrmann, *Fundamentals of Analytical Flame Spectroscopy*, Wiley, New York, 1979.
26. L. de Galan and H. C. Wagenaar, *Methodes Phys. Anal.* 3 (1971) 10.
27. C. F. Bruce and P. Hannaford, *Spectrochim. Acta* 26B (1971) 207.
28. P. L. Larkins, *Spectrochim. Acta* 39B (1984) 1365.
29. H. Kopfermann, *Nuclear Moments* (English translation by E. E. Schneider), Academic Press, New York, 1958.
30. G. J. De Jong and E. H. Piepmeier, *Spectrochim. Acta* 29B (1974) 179.
31. E. H. Piepmeier and L. de Galan, *Spectrochim. Acta* 30B (1975) 263.
32. H. C. Wagenaar and L. de Galan, *Spectrochim. Acta* 28B (1973) 157.
33. P. L. Larkins, *Spectrochim. Acta* 43B (1988) 1175.
34. J. W. Hosch and E. H. Piepmeier, *Appl. Spectrosc.* 32 (1978) 444.
35. E. H. Piepmeier, *Spectrochim. Acta* 44B (1989) 609.
36. V. Svoboda, R. F. Browner, and J. D. Winefordner, *Appl. Spectrosc.* 26 (1972) 505.
37. J. A. Goleb and Y. Yokoyama, *Anal. Chim. Acta* 30 (1964) 213.
38. J. D. Ingle, Jr., *Anal. Chem.* 46 (1974) 2161.
39. S. B. Smith, Jr., and G. M. Hieftje, *Appl. Spectrosc.* 37 (1983) 419.
40. J. V. Sullivan, *Anal. Chim. Acta* 105 (1979) 213.
41. P. B. Farnsworth and J. P. Walters, in: *Improved Hollow Cathode Lamps for Atomic Spectroscopy* (S. Caroli, ed.), Wiley, New York, 1985.
42. S. Caroli, A. Alimonti, and F. Petrucci, in: *Improved Hollow Cathode Lamps for Atomic Spectroscopy* (S. Caroli, ed.), Wiley, New York, 1985.
43. H. C. Wagenaar, I. Novotny, and L. de Galan, *Spectrochim. Acta* 29B (1974) 301.
44. D. C. McDonald, *Anal. Chem.* 54 (1982) 1052.
45. C. Van Dijk, B. W. Smith, and J. D. Winefordner, *Spectrochim. Acta* 37B (1982) 759.
46. N. P. Ferreira and H. G. C. Human, *Spectrochim. Acta* 36B (1981) 215.
47. A. J. Stirling and W. D. Westwood, *J. Phys. D.* 4 (1971) 246.
48. A. J. Stirling and W. D. Westwood, *J. Appl. Phys.* 41 (1970) 742.
49. D. S. Gough, P. Hannaford, and R. Martin Lowe, *Anal. Chem.* 61 (1989) 1652.
50. P. W. J. M. Boumans, *Theory of Spectrochemical Excitation*, p. 297, Plenum Press, New York, 1966.
51. H. J. Kim and E. H. Piepmeier, *Anal. Chem.* 60 (1988) 2040.
52. C. L. Chakrabarti, K. L. Headrick, J. C. Hutton, B. Marchand, and M. H. Back, *Spectrochim. Acta* 44B (1989) 385.
53. C. L. Chakrabarti, K. L. Headrick, J. C. Hutton, and M. H. Back, *XXVI Colloquium Spectroscopicum Internationale, Invited Paper No.* 4 (1989) 10.
54. C. L. Chakrabarti, K. L. Headrick, P. C. Bertels, and M. H. Back, *J. Anal. At. Spectrom.* 3 (1988) 713.
55. A. E. Bernhard, personal communication.
56. Y. Demers, J.-M. Gagné, and P. Pianarosa, *J. Anal. At. Spectrom.* 2 (1987) 59.
57. C. H. Corliss and W. R. Bozman, *Experimental Transition Probabilities for Spectral Lines of Seventy Elements*, National Bureau of Standards Monograph 53, U.S. Government Printing Office, Washington, D.C., 1962.
58. P. L. Larkins, *Anal. Chim. Acta* 132 (1981) 119.
59. D. Fang and R. K. Marcus, *Spectrochim. Acta* 43B (1988) 1451.
60. G. S. Anderson, *J. Appl. Phys.* 40 (1969) 2884.

61. A. I. Drobyshev and Y. I. Turkin, *Spectrochim. Acta* 36B (1981) 1153.
62. E. Gillam, *J. Phys. Chem. Solids* 11 (1959) 55.
63. W. L. Patterson and G. A. Shirn, *J. Vac. Sci. Technol.* 4 (1967) 343.
64. D. Fang and R. K. Marcus, *J. Anal. At. Spectrom.* 3 (1988) 873.
65. Y. Yokoyama and S. Ikeda, *Spectrochim. Acta* 24B (1969) 117.
66. D. S. Gough and R. J. Meldrum, *Anal. Chem.* 52 (1980) 642.
67. M. R. Winchester, S. M. Hayes, and R. K. Marcus, *Spectrochim. Acta* 46B (1991) 615.
68. B. M. Patel and J. D. Winefordner, *Spectrochim. Acta* 41B (1986) 469.
69. D. C. McDonald, *Anal. Chem.* 54 (1982) 1057.
70. D. C. McDonald, *Anal. Chem.* 49 (1977) 1336.
71. B. M. Patel and J. D. Winefordner, *Appl. Spectrosc.* 40 (1986) 667.
72. J. D. Ingle, Jr., and S. R. Crouch, *Spectrochemical Analysis*, Prentice–Hall, Englewood Cliffs, N.J., 1988.
73. B. Welz, *Atomic Absorption Spectrometry*, 2nd ed., VCH Verlagsgesellschaft, Weinheim, 1985.
74. L. de Galan and M. T. C. de Loos-Vollebregt, *Spectrochim. Acta* 39B (1984) 1011.
75. M. T. C. de Loos-Vollebregt and L. de Galan, *Spectrochim. Acta* 41B (1986) 597.
76. G. Tie-Zheng and R. Stephens, *J. Anal. At. Spectrom.* 1 (1986) 355.
77. E. H. Piepmeier and L. de Galan, *Spectrochim. Acta* 31B (1976) 163.
78. K. Ohls, *Fresenius Z. Anal. Chem.* 327 (1987) 111.
79. E. H. Piepmeier, *Anal. Chem.* 48 (1976) 1296.
80. E. D. Salin and J. D. Ingle, Jr., *Appl. Spectrosc.* 32 (1978) 579.
81. E. D. Salin and J. D. Ingle, Jr., *Anal. Chem.* 50 (1978) 1737.
82. R. Mavrodineanu and R. C. Hughes, *Appl. Opt.* 7 (1968) 1281.
83. W. Niu, R. Haring, and R. Newman, *Am. Lab.* 19 (1987) 40.
84. J. V. Sullivan and A. Walsh, *Appl. Opt.* 7 (1968) 1271.
85. A. E. Bernhard, E. H. Piepmeier, H. J. Kim, and H. L. Kahn, 39th Pittsburgh Conference and Exposition on Analytical Chemistry and Applied Spectroscopy, 1988, Paper No. 809.
86. E. H. Piepmeier, H. J. Kim, C. E. Crandall, and A. E. Bernhard, 16th Annual Meeting of the Federation of Analytical Chemistry and Spectroscopy Societies, 1989, Paper No. 920.
87. H. Jäger, *Anal. Chim. Acta* 60 (1972) 303.
88. Analyte Corp., Medford, Oregon, brochure, 1990.
89. M. R. Winchester and R. K. Marcus, *Appl. Spectrosc.* 42 (1988) 941.
90. D. A. McCamey and T. M. Niemczyk, *Appl. Spectrosc.* 34 (1980) 692.
91. P. Hannaford and A. Walsh, *Spectrochim. Acta* 43B (1988) 1053.
92. J. A. Goleb, *Anal. Chim. Acta* 34 (1966) 135.
93. T. W. Hänsch, I. S. Shahin, and A. L. Schawlow, *Phys. Rev. Lett.* 27 (1971) 707.
94. D. S. Gough and P. Hannaford, *Anal. Chem.* 55 (1985) 91.
95. D. S. Gough, P. Hannaford, R. M. Lowe, and A. P. Willis, *J. Phys. B* 18 (1985) 3895.
96. R. J. McLean, P. Hannaford, H.-A. Bachor, P. T. H. Fisk, and R. J. Sandeman, *Z. Phys. D* 1 (1986) 253.
97. D. S. Gough, P. Hannaford, and R. J. McLean, *J. Phys. B* 21 (1988) 547.
98. P. Hannaford, *Contemp. Phys.* 24 (1983) 251.
99. D. C. Duckworth and R. K. Marcus, *Anal. Chem.* 61 (1989) 1879.

4

Atomic Emission Spectrometry

J. A. C. Broekaert

4.1. Optical Emission Spectrometry

4.1.1. Principles

4.1.1.1. Optical Spectra

Optical emission spectrometry is one of the oldest physical methods of analysis, enabling multielement determinations at the level of major elements, minor elements, as well as trace elements. Its historical development is closely related to the progress in basic understanding of the nature of atomic spectra and the structure of matter. The optical emission spectra of atoms and ions (for a discussion see Refs. 1, 2) stem from transitions between the outer electron shells of the chemical elements which give rise to line spectra where the wavelength of the lines relates to the energy difference of the levels concerned according to:

$$\Delta E = hc/\lambda \tag{4-1}$$

The atomic levels as known from Bohr's atomic model relate to the atomic quantum numbers. The main quantum number is described by

$$E = -\frac{2\pi\mu Z^2 e^4}{n^2 h^2} \tag{4-2}$$

J. A. C. Broekaert • Department of Chemistry, University of Dortmund, D-4600 Dortmund 50, Germany.

Glow Discharge Spectroscopies, edited by R. Kenneth Marcus. Plenum Press, New York, 1993.

with μ being the reduced mass of the electron and the nucleus, e the charge of the electron, h Planck's constant, and $n = 1, 2, 3, \ldots$. The l quantum number is defined by the quantization of the impulse moment of the electron as:

$$|\mathbf{L}| = \frac{h}{2\pi}\sqrt{l(l+1)} \tag{4-3}$$

and has values of $l = 0, 1, 2, \ldots, n-1$.

The magnetic quantum number is related to the quantization of the impulse momentum in the orientation of an external field and is defined by:

$$L_z = \frac{h}{2\pi} m_1 \tag{4-4}$$

with $m_1 = \pm 1, \pm(l-1), \ldots, 0$.

The spin quantum number is introduced for describing the splitting of the atomic spectral lines in a magnetic field and is defined by:

$$|\mathbf{S}| = \frac{h}{2\pi}\sqrt{S(S+1)} \tag{4-5}$$

with S the spin momentum and for the component in the field direction s_z:

$$s_z = \frac{h}{2\pi} m_s \qquad \text{with } m_s = \pm\tfrac{1}{2} \tag{4-6}$$

The impulse momentum **L** and the spin momentum **S** combine to a total angular momentum **J** of which

$$|\mathbf{J}| = \frac{h}{2\pi}\sqrt{j(j+1)} \tag{4-7}$$

with $j = l \pm s$ the internal quantum number.

The atomic energy levels can be described by the quantum numbers and indicated by a term symbol:

$$n^{m} l_{j} \tag{4-8}$$

When two valence electrons are present, coupling of the angular momentum of both electrons gives rise to a total angular momentum and the coupling of the spin momentum of both electrons to a total spin momenum ($L - S$ or Russell Saunders coupling) and total quantum numbers are defined as:

$$L = l_1 + l_2 \tag{4-9}$$

and

$$S = \sum_{i} s_i \qquad \text{(being 0 or 1)} \tag{4-10}$$

and

$$|\mathbf{J}| = \frac{h}{2\pi}\sqrt{J(J+1)} \quad \text{with } J = L - S, \ldots, L + S \tag{4-11}$$

Also here the terms can be indicated as

$$n\,{}^{m}L_{J} \tag{4-12}$$

and radiative transitions are only allowed when Δn is an integer number, $\Delta L = 0, \pm 1$, and $\Delta J = 0, \pm 1$, but not for $J = 0$. The complexity of the atomic spectra is reflected by the number of terms relevant for the element concerned.

The spectral lines of relevance in atomic spectrometry occur between 160 and 800 nm. As the line spectra are characteristic of the emitting element, the presence of lines of known wavelength in a spectrum is a criterion for the presence of the element above a detectable concentration level. This forms the base for emission spectrometry as a tool for qualitative analysis.

4.1.1.2. Line Intensities

The line intensities emitted by the so-called radiation source, the glow discharge in this work, directly relate to the number density of the atoms or ions present, as given by the expression for the radiance in terms of the energy emitted per unit of space angle:

$$I_{nm} = \frac{1}{2\pi} n_m A_{nm} h \nu_{nm} \tag{4-13}$$

where I_{nm} is the radiation density ($W/cm^2 \cdot sr$), n_m is the number density of the excited-state atoms or ions, h is Planck's constant (6.623×10^{-27} erg · s), A_{nm} is the Einstein transition probability for spontaneous emission (s^{-1}), and ν_{nm} is the frequency of the emitted line (s^{-1}).

In the radiation source, free atoms which are generated by thermal dissociation of the sample material or entrained from a separated atomizer, are excited or ionized and excited additionally. Several collisional or other processes may be responsible for delivering the required energy to the analyte species. When the excitation energy of the excited state is E_m, the population of this level is given by the Boltzmann equation provided the source is in so-called thermal equilibrium and is given by:

$$\frac{n_m}{n_0} = \frac{g_m}{g_0} e^{-E_m/kT} \tag{4-14}$$

where n_m and n_0 are the populations of the excited level and the ground state, respectively, g_m and g_0 are their statistical weights, k is Boltzmann's constant (1.38×10^{-16} erg/K), and T is the temperature in degrees Kelvin. When, instead of n_0, a summation for all levels is made, the Boltmann equation reduces to:

$$n_m = n \frac{g_m}{Z(T)} e^{-E_m/kT} \tag{4-15}$$

where $Z(T)$ is the partition function and n is the total number density for all levels of the analyte atoms. Accordingly, the expression for the line intensities becomes:

$$I_{nm} = n \frac{g_m}{Z(T)} e^{-E_m/kT} A_{nm} h\nu_{nm} \tag{4-16}$$

However, radiation sources often are very hot or they are electrical discharges in which the analyte substance is partially ionized. The degree of ionization α_j for a certain element j according to the reaction:

$$n_{aj} \leftrightharpoons n_{ij} + n_e \tag{4-17}$$

where n_{aj}, n_{ij}, and n_e are the atom, the ion, and the electron number densities, respectively, can be expressed as:

$$\alpha_j = \frac{n_{ij}}{n_{aj} + n_{ij}} \tag{4-18}$$

and $n_{ij} = \alpha_j n_j$ and $n_{aj} = (1 - \alpha_j) n_j$.

Accordingly, the intensities for atom and ion lines are given by:

$$I^{+}{}_{nm} = \alpha_j n_j \frac{g_m}{Z(T)} e^{-E_m/kT} A_{nm} g_m h\nu_{nm} \tag{4-19}$$

and

$$I_{nm} = (1 - \alpha_j) n_j \frac{g_m}{Z(T)} e^{-E_m/kT} A_{nm} g_m h\nu_{nm} \tag{4-20}$$

According to the Saha equation, the electron partial pressure p_e in a plasma which is in so-called thermodynamical equilibrium is given by:

$$S_{pj}(T) = \frac{p_{ij}p_e}{p_{aj}} = \frac{(2\pi m)^{3/2}(kT)^{5/2}}{h^3} \frac{2Z_{ij}}{Z_{aj}} e^{-E_{ij}/kT} \tag{4-21}$$

where p_{ij} and p_{aj} are the partial pressures of the ions and the atoms of the element in the plasma, respectively. With the aid of Eq. (4-21) and the temperature, the degree of ionization of an element can be calculated from the intensity ratios of ion and atom lines as expressed by Eqs. (4-19) and (4.20). This, however, only applies provided the plasma is in local thermal equilibrium.

In electrical discharges under reduced pressure, as schematically shown in Fig. 4-1, free electrons are accelerated in the vicinity of the cathode. They ionize by electron impact and may cause secondary emission. Energy exchange between electrons and ions formed may result from elastic collisions, charge transfer, and recombination according to the reactions:

$$A + e \rightarrow A^* + e \quad \text{electron impact} \tag{4-22}$$

$$A + e \rightarrow A^{+*} + 2e \tag{4-23}$$

$$A^* + B \rightarrow A + B^* \quad \text{collision of the second kind} \tag{4-24}$$

$$A^* + B \rightarrow A + B^{+*} \quad \text{Penning ionization} \tag{4-25}$$

$$A^+ + B \rightarrow A + B^{+*} \quad \text{charge transfer} \tag{4-26}$$

$$A^+ + e \rightarrow A^* \quad \text{recombination} \tag{4-27}$$

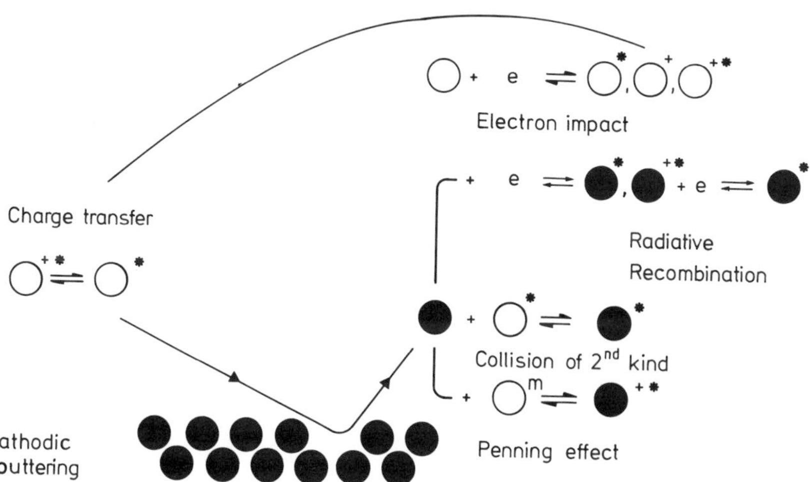

Figure 4-1. Excitation processes in discharges under reduced pressure.

All of these collision-dependent processes are hampered by the low number densities of all partners as well as by their high energies, resulting from the high electrical field in the relevant parts of the discharge as illustrated by Fig. 4-2. The collision processes occur especially in the vicinity of the cathode, as well as in the negative glow. Absence of local thermal equilibrium will be found, as the distribution of the kinetic energies due to the high electrical field can no longer be described by a Maxwell distribution function and as the mean kinetic energies of the different plasma components will be different. Thus, one will have to distinguish between electron temperatures, which have to be determined by probe measurements, kinetic temperatures of atoms and ions, and excitation temperatures depending on the lines (transitions) considered. The process (4-25) is particularly interesting in the case when noble gases such as argon and helium are employed as the discharge gas. They have metastable states (e.g., 11.5 and 11.7 eV for argon) which can only decay by collisions and may ionize and excite atoms released from the cathode in one step.

The energy distribution of the charged particles in a glow discharge has been described by a Druyvenstein function:

$$\frac{dn}{n} = 1.039\sqrt{u'}\, e^{-0.548u'^2}\, du' \tag{4-28}$$

where $u' = E/kT$. In the case of glow discharges, even the existence of two groups of electrons has been proposed, namely, a group of slow electrons involved in recombination processes and a group of highly energetic electrons mainly involved in excitation. Also, the Saha equation has to be replaced by the much more complex Corona equation. Accordingly, intensity ratios of atomic and ionic lines, as well as the relations with the electron number densities become complex.

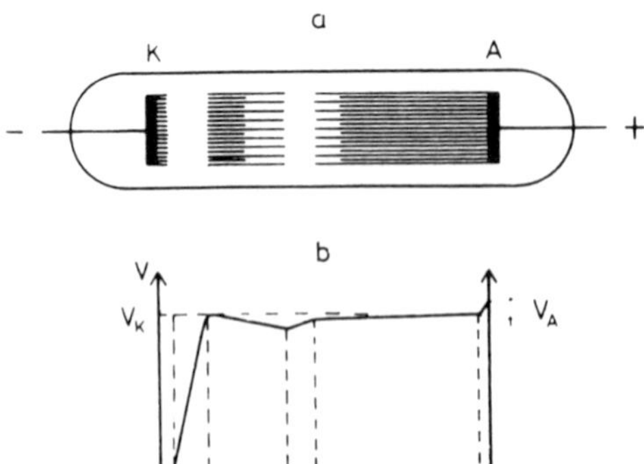

Figure 4-2. (a) Geometry and (b) potential distribution of a dc electrical discharge under reduced pressure. (1) Aston dark space; (2) Hittorf dark space; (3) negative glow; (4) Faraday dark space; (5) positive column; (6) anode region.

In the excitation of molecular spectra, however, the Boltzmann function still reasonably well describes the population of the rotational–vibrational states in the different electronic levels. Both for oxides such as AlO^+, TiO^+, Y0, . . . and radicals stemming from molecular impurities (OH, CN_2^+, N_2^+, . . .) the electronic states ($^1\Sigma$, $^2\Sigma$, $^2\Pi$, . . .) have energy differences of 1–10 eV. They have a vibrational fine structure ($v = 0, 1, 2, 3, \ldots$) and a rotational hyperfine structure ($j = 0, 1, 2, 3, \ldots$). The energy difference between two vibrational states of the same electronic level is ~0.25 eV and between two rotational levels of the same vibrational states ~0.005 eV. Accordingly, the radiation emitted from transitions between rotational and vibrational levels of the same electronic states lies in the infrared region. However, when two different electronic levels are involved, the emission of radiation is at visible or UV wavelengths. This is the case for the $^2\Sigma$–$^2\Sigma(0, 0)$ CN bands, different OH bands, N_2^+ bands, etc. They consequently will complicate the optical spectra of glow discharges, when these gases are used as discharge gas or when they are present as impurities in the noble gases used as discharge gas.

Transitions between rotational levels are possible when $\Delta j = 0, \pm 1$ and the P, Q, and R branches correspond to $+1$, 0, and -1, respectively. For a transition from an upper level m to a lower level n, the intensity of the corresponding rotational line is given by:

$$I_{\mathrm{nm}} = N_{\mathrm{m}} A_{\mathrm{nm}} h \nu_{\mathrm{nm}} \frac{1}{2\pi} \tag{4-29}$$

where N_{m} is the population of the level m and ν_{nm} is the frequency of the rotational line. A_{nm}, being the transition probability for dipole radiation, is given by:

$$A_{\mathrm{nm}} = \frac{64\pi^4 \nu_{\mathrm{nm}}^3}{3k} \frac{1}{g_{\mathrm{m}}} \Sigma|^{R} n_{\mathrm{i}} m_{\mathrm{k}}|^2 \tag{4-30}$$

where i and k indicate the degeneracy of the levels n and m, $^{R}n_{\mathrm{i}}m_{\mathrm{k}}$ is a matrix element of the electrical dipole momentum, and g_{m} is the statistical weight of the vibrational level. N_{m} is given by the Boltzmann equation

$$N_{\mathrm{m}} = N \frac{g_{\mathrm{m}}}{Z(T)} e^{-E_{\mathrm{r}}/kT} \tag{4-31}$$

where E_{r} is the rotational energy of the excited electronic level and relates to the quantum number of the upper level as:

$$E_{\mathrm{r}} = hcB_{v'}J'(J' + 1) \tag{4-32}$$

where $B_{v'}$ is the rotational constant of the excited electronic and vibrational level. For a $^2\Sigma_g$–$^2\Sigma_u$ transition, $|^R n_i m_k|^2$ is given by $(J' + J'' + 1)$ where J'' is the rotational quantum number of the lower level. Accordingly,

$$\ln \frac{I_{nm}}{J' + J'' + 1} = C - \frac{hcB_{v'}J'(J' + 1)}{kT} \tag{4-33}$$

By plotting $\ln[I_{nm}/(J' + J'' + 1)]$ versus $J'(J' + 1)$ for a series of rotational lines, a rotational temperature can be determined which in the case of low-pressure discharges is a good approximation for the gas temperature. As shown for end-on measurements in a hollow cathode glow discharge, values of 600–1000 K are obtained (Fig. 4-3).[3]

4.1.1.3. Spectral Linewidths

The linewidths of atomic emission lines result from several broadening mechanisms.[4] The natural width of spectral lines results from the finite

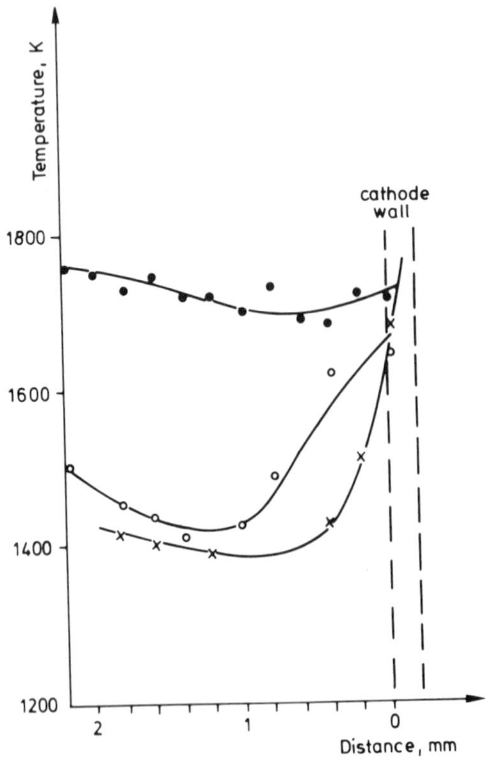

Figure 4-3. Rotational temperature distribution in a hollow cathode glow discharge.[3] i: 150 mA; graphite hollow cathode with 4.71-mm internal diameter; ×: 530 Pa, ○: 800 Pa, ●: 1060 Pa argon; $^2\Sigma$–$^2\Sigma(0, 0)$ CN 388.3 nm.

lifetime of excited levels from which radiative transitions to lower levels are allowed (about 10^{-8} s) and is given by:

$$\Delta \nu_{\mathrm{N}} = \frac{1}{2\pi\tau} \tag{4-34}$$

Natural linewidths are on the order of 10^{-2} pm.

As the emitting atoms and ions have a velocity component in the direction of observation, spectral linewidths also include a Doppler broadening component. Its full width at half-maximum (FWHM) in terms of the frequency is given by:

$$\Delta \nu_{\mathrm{D}} = \frac{2\sqrt{\ln 2}}{c} \nu_0 \sqrt{\frac{2RT}{M}} \tag{4-35}$$

where c is the velocity of light, ν_0 is the frequency of the line maximum, R is the gas constant, and M is the atomic mass. The Doppler broadening strongly depends on the gas temperature. For a hollow cathode glow discharge the Doppler linewidth of the Ca(I) 422.5-nm line for instance ranges from 0.8 to 2 pm in the case of temperatures between 300 and 2000 K.

Pressure, or Lorentz broadening results from the interaction of emitting and foreign atoms. Its contribution to the frequency FWHM is given by:

$$\Delta \nu_{\mathrm{L}} = \frac{2}{\pi} \sigma_{\mathrm{L}}^2 N \sqrt{2\pi RT\left(\frac{1}{M_1} + \frac{1}{M_2}\right)} \tag{4-36}$$

where M_1 and M_2 are the atomic masses, N the concentration of the foreign atoms, and σ_{L} the collision cross section. The pressure broadening is low in the case of low-pressure discharges. For instance, in a discharge operated at 9 mbar, in which the kinetic temperature is 300 K, the pressure broadening of the Ca(I) 422.6-nm line is only 0.02 pm, and negligible compared with the Doppler broadening. From the latter, the gas temperature can be determined in the case of glow discharges. The values are on the same order of magnitude as the rotational temperatures found from the molecular spectra.

As in every plasma discharge, self-absorption occurs in glow discharges. However, because of the optically thin plasma, it is much less than in atmospheric-pressure plasmas. With some anode–cathode geometries, a cloud of ground-state atoms may be present in front of the emitting zones. Accordingly, self-reversal may occur especially for matrix resonance lines, as shown by the profile of a copper line measured for a Grimm-type glow discharge

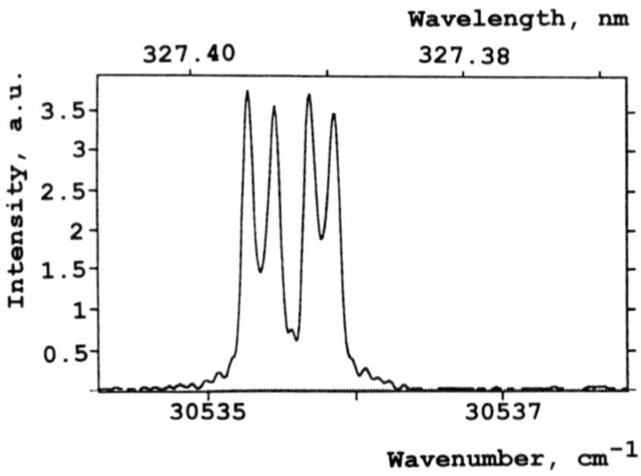

Figure 4-4. Spectral line profile of the Cu(I) 324.7-nm line emitted by a Grimm-type glow discharge and recorded by Fourier transform atomic emission spectroscopy.[5] Grimm-type glow discharge lamp with floating restrictor (diameter: 8 mm); end-on observation; copper sample; 3.5 Torr argon pressure; 50 mA, 1020 V.

lamp with the aid of high-resolution Fourier transform spectrometry (Fig. 4.4).[5]

The optical spectra in glow discharges, according to the above-described principle, will contain mainly atomic and less ionic lines. Because of the absence of local thermal equilibrium, the kinetic gas temperatures will be low (1000–2000 K) relative to the electron temperatures measured with probes (above 20,000 K). Molecular bands may occur as a result of water vapor or molecular gases present in the filler gases, which in most cases are noble gases. They may lead to considerable spectral interferences and their removal calls for the use of high-purity gases and effective evacuation of the discharge lamp subsequent to sample change. The analyte lines have low physical widths as pressure broadening is minimal, by which spectral interferences of atom and ion lines are minimal. For matrix atomic resonance lines, self-reversal may occur, but only at rather high analyte number densities. Accordingly, the linear dynamic range obtained still will be considerable.

4.1.2. Radiation Sources

For atomic emission spectrometry a wide variety of radiation sources have been developed and used for practical analysis (Fig. 4-5). For the analysis of liquids, *flames* were used as emission sources beginning with the early work of Bunsen and Kirchhoff, and still play a role in the determination of the alkalines. This work led to the development of electrical plasmas with a geometry and stability similar to those of flames but with much higher

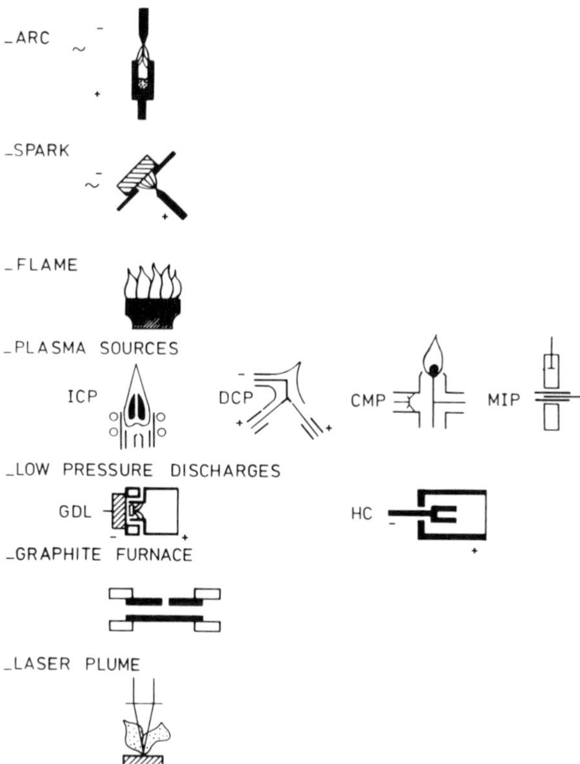

Figure 4-5. Radiation sources for atomic spectrometry.[6]

temperatures. These *plasma sources* [inductively coupled plasma (ICP) and microwave-induced plasma (MIP) sources] now are routine tools for atomic emission spectrometric[6] and mass spectrometric analysis of liquids and solids subsequent to sample disssolution. *Dc Arc* and *spark* sources have developed into powerful sources for the direct analysis of solids.

4.1.2.1. dc Arcs

These sources are operated in the so-called abnormal part of the current–voltage characteristics (see Fig. 4-6). Here, an intensive heating of the sample contained in the anode takes place and volatilization results from thermal evaporation. In the cathode region there may even be an enrichment of analyte material by which an intensification of the analytical signal may occur (cathode-layer effect). The volatilization may be effective for distilling volatile elements selectively from a refractory matrix. However, one also has

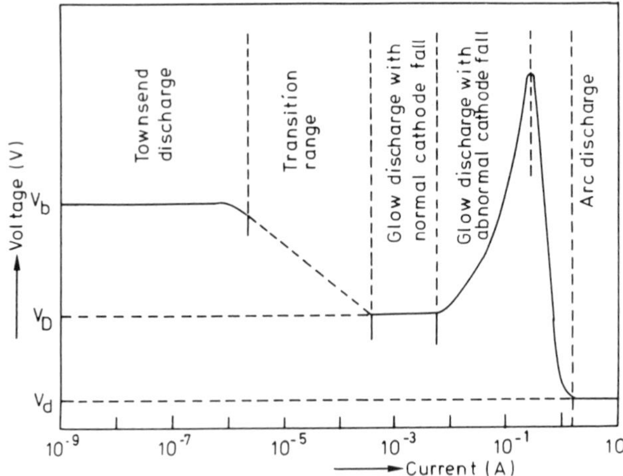

Figure 4-6. Current–voltage characteristics of dc electrical discharge.[32] (Courtesy of the American Chemical Society.)

to take into account the matrix effects, which occur as a result of anions (such as halogens) producing volatile compounds and causing volatilization interferences or carrier effects because of the presence of volatile matrix elements. Accordingly, the accuracy of DC arc emission spectroscopy is rather limited. Nevertheless, for survey analysis it still is one of the most sensitive multielement methods, as shown by the example of trace analysis in U_3O_8 (Table 4-1).[7]

4.1.2.2. Spark Sources

Spark emission spectrometry is a powerful method for multielement analysis of metals and alloys as required for production control and product control in the steel industry.[8] Detection limits for modern medium-voltage unidirectional sparks are at the microgram per gram level. Especially for light elements, spark atomic emission spectrometry is a necessary complement to x-ray fluorescence spectrometry for metallurgical analysis. In sparks, a capacitor is charged by a tank circuit and discharged across an analytical gap. The samples are mounted as flat disks opposite a tungsten electrode (point-to-plane geometry) with an interelectrode distance of 2–4 mm. Often the diameter of the spark impact on the sample is limited to avoid wandering across the sample, which would lead to irreproducible results. Mostly unidirectional sparks are used. Whereas earlier on high-voltage sparks (voltage 2–5 kV) were used, modern sparks almost always are operated at medium

Table 4-1. Detection Limits Obtained by dc Arc Analysis Atomic Emission Spectrometry of U_3O_8 (Cathode Region Method)[7]

Element	c_L (μg/g)
B	0.1
Ba	1.0
Bi	0.3
Cd	0.1
Co	1.0
Ga	1.0
Ge	0.3
In	0.3
K	1.0
Li	1.0
Mg	0.1
Mo	0.1
Na	1.0
Nb	4.0
Pb	0.5
Si	1.0
Sr	1.0
Zn	10.0

voltage (500 V). The repetition rate (100–1000 Hz) can be varied electronically but also the energy dissipation can be changed by changing the resistance, the capacitances, and the inductances in the discharge circuit. Material ablation in spark discharges partially results from cathodic sputtering due to the highly energetic particles produced in the spark channel. However, because of the intense bombardment, the sample is locally molten (Fig. 4-7), by which thermal evaporation contributes considerably to material volatilization. The contribution of both can be varied by changing the circuitry parameters. Regardless, the contribution of thermal evaporation leads to selective volatilization and related matrix effects. Accordingly, matrix correction calculations with extensive software as well as a careful selection of the standard samples for calibration are required. With sparks, analysis times down to 20 s can be obtained. This is possible by the application of high-energy presparking, by which the metallograpical structure of the sample is destroyed. When using internal standardization, a precision of 1% in the concentration range of the minor elements can be realized. Considerable innovation in this method, however, is still taking place. For instance, by a proper selection of the analytical zone, as is possible with the aid of quartz optical fibers, the matrix effects can be greatly reduced. Electronic control of the waveform may also lead to improvements of the precision and

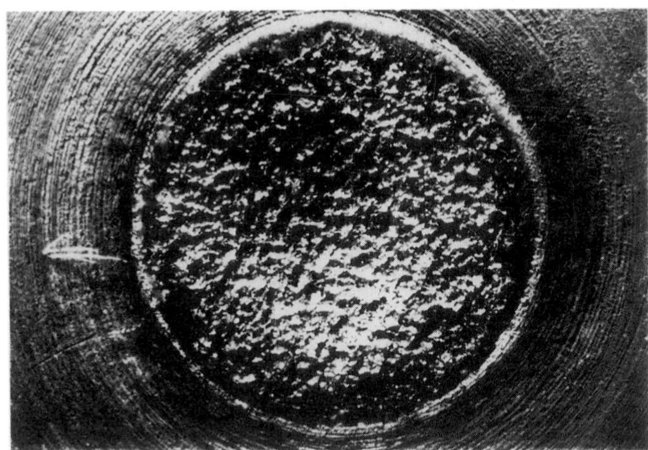

Figure 4-7. Burning spot obtained in spark emission spectroscopy. 500 Hz monoalternance spark in argon; sample: aluminum.

of the power of detection. The latter also is still attainable by using cross-excitation with high-frequency or microwave energy.[9]

Furnaces and electrothermal devices are of special interest as thermal atomizers in combination with other sources for signal generation, such as glow discharges. This combination is known as furnace atomic nonresonance emission spectrometry (FANES)[10] (Fig. 4-8) and is discussed in a separate chapter in this book. Combinations with plasma sources at atmospheric pressure also have been studied extensively and shown to be useful for

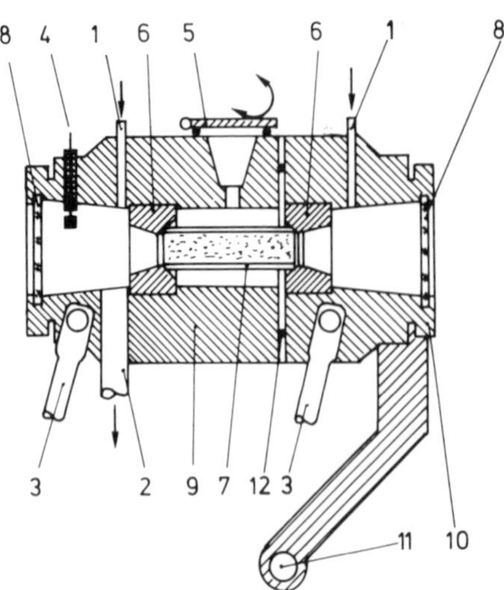

Figure 4-8. Principle of FANES (furnace atomic nonresonance emission spectrometry.[10] (Courtesy of Pergamon Press.) (1) carrier-gasport; (2) pump port; (3) electrical connector; (4) anode; (5) removable lid for sample injection; (6) graphite electrode; (7) graphite furnace and hollow cathode; (8) window; (9) water-cooled vacuum vessel; (10) water-cooled part of the vacuum vessel and rotation arm for changing the graphite tube; (11) pivot; (12) gasket.

analyzing microaliquots of solutions as well as powdered organic samples. Only in a limited sense can they also be used as direct emission source.[11] In this case, only elements of which the most sensitive lines have low excitation energies, such as the alkali elements, can be determined down to the trace element level.

4.1.2.3. Laser Plumes

Laser-generated plumes have been used both as atomizers and as direct emission sources in atomic emission spectrometry. A review has recently appeared.[12] These techniques, of which the principles stem from the 1960s, have gained new interest since stable and reliable Nd-YAG laser sources with high modal purity have recently become availabe. Both atomic emission and dedicated mass spectrometers for work with pulsed sources are available. The fact that there is a large need for the direct and microdistributional analysis of compact electrically nonconductive samples such as ceramics has spurred this development. Volatilization by laser ablation in combination with other signal generation techniques such as ICP-OES and ICP-MS, and also direct emission spectrometry (e.g., see Ref. 13) enables it to obtain detection limits down to the microgram per gram range. Laser-induced atomic fluorescence at the laser plume leads to an increase of the power of detection by at least a factor of 10.[14]

4.1.2.4. Hollow Cathode Devices

Glow discharge sources have become of special interest as sources for atomic emission spectrometry. This applies especially to hollow cathode sources as described by Schüler and Gollnow in 1935.[15] They still are important as emission sources and are treated in another chapter in this book. The sample here is placed in a hollow cathode and volatilized partly by cathode sputtering, as well as by thermal evaporation when the cathode is not cooled. Its unique features lie in the fact that the analyte vapor has a very long residence time in the negative glow of the plasma which fills the cathode cavity and thus is very efficiently excited. Moreover, background intensities are low because the electron number densities are low and accordingly high-intensity ratios of line to background are obtained.[16] Hollow cathodes for this reason are the most sensitive glow discharge emission sources. The absolute detection limits are in the subnanogram range. However, no steady-state analyte volatilization is obtained. This may be a drawback with regard to a flexible multielement determination. However, in the case of a hot hollow cathode, a selective volatilization occurs and this fact may be analytically used for performing a sensitive determination of volatile elements in refractory matrices, as shown by Thornton.[17] This approach is still very useful for the determination of P, As, Se, Bi, etc. in high-temperature alloys,[18] for which atomic emission spectrometry using a hot hollow cathode is still a unique approach for routine use.

4.1.2.5. Planar Glow Discharges

Glow discharges with flat cathodes first became of interest for atomic emission spectrometry with the lamp introduced by Grimm in 1968.[19] The Grimm-type lamp (Fig. 4-9) is a constricted type of glow discharge, of which the discharge is confined to the sample which is taken as the cathode and is water-cooled. Confinement is achieved in the classical Grimm-type lamp by keeping the distance between the anode tube and the cathode block below the mean free path length of the electrons, which is at the pressure of operation (some 100 Pa) below a few tenths of a millimeter. Moreover, the vacuum in the anode–cathode interspace is lower than in the discharge region itself, which is achieved by the use of a dual outlet pump with a larger throughput for the interelectrode space. Accordingly, the sample, which must be vacuum-tight and electrically conducting, is volatilized by cathodic sputtering only and volatilization interferences do not occur. The sample material is excited in the negative glow of the low-pressure discharge. Because of the absence of thermal evaporation as well as the separation of analyte volatilization and the excitation in the negative glow (which is away from the most energetic cathode fall), interelement effects are low and it may be possible to analyze a large variety of samples with the same calibration

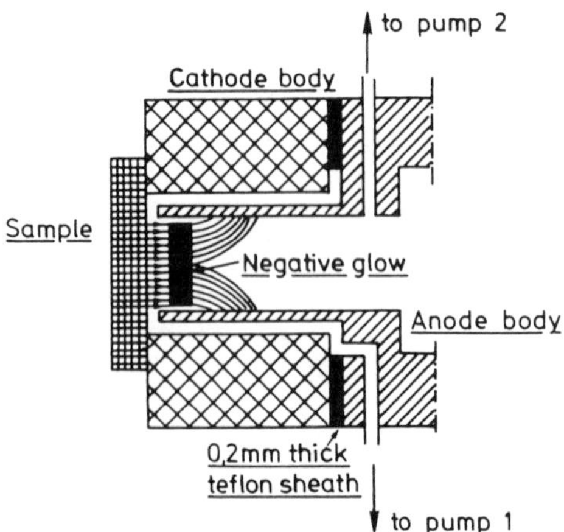

Figure 4-9. Principle of the Grimm-type glow discharge lamp.[32] (Courtesy of the American Chemical Society.)

function. Because of the consistency of the material volatilization and excitation processes with time, rapid simultaneous but also very flexible multielement determinations of bulk constituents with glow discharge sources are possible. Moreover, because of cathodic sputtering the sample is ablated layer by layer and the source is suitable for performing in-depth profiling, as is discussed in a separate chapter in this book. The Grimm-type glow discharge is now a routine tool for the direct atomic emission spectrometric analysis of solids and will be treated in detail.

4.1.3. Spectrometers

4.1.3.1. Dispersive Spectrometers

Dispersive spectrometers used in atomic emission spectrometry with glow discharges normally make use of a diffraction grating as the dispersive element. For sequential multielement determinations, monochromators are used whereas simultaneous multielement analyses are performed with polychromators. As the physical widths of the analytical lines in glow discharge atomic emission spectrometry are on the order of 2 pm, optical spectrometers with a high power of detection are required for resolving the physical linewidths and obtaining maximum signal-to-background ratios. Therefore, a practical resolution of 200,000 at 300 nm would be required. Accordingly, the use of high-resolution gratings as well as spectrometers with sufficient focal length f and narrow slit widths is required. The theoretical resolution obtained with a diffraction grating is given by:

$$R_0 = Nm \tag{4-37}$$

where N is the total number of grooves (grating constant, lines/mm × width) and m is the order. Normally, holographic gratings, with which a uniform response over a broad spectral range can be obtained, are used; the groove density is beyond 1800/mm and the width is at least 60 mm. The reciprocal linear dispersion of the spectrometer is given by:

$$(dz/d\lambda)^{-1} = fm/a \cos\theta \cos\varphi_2 \tag{4-38}$$

where φ_2 is the angle between the grating normal and the diffracted beam and θ is the angle between the diffracted beam and the radiation detector. In order to keep the spectral slit width below the contributions of diffraction and of the physical widths, the entrance and exit slits of the spectrometer should often be in the micrometer range. However, this is not possible

Table 4-2. Background Radiances in a Grimm-Type Glow Discharge Lamp

Wavelength (nm)	Photocurrent (A) ($\times 10^{-9}$)
200	2.8
220	8
240	5
260	4
280	4.5
300	5
320	5.5
340	8.4
360	7
380	5
400	5
420	7
440	5.5
460	4

[a]Grimm-type lamp with floating restrictor (diameter: 8 mm), 0.35-m McPherson monochromator (Czerny–Turner mounting, grating: 1200 lines/mm, width: 50 mm, slit widths: 15 μm, 1P28 photomultiplier with dark current of 2×10^{-9} A), copper sample, 4 Torr, 50 mA, 1 kV.

because of stability requirements. Moreover, as shown by the results in Table 4-2, the background radiances of a glow discharge lamp measured with a rather quick (high throughput) spectrometer are already on the same order of magnitude as the dark current of a good photomultiplier and the signal-to-background ratios would be determined by dark current limitations. Therefore, typical slit widths are at least 30 μm.

For sequential spectrometers both the Ebert (Fig. 4-10a) and the Czerny–Turner (Fig. 4-10b) mountings are used. In the first case, optical aberrations especially when using narrow slits and a focal length >0.5 m are minimized by using curved slits. With a Czerny–Turner mounting this problem does not occur as two mirrors with slightly different focal length and slightly different incident angles can be used. For measurements at VUV wavelengths purging the spectrometer with nitrogen or argon can be applied to yield access to wavelengths down to 165 nm. Below these wavelengths, quartz lenses and windows have to be replaced by LiF. In order to minimize the number of reflecting surfaces, a Seya–Namioka setup with a concave diffraction grating (Fig. 4-10c) may be useful. Wavelength selection is performed by turning the grating in all cases. By using high-precision grating drives which make use of optical encoders, an angular precision of 10^{-4} degree becomes possible. For a grating with 2400 grooves/mm, this results

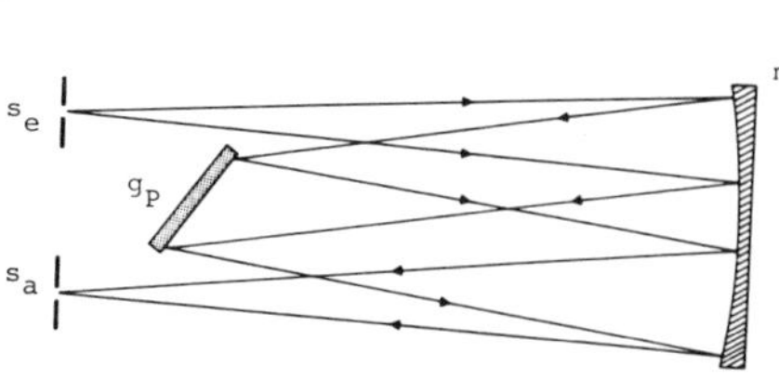

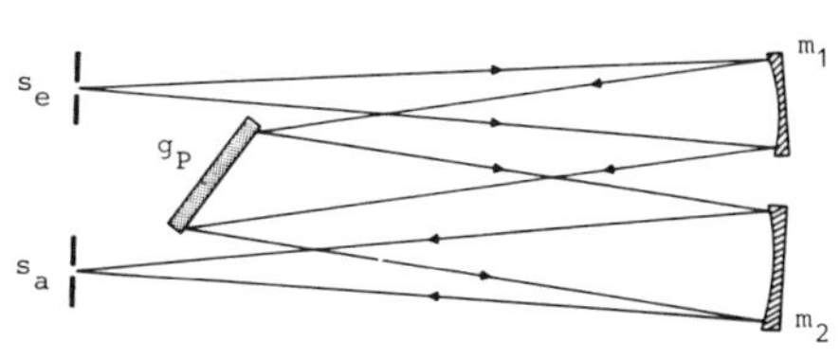

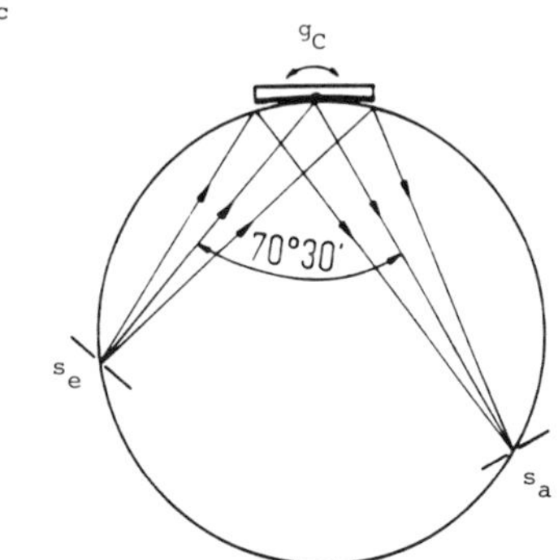

Figure 4-10. (a) Ebert, (b) Czerny–Turner, and (c) Seya–Namioka mounting for monochromators. s_e : entrance slit; s_a : exit slit; m: mirror; m_1 : mirror of entrance collimator; m_2 : mirror of camera; g_P : plane grating; g_c : concave grating.

in a wavelength presetting precision of 1 pm according to:

$$\sin \varphi = \lambda / 2a \cos \varphi \tag{4-39}$$

where $\varphi = 10^{-4}$ degree, $m = 1$, and $a = (2400)^{-1}$ mm. In order to accurately peak wavelengths, searching procedures are used where a scan across the line is made, the intensities measured are digitized, and the apparatus is automatically set at the maximum. This of course presumes that line intensities sufficiently differ from the background features, which is not the case in trace analysis. For the latter aim, spectral apparatus with fixed optical elements, as is the case with polychromators, should be used.

Polychromators used for simultaneous measurements have a Paschen–Runge setup (Fig. 4-11) where the concave grating, the entrance slit, and many exit slits are on the Rowland circle. These instruments have a focal length between 0.5 and 1 m and are used in most commercial equipment. By providing a computer-controlled displacement of the entrance slit, spectral windows around the analytical lines provided can be scanned. Accordingly, background and interference correction is possible. As a displacement of the entrance slit can be made without mechanical loss, rapid sequential analyses can be performed by displacement of the detector in the focal plane. Then, random access to the provided slits or access to any wavelength is possible. In the latter case, one can provide equidistant exit slits in a mask at every 2-nm wavelength difference for instance and adjust the wavelength in between by displacement of the entrance slit. By performing computer-controlled displacement of detector and entrance slit, rapid access to any wavelengths is possible.

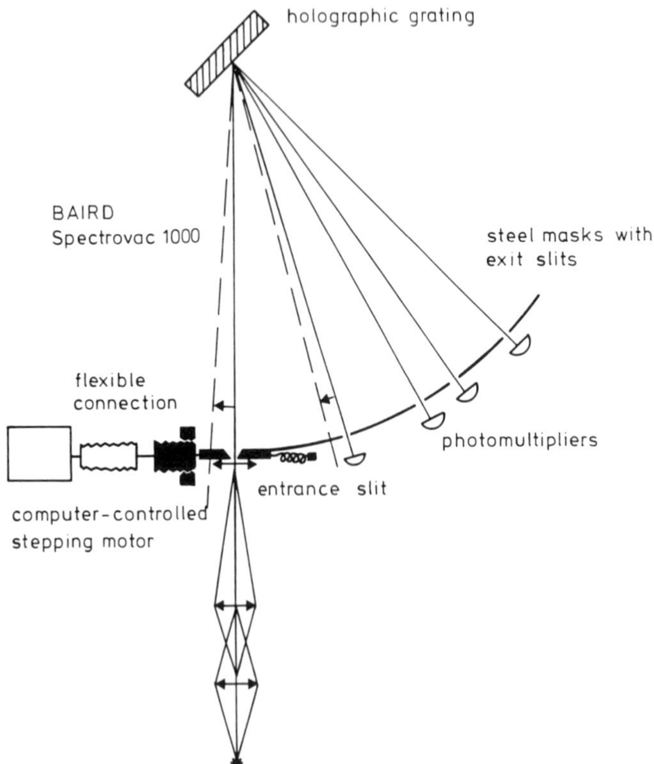

Figure 4-11. Paschen–Runge mounting with concave grating and background measurement facilities for polychromator.

Instead of one dispersive element, multiple dispersion elements can be applied so as to realize high spectral resolution with compact instruments. When using an echelle grating (with a grating constant of up to 150 lines/mm) a prism can be used and both crossed dispersion and parallel dispersion can be applied. An echelle spectrometer with a 0.5-m Czerny–Turner monochromator, a quartz prism for cross-dispersion, and an echelle grating with 79 lines/mm being used in the 75 order and higher, allows a reciprocal linear dispersion of 0.15 nm/mm and is commercially available. The entrance slit heights are only some 0.1 mm, but nevertheless a sufficient radiant throughput for many purposes is still obtained. Also, instruments with parallel dispersion are commercially available.

4.1.3.2. Multiplex Spectrometers

Apart from dispersive spectrometers, multiplex spectrometers such as Hadamard transformation spectrometers[20] and especially interferometers using Fourier transform processing[21] are useful for work with glow discharges. The latter especially applies to glow discharges delivering time-stable signals, as will be discussed further. This especially relates to the noise properties as well as to the relatively line-poor emission spectra of these sources by which multiplex disadvantages are much lower than in the case of inductively coupled plasmas, for instance.[100] For the latter aim a Michelson interferometer (Fig. 4-12) can be used to produce the interferogram. With the aid of a beam splitter the radiation is split into two parts which are each directed to a mirror. When shifting the mirror in one of the side

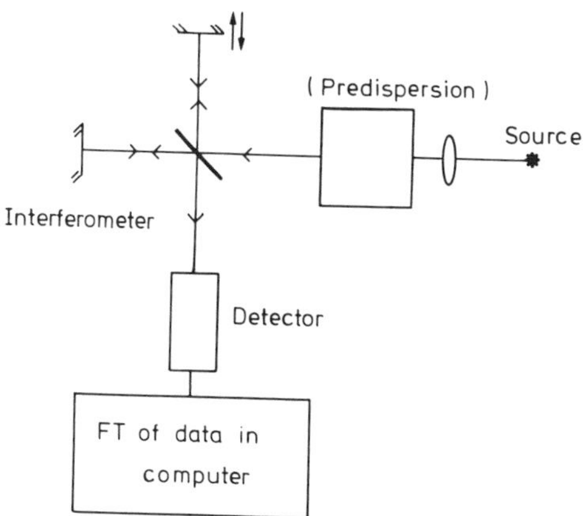

Figure 4-12. Principle of Fourier transform emission spectrometry.

arms, the interference for each wavelength is given by:

$$I(x) = B(\sigma)[1 + \cos(2\pi\sigma x)] \tag{4-40}$$

where x is the optical path difference, σ is the wave number of the radiation (cm^{-1}), I is the intensity measured with the detector, and B is the radiation density of the source. The frequency f of the signal on the detector for a radiation with frequency ν is a function of the velocity v with which x is changed:

$$f = \sigma v \tag{4-41}$$

by which optical frequencies are shifted to the audio range and their variations can be measured with conventional detectors. A polychromatic radiation source accordingly gives an interferogram where the intensity of each point is the sum of all values resulting from Eq. (4-40). The central part contains the information at low resolution and the ends contain information at high resolution. The latter also depends on the registration time, the spectral bandwidth and the number of repetitive scans. By applying a Fourier transform on the signal for each point of the interferogram

$$I(x) = \int_{-\infty}^{+\infty} B(\sigma)\, d\sigma + \int_{-\infty}^{+\infty} B(\sigma) \cos(2\pi\sigma x)\, d\sigma \tag{4-42}$$

$$= C + \int_{-\infty}^{+\infty} B(\sigma) \cos(2\pi\sigma x)\, d\sigma \tag{4-43}$$

with C a constant term being subtracted before the transformation one obtains:

$$B(\sigma) = \int_{-\infty}^{+\infty} I(x) \cos(2\pi\sigma x)\, dx \tag{4-44}$$

The spectrum thus is obtained in wave numbers. By digitizing the interferogram, a quick Fourier transform is possible and by multiplying with a block function a limited but highly accurate displacement L of the arm allows one to collect sufficient information for obtaining a resolution which is proportional to $(2L)^{-1}$ as well as with the number of measurement steps. In addition, repetitive scanning is applied where a reference laser is used to obtain sufficient precision in mirror positioning. The technique has been applied at IR wavelengths for some time and is now used for spectra in the visible and UV region as systems for a sufficiently precise mechanical displacement and high-capacity computers for the calculation of complex spectra have just recently become available. Fourier transform spectrometry is attractive for sources with low radiance and rather line-poor spectra but high stability, and therefore is uniquely suitable for use in combination with glow discharges.

4.1.4. Radiation Detectors

4.1.4.1. Photomultipliers

For atomic emission spectrometry, photomultipliers are most commonly used for radiation detection. In a photomultiplier the photons strike a photocathode, from which photoelectrons are released and accelerated to a number of dynodes mounted in a cascade geometry and with an anode as the last stage. The current obtained at the anode (I_a) depends on the radiation flux N_ϕ as:

$$I_a = I_c\theta(V) \tag{4-45}$$

with

$$I_c = N_e e \qquad (e = 1.6 \times 10^{-19}\ \text{A} \cdot \text{s}) \tag{4-46}$$

in which

$$N_e = N_\phi \text{QE}(\lambda) \tag{4-47}$$

$\theta(V)$ is the amplification factor (to up to 10^6) and QE(λ) (up to 20%) the quantum efficiency. Together with the dark current, being I_a when no radiation comes on the detector ($\approx$1 nA), they are characteristic of the type of photomultiplier. Photomultiplier currents are amplified and integrated on a capacitor, which is read out prior to analog-to-digital conversion. By using a preamplifier with several capacitors, a high dynamic range can be realized, as is required for glow discharge atomic emission sources having large linear response functions (Fig. 4-13). Repeated readout with the aid of rapid A/D conversion under control of a personal computer now is standard in atomic emission spectrometric instrumentation.

Photomultipliers which can be used at wavelengths between 160 and 500 nm often have an S-20 photocathode having a quantum efficiency of 5–25%, the multiplication factors are 3–5 at dynode voltages between 50 and 100 V, and they have 9–15 dynodes. For selected types, the dark current may be below 100 photoelectrons/s ($I_d < 10^{-10}$ A). Red-sensitive photomultipliers have a much higher dark current which can be lowered by

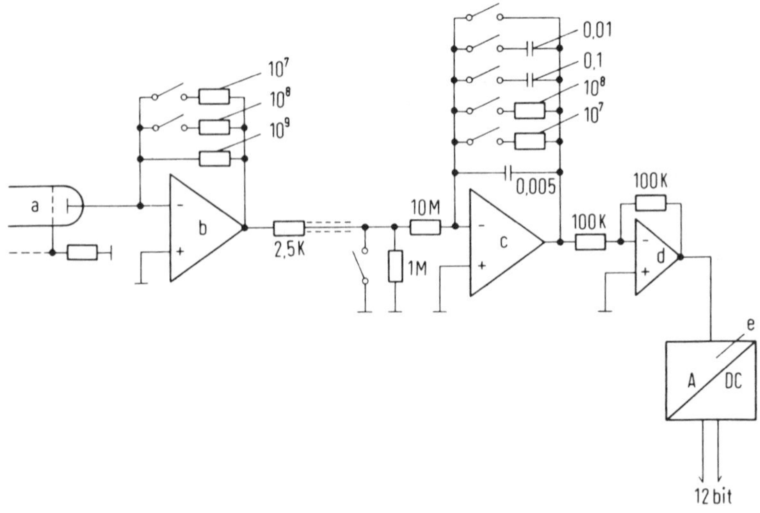

Figure 4-13. Photomultiplier tube and measurement system.[129] (Courtesy of Verlag Chemie GmbH.) a: photomultiplier tube; b: current amplifier; c: measurement amplifier for dc signal and for signal integration; d: conversion of sign; e: analog-to-digital conversion.

cooling. For wavelengths below 160 nm, MgF_2 must be used instead of quartz as window material. For the low UV, "solar blind" photomultipliers which are not sensitive to radiation of longer wavelengths (>350 nm) are often used. They are particularly useful in polychromators where the analytical lines at longer wavelengths are measured in the first order and the VUV and UV lines below 300 nm in the second order.

4.1.4.2. Multichannel Detectors

Multichannel detectors with parallel input have made considerable progress for atomic emission spectrometry. They include vidicon, SIT-vidicon, photodiode array detectors, and image dissector tubes.[22] However, more recently, charge-coupled devices (CCD) and charge injection devices (CID)[23] have been introduced. Photodiode arrays (PDA) may comprise matrices of up to 512 and even 1024 individual diodes (e.g., Reticon) which have individual dimensions of 10 μm in width and up to 20 mm in height. The charge induced by photoelectric effects gives rise to photocurrents which are sequentially fed into a preamplifier and a multichannel analyzer. Therefore, they are rapid sequential devices. They have the advantage that memory effects due to the incidence of high radiant densities on individual diodes (lag) as well as cross talk between individual diodes (blooming) are low.

Their sensitivity in the UV and the VUV range is low compared with photomultipliers. They nevertheless presently are of interest for atomic emission spectrometry as new scintillators with sufficient stability are available and as they can be coupled with microchannel plate detectors giving an additional amplification. Signal-to-noise ratios can be considerably improved by cooling the array with the aid of a Peltier element or liquid nitrogen. Commercial instrumentation is now available and has been successful (e.g., EG&G: OMA 4; Spectroscopy Instruments Inc.). They have replaced silicon intensified target (SIT) vidicon detectors, where the charge is integrated in the individual pixels and the simultaneously collected information is read out sequentially with the aid of an electron beam. When using an electrostatic image intensifier (Fig. 4-14), these devices for many purposes are quite sensitive and permit bidimensional resolution, and thus allow spatially resolved measurements of spectral segments (for an example in the case of microwave-induced plasmas, see Ref. 24). Photodiode arrays have been used successfully in a segmented echelle spectrometer (Plasmarray, Leco Co.)[25] where spectral segments around the analytical lines are sorted out by primary masks subsequent to spectral dispersion with an echelle grating and are detected after a second dispersion onto a diode array. Accordingly, more than ten analytical lines and their background contributions can be measured simultaneously. The spectrometer is used succesfully in the case of inductively coupled plasmas and glow discharges (e.g., see Fig. 4-15),[26] but with restrictions due to detector noise limitations at low wavelengths.

Image dissector tubes (Fig. 4-16) make use of an entrance aperture behind the photocathode, by which the photoelectrons stemming from different locations of the photocathode can be scanned and measured after amplification in the dynode train, as is the case in a conventional photomultiplier. The system has been used in combination with an echelle spectrometer with crossed dispersion for flexible rapid sequential analysis.[27] However, due to the limited size of the photocathodes available and stability problems, the technique has not made a breakthrough.

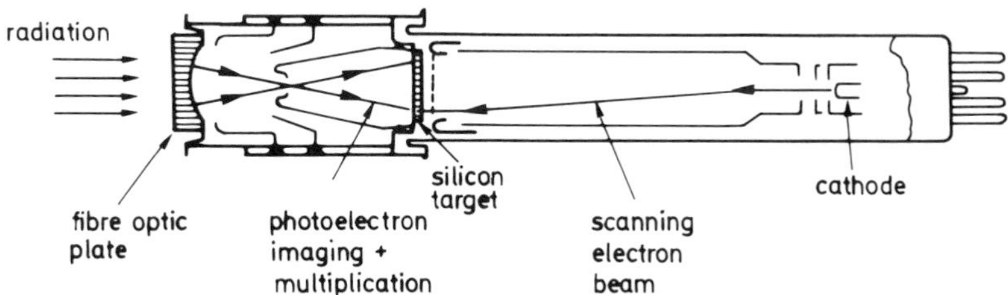

Figure 4-14. Early SIT vidicon system. (Courtesy of SSR Instruments, Inc.)

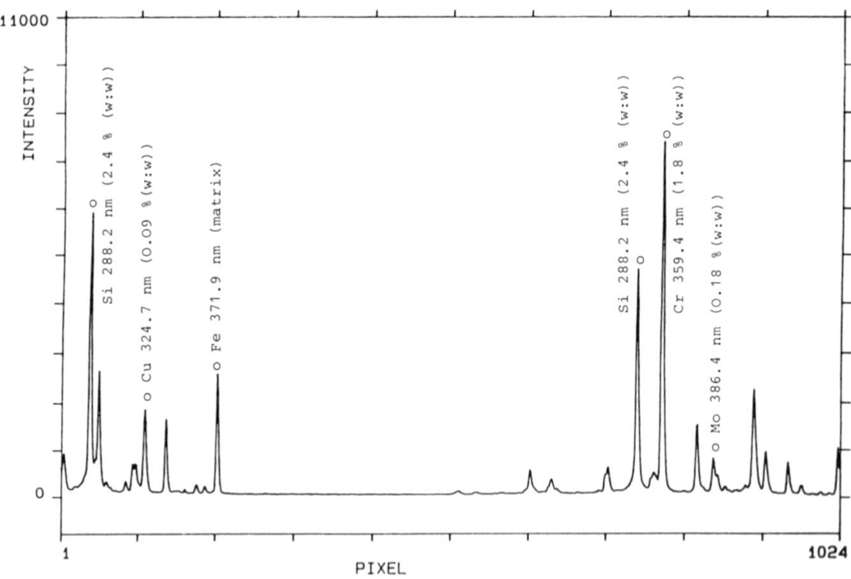

Figure 4-15. Spectral-segmented spectrum of glow discharge.[26] Grimm-type glow discharge lamp with floating restrictor (8-mm diameter): 50 mA, 3.5 Torr, 2 kV; Plasmarray spectrometer (Leco); steel sample: 217A (Research Institute, CKD, Prague, Czech Republic).

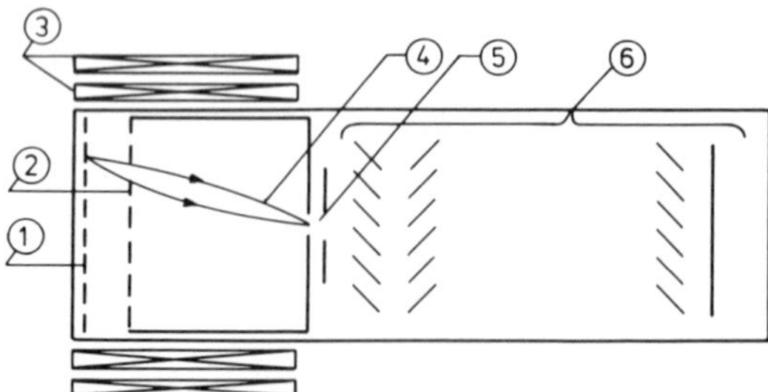

Figure 4-16. Image dissector tube. (1) photocathode; (2) accelerating electrode (mesh); (3) deflection coils; (4) photoelectron beam; (5) aperture; (6) electron multiplier.

Charge-coupled devices (Fig. 4-17) allow the readout of the induced charge after it passes a preset value and can be used as bidimensional detectors[28] by which they are ideally suited for coupling with an echelle spectrometer. As their sensitivity is high, they may be expected to become real alternatives to photomultipliers for atomic emission spectrometry. An evaluation of the different radiation detectors with respect to their feasibility for work with glow discharges is given in Table 4-3.

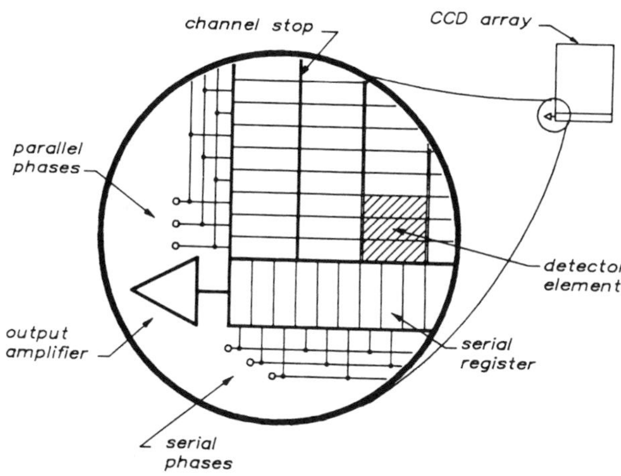

Figure 4-17. Charge-coupled device (CCD) detector.[28] (Courtesy of the Society for Applied Spectroscopy.) Layout of a typical three-phase CCD. Photogenerated charge is shifted in parallel to the serial register. The charge in the serial register is then shifted to the amplifier.

Table 4-3. Evaluation of Some New Radiation Detectors for Atomic Emission Spectrometry

Detector	Dimensions	Spectral region	Sensitivity versus photomultiplier	Ref.
Vidicon	12.5 × 12.5 mm bidimens.[a]	With scintillator down to 200 nm	Poorer especially at < 350 nm dynamics < 10^2 lag/blooming	22
diode array	Up to 25 mm, up to 1024	Especially for >350 nm	Poor in UV dynamics 10^3	25
Diode array – MCP	12.5 mm unidimens.	200–800 nm	Similar to PM dynamics > 10^3	22
Dissector tube	Up to 60 mm bidimens.[a]	200–800 nm	Similar to PM dynamics > 10^3	27
CCD	Up to 13 × 2 mm bidimens.[a]	200–800 nm	Similar to PM dynamics > 10^5	23

[a]Coupled with crossed-dispersion echelle spectrometer possible.

4.2. Glow Discharges with Planar Cathodes

The Grimm-type glow discharge source mentioned in Section 4.1.2.5 and its further developments are the most important types of glow discharges employing planar cathodes now in use. Their analytical performance is closely related to their electrical properties, the nature of material volatilization and of the relevant excitation processes.

4.2.1. Electrical Characteristics

The classical glow discharge is operated with argon as working gas. Its pressure is maintained at 150 to 600 Pa and it is continuously bled into the lamp through a needle valve and pumped away. As the working gas flow and pressure are of paramount importance both for the sample volatilization and for the signal generation processes, the pressure measurement should be made as close as possible to the discharge plasma. A capacitance transducer as available from several manufacturers (e.g., MKS Baratron, Phillips–Granville) should be used to be as independent as possible from the nature of the working gas used. The electrical characteristics of a classical glow discharge lamp are abnormal because of the obstructed nature of the discharge. They depend on the anode–cathode interspace (on the order of some 0.1 mm) but also on the burning spot diameter. For a burning spot diameter of 8–10 mm, and argon as working gas, burning voltages at discharge currents between 40 and 200 mA range from 0.5 to 2 kV but depend considerably on the nature of the gas, its pressure, and the cathode material (Fig. 4-18).[29] With krypton and neon the burning voltage is higher than in the case of argon, which in the former case is due to the lower mobility of the heavier krypton ions and in the latter case to the higher ionization energy of neon compared with argon. Further, the burning voltage depends on the filler gas pressure, decreasing with increased pressure as the ion production due to collisions becomes more effective and more charge carriers become available. The dependence of the burning voltage on the sample matrix analyzed is considerable, as shown by the diagrams obtained with steel, copper, and aluminum samples (Fig. 4-19).[30] This is due to the fact that a considerable part of the current is carried by analyte species and the ablation rate depends on the sample matrix. Indeed, at an ablation rate of some milligrams per minute, the seeding of the plasma with analyte atoms or ions is on the order of 1%. This is considerably higher than in the case of atmospheric-pressure discharges such as inductively coupled plasmas for solution analysis where analyte concentration is on the order of 0.01% at maximum.

As the sputtered material is deposited at the edge of the burning crater, where the field strength is somewhat higher and suction of the pump is effective, the burning time of a classical Grimm-type lamp is limited to about 15 min before electrical instabilities occur. The inner wall of the anode tube as well as the anode–cathode interspace have to be cleaned regularly because of sputtered atom deposition. For a given discharged lamp geometry and type of sample matrix, two of the parameters (current, voltage, and pressure)

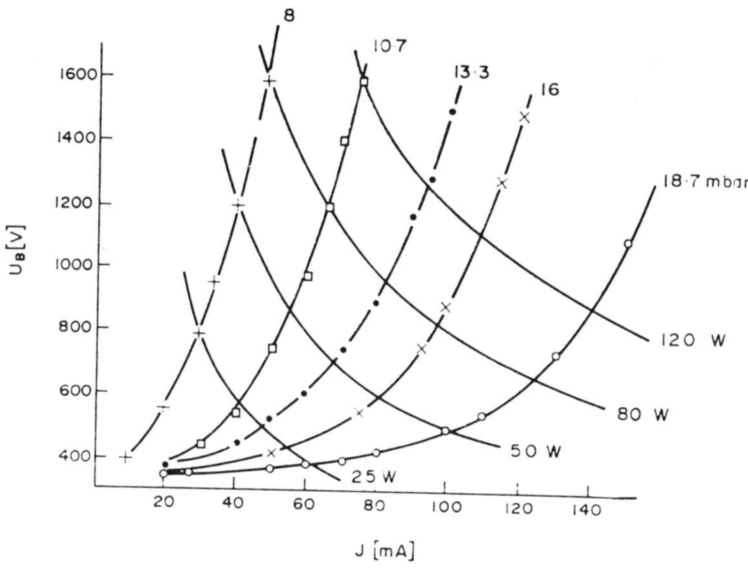

Figure 4-18. Current–voltage characteristics of Grimm-type glow discharge lamp.[29] (Courtesy of Pergamon Press.) Burning voltage U_B as a function of the current J with argon as discharge gas and a steel sample. Argon pressure in mbar. Curves indicating equal power are drawn.

may be selected and the third one then is fixed. Normally the pressure is kept at a preselected value and constant-current or constant-voltage operation, where the voltage or the current respectively then vary, is selected. One also can operate the lamp at constant power, by imposing small voltage corrections when current changes occur. It should also be mentioned that at the initiation of the discharge the voltage and current may oscillate for a certain time, which is due to the release of molecular gases from the sample surface. Eventual breakdowns of the discharge can easily be prevented by providing a large inductor (1–2 H) in the discharge circuit.

4.2.2. Sample Volatilization

In the case of a restricted type of glow discharge with a cooled and flat surface, the sample volatilizes as a result of cathodic sputtering only. The theory of cathodic sputtering (e.g., see Ref. 31) can be well applied so as to qualitatively describe the influence of various discharge conditions on the sputtering. A quantitative description, however, is hampered by the fact that sputtering models have been developed for high-vacuum sources where the impacting particles exist as monoenergetic beams with a well-defined angle

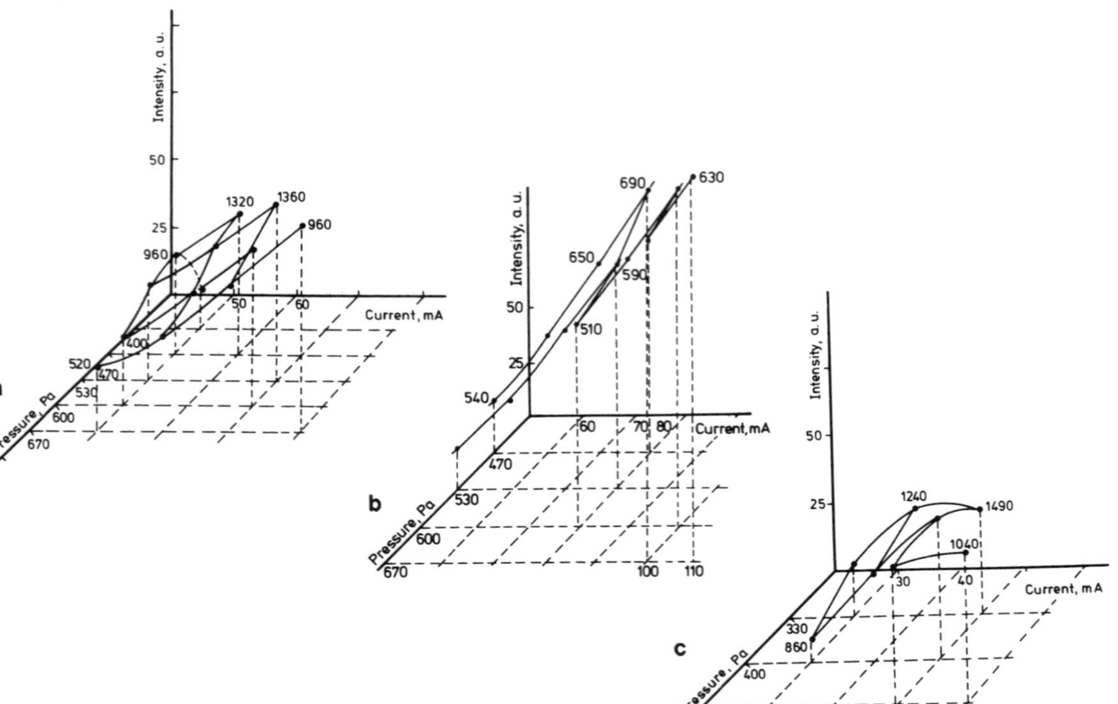

Figure 4-19. Response surfaces of GD-AES for different metal matrices.[30] Glow discharge lamp with floating restrictor diameter of 8 mm. (a) Steel matrix: spectral line Fe(I) 371.9 nm; (b) aluminum matrix: spectral line Al(I) 396.1 nm; (c) copper matrix: spectral line Cu(I) 324.7 nm. The values listed near each data point indicate the GDS operating voltage at that point on the response surface.

with respect to a monocrystalline target. In an analytical glow discharge, the pressure is on the order of some 100 Pa and accordingly, due to collisions, ions and also neutrals impact under different angles. Backdiffusion of sputtered atoms is also considerable. Furthermore, the sample material is mostly polycrystalline and structural defects as well as chemically inhomogeneous grains may be present.

The general theory of cathodic sputtering has been covered extensively in Chapter 2 and this treatment will be confined to its use for understanding sample volatilization in a glow discharge lamp with a planar and cooled cathode. The cathodic sputtering in a glow discharge with a flat cathode can be well explained by the model of impulse theory. After the removal of the molecular gases and impurities which are adsorbed on the surface and the formation of a stable burning crater surface, we obtain sputtering equilibrium. At this point, the discharge penetrates with a constant velocity into the sample and the composition of the removed material equals the bulk composition of the sample. From this time onwards, reliable bulk analyses can be made. The preburn required depends on the nature and the pretreatment of the sample. For instance, at 90 W with an argon discharge gas the preburn times required are: for zinc, 6 s; brass, 3–5 s; steel, 20 s; aluminum, 40 s.[29] Accordingly, spectrometry with glow discharges normally is slower than spark emission spectrometry and thus less suitable for rapid production control. However, when applying high-energy preburns the speed of analysis can be considerably increased. The preburn time also depends on the filler gas used: it has been found to be shorter in krypton than in argon which again is shorter than in neon. The burning crater has a structure (Fig. 4-20)

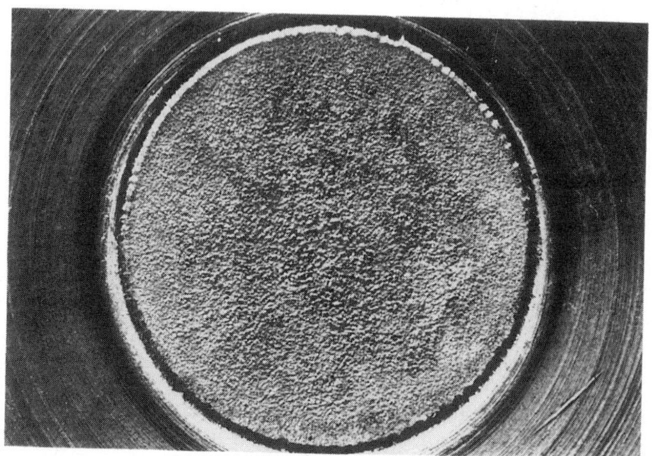

Figure 4-20. Burn crater obtained by glow discharge optical emission spectrometry of a compact metal sample (5 min burning at 50 W).

which results from the presence of different crystals with different orientation and thus different ablation by sputtering and also from chemical inhomogeneities. This structure limits the in-depth resolution when profiling is performed. This is discussed in detail in Chapter 8.

The material volatilization depends on the different working parameters. The dependence on the sample material can be explained by impulse theory. At 100 W and 1330 Pa in argon, the sputtering rate increases in the order C < Al < Fe < steel < Cu < brass < Zn. As the discharge characteristics depend on the filler gas, it is difficult to compare results obtained with different filler gases. O_2 and N_2 are hardly used as discharge gases as they give rise to oxides and nitrides, respectively, and molecular band emission would complicate the spectra. For noble gases, the sputtering rate (for aluminum at 90 W) increases in the order neon < argon < krypton, which again agrees with impulse theory. Helium gives a very poor ablation and can only be used as an additive. In the case of argon the sputtering rate is on the order of 0.3–0.5 mg/min and the influence of pressure on the sputtering rate at constant power is given by:

$$q = c/\sqrt{p} \tag{4-48}$$

where c depends on the cathode geometry, the power used, and the sample material, as investigated by Boumans.[(32)] At low pressure, the burning voltage indeed is high and energy losses due to collisions are minimal.

4.2.3. Excitation Processes

In a Grimm-type glow discharge the plasma is not in local thermal equlibrium. Excitation temperatures of 5000 and >10,000 K are cited, depending on the lines used for temperature measurement. This would refer to the existence of two groups of electrons, namely, high energetic electrons for excitation and slow electrons involved in radiative recombination. However, these values result from end-on measurements which give no account of spatial inhomogeneities in the plasma. The gas temperatures in the plasma, as derived from Doppler profile measurements or estimated from rotational temperatures measured, are below 1000 K. The Doppler broadening of the lines thus is low and the physical widths of the analytical lines are 1–2 pm, as obtained by interferometric measurements[(33)] and by recent measurements using Fourier transform spectrometry (Table 4-4).[(30)] The measurement error in these data is about 0.5 pm. A tendency toward increasing linewidths can be observed as the glow discharge source power is increased, especially for the Fe(I) 371.9-nm and the Cr(I) 425.4-nm lines. The glow discharge source power can be increased either by lowering the pressure while maintaining the current constant or by raising the current while maintaining the

Table 4-4. Physical Widths of Atomic Lines Emitted by Glow Discharges Operated under Different Working Conditions and Obtained by (a) Fourier Transform Spectrometry[30] and (b) Interferometry[33]

(a) Steel sample (CKD Research Institute) with 23.9 mg/g Cr and 1.3 mg/g Mo

Glow discharge working conditions			Physical linewidth (pm)			
Pressure (Pa)	Current (mA)	Power (W/cm^2)	Fe 371.9 nm	Mo 379.8 nm	Cr 359.3 nm	Cr 425.4 nm
430	50	200	2.1	1.6	1.6	1.9
530	50	120	1.8	1.7	1.7	1.6
600	50	115	1.5	1.7	1.5	1.4
	70	270	1.6	1.6	1.9	1.7
	60	190	1.6	1.6	1.7	1.6
	30	50	1.2	—	1.3	1 0
	15	20	1.2	—	—	—

(b) Glow discharge lamp operated at 1 kV, argon pressure variable; line: Cr 425.4 nm

Cr in steel (mg/g)	Current (mA)	Physical linewidth (pm)
5.6	25	1.3
	50	1.6
	100	1.65
	200	2.1
21	25	1.4
	50	1.65
250	25	1.4
	50	1.6
	100	1.7
	200	2.15

pressure constant; both seem to produce equivalent results and both are known to increase the sample ablation rates.[32] The observed increase in linewidth may be due to self-absorption, which is very strong in the case of resonance lines.[5]

Excitation mostly takes place in the negative glow where the number density of free electrons is high. Mainly electron impact, but possibly also Penning ionization, contributes to the excitation. The number density of metastables in the negative glow is on the order of 10^{12} cm^{-3} [34] as determined by absorption measurements, whereas 10^{16} cm^{-3} argon atoms and 10^{11}–10^{13} cm^{-3} sputtered analyte atoms are present.[35] Steers and Leis[36] recently showed that charge transfer processes may also play a role in excitation in a Grimm-type lamp.

As the electron number densities are considerably lower than in atmospheric-pressure plasmas ($n_e = 10^{14}$ versus 10^{16}), the radiation density of the continuum, which is caused by the interaction of free electrons (bremsstrahlung) and mainly by the interaction of free and bound electrons (recombination continuum), is given by:

$$I_\nu \, d\nu = K n_e n_r \frac{r^2}{T_e^{1/2}} e^{-h\nu/kT_e} \, d\nu \qquad \text{(free–free)} +$$

$$K' \frac{1}{j^3} n_e n_Z \frac{Z^4}{t_e^{3/2}} e^{(U_j - h\nu)/kT_e} \qquad \text{(free-bound)} \tag{4-49}$$

where n_r is the concentration of ions with charge r, Z is the charge of the nucleus, and U_j is the ionization energy of the term with quantum number j. As n_e is considerably lower than in atmospheric-pressure plasmas, the background continuum will be accordingly lower and high signal-to-background ratios can be expected.

4.3. Analytical Performance

Because of the material volatilization by cathodic sputtering and the excitation in a low-pressure discharge, glow discharges with a flat cathode are very suitable sources both for sequential and for simultaneous multielement determinations in electrically conducting solids. The optimization will include a careful consideration of the electrical parameters, the working gas and its pressure as well as the sample preparation. The optimization, however, must be made in view of the analytical performance sought, and accordingly, differences can be expected when optimization aims at the highest power of detection, minimal interferences, or the highest analytical precision.

4.3.1. Power of Detection

Optimization toward a high power of detection in the case of the Grimm-type glow discharge requires measures regarding high radiant output. Indeed, as reflected by background measurements performed with a high-speed monochromator (Table 4-2), especially at low wavelengths the background signals obtained are hardly higher than the dark current of common photomultipliers. Accordingly, dark current noise as well as counting statistics limitations may curtail the power of detection. Indeed, the total statistical error $\sigma(I_U)$ is given by:

$$\sigma^2(I_U) = \sigma_P^2 + \sigma_f^2 + \sigma_D^2 + \sigma_V^2 + \sigma_c^2 \tag{4-50}$$

where σ_P is the photon noise, σ_D the dark current noise, σ_f the flicker noise, σ_V the amplification noise, and σ_c the contribution from the counting statistics. A photon noise contribution from the counting statistics. A photon noise contribution of 5×10^{-4} at 200 nm is only slightly beyond the contribution of the dark current (3×10^{-4}) and a decrease of the latter by a factor of five would improve the noise level by 30% and the detection limit accordingly. This applies when taking smooth spectral background into account. When blank contributions or spectral line wing contributions become important, the detection limit is deteriorated according to:

$$c_L = (I_X/I_U \cdot 1/c)^{-1} 3\sqrt{2}\left(\frac{I_B + I_U}{I_U}\right)\left(\frac{\sigma_r(I_B + I_U)}{\sigma_r(I_U)}\right) \tag{4-51}$$

where I_X is the net signal for a concentration c, I_U the total signal for background and dark current, I_B the net blank contribution (i.e., the contribution stemming from interfering lines) and $\sigma_r(I_B + I_U)$ and $\sigma_r(I_U)$ the respective standard deviations.[37] From Eq. (4-50), it can be seen that blank contributions or signals from interfering lines do influence the detection limits even when these contributions are highly stable.

The principal part of optimization toward a high power of detection, however, is related to optimization for the highest line-to-background intensity ratio. In order to get maximum line intensities: (1) the material ablation should be high; this is achieved as discussed earlier at high burning voltage, and can be realized at relatively low gas pressure, which in addition has the advantage of minimizing collisions by which the energy of the incident species, and thus also the ablation rate, would decrease; and (2) the excitation efficiency, however, also should be as high as possible, which is achieved at higher working gas pressures, as then the number densities for the exciting species and thus the number of collisions will be high.

The significance of the working pressure is clearly shown in many studies (e.g., Fig. 4-21).[29] Both the material ablation and the excitation efficiencies increase with current density (Fig. 4-22).[32] Apart from the line intensities, both optimization parameters also considerably influence the background intensities, as both the electron number densities and the analyte number densities change.[38] Furthermore, each sample matrix has its own current–voltage characteristics, which in addition depend on the construction of the lamp (anode tube diameter and anode–cathode interspace as well as gas throughput of the lamp), and accordingly further influences the optimum values for the working parameters. Apart from these parameters, the possibility of selective excitation processes (e.g., for copper see Ref. 38) makes an optimization for each particular line necessary. This also applies when using different working gases.

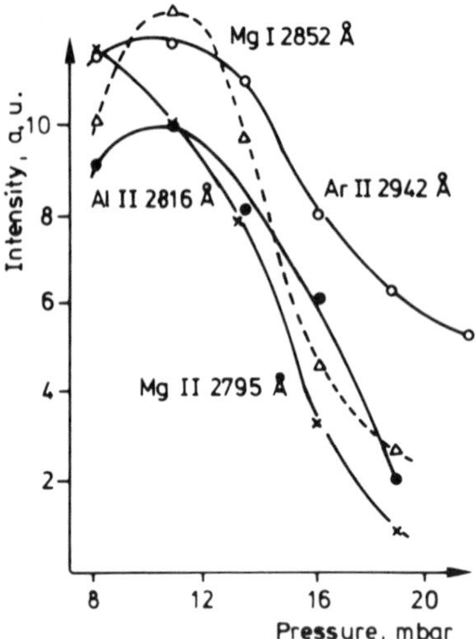

Figure 4-21. Influence of working pressure on line intensities.[29] (Courtesy of Pergamon Press.) Conventional Grimm-type glow discharge lamp operated in argon; 70 mA, aluminum sample.

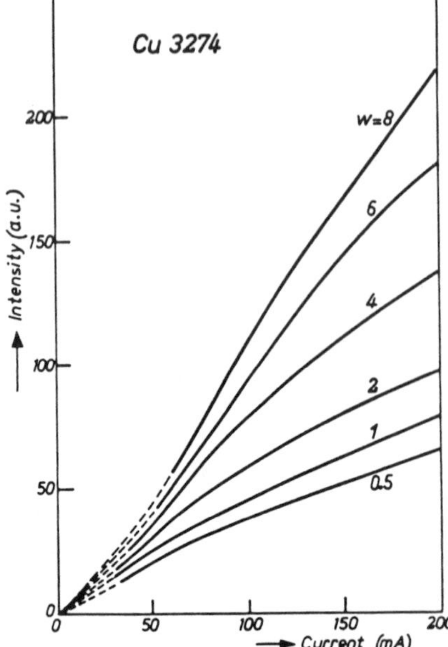

Figure 4-22. Influence of current density on intensity of Cu 324.7 nm at different penetration rates (μm/min) for a copper target.[32] (Courtesy of the American Chemical Society.)

Detection limits obtained with a Grimm-type glow discharge in the case of metallic samples are in the microgram per gram range, as shown for steels and copper samples (Table 4-5).[39,44,48] The values found in the literature often are higher than expected, which may be due to the fact that polychromators are used on which the most sensitive lines are not always available or

Table 4-5. Detection Limits of Glow Discharge Optical Atomic Emission Spectrometry for the Analysis of Steel[44] and Copper Samples[57]

Element/line (nm)		Detection limits (μg/g)		
		Steel with microwaves	Steel without microwaves	Copper
Ag(I)	338.3			1.8
Al(I)	396.2	0.1	0.4	
As(I)	190.0			<10
B(I)	209.1	0.3	0.8	
	208.9	0.4	0.9	
Cr(I)	425.4	0.05	0.2	
Cu(I)	327.4	0.3	1.5	
	324.8	0.9	2	
Mg(I)	285.2	1.5	2	
Mg(II)	279.6	0.9	1.3	
Mn(I)	403.1	0.2	1	
Mo(I)	386.4	0.8	1.5	
Nb(I)	405.9	0.6	4	
Ni(I)	232.0	0.1	0.5	
Si(I)	288.2	0.4	3	
Ti(I)	364.3	0.6	3	
V(I)	318.4	1	3	
Zr(I)	360.1	1.5	8	

of which the practical resolution is low because of stability reasons (e.g., for steels, see the values in Ref. 40).

Several means are available to improve the power of detection. They include

1. Innovation in the lamp concept itself. Here, measures that improve the ablation rate as well as those that increase the excitation efficiency can be taken. For the first part, the use of magnetic fields to increase the density of the sputtering beam[41] increases the current density by further constricting the burning spot, and improving the cooling of the lamp can be advantageous. For the second part, the use of cross-excitation by a dc discharge,[42] a high-frequency[43] or a microwave discharge[43,44] (see Section 4.5.1) has been shown to be effective.
2. Improvement of the speed of the spectrometer. This is particularly useful to overcome detector noise limitations which may become limiting at low wavelengths as already discussed. As the spectra of glow discharges are almost less line-rich than those of classical sources such as the dc arc, this can be achieved by employing spectrometers of rather low practical resolution which have large spectral

bandwidths, insofar as spectral interferences do not take place. A straightforward approach is the use of Fourier transform spectrometry, as discussed in Section 4.5.2.

4.3.2. Matrix Interferences

Because of the sample volatilization by cathodic sputtering, matrix interferences due to different volatilities of the elements are absent. Therefore, matrix effects may result from differences in elemental sputtering rates from one alloy to another and even from the influence of differences in the metallographic structure from one sample to another. For a minimization of these matrix effects, measurements should be performed after a sputtering equilibrium is obtained. When igniting the discharge, the impurities present on the sample surface enter the discharge plasma and may influence the excitation conditions. Further, during the initial discharge stage the preferentially sputtered elements are enriched in the material leaving the surface until, as a result of a drop in the bulk concentration for these elements and of backdiffusion, a dynamic equilibrium is reached. At this point, the line intensity signals become constant and matrix influences are minimal.

The preburn time in the case of a Grimm-type glow discharge is on the order of 20–30 s for metal samples.[45] Accordingly, and because of the more difficult sample change as compared with sparks, glow discharges are less useful for rapid production control as required in the steel industry. They are excellent for product control. The preburn time can be decreased by all measures which increase the ablation rate. This can be achieved by high-energy preburn which can be done by a controlled decrease of the discharge gas pressure, by which the burning voltage at constant current increases, or by an increase of the discharge current at constant discharge gas pressure. It has been found that for the different matrices the preburn times correlate with the sputtering rates. Electrically nonconducting powder samples can be analyzed by glow discharge lamp spectrometry by mixing them with a metal powder such as copper and briquetting vacuum-tight, electrically conducting pellets, as described by El Alfy *et al.*[46] As reported in the literature, the preburn times in this case are considerably longer (>1 min) but also show the same dependence on the discharge parameters.[47]

The magnitude of the matrix effects in glow discharge emission spectrometry is exemplarily shown for the case of metals in Fig. 4-23[48] (aluminum alloys) and for the case of electrically nonconducting powders in Fig. 4-24[49] (silicate rocks). For aluminum alloys, matrix effects were found to be considerably lower than in spark emission spectrometry. In the latter case, a strong curvature and even a splitting into several curves occurred. It should be emphasized that, in any case, referring to a matrix line is required to compensate for differences in sample ablation rates. When analyzing pul-

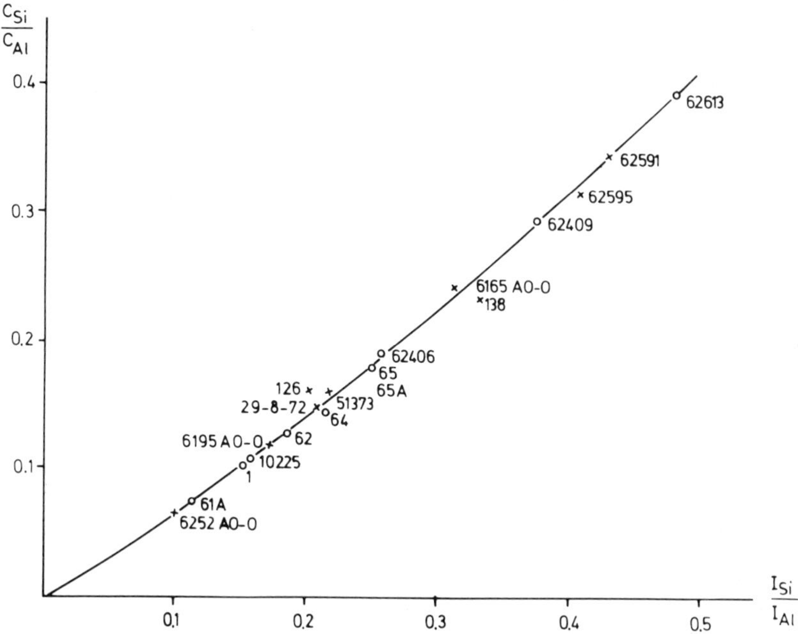

Figure 4-23. Analysis of aluminum samples by glow discharge emission spectrometry.[48] (Courtesy of GAMS.) Conventional Grimm-type glow discharge lamp; restrictor diameter: 7 mm; 125 W, 1.1 kV; 11 mbar argon; analytical lines: Si(I) 251.6 nm and Al(I) 236.7 nm.

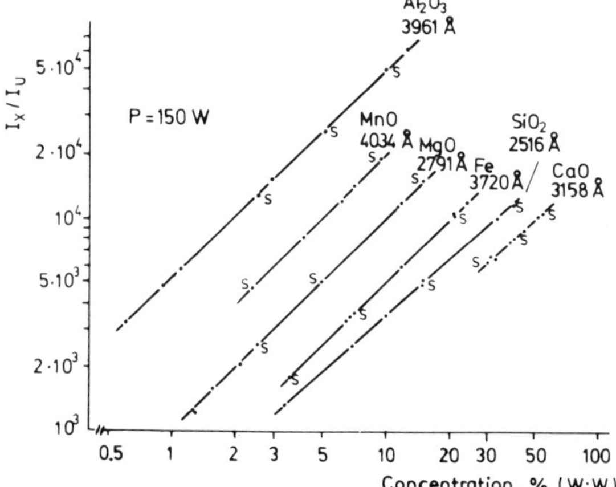

Figure 4-24. Calibration curves for the determination of minor and major elements in slags after pulverization and mixing with copper powder by glow discharge optical emission spectrometry.[49] (Courtesy of Dr. S. El Alfy.) S: synthetic powder mixtures.

verized mineralogical samples, it has been found possible to calibrate using synthetic mixtures of metal oxides.

4.3.3. Analytical Precision

The types and sources of noise occurring in the case of glow discharges have to be investigated to discover the ultimate precision achievable, and to find ways for its further improvement. Noise power spectra reflect the types of noise present in an analytical system, and results of rapid Fourier transform of the direct signal of a Grimm-type glow discharge have been recorded.[30] The authors found that most noise is white noise and that flicker noise displaying a $1/f$ frequency dependence is hardly present. However, discrete noise frequency bands also occur. It was found that some contributions may stem from the vacuum system and others from the power line frequency (Fig. 4-25). Thus, both of the last points require care with respect to the instrument design. From comparative analyses of Cu/Zn samples, it was found that the analytical precision of glow discharge emission spectroscopy approximates that of x-ray fluorescence spectrometry and is similar to solution analysis for major elements in the same samples with ICP optical emission spectrometry, provided that line intensity ratios are used as analytical signals (Table 4-6).[50]

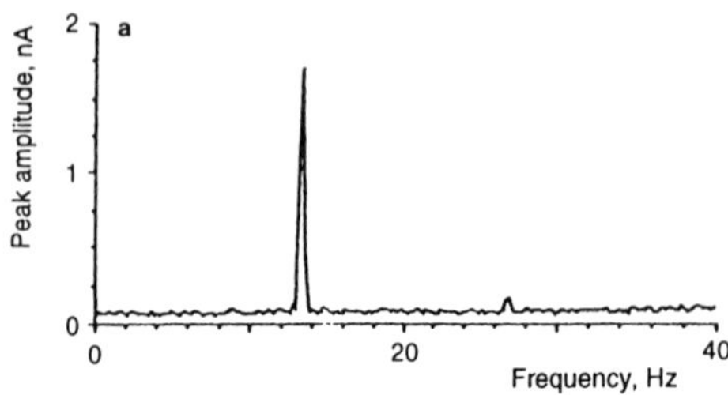

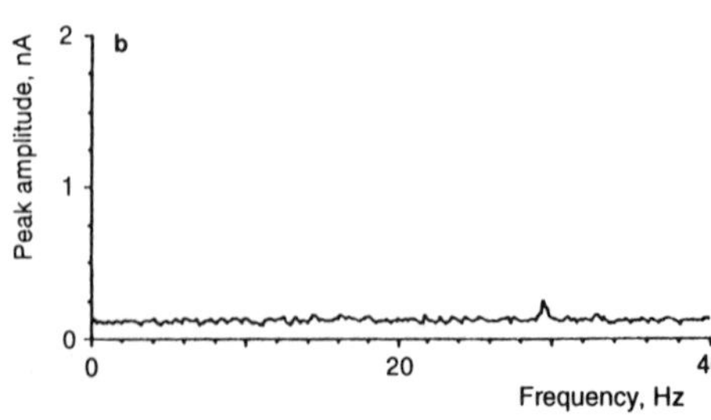

Figure 4-25. Amplitude noise spectra of glow discharge optical emission spectrometry with a Grimm-type glow discharge lamp.[30] Steel standard sample 218A (Research Institute CKD, Czech Republic); *i*: 50 mA; argon pressure: 3.5 Torr; burning voltage: 900 V; 0.35-m McPherson monochromator; line: Fe(I) 371.9 nm. (a) Without needle valve between glow discharge lamp and vacuum pump; (b) with needle valve between glow discharge lamp and pump.

Table 4-6. Analytical Precision Obtainable in Glow Discharge Atomic Emission Spectrometry, X-ray Fluorescence Spectrometry, and ICP Atomic Emission Spectrometry[(50)a]

Method	Samples (brass)	Concentration range for Cu (mg/g)	Standard deviation over the regression (mg/g)
GD-AES	Alloy	500–900	2.1
XRF	Alloy	500–900	0.6
ICP-AES	Solution	100–900	2.1

[a]Calibration according to the method of binary ratios.

The use of an internal standard is known from the early days of analytical emission spectrometry. With I_{qp} and I_{sr} the intensities of an analytical line of the element X and a reference line of the element R, respectively, the intensity ratio can be written as:[(51)]

$$\frac{I_{qp}}{I_{sr}} = \frac{n_{(X)} A_{qp} \nu_{qp} g_q Q_{(R)}}{n_{(R)} A_{sr} \nu_{sr} g_s Q_{(X)}} \exp(E_s - E_q)/kT \tag{4-52}$$

In order to compensate efficiently for fluctuations in sample introduction as well as for changes in excitation temperature, the excitation energies E_s and E_q, and the partition functions Q should be similar. In any case, either two atom or two ion lines should be selected, for otherwise ionization would falsify the number densities of the radiating analytical or reference species. For the same reason the ionization energies of both should be as similar as possible.

The use of reference signals is indispensible when the highest analytical precision is required. It compensates both for differences in ablation rate from one sample to another as well as for small differences in excitation conditions and small fluctuations of the source. This also applies to the determination of major elements as shown by the procedure used by Jäger for the analysis of gold.[(52)]

4.4. Analytical Applications

4.4.1. Analysis of Metals

For both bulk and in-depth analysis of metal samples, glow discharge atomic emission spectrometry with a Grimm-type glow discharge lamp is very powerful and many systematic studies and applications to diverse matrices have been described in the literature.[(108)]

4.4.1.1. Bulk Analysis

Multielement analyses for major, minor, and trace elements have been described for most industrially important matrices. In methodological studies, the sputtering and ablation rates of various metals in a Grimm-type glow discharge have been investigated.(32) Under analytically relevant conditions, ablation rates range from 0.1 (Al) to 3 mg/min (Zn) for a target area of 0.5 cm^2, a current of 180 mA, and a voltage of 500–600 V. Accordingly, preburn times of at least 30 s are required to reach sputtering equilibrium. In the first paper of Grimm,(19) the strong dependence of the voltage on the metal matrix was shown and therewith the necessity to relate relative intensities to concentration ratios in the case of the determination of major elements, as shown by the example of determination of Ni in binary Fe–Ni alloys (250–1000 mg/g).

The working pressure and the related sputtering and excitation properties of the source have also been shown to strongly influence the analyte line intensities, whereas the low spectral background is much less influenced. The properties of the source are also known to vary considerably with the discharge gas used. As is known from sputtering studies (Chapter 2), argon represents a good compromise as sputtering gas. However, neon or argon–helium mixtures have also been used.(53,54) In the latter case, the sputtering rate decreases with increasing amounts of helium, although the net line intensities for many lines increase.(55) The properties of glow discharge lamp spectrometry were critically compared with those of spark emission by Durr and Vandorpe.(56) Whereas in the first case rectilinear calibration curves were obtained for nickel and sulfur in steel, the latter suffered from curved calibration curves (Ni) or even from splitting. For Sn, however, the glow discharge calibration was also curved. With a demountable type of glow discharge lamp, where different anode–cathode geometries and a floating restrictor can be used,(57) fine-tuning of the pressure and flow conditions in the lamp measurably influences the freedom from interferences stated above. This has been shown mainly for Cu/Zn and Cu/Zn/Pb alloys. For steel analysis, the Grimm-type lamp can be used to advantage product control, as shown by Rademacher and de Swardt(45) who reported preburn times on the order of 20–30 s for Si, Mn, P, C, and S. For C, 156.1 nm was found most advantageous. RSDs from 0.5 to 2.5% were found in the 0.01 to 10 mg/g range. With a magnetically focused glow discharge lamp, Kruger *et al.*(40,41) succeeded in increasing the power of detection by a factor of 2–5. This is partly due to the increase in ablation rate (up to twofold). However, it may also be that the residence time of electrons, and accordingly their number density, in the negative glow, increases by which an improvement in excitation takes place. The lamp, however, also showed promise for nonferrous metals as here no rinsing problems occur.(128) In the investigation of iron-based alloys, Wagatsuma and Hitokawa[58] reported that chromium oxide films on the surface influence the sputtering yield, as reflected by the change of chromium and iron line intensities. Results for ternary Fe/Ni/Co alloys are presented. By applying voltage modulation of the lamp and lock-in amplification of the analytical signals, the signal-to-noise ratios can be improved by a factor of 20–50 in the case of low-alloyed steels.[59] Apart from steels, cast iron can also be analyzed. For C, here, the C 247.86-nm line is proposed. However, for cast irons with different metallographic structure, not only the signals but also the intensity ratios are different. Further working at increased burning voltage was shown to shorten the preburn time considerably.[60]

For the analysis of noble metals by glow discharge emission spectrometry, techniques for increasing the precision were developed; they are based on the use of the sputtering rate as reference signal and allow it to obtain in gold analyses RSDs down to 0.1% in the major element range (for Au). Further cast and rolled samples could then be analyzed with the same calibration curves.[61] The same principle allowed RSDs below 0.09% to be obtained for silver in silver analyses, whereas for Au, Pd, Pt, and Cu, the RSDs were on the order of 5% at the 1–2% (w/w) range.

For Cu–Al alloys, Dogan[62] found an RSD at a concentration of 0.06% $(1 - w)$, with w the concentration for a concentration range of 0.5–99.5% (w/w) of Cu or Al in binary alloys. This shows that, especially in the case of simultaneous emission spectrometry, a high analytical precision for major elements is attainable. Also, in other papers on the analysis of Cu alloys,[63] RSDs of 0.2% have been reported for the major elements. Nevertheless, proper measures must be taken to correct for the influence of minor elements on the emission intensities. Yamada *et al.*[64] reported that for copper alloys, the metallurgical structure considerably influenced the sputtering rate; changes with the concentration of aluminum [1–12% (w/w)] were discussed. It was found that after a preburn time of 100 s, however, structural effects no longer affected the Al(I)/Cu(I) line intensity ratios.

4.4.1.2. In-Depth Profiling

Due to the layer-by-layer ablation in glow discharges with flat cathodes, in-depth profiling was found at an early stage to be one of the most important analytical features of glow discharges, as shown in 1973 by Belle and Johnson[65] and in 1978 by Berneron.[66] This important feature is discussed in detail in Chapter 8, so only a brief outline will be given here.

For in-depth profiling, it is important to calibrate the penetration rate as a function of the working conditions as well as to know its dependence on the material composition. The latter point may be difficult especially as the main composition in the case of coatings may vary considerably with the sampling depth. For solving this problem, approaches that make use of reduced sputtering coefficients, as discussed and developed by Bengtson,[67] may be helpful. As shown by Berneron[68] the problem of converting the measured intensities of in-depth resolved concentrations is still more difficult. Here, an attempt could be made to calibrate intensity ratios by comparing them with values obtained for signals from bulk analysis of samples of known composition. Adequate software, which uses these data as a function of the sputtering conditions (current density, working gas pressure, and voltage) to calculate the in-depth variations of elemental concentrations, is under development (see Chapter 8).

For obtaining sufficient resolution in in-depth profiling, the radiant density per ablated layer should be high.[69] This necessitates operating the lamp at relatively low burning voltage, to achieve low sputtering rates, and at relatively high working gas pressure, to yield a high excitation efficiency. Typical values are 400–500 V burning voltage at a pressure of 4–5 mbar argon. At a discharge current of some 10 mA, ablation rates for steels were 4 nm/s, as reported by Waitlevertch and Hurwitz[70] in early glow discharge work. De Gregorio and Savastano[71] reported the use of discharge currents of 20–100 mA, a burning voltage of 1 kV, and a time resolution of 0.1 s for steels. However, it should be clear that the optimum conditions strongly depend on the construction of the lamp used. The analytical figures of merit of glow discharge emission spectrometry for in-depth profiling have to be compared with those of other techniques for in-depth profiling such as secondary ion mass spectrometry (SIMS), as shown by Takimoto *et al.* in the case of galvanized steels.[72] For SIMS, the in-depth resolution is much better than for GD-OES (in the nm range). However, in the production control of technical surfaces, it is often necessary to investigate thicker layers, for which SIMS is too expensive and time-consuming. Often it is sufficient to accurately control the thickness of the layer to have semiquantitative information on its elemental composition.

Most applications reported stem from the steel industry. Koch *et al.*[73] reported that for bulk analysis, glow discharges have not made a breakthrough in the steel industry. However, for surface characterization of steels involving the investigation of oxide and nitride layers with a thickness greater than 0.1 μm, GD-OES is of use.[74] In their work, they used a burning voltage of 1000 V and a discharge current of 100 mA and reported sputtering rates of 25 nm/s for iron oxide layers. The problem of quantization was experimentally attacked by Wagatsuma and Hirokawa,[75] who studied the sputtering behavior of binary alloys so as to obtain sputtering constants that could be used in calibration programs. In the case of silver–copper alloy surfaces,[54] they found that the metallurgical structure of the samples played an important role regarding the length of the transition times prior to steady-state sputtering. They also observed that redeposition of the silver grains with copper may influence the sputtering yield after some time, which further may falsify in-depth determination from sputtering times. Ohashi *et al.*[76] used a glow discharge with floating restrictor and remote electrode to record in-depth profiles of steels. In the case of measurements at 4 mbar, 50 mA, and 200–400 V for samples of differing nature, layers between 1 and 10 nm still could be characterized using a time resolution of 1 s for the measurements.

Progress in the use of glow discharges for in-depth profiling will result from an improvement of the flatness of burning craters. As shown by investigations with copper alloys in a demountable glow discharge lamp by Ko,[57] the use of a floated restrictor for this aim is very helpful. Furthermore, the radiant output per ablated layer in the glow discharge still can be increased considerably. This is indicated by the self-reversal found in the line profiles of copper resonance lines measured with Fourier transform spectrometry[77] which relates to a high number density of ground-state atoms in the glow discharge plasma. As shown by recent measurements, the self-reversal of the resonant lines emitted by the glow discharge vanishes when applying cross-excitation with microwaves as described by Leis *et al.*[44] This was indeed found to considerably enhance the net line intensities (Table 4-7).

4.4.2. Analysis of Dry Solution Residues

The analysis of dry solution residues is known from work with hollow cathodes enabling extremely low absolute detection limits. This especially

Table 4-7. Signal Enhancement in Glow Discharge Atomic Emission Spectrometry Resulting from Microwave Boosting[44][a]

Element/line (nm)	dc current (mA)	Gain factor		
		300 Pa	420 Pa	600 Pa
Cr 425.4	10	27	21	33
	20	23	10	23
	30	23	14	4
	40	—	9	—
Cu.327.4	10	58	124	—
	20	43	83	81
	30	—	56	48
	40	—	—	32
	50	—	—	27
	60	—	—	25
Si 288.2	10	36	44	39
	20	22	14	16
	30	11	—	10

[a]Steel samples, 40 W microwave power.

applies when the technique is used as the determination step in a multistage combined procedure where the analyte is isolated by chemical procedures and presented as a small-volume multielement concentrate.

Büger and Fink[(78)] tried to directly introduce liquids in the hollow cathode glow discharge plasma but found strong interferences of band spectra. Harrison and Prakash[(79)] succeeded in analyzing liquids by hollow cathode emission spectrometry subsequent to their evaporation in steel or graphite cathodes and obtained detection limits down to 0.1 μg/ml, requiring less than 0.5 ml of sample. For rare earths, similar results were obtained with copper[(80)] and graphite cathodes.[(81)] For the alkali elements, extremely low absolute detection limits (pg range) have been reported recently.[(82)] However, the dispensing of small liquid aliquots to plane surfaces is much easier as reported by Brackett and Vickers.[(83)] Provided the necessary optimization and choice of carriers is made, a handy method with high power of detection and multielement capacity can be developed, as has been shown for the case of a Grimm-type glow discharge used as ion source for mass spectrometry recently.[(84)]

4.4.3. Analysis of Electrically Nonconducting Samples

Electrically nonconducting powder samples have for some time been analyzed with glow discharges after mixing with a metal powder and briquetting compact and vacuum-tight pellets of good thermal conductivity. Toward this aim, El Alfy *et al.*[(46)] proposed a mixture of 3–4 parts of ductile copper powder with the powder sample. In order to keep sample consumption low (25 mg), a small amount of this mixture is embedded in pure copper powder and pressed at 10^4 N/cm^2 with a hydraulic press (Fig. 4-26). It should be mentioned that samples such as Al_2O_3 can be easily analyzed, provided no crystal (adsorbed) water is present, but that materials such as SiO_2 tend to swell subsequent to pressing and make it difficult to get stable pellets. The powder particle size must be small enough to effect a stable material ablation. From sieving analysis, El Alfy[(49)] reported no problems with pellet stability when the powder particle size was below 25 μm.

The behavior and optimization of the Grimm-type glow discharge in the case of pellets for powder analysis have been investigated by a number of groups. As this is covered in detail in Chapter 7, only some lines of research and their trends will be mentioned here. El Alfy *et al.*[(46)] described the determination of the minor compounds in ores and minerals and showed that a calibration could be performed with synthetic standards prepared by mixing the analyte oxide powders in a calcium carbonate matrix (Fig. 4-24). The preburn times, however, were on the order of 2 min. Si, Ca, Mg, Mn, Fe, Al, Ti, Na, K, P, S, and C have been determinated in a wide variety of electrically nonconducting substances (e.g., rocks, minerals, ores, glasses,

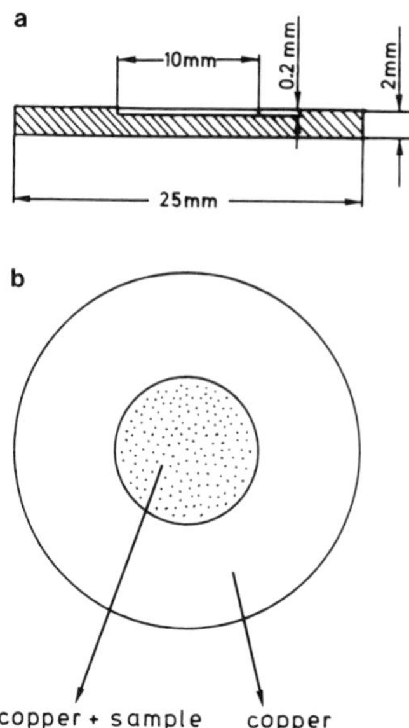

Figure 4-26. Briquetting of pellets for the analysis of electrically nonconducting powders by glow discharge atomic emission spectrometry.

slags, cement). It was found that pronounced structures occurred in the burning craters. Different theories were developed to explain their form. Jäger and Blum[(85)] proposed the distortion of the local electrical field in the vicinity of electrically nonconducting grains to be responsible for changes in the local sputtering rates. However, preferential sputtering due to the orientation of the crystallites with respect to the incident ions and atoms, as well as redeposition at favored places, may play a role. This has been studied in more detail by electron probe microanalysis[(86)] and it was suggested that partial melting at the sample surface may occur but did not influence the volatilization of the nonconducting grains. Mai and Scholze[(47)] mentioned that the preburn times required in the case of the discharge power of 100 W are below 1 min. They also suggested that diffusional transport to analyte from the nonconducting grains to their surface might play a role in the material volatilization. Sputtering of these surfaces indeed could occur as a result of surface conduction of redeposited material. XPS signals for CuO and Cu supported this theory.

For the analysis of electrically nonconducting powders, glow discharge emission spectrometry is a viable alternative to x-ray fluorescence spectrometry, as the light elements such as B, Be, and Li are easily determined. Furthermore, the power of detection for these elements extends to the microgram per gram range and the precision in terms of the RSD relative to a reference line is at the 1% level or better within a large range of concentration. As the analysis of ceramic powders becomes more and more important, many elements (including the light ones) have to be determined in Al_2O_3, AlN, SiC, Si_3N_4, and ZrO_2 matrices[87] and these applications of glow discharges will become more important.

Recently, the use of boosted glow discharges for the analysis of electrically nonconducting powders has also been described. Lomdahl *et al.*[88] used the dc-boosted glow discharge lamp developed by Gough[89] to determine down to 0.002% (w/w) NiO and 0.0015% (w/w) CuO in copper and nickel binders, respectively, and achieved an RSD of 1–2%. Minerals and rocks[90] were also analyzed. Samples were mixed 5%–95% (oxide to host) with copper powder and pellets of 6-mm diameter were pressed at 10^9 N/cm^2. The method was used to determine major to trace compounds [e.g., down to 0.004% (w/w) K] in rocks. The intensity enhancement realized by the auxiliary dc discharge is essential for penetrating the trace element concentration level.

Further progress lies in the use of an rf glow discharge with which compact electrically nonconducting samples such as ceramics can be analyzed directly, as recently proposed by Duckworth and Marcus[91] first for mass spectrometry and later for optical emission spectrometry[92] as discussed in detail in Chapter 7.

4.5. Lines of Research for Further Improvement

Further improvement in the analytical figures of merit of glow discharge optical emission spectrometry may be realized by improving the glow discharge itself and also by improved use of the spectral information.

4.5.1. Application of Cross-Excitation

The idea of boosting a glow discharge to achieve a higher radiation output, overcoming shot-noise limitations and also increasing the line-to-background intensity ratios, has been pursued for some time. Indeed, the need for high-intensity primary line sources for atomic absorption and atomic fluorescence spectroscopy forced the research on increasing the radiation output of hollow cathode lamps. This could be achieved by pulsing the sources and also by boosting the dc hollow cathode discharges. Indeed, plasma spectroscopic investigations revealed that in glow discharges a lot of the volatilized material is present in the form of ground-state atoms, and by increasing the excitation efficiency the number densities of excited atoms still can be substantially increased.

For the glow discharge sources in emission spectrometry, several methods for cross-excitation have been investigated. First, a dc discharge has been put across the glow discharge plasma, an idea that was mainly pursued in Australia by Gough, Sullivan, and others. In the work of Lowe,[42] a dc discharge of several hundred milliamperes was put across the glow discharge plasma and found to produce a considerable increase in line intensities. McDonald[93] found excitation temperatures for Fe(I) and Fe(II) lines in the range of 4000 K and electron densities of 5×10^{13} while the degrees of ionization were 81.3% for Ca, 19.5% for Cr, and 5.5% for Mg. Gough and Sullivan[94] combined a source using a secondary discharge (up to 750 mA at 120 V) on a normal glow discharge (up to 100 mA at 900 V) with a polychromator and analyzed steels. When using mainly resonant lines for analysis, a series of elements could be determined down to the microgram per gram level. The secondary discharge made use of a heated filament (0.3 V, 10 A) as an electron source. This emitter, however, as well as crystals of rare earth compounds used later are easily poisoned by the air contact during sample change as well as by deposition, which has made the lamp difficult to operate. Other approaches make use of electrodeless discharges for boosting. Ferreira *et al.*[43] used a high-frequency discharge and reported an enhancement between 10- and 20-fold when superimposing a 13.6-MHz discharge on a 25-mA glow discharge. However, when the current was increased to 100 mA, the difference became negligible.[95]

The use of microwave boosting seems to be more effective. Coupling with an antenna was used by Ferreira *et al.*[43] for a Grimm-type glow discharge and by Caroli *et al.*[96] for a hollow cathode glow discharge. Later it was recognized that the efficiency of cross-excitation could be increased greatly by integrating a Beenakker TM_{010} cavity with the low-pressure discharge, as reported for the hollow cathode discharge by Caroli *et al.*[97] and for the Grimm-type glow discharge lamp by Leis *et al.*[44]

The lamp used by Leis *et al.*[44] is schematically represented in Fig. 4-27. The GDL part is essentially the same as Grimm's original lamp with the exception that it is relatively thin so that the integrated microwave cavity part is as close to the cathode as possible. However, for construction reasons (water cooling, gas inlet and outlet, O-rings) the distance from the cathode to the inner wall of the microwave resonator is still 16 mm. The exchangeable anode tube has an internal diameter of 8 mm and is at ground potential by electrical contact with the walls of the cavity. A quartz tube, also of 8-mm internal diameter, separates the inner low-pressure region from the outer part of the cavity, which is filled with air at atmospheric pressure. The gas inlet, the illumination of the 1-m sequential spectrometer, and readout system are described in Ref. 44.

The lamp can be operated with or without microwaves and has always been used in a current-regulated mode. When microwaves are superimposed,

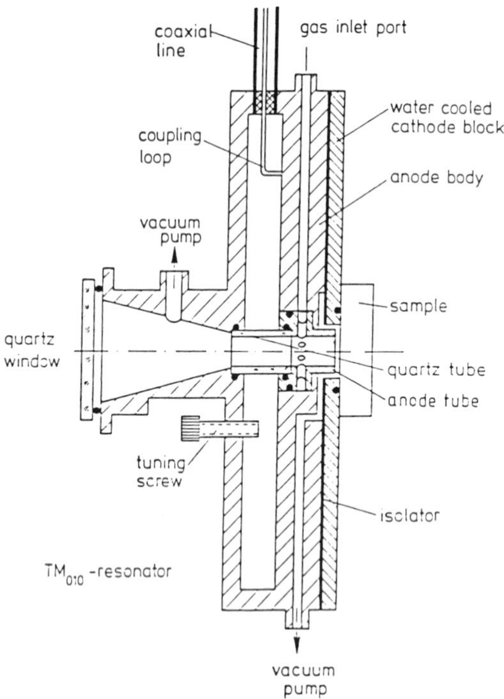

Figure 4-27. Glow discharge lamp with integrated microwave cavity.[44]

a microwave discharge which forms one optically thin plasma with the positive column can be observed. A microwave discharge occurs, depending on pressure and current of the dc discharge, at 30 W (at 300 Pa and 10 mA) and from 60 W onwards, it can be sustained without the glow discharge. The addition of the microwave discharge considerably decreases the burning voltage and a most favorable tuning is obtained at a forward power of 40 W (reflected power 15 W). The long-term stability (after more than one day) is limited by deposits of sputtered material on the quartz tube, influencing the microwave coupling, and on the observation window.

The burning craters obtained with microwaves tend to be somewhat deeper in the center, their depth determined by changes in burning voltage, which is a function of the microwave power applied. The erosion rates are on the order of 10–30 nm/s, increasing with the applied anode currents. These values are 2–5 times lower than in a conventional glow discharge. Accordingly, the preburn times will be long (the signals fully reach their equilibrium value after 3 min). Considerable enhancement in net line intensity especially for lines with rather low excitation energies (resonant lines) has been found. For phosphorus and sulfur lines, for instance, the enhancement was lower. At low glow discharge currents, the enhancement was highest, which is similar to the observations of Walters and Human for the rf-boosted lamp.(95) Also, when comparing the background equivalent concentrations obtained under optimized conditions (16 mA, 40 W, 300 Pa) with and without supplementary microwave excitation, a gain is still obtained. The RSDs both for the line intensities and for the spectral background at integration times of 8 s are 1% or lower. The detection limits obtained for steel samples in the case of the microwave-boosted discharge are 3–5 times lower than in the case of a conventional glow discharge lamp operated under optimized conditions (Table 4-5). The linearity of the calibration curves extends over three decades of concentration with no curvature observed at the high-concentration side (Fig. 4-28). It is interesting to remark that the self-reversal, which can be observed for resonance matrix lines [see Fourier transform spectra for Cu(I) 327.4 nm], is no longer observed in the case of the microwave-boosted lamp.(77) The glow discharge lamp with integrated microwave cavity thus is a suitable tool for the analysis of electrically conducting solid samples. For many elements in steel, this constitutes progress in the power of detection down to the submicrogram per gram level. Further measurements showed that this also applies to aluminum and copper

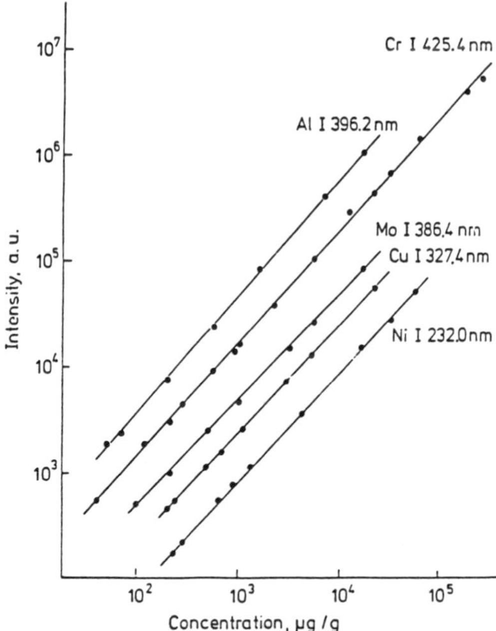

Figure 4-28. Calibration curves for five elements in steel samples measured with the microwave-supported glow discharge lamp.(44) Each sample was measured only once with an integration time of 3 s.

matrices.[98] The high radiance, which is achieved even at a reduced erosion rate, also is favorable for in-depth profiling, although under the conditions used, the crater is not as flat as with a conventional lamp. In addition, the low heating of the sample is a suitable basis for the analysis of metals and alloys of low melting point such as lead.

Another approach is furnace nonthermal emission spectrometry, developed by Falk *et al.*[10] and discussed in a separate chapter of this book. Here, progress lies in the complete separation of sample volatilization in a furnace and excitation in the negative glow of a dc discharge. Liang and Blades[99] follow a similar approach using capacitive powering of the plasma. Further lines of improvement in the power of detection of glow discharge emission sources may be found in side-on observation of the glow discharge plasma. This can be done easily with the aid of quartz fibers. Indeed, preliminary measurements performed with a glow discharge operated at 25 W/cm^2, 350 Pa for the Fe 371.9-nm line revealed that the signal-to-background ratio for side-on measurements at 2 mm above the cathode was 12 compared with 7.5 for end-on measurements. Of course, optimization of the working conditions here may bring still further progress. Time-resolved measurements in pulsed glow discharges may yield progress in the power of detection. Indeed, especially due to the long lifetime of the exciting species (especially of the argon metastables) the excitation capacity may remain high while a decrease in background continuum may take place.

4.5.2. Improvements in the Use of the Spectral Information

The radiant output of the Grimm-type glow discharge lamp with a flat cathode is very stable. Moreover, the optical emission spectra consist primarily of neutral atomic lines, of which the resonance transitions dominate. Accordingly, emission spectra of the glow discharge source ordinarily contain fewer emission lines than do spectra of such sources as the inductively coupled plasma (ICP). Furthermore, the relatively high analyte-vapor concentration in the source (up to 1% relative to the argon density) caused by the high ablation rate from the cathode surface (on the order of a few milligrams per minute for a Cu matrix) leads to favorable analyte-to-argon line emission intensity ratios.

These strengths make the glow discharge source particularly attractive as a potential source for Fourier transform spectrometry (FTS). In previous studies of FTS-AES, where an ICP was used as a source of atomic emission, a loss in the power of detection and a substantial dependence of sensitivity on the type and concentration of the matrix were observed.[100] In contrast, the stability of the GDS and the line-poor character of its spectra should yield signal-to-noise ratios in FTS–AES that are comparable to those realized in wavelength-dispersive spectrometers. Finally, the high optical conductance of the FTS and the advantage of working with physically resolved linewidths (readily available with FTS) guarantee the highest degree of sensitivity[101] and the lowest level of systematic errors. Initial experiments have been performed at the Los Alamos Fourier Transform Spectrometer Facility[102,103] with a Grimm-type glow discharge lamp using a floating restrictor. The glow discharge plasma was imaged 1 : 1 on the 8-mm entrance aperture of the FTS instrument. Emission spectra over the spectral range 300–600 nm were calculated from interferograms recorded during a mirror displacement of approximately 12 cm. The sampling rate of the interferometer was 5000 Hz and approximately 7 min was required for each scan. The signal was filtered by low- and high-pass filters before digitalization. The FTS operating conditions provided a spectral resolution of 0.043 cm^{-1} (at 300 nm, the spectral bandwidth was 0.39 pm).

The high resolution of the FTS provides a means by which the widths (FWHM) of matrix and minor elements lines can be measured directly and simultaneously. Moreover, the dynamic range of the FTS is sufficient to measure at the same time matrix lines, argon lines, and trace element lines. The spectral widths of a number of analytical lines under different GDS operating conditions are listed in Table 4-4. At discharge currents of 15–70 mA and argon pressures of 430–600 Pa, the observed linewidths were on the order of 1–2 pm, in good agreement with results reported by West and Human,[33] and the measurement error is about 0.5 pm. Understandably, the linewidths are lower than those obtained with sources operated at atmospheric pressure.[101,104,105] A tendency toward increasing spectral linewidths can be observed as the glow discharge operating power is increased, especially for the Fe(I) 371.9-nm and the Cr(I) 425.4-nm lines. The GDS power can be increased either by lowering the pressure while maintaining the current constant or by raising the current while maintaining the pressure constant; both seem to produce equivalent results and both are known to increase the sample ablation rates.[32] The observed increase in linewidth with operating power may be due to self-absorption.

With spectrometers of sufficiently high resolution, matrix and spectral interferences are ordinarily not significant in glow discharge spectrometry. Consequently, it is possible to estimate detection limits by extrapolating an analyte atomic line intensity (I) to a level equal to three times the standard deviation of the spectral background adjacent to the line. Because of the flat noise distribution in FTS, a valid estimate of the noise can be obtained as the root mean square (N_{rms}) noise in a spectral region close to, but distinct from, the spectral line of interest. Following this procedure (see Fig. 4-29), a detection limit of 30 μg/g is obtained for the FTS determination of molybdenum [Mo(I) 379.8 nm] in low-alloy steel ($I_{Mo} = 7.74 \times 10^6$, $I_{avg} = 2.00 \times 10^6$, $N_{rms} = 1.00 \times 10^6$, and [Mo] = 50 μg/g).

It should, however, be emphasized that the value stated for Mo is only an estimate as no serious attempts were made to optimize the Mo signal. Nevertheless, it suggests that the sensitivity of GDS–FTS approaches that

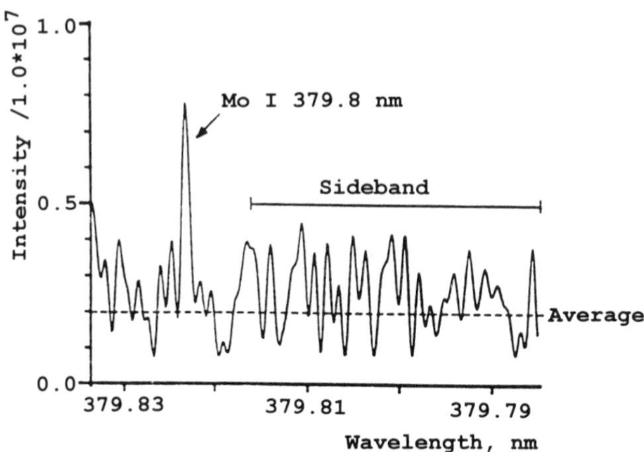

Figure 4-29. High-resolution record of spectral structure in the vicinity of Mo(I) 379.8 nm.[30] Sample: steel (216A) (c_{Mo}: 50 μg/g); i: 50 mA; argon pressure 470 Pa; burning voltage: 1500 V; high-resolution FTS (0.043 cm^{-1}). Detection limit for Mo: 30 μg/g.

obtainable from a GDS in combination with a high-resolution wavelength-dispersive monochromator.[44] The latter might be limited by dark current rather than by line-to-background intensity ratios. This behavior is verified by the spectral background signals obtained with a small (0.35 m, slit width: 15 μm, spectral bandpass: 0.03 nm) monochromator (at 200 nm: 2.8×10^{-9} A; at 300 nm: 5×10^{-9} A; at 400 nm: 4×10^{-9} A), which are hardly above the dark-current level of the photomultiplier (2×10^{-9} A). In addition, these results indicate that the multiplex disadvantage is not a serious limitation for FT-GDS-AES. Moreover, because the entire spectrum is available at high resolution, multiline calibration could further enhance the precision and the reliability of the analytical data. Figure 4-30 shows the molybdenum/iron pair observed in the analysis of low-alloy steel and documents the analytical performance of the technique.

The use of new radiation detectors is of interest for improving the analytical performance of GD-OES. The use of silicon photodiode arrays as detectors in GD-OES was described by Bubert and Hagenah.[106] They recognized the strengths of the available arrays for work in the visible and IR region and determined the alkaline elements as well as Ba, La, and Tl having their most sensitive lines in this region. An array with 20 individual diodes with the corresponding electronics enabled the simultaneous measurement of the line and background intensities and thus allowed rapid sequential analyses. In the spectral region investigated, both detector noise and fluctuations of the radiation source were found to limit the power of detection. With the modern arrays which have up to 1024 diodes, the spectral range

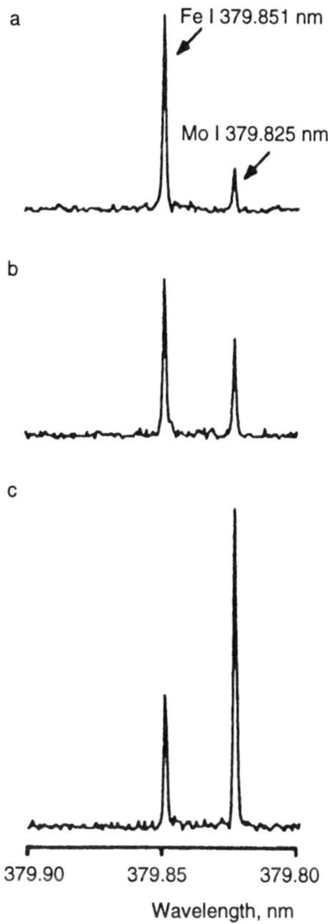

Figure 4-30. Iron and molybdenum peaks from steel samples (Research Institute CKD, Czech Republic) obtained with FTS.[30] i: 50 mA; argon pressure: 470 Pa; burning voltage: 1340 V; (a) sample 218A (c_{Mo} : 1.3 mg/g); (b) sample 224A (c_{Mo} : 4.1 mg/g); (c) sample 221A (c_{Mo} : 14.8 mg/g).

covered and the number of pixels available for simultaneous data acquisition are greatly enhanced. These types of detectors have especially regained interest since spectrally segmented spectrometers have become commercially available (PRA Inc.).[107] Here a primary dispersion element with the aid of an echelle grating is applied and with an exchangeable mask the spectral regions around the analytical lines of interest are isolated and recombined. With a second grating the radiation is spectrally resolved again with the lines and their spectral environment imaged on the photodiode array. In this way, up to ten analytical lines can be simultaneously detected together with their spectral background intensities. With the array used, detector noise limitations may still occur, as shown by Hieftje and Brushwyler[109] for the ICP and found in initial experiments with the GDS (Fig. 4-15).[26] Here, progress may be expected both from the detectors combining a photodiode array and a microchannel plate as well as from CCD and CID devices.[110]

4.5.3. Glow Discharges as Atom Reservoirs

Glow discharges with flat cathodes not only are suitable sources for atomic emission spectrometry but also perform well as atom reservoirs. Their use in atomic fluorescence dates back to combinations with a glow discharge as primary source, as in the resonance fluorescence detector studied by Butler *et al.*,[111] Human *et al.*,[112] and Bubert.[113] The full power of detection of this approach, however, is realized only when laser-induced fluorescence spectrometry is employed. For lead, detection limits of 20 pg have been found by Smith *et al.*[114] when dispensing 5-μl aliquots on a flat copper cathode. In hollow cathodes, the feasibility of performing single atom detection has been described.[115] When soft discharge conditions are used, it is even possible to detect diatomic molecules.[116] The use of glow discharges as atom reservoirs for laser-enhanced ionization (LEI)[117] and resonance ionization mass spectrometry (RIMS)[118] has been described as well.

Special interest has been generated for the use of glow discharges as sputtering cells enabling direct atomic absorption spectrometric analysis of metals and alloys (see Chapter 3). Gough[89] described the use of jet-assisted sputtering to produce high analyte number densities in a glow discharge plume and determined minor and trace elements in steel [0.08% (w/w) Cr]. Detection limits were shown to be at the 0.0001 to 0.001% (w/w) level. Larkins[119] improved the gas-control unit and reduced the moisture introduction in the cell by which both the analytical precision and the power of detection could be further improved. McDonald[120] further discussed the use and selection of an internal standard to increase the analytical precision. He also studied the diffusion of analyte atoms in a glow discharge source by atomic absorption measurements.[121] The technique recently became of interest when a commercial system was introduced by the Analyte Corporation. Kim and Piepmeier[122] studied the influence of the gas-jet parameters on the material ablation and used electron microscopy to investigate the analyte removal mechanism. Sputtering studies of metal[123] and also of electrically nonconducting powders[124] for atomic absorption purposes will promote this technique further, which Ohls *et al.*[125] have recently shown to be of use for multielement analyses in steel samples, including refractory elements such as Zr. Banks and Blades[126] and Broekaert *et al.*[130] also studied the possibilities of the source for optical emission spectrometry. It was found that despite an enhancement of the sputtering rates, the intensities of Cu(I) resonant lines are not enhanced in contrast to those of Cu(II) and Zn(I) lines, which may relate to self-reversal.

The experience gained with glow discharges using flat cathodes as sources for optical atomic spectrometry has also been very valuable regarding progress in their use as ion sources for mass spectrometry (see Chapter 5).

4.6. Comparison with Other Methods for the Direct Analysis of Solids

For *bulk analysis*, glow discharges have their strength in the material volatilization mechanism of cathodic sputtering in which the volatilization interferences are low compared with classical spectrochemical sources such as arcs and sparks. In comparison with x-ray fluorescence methods, GD-AES has the advantage of low matrix effects and in addition allows trace determinations even for the light elements down to the microgram per gram range. Here, improvement through the use of cross-excitation and side-on observation still might shift the limits by one order of magnitude. GD-AES also enables high-precision work from the trace element to the major element

Table 4-8. Comparison of Glow Discharge Atomic Emission Spectrometry with Other Methods for the Direct Analysis of Solid Samples

	Bulk analysis			
Method	Power of detection	Precision	Matrix effects	Analysis throughput
dc arc AES	+ + +[a]	+	+	+ +
Spark AES	+	+ +	+	+ + +
Spark ablation ICP-AES	+ +	+ +	+ + +	+ +
Glow discharge AES	+	+ +	+ + +	+
Glow discharge mass spectrometry	+ + +	+ +	+ + +	+
Spark mass spectrometry	+ + +	+	+	+
Laser AES	+ +	+ +	+	+ +
X-ray fluorescence	+	+ + +	+	+ + +
	In-depth profiling			
Method	Power of detection	Precision	Resolution	Analysis throughput
Secondary ion mass spectrometry	+ + +	+ + +	+ + +	+
Auger electron spectrometry	+	+ +	+ + +	+
Sputtered neutrals mass spectrometry	+ + +	+ + +	+ + +	+
Glow discharge AES	+ +	+ +	+	+ + +
Glow discharge mass spectrometry	+ + +	+ + +	+ + +	+ +

[a] +, poor; + +, good; + + +, very good.

level. The use of internal standardization allows it to approach the performance of x-ray fluorescence spectrometry, with the improvement of the spectral information acquisition by multiplex detectors with parallel input allowing further improvement. Glow discharges, however, are more complex to operate than the above-mentioned sources. This stems from their operation at reduced pressure, which complicates sample preparation and interchange, as well as the instrumentation, and decreases the speed of analysis. The latter has until now hampered the use of GD-AES for product control in metallurgy. High-energy preburning might lower the analysis times, but only to a small extent. The strengths of glow discharges for bulk analyses therefore lie in high-precision product control for a wide variety of samples. Particular fields of application here are copper analysis, which at the trace element level in some cases may present difficulties with modern spark emission spectrometry, precious metals analysis, and the forthcoming interest in powder analysis.

With respect to *microdistributional analysis*, GD-AES offers possibilities for in-depth profiling, but not for laterally resolved measurements. The latter is the field of surface scanning techniques such as Auger electron spectroscopy, laser ablation in combination with different types of spectrometry [e.g., mass spectrometry (LAMMS)], proton-induced x-ray emission (PIXE), electron probe microanalysis (EPMA), secondary ion mass spectrometry (SIMS), and scanning Raman spectrometry, which all differ in principle, lateral resolution, and sample requirements.[127] GD-AES, however, enables in-depth profiling with a depth resolution of 1–10 nm, a good power of detection (down to the trace element level), and a fair precision (RSD below 10%). Quantization with respect to both the in-depth scale as well as the determination of in-depth concentration profiles is making considerable progress. It can be used for a control of thicknesses of coatings as required by the metallurgical industries producing technical surfaces. It is quicker and easier to handle than SIMS and has found its way to the routine industrial laboratories. Here, alternative techniques such as laser sources enabling reproducible in-depth profile recording also exist. The latter are also of use for analyses of compact ceramics, by direct emission spectrometry[13] as well as the more sensitive laser-induced fluorescence technique.[14] Important progress in this area of work is possible with glow discharges operating as rf plasmas, enabling a reproducible and sufficiently intensive ablation of electrically isolating solids.[91,92]

4.7. Conclusions

Optical emission spectrometry with glow discharges is now an established method for the direct analysis of solid samples of widely different origin and composition. It will further develop together with the use of glow discharges as atom reservoirs for atomic absorption or atomic fluorescence spectroscopy and as ion sources for mass spectroscopy.

In particular, the improvements in sample ablation and excitation as well as in the use of the spectral information are likely to give rise to a wide diversity of instrumentation for practical spectrochemical analysis and to contribute to the solution of analytical tasks in the most widely different fields of work.

Acknowledgments

This study has been supported by the Bundesministerium für Forschung und Technologie, Bonn and by the Ministerium für Wissenschaft und Forschung des Landes Nordrhein-Westfalen, Düsseldorf.

References

1. P. W. J. M. Boumans, *Theory of Spectrochemical Excitation*, Hilger & Watts, London, 1966.
2. J. D. Ingle, Jr. and S. R. Crouch, *Spectrochemical Analysis*, Prentice–Hall, Englewood Cliffs, N.J., 1988.
3. J. A. C. Broekaert, *Bull. Soc. Chim. Belg.* 86 (1977) 895.
4. J. Junkes and E. W. Salpeter, *Ric. Spettrosc.* 2 (1961) 255.
5. J. A. C. Broekaert, K. R. Brushwyler, and G. Hieftje, unpublished work.
6. J. A. C. Broekaert, *Anal. Chim. Acta* 196 (1987) 1.
7. R. Avni, in: *Applied Atomic Spectroscopy* (E. L. Grove, ed.), Vol. 1, Chap. 4, Plenum Press, New York, 1978.
8. K. Slickers, *Automatic Emission Spectroscopy*, Brühl Druck & Pressehaus, Giessen, 1977.
9. D. M. Coleman, M. A. Sainz, and H. T. Butler, *Anal. Chem.* 52 (1980) 746.
10. H. Falk, E. Hoffmann, and C. Lüdke, *Spectrochim. Acta* 39B (1984) 283.
11. H. Falk, E. Hoffmann, C. Lüdke, J. M. Ottaway, and S. K. Giri, *Analyst* (*London*) 108 (1983) 1459.
12. L. Moenke-Blankenburg, *Laser Microanalysis*, Wiley, New York, 1989.
13. F. Leis, W. Sdorra, J. B. Ko, and K. Niemax, *Mikrochim. Acta* II (1989) 185.
14. W. Sdorra, A. Quentmeier, and K. Niemax, *Mikrochim. Acta* II (1989) 201.
15. H. Schüler and H. Gollnow, *Z. Phys.* 93 (1935) 611.
16. S. L. Mandelstam and V. V. Nedler, *Spectrochim. Acta* 17 (1961) 885.
17. K. Thornton, *Analyst* (*London*) 94 (1969) 958.
18. B. Thelin, *Appl. Spectrosc.* 35 (1981) 302.
19. W. Grimm, *Spectrochim. Acta* 23B (1968) 443.
20. P. J. Treado and M. D. Morris, *Anal. Chem.* 61 (1989) 723A.
21. L. M. Faires, *Anal. Chem.* 58 (1986) 1023A.
22. Y. Talmi, *Multichannel Image Detectors*, American Chemical Society Symposium Series 102 (1979).
23. J. V. Sweedler, R. F. Jalkian, and M. B. Denton, *Appl. Spectrosc.* 43 (1989) 953.

24. G. Heltai, J. A. C. Broekaert, F. Leis, and G. Tölg, *Spectrochim. Acta* 45B (1990) 301.
25. N. Furuta, K. R. Brushwyler, and G. M. Hieftje, *Spectrochim. Acta* 44B (1989) 349.
26. J. A. C. Broekaert, K. R. Brushwyler, and G. M. Hieftje, unpublished work.
27. A. Danielson, P. Lindblom, and E. Södermann, *Chem. Scr.* 6 (1974) 5.
28. R. B. Bilhorn, P. M. Epperson, J. V. Sweedler, and M. B. Denton, *Appl. Spectrosc.* 41 (1987) 1114.
29. M. Dogan, K. Laqua, and H. Maßmann, *Spectrochim. Acta* 26 (1971) 631.
30. J. A. C. Broekaert, C. A. Monnig, K. R. Brushwyler, and G. M. Hieftje, *Spectrochim. Acta* 45B (1990) 769.
31. M. Kaminsky, *Atomic and Ionic Impact Phenomena on Metal Surfaces*, Springer, Berlin, 1965.
32. P. W. J. M. Boumans, *Anal. Chem.* 44 (1972) 1219.
33. C. D. West and H. G. C. Human, *Spectrochim. Acta* 31B (1976) 81.
34. N. P. Ferreira, J. A. Strauss, and H. G. C. Human, *Spectrochim. Acta* 37B (1982) 273.
35. N. P. Ferreira and H. G. C. Human, *Spectrochim. Acta* 36B (1981) 215.
36. E. B. M. Steers and F. Leis, *J. Anal. At. Spectrom.* 4 (1989) 199.
37. A. Aziz, J. A. C. Broekaert, and F. Leis, *Spectrochim. Acta* 36B (1981) 251.
38. E. B. M. Steers and R. J. Fielding, *J. Anal. At. Spectrom.* 2 (1987) 239.
39. M. Dogan, K. Laqua, and H. Maßmann, *Spectrochim. Acta* 27B (1972) 65.
40. R. A. Kruger, R. M. Bombelka, and K. Laqua, *Spectrochim. Acta* 35B (1980) 589.
41. R. A. Kruger, R. M. Bombelka, and K. Laqua, *Spectrochim. Acta* 35B (1980) 581.
42. R. M. Lowe, *Spectrochim. Acta* 31B (1976) 257.
43. N. P. Ferreira, J. A. Strauss, and H. G. C. Human, *Spectrochim. Acta* 38B (1983) 899.
44. F. Leis, J. A. C. Broekaert, and K. Laqua, *Spectrochim. Acta* 42B (1987) 1169.
45. H. W. Rademacher and M. C. de Swardt, *Spectrochim. Acta* 30B (1975) 353.
46. S. El Alfy, K. Laqua, and H. Maßmann, *Fresenius Z. Anal. Chem.* 263 (1973) 1.
47. F. Mai and H. Scholze, *Spectrochim. Acta* 41B (1986) 797.
48. J. B. Ko and K. Laqua, *18th Colloq. Spectrosc. Int.*, Grenoble, Conférences et Tables Rondes, Vol. II, pp. 543–548, 1975.
49. S. El Alfy, Ph.D. dissertation, University of Dortmund, 1978.
50. J. A. C. Broekaert, R. Klockenkämper, and J. B. Ko, *Fresenius Z. Anal. Chem.* 316 (1983) 256.
51. W. B. Barnett, V. A. Fassel, and R. N. Kniseley, *Spectrochim. Acta* 23B (1968) 643.
52. H. Jäger, *Anal. Chim. Acta* 60 (1972) 303.
53. K. Wagatsuma and K. Hirokawa, *Anal. Chem.* 56 (1984) 2024.
54. K. Wagatsuma and K. Hirokawa, *Anal. Chem.* 58 (1986) 1112.
55. K. Wagatsuma and K. Hirokawa, *Anal. Chem.* 60 (1988) 702.
56. J. Durr and B. Vandorpe, *Spectrochim. Acta* 36B (1981) 139.
57. J. B. Ko, *Spectrochim. Acta* 39B (1984) 1405.
58. K. Wagatsuma and K. Hirokawa, *Anal. Chem.* 56 (1984) 908.
59. K. Wagatsuma and K. Hirokawa, *Bunko Kenkyu* 33 (1984) 320.
60. M. Fujita, J. Kashima, and K. Naganuma, *Anal. Chim. Acta* 124 (1981) 267.
61. H. Jäger, *Anal. Chim. Acta* 58 (1972) 57.
62. M. Dogan, *Spectrochim. Acta* 36B (1981) 103.
63. R. A. Kruger, L. R. P. Butler, C. J. Liebenberg, and R. G. Böhmer, *Analyst* (*London*) 102 (1977) 949.
64. T. Yamada, J. Kashima, and K. Naganuma, *Anal. Chim. Acta* 124 (1981) 275.
65. C. J. Belle and J. D. Johnson, *Appl. Spectrosc.* 27 (1973) 118.
66. R. Berneron, *Spectrochim. Acta* 33B (1978) 665.
67. A. Bengtson, *Spectrochim. Acta* 40B (1985) 631.
68. R. Berneron and J. C. Charbonnier, *Anal. Proc.* 17 (1980) 488.

69. A. Quentmeier and K. Laqua, in: *13. Spektrometertagung, Düsseldorf* (K.-H. Koch and H. Massmann, eds.), pp. 37–49, de Gruyter, Berlin, 1981.
70. M. E. Waitlevertch and J. K. Hurwitz, *Appl. Spectrosc.* 30 (1976) 510.
71. P. De Gregorio and G. Savastano, *Ann. Ist. Super. Sanita* 19 (1983) 613.
72. K. Takimoto, K. Suzuki, K. Nishizaka, and T. Ohtsubo, *Quantitative Analysis of Zinc–Iron Alloy Galvanized Coatings by Glow Discharge Spectrometry and Secondary Ion Mass Spectrometry*, Nippon Steel Technical Report No. 33 (1987).
73. K.-H. Koch, M. Kretschmer, and D. Gruenenberg, *Arch. Eisenhüttenwes.* 54 (1983) 395.
74. K.-H. Koch, D. Sommer, and D. Grünenberg, *Mikrochim. Acta* II (1985) 1.
75. K. Wagatsuma and K. Hirokawa, *Anal. Chem.* 56 (1984) 412.
76. Y. Ohashi, Y. Yamamoto, K. Tsunoyama, and H. Kishidaka, *Surf. Interface Sci.* 1 (1979) 53.
77. F. Leis, J. A. C. Broekaert, and E. B. M. Steers, 1990 Winter Conference on Plasma Spectrochemistry, St. Petersburg, Fla., Paper WP18.
78. P. A. Büger and W. Fink, *Fresenius Z. Anal. Chem.* 244 (1969) 314.
79. W. W. Harrison and N. J. Prakash, *Anal. Chim. Acta* 49 (1970) 151.
80. J. Mierrzwa and W. Zyrnick, *Anal. Lett.* 21 (1988) 115.
81. J. A. C. Broekaert, *Spectrochim. Acta* 35B (1980) 225.
82. J. Y. Ryu, R. L. Davies, J. C. Williams, and J. C. Williams, Jr., *Appl. Spectrosc.* 42 (1988) 1379.
83. J. M. Brackett and T. J. Vickers, *Spectrochim. Acta* 37B (1982) 841.
84. N. Jakubowski, D. Stüwer, and G. Toelg, *Spectrochim. Acta* 46B (1992) 155.
85. H. Jäger and F. Blum, *Spectrochim. Acta* 29b (1974) 73.
86. I. Brenner, K. Laqua, and K. Dvorachek, *J. Anal. Atm. Spectrom.* 2 (1987) 623.
87. J. A. C. Broekaert, T. Graule, H. Jenett, G. Tölg, and P. Tschöpel, *Fresenius Z. Anal. Chem.* 332 (1989) 825.
88. G. S. Lomdahl, R. McPherson, and J. V. Sullivan, *Anal. Chim. Acta* 148 (1983) 171.
89. D. S. Gough, *Anal. Chem.* 48 (1976) 1926.
90. G. S. Lomdahl and J. V. Sullivan, *Spectrochim. Acta* 39B (1984) 1395.
91. D. C. Duckworth and R. K. Marcus, *Anal. Chem.* 61 (1989) 1879.
92. M. R. Winchester and R. K. Marcus, *J. Anal. At. Spectrom.* 5 (1990) 575.
93. D. C. McDonald, *Spectrochim. Acta* 37B (1982) 747.
94. D. S. Gough and J. V. Sullivan, *Analyst (London)* 103 (1978) 887.
95. P. E. Walters and H. G. C. Human, *Spectrochim. Acta* 36B (1981) 585.
96. S. Caroli, A. Alimonti, and F. Petrucci, *Anal. Chim. Acta* 136 (1982) 269.
97. S. Caroli, O. Senofonte, N. Violante, and L. Di Simone, *Appl. Spectrosc.* 41 (1987) 579.
98. F. Leis, J. A. C. Broekaert, and E. B. M. Steers, *Spectrochim. Acta* 46B (1991) 243.
99. D. C. Liang and M. W. Blades, *Spectrochim. Acta* 44B (1989) 1049.
100. L. M. Faires, *Spectrochim. Acta* 40B (1985) 1473.
101. K. Laqua, W.-D. Hagenah, and H. Waechter, *Fresenius Z. Anal. Chem.* 225 (1967) 142.
102. B. A. Palmer, Los Alamos Technical Bulletin, Los Alamos National Laboratory, Los Alamos, N.M., August 1985.
103. M. L. Parsons and B. A. Palmer, *Spectrochim. Acta* 43B (1988) 75.
104. J. A. C. Broekaert, F. Leis, and K. Laqua, in: *Developments in Atomic Spectrochemical Analysis* (R. M. Barnes, ed.), pp. 84–93, Heyden, London, 1981.
105. P. W. J. M. Boumans and J. J. A. M. Vrakking, *Spectrochim. Acta* 41B (1986) 1235.
106. H. Bubert and W.-D. Hagenah, *Spectrochim. Acta* 36B (1981) 489.
107. G. M. Levy, A. Quaglia, R. E. Lazure, and S. W. McGeorge, *Spectrochim. Acta* 42B (1987) 341.
108. J. A. C. Broekaert, *J. Anal. At. Spectrom.* 2 (1987) 537.
109. G. M. Hieftje and K. R. Brushwyler, *Appl. Spectrosc.* 45 (1991) 582.

110. G. R. Sims and M. B. Denton, *Opt. Eng.* 26 (1987) 1009.
111. L. R. P. Butler, K. Kroger, and C. D. West, *Spectrochim. Acta* 30B (1975) 489.
112. H. G. C. Human, N. P. Ferreira, R. A. Kruger, and L. R. P. Butler, *Analyst* (*London*) 103 (1978) 469.
113. H. Bubert, *Spectrochim. Acta* 39B (1984) 1377.
114. B. W. Smith, N. Omenetto, and J. D. Winefordner, *Spectrochim. Acta* 39B (1984) 1389.
115. B. W. Smith, J. B. Womack, N. Omenetto, and J. B. Winefordner, *Appl. Spectrosc.* 43 (1989) 873.
116. B. M. Patel and J. D. Winefordner, *Appl. Spectrosc.* 40 (1986) 667.
117. T. Masaki, Y. Adachi, and C. Hirose, *Appl. Spectrosc.* 42 (1988) 51.
118. P. J. Savickas, K. R. Hess, R. K. Marcus, and W. W. Harrison, *Anal. Chem.* 56 (1984) 819.
119. P. L. Larkins, *Anal. Chim. Acta* 132 (1981) 119.
120. D. C. McDonald, *Anal. Chem.* 54 (1982) 1052.
121. D. C. McDonald, *Anal. Chem.* 54 (1982) 1057.
122. H. J. Kim and E. H. Piepmeier, *Anal. Chem.* 60 (1988) 2040.
123. S. L. Tong and W. W. Harrison, *Anal. Chem.* 56 (1984) 2028.
124. M. R. Winchester and R. K. Marcus, *Appl. Spectrosc.* 42 (1988) 941.
125. K. Ohls, J. Flock, and H. Loepp, *Fresenius Z. Anal. Chem.* 332 (1989) 456.
126. P. R. Banks and M. W. Blades, *Spectrochim. Acta* 44B (1989) 1117.
127. W. A. Van Borm and F. C. Adams, *Anal. Chim. Acta* 218 (1989) 185.
128. J. B. Ko and K. Laqua, in: *13. Spektrometertagung Düsseldorf* (K.-H. Koch and H. Maßmann, eds.), pp. 51–68, de Gruyter, Berlin, 1981.
129. K. Laqua, in: *Ullmanns Enzyklopädie der technischen Chemie*, 4th ed., vol. 5, p. 460, Verlag Chemie, Weinheim, 1980.
130. J. A. C. Broekaert, T. Bricker, K. R. Brushwyler, and G. M. Hieftje, *Spectrochim. Acta* 47B (1992) 131.

5

Glow Discharge Mass Spectrometry

F. L. King and W. W. Harrison

5.1. Introduction

Glow discharge mass spectrometry (GDMS) is associated with two closely related techniques for the characterization of ion populations in glow discharge plasmas. The materials scientist is familiar with GDMS as a tool for plasma processing diagnostics, whereas the analytical chemist is familiar with GDMS as a method for direct solids analysis. The present chapter will focus on GDMS as a maturing technique in elemental analysis. The reader interested in GDMS as a plasma diagnostic tool is referred to reviews of the topic by Aita[1] and Coburn.[2]

The relative simplicity of mass spectra compared with optical spectra makes mass spectrometry an attractive alternative to optical spectrometry for trace element analysis. By providing isotopic abundance information, mass spectrometry permits the use of isotope dilution techniques and isotope tracer studies. The ability to generate a stable analyte ion population directly from a solid sample, thereby precluding the problems of dissolution, dilution, and contamination, attendant to techniques requiring solution samples, makes the glow discharge an attractive ion source for elemental mass spec-

F. L. King • Department of Chemistry, West Virginia University, Morgantown, West Virginia 26506. *W. W. Harrison* • Department of Chemistry, University of Florida, Gainesville, Florida 32611.

Glow Discharge Spectroscopies, edited by R. Kenneth Marcus. Plenum Press, New York, 1993.

trometry. Coupling a glow discharge atomization/ionization source with a mass spectrometer yields a state-of-the-art technique for the direct elemental analysis of solid-state materials, GDMS.

The division of the chemical analysis process into three distinct steps, representation, speciation, and quantitation,[3] permits a parallel division of the functions performed in GDMS, i.e., atomization/ionization, mass analysis, and data acquisition/interpretation. In this chapter, we seek to develop both an understanding of the fundamental principles underlying these functions and an appreciation of the broad analytical potential presented by GDMS. The discussion focuses on general principles rather than any of the various turnkey and add-on systems available from commercial vendors.[4] Initially, the principles underlying the operation of various mass analyzers will be reviewed. The merits of each type of mass analyzer for use in elemental mass spectrometry will be evaluated. Atomization and ionization processes in glow discharges will be discussed in the context of their impact on quantitative interpretation of GDMS data. The reader interested in a complete description of the atomization and ionization processes is referred to other chapters of this text where these processes are described in detail. The configurations and attributes of the various glow discharge ion source configurations will be described. A sampler of the analytical results obtained with GDMS will conclude this chapter.

5.2. Mass Spectrometry

Traditionally, elemental analysis has been dominated by optical spectrometric techniques, such as atomic emission spectrometry and atomic absorption spectrometry, that probe the transitions between energy levels characteristic of an analyte element. Mass spectrometry originated as a tool for the characterization of ions present in gas discharge plasmas. The desire to extract further information concerning the nature of matter from "positive rays" emitted by electric discharges led Thompson, Aston, and Bainbridge to develop a new type of spectrograph.[5] These mass spectrographs separated the components of these "positive rays" and provided direct information regarding the distribution of isotopic constituents in these discharge plasmas. Unlike the complex optical spectra examined previously, these new mass spectra were interpreted readily because each elemental isotope yielded only a single line at a characteristic mass-to-charge ratio (m/z).[5] Characterization of an atom on the basis of its mass-to-charge ratio leads to an immediate simplification; because there are only a few different isotopes of any given element, and in general only the singly ionized species will be detectable under the conditions employed, the resulting mass spectrum will be less complex and isobaric interferences will be reduced. It is this spectral simplicity coupled with the sensitivity and dynamic range of modern ion detection techniques that makes mass spectrometry a powerful technique for elemental analysis.

5.2.1. Fundamentals

Although mass spectrometry is fundamentally different from other analytical spectrometric techniques in that the absorption, or emission, of radiation is not necessarily the physical phenomenon by which the analytical species are distinguished,* there are certain analogies from which much of the terminology of mass spectrometry is derived. We begin with a source of ions, rather than photons, that are characteristic of the analyte. The composition of the ion beam emitted by the source should be a steady and reproducible reflection of the original sample composition. Ion optics serve to collimate and condition the ion beam optimizing the transmission of the beam into the mass analyzer. Mass analyzers permit the temporal or spatial isolation of different m/z ions in the ion beam much the same as a wavelength-selective device separates photons of different energy in a beam of light. The beam emerging from the mass analyzer is then of a single m/z with an intensity that is characteristic of the isotopic composition of the original sample. The beam is detected either by direct measurement of the ion current or by conversion to electrons or photons that are amplified within the detector prior to current measurement. A plot of the signal intensity as a function of transmitted m/z value is the mass spectrum. Mass spectral resolution, $(m/z)/(\Delta m/z)$, is a measure of the ability to separate ions having small m/z differences just as optical spectral resolution is a measure of the ability to separate photons having small energy differences. The signal recorded in mass spectrometry is a direct measure of the ion flux at a selected m/z; as in optical spectrometry it is a measure of the photon flux at a selected energy.

5.2.2. Mass Analysis

Mass analyzers can be classed loosely into three general categories: continuous dispersion instruments, sequential filtration instruments, or ion trapping instruments. At the present time, sector or quadrupole-based instruments are the only commercially available GDMS instruments although the utility of other mass analyzers is under investigation. A general comparison of mass analyzers for GDMS appears in Table 5-1.

*Photoselective ionization techniques, such as RIMS, actually employ classical spectroscopy in concert with mass spectrometry to effect a two-dimensional separation of species.

Table 5-1. Comparison of Mass Analyzers for Elemental MS

Mass Analyzer	Advantages	Disadvantages
Double focussing sector	>Resolution adequate to discriminate against some isobaric interferences >Commercial systems available	>Complexity >Cost >Scan speed
Time-of-flight	>Mass analyzer simplicity >Speed >Good resolution with use of reflection	>Complicated ion extraction >No results yet
Quadrupole mass filter	>Scan speed >Peak hopping >Commercial systems available	> Limited resolution
Quadrupole ion trap	>*In situ* CID to remove polyatomic isobaric interferences	>Problems attendant to external ion generation
FT-ICR	>High resolution >CID possible	>Cost >Complexity >Space charge limited dynamic range >Problems attendant to external ion generation

5.2.2.1. Dispersion in Space (Sector MS)

The prototype dispersion instrument was Aston's original mass spectrograph, which employed an electromagnetic field to disperse different m/z ions spatially.[(5)] The descendants of this instrument are the sector mass analyzers employed today. Sector instruments employ electrostatic and magnetic fields to effect ion separation. The mathematics essential to the explanation of ion trajectories in such fields is relatively straightforward. A magnetic field acts as a prism, dispersing monoenergetic ions of differing m/z values across a focal plane (Fig. 5-1).[(6)] The radius of the curved flight path of an ion through a magnetic field is given by

$$r_{\rm m} = (144/B)(mV/z)^{1/2}$$

where B is the field strength in gauss, m is the atomic mass of the ion, V is the voltage with which the ion is accelerated into the magnetic field, and z is the charge on the ion.[(6)] Such instruments can provide a simultaneous mass spectrogram of all the species of interest or can scan a mass spectral region of interest.

Two modes of detection can be employed with sector instruments: (1) simultaneous detection of a range of m/z ions by a photographic plate or array detector located in the focal plane or (2) sequential detection by an

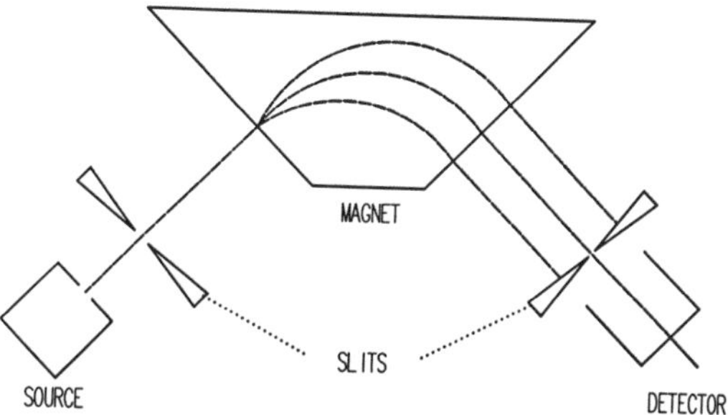

Figure 5-1. Magnetic sector mass analyzer.

electronic detector on which different m/z ions are brought into focus by scanning B or V. Resolution is determined by ion beam focusing in the detection plane. Two types of focusing are possible with sector instruments: direction and velocity focusing. In single-focusing instruments, increased resolution is gained by narrowing the entrance and exit slits.(6) Unfortunately, such instruments depend on the availability of an ion beam having a very narrow ion energy spread to prevent the occurrence of aberrations arising from a lack of energy focusing. More complex instruments eliminate aberrations resulting from wide ion energy spreads by employing double (velocity and direction) focusing.(6) In such instruments, resolution is determined largely by the width of the entrance slit, which determines the physical size of the ion beam image in the detection plane. In both, increased resolution comes at the expense of ion beam attenuation, resulting in decreased sensitivity.

The first commercially available GDMS system employed a double-focusing sector mass analyzer.(7) The practical resolution attainable with such a mass analyzer is on the order of 4000 while providing suitable sensitivity to the ppm level. The system was equipped with both a Faraday detector for matrix and discharge gas ion signals and a Daly-type detector for trace constituents. An automatic detector switching algorithm permitted a dynamic range of nine orders of magnitude during a single scan. The complexity and cost of these instruments limited the consideration of GDMS to a small segment of the analytical community whose specialized needs warranted the purchase of such a system.

5.2.2.2. Dispersion in Time (Time-of-Flight MS)

Perhaps the simplest form of mass spectrometry is time-of-flight mass spectrometry (TOFMS). Recalling that kinetic energy, E, is a function of mass, m, and velocity, v; or charge, z, and acceleration potential, V,

$$E = mv^2/2 = zV$$

it is obvious that monoenergetic ions of different m/z accelerated to the same kinetic energy will have different velocities. A TOF instrument operates on the principle that if all ions exit the extraction lens with equal kinetic energies, they will travel with different velocities, v_x, proportional to the reciprocal square root of their respective masses, m_x:

$$V_x = (2E/m_x)^{1/2}$$

The time required to traverse the flight distance D and to arrive at the detector is related to the m/z of the ion (Fig. 5-2):[8]

$$t = D(m/2Vz)^{1,2}$$

Monitoring the current at the detector yields a time-dependent signal. Conversion of the time axis in the intensity versus time plot to m/z values yields a mass spectrum. The mass resolution is determined directly by the temporal resolution, $\Delta t/t$, which is limited by the initial energy spread of the ions and the speed of the detection electronics.

The TOF mass analyzer is suited to short pulses of ion signal such as generated by pulsed laser desorption or pulsed ion beam sputtering. Secondary ion mass spectrometry employing TOF mass analysis[9] has developed into a successful technique for the characterization of a variety of materials. The potential applicability for TOF with the glow discharge is related to the use of either a pulsed glow discharge ion source or resonance-enhanced photoionization of neutrals by a pulsed laser (RIMS employing the GD as an atom reservoir).

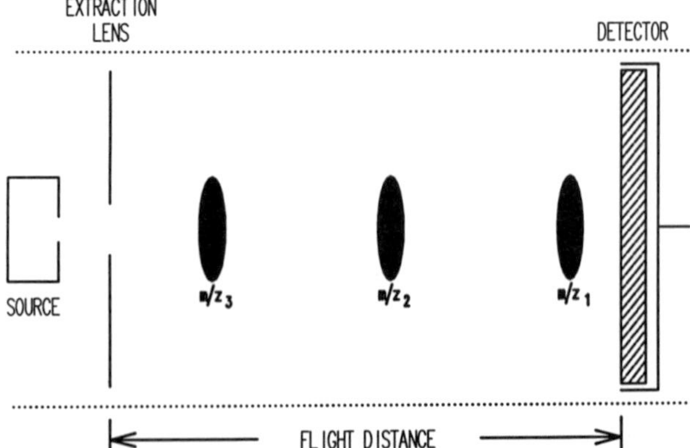

Figure 5-2. Time-of-flight mass analyzer.

5.2.2.3. Selective m/z Transmission (Quadrupole Mass Filters)

Whereas a sector mass analyzer is analogous to an optical monochromator, the quadrupole mass analyzer (QMA) is a variable bandpass filtering ion optic analogous to an optical bandpass filter. These devices are capable of transmitting ions of all masses simultaneously, or of transmitting ions of a selected mass window only.(10) By nature, they are capable of only the sequential measurement of various isotopes; however, their advantage over magnetic sector analyzers lies in the ability to scan very rapidly or to "peak-hop" among a series of isotopes.(11)

The QMA consists of four cylindrical rods arranged in a square array to which a combination of dc and rf potentials is applied (Fig. 5-3).(12) For optimal approximation of hyperbolic electric fields about each rod, the case of the ideal quadrupole, the ratio of the rod radius, r, to the radius enclosed by the electrodes, r_0, should equal a constant value:(13)

$$r/r_0 = 1.148$$

In order to provide a qualitative feel for the operation of the quadrupole mass filter, the treatment of this subject by Miller and Denton(8) will be followed. When utilized as a mass spectrometer, the quadrupole mass filter operates as a bandpass filter created by the combination of a high-pass filter and a low-pass filter (Fig. 5-4).(8) The quadrupole field is established by the application of potentials, dc and rf, to the four quadrupole electrodes. Opposite pairs of electrodes are electrically connected and lie in the xz or yz planes of the coordinate diagrams as shown in Fig. 5-4. The xz electrode pair have a positive dc potential applied along with an rf potential, while

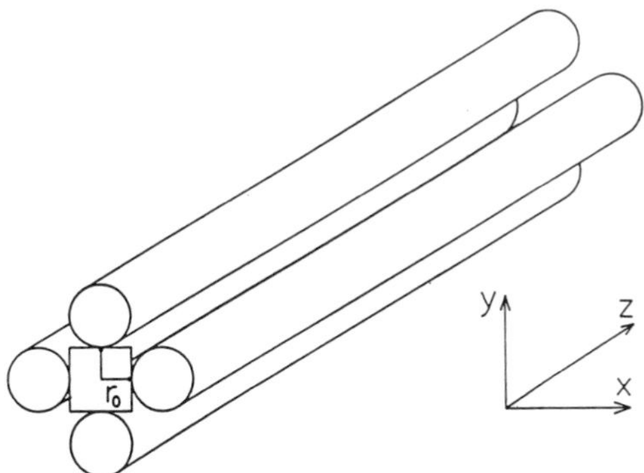

Figure 5-3. Quadrupole mass filter. Reprinted with permission from R. E. March and R. J. Hughes, *Quadrupole Storage Mass Spectrometry*, p. 43, copyright 1989, Elsevier Science Publishers.

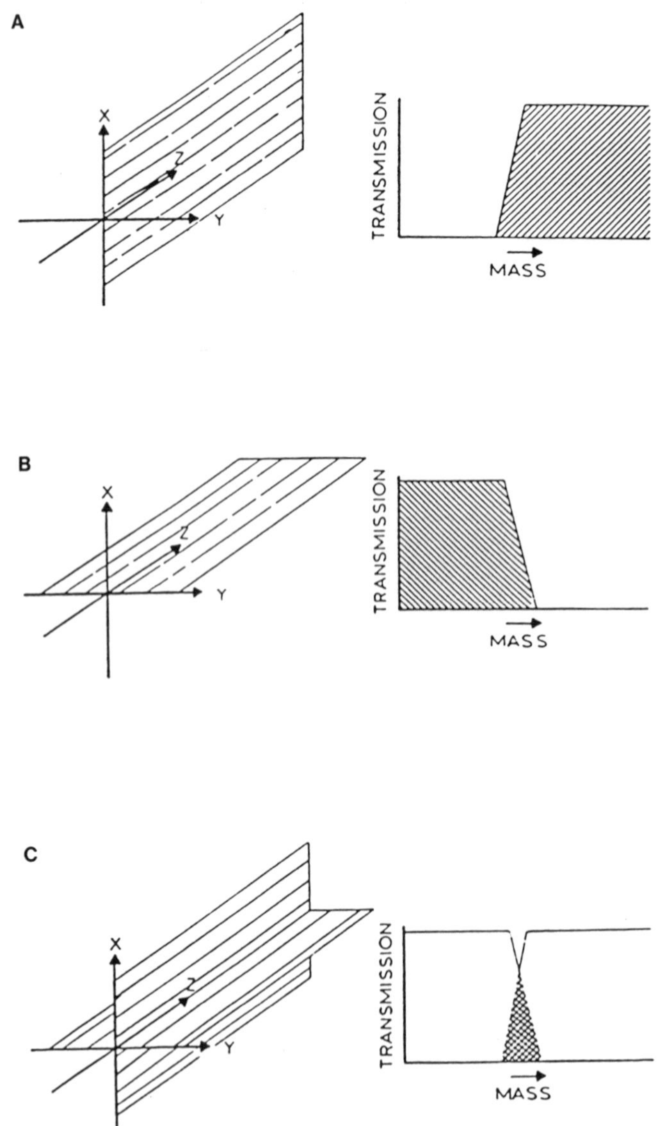

Figure 5-4. (A) High-pass mass filter; (B) low-pass mass filter; (C) bandpass mass filter. Reprinted with permission from P. E. Miller and M. B. Denton, *J. Chem. Educ.* 63 (1986) 619, copyright 1986, Division of Chemical Education, American Chemical Society.

the *yz* electrode pair have a negative dc potential applied along with an rf potential that is 180° out of phase with that applied to the *xz* pair. Ions are injected into the field along the *z* axis. In the *xz* plane, ions of larger masses are focused along the *z* axis by the positive dc field, whereas ions of smaller masses are subject to destabilization by the rf field. In the *yz* plane, ions of larger masses are deflected away from the *z* axis by the negative dc field. Ions of smaller masses in the *yz* plane will be stabilized by the ac field. The net result is the coexistence of a high-pass mass filter in the *xz* plane, A of Fig. 5-4, along with a low-pass mass filter in the *yz* plane, B of Fig. 5-4, creating a narrow bandpass mass filter, C of Fig. 5-4.

A more rigorous quantitative explanation of quadrupole mass filtering employs equations of motion for a charged particle in an oscillating electric field.[13] The electrical potential, Φ, of the quadrupole field at any given time, t, is given by the equation

$$\Phi = [U + V\cos(\omega t)](x^2 - y^2)/2r_0^2$$

where U is the dc potential, V is the rf magnitude, ω is the angular frequency ($2\pi f$) of the rf potential, x and y are the linear displacements along the respective axes, and r_0 is the radius encompassed by the four electrodes that encircle the z axis. The partial derivatives of this equation yield the electric field magnitudes, E_i, along the i axis. Multiplication of the partial derivative by the particle charge yields the force, F_i, applied to the particle along the i axis. From this, equations can be derived that describe the path of any ion through the field. Parameters a, q, and ϕ

$$a = 4zU/m\omega^2 r_0^2$$

$$q = 2zV/m\omega^2 r_0^2$$

$$\phi = \omega t/2$$

can be substituted yielding a Mathieu differential equation of the form

$$d^2u/d\phi^2 + (a_u + 2q_u \cos 2\phi)u = 0$$

which is solved for values of a and q yielding u, such that $0 < u < r_0$. Under these conditions the ion is never outside of the radius bounded by the quadrupole rods along axis x or y and its trajectory is said to be stable. In Fig. 5-5,[14] these stable ion trajectory conditions are met for those values of a and q contained under the curve. Imposed upon this stability diagram is the operational condition that results in the mass filtering ability of the quadrupole: the ratio between dc and rf potentials is set to a constant value, $a/q \propto U/V = \text{constant}$. The resulting mass scan line intersects the region of

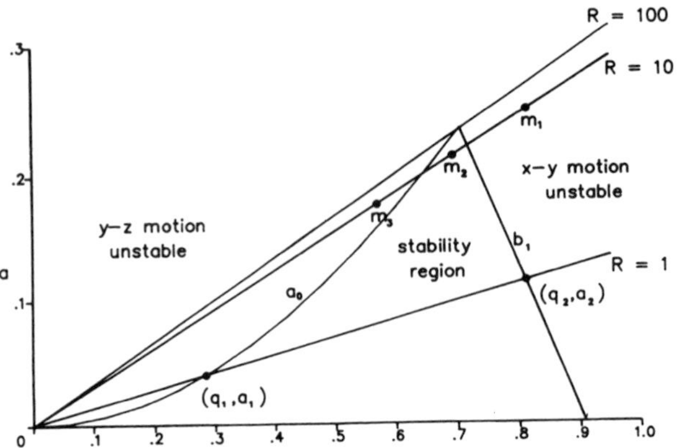

Figure 5-5. Quadrupole mass filter stability diagram. Reprinted with permission from J. E. Campana, *Int. J. Mass Spectrom. Ion Phys.* 33 (1980), 108, copyright 1980, Elsevier Science Publishers.

a–q space that corresponds to a stable ion trajectory, and only ions with those values of m/z that fall in this region can successfully traverse the quadrupole. As the potentials U and V are ramped, the m/z values falling within this window are changed. The width of this window, and thus the mass resolution, is determined by the slope of the mass scan line.

Now that commercial vendors have begun marketing quadrupole-based systems, GDMS should begin to receive the same level of acceptance and use as ICPMS has enjoyed. The principal limitation faced by these systems is their inability to resolve analyte isotopes from overlapping polyatomic interferences. Judicious choice of operating conditions and the implementation of collision-induced dissociation can minimize or remove interferences arising from polyatomic ions.[15]

5.2.2.4. *Selective m/z Ejection (Quadrupole Ion Traps)*

Quadrupole ion traps are similar to quadrupole mass filters in their operation. Essentially, two opposing rods in the QMA are connected to form a ring and the remaining two rods are converted into hyperbolic end caps, (Fig. 5-6).[16] The result is a three-dimensional quadrupole field that is symmetric with respect to rotation about the center. The end caps are taken to be along the former z axis and the former x and y axes now become a plane, r, symmetric about z. The a and q parameters employed in describing the quadrupole mass filter are modified to fit into the same Mathieu equation[13]:

$$a = -8zU/m\omega^2 r_0^2$$

$$q = -4zV/m\omega^2 r_0^2$$

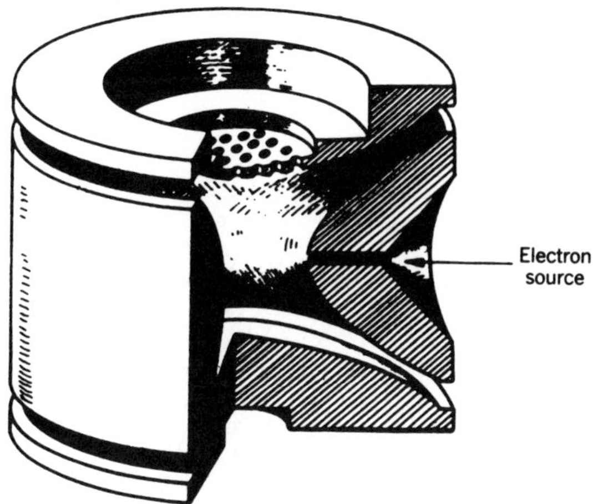

Figure 5-6. Quadrupole ion trap. Reprinted with permission from R. E. March and R. J. Hughes, *Quadrupole Storage Mass Spectrometry*, p. 4, copyright 1989, Elsevier Science Publishers.

The resulting stability diagram is shown in Fig. 5-7.[16] Mass analysis is achieved by scanning through the regions of ion instability, as opposed to scanning ion stability with the QMA.[17] Ions become unstable and are ejected selectively from the trap through orifices in one of the end caps. An electron multiplier located outside of the trap then detects the ejected ions. Quadrupole ion traps are small, relatively simple, and comparatively inexpensive. Although resolution of singly charged ions is limited in the trap, ion traps permit the use of collision-induced dissociation to remove polyatomic interferences. The ability to store particular ions for long periods of time permits the effective integration of low-abundance ions with trap technology. The limitation at the present time is the efficiency with which ions can be admitted to the trap from external sources.

The use of a glow discharge ion source combined with a quadrupole ion trap mass spectrometer (QITMS) was employed for atmospheric sampling by McLuckey *et al.*[18] The principal aim was the determination of airborne trace organic species. Hemberger and colleagues investigated the utility of a GD ion trap instrument for field elemental analysis.[19] Further results from investigations of GDMS using QITMS are awaited anxiously.

5.2.2.5. Fourier Transform Ion Cyclotron Resonance MS

Ion trapping within a cubic cell by a combination of electric and magnetic fields is also employed as a mass spectrometric method. In the presence

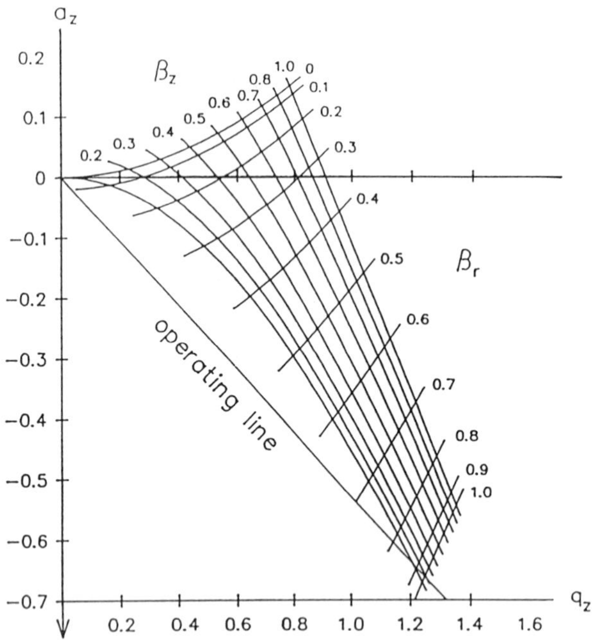

Figure 5-7. Quadrupole ion trap stability diagram. Reprinted with permission from R. E. March and R. J. Hughes, *Quadrupole Storage Mass Spectrometry*, p. 118, copyright 1989, Elsevier Science Publishers.

of a strong magnetic field, ions can be trapped in circular orbits with the application of low voltages. The ions orbit perpendicular to the applied magnetic field at characteristic cyclotron frequencies, ω_c, that are inversely proportional to their m/z values and proportional to the magnetic field strength, B (Fig. 5-8):[20]

$$\omega_c = zB/m$$

Application of an rf-voltage pulse to the transmitter plates of the cubic cell brings the orbiting ions into coherent motion. The ions are then detected as image currents generated at the receiver plates of the cell. This image current arises from the field induced as the ions come into close proximity of the receiver plate. The image currents rise and fall with the orbit of the ions yielding a transient that can be Fourier transformed to give the characteristic frequencies of stored ions and the relative signal contribution at each frequency.

Workers at IBM and the University of Wisconsin first reported FT-ICR MS investigations of an rf glow discharge plasma in 1989.[21] Although

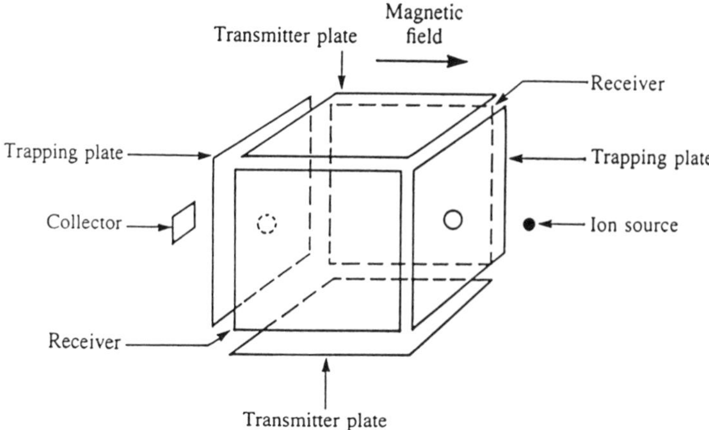

Figure 5-8. Ion cyclotron resonance cell employed in Fourier transform mass analysis. Reprinted with permission from H. H. Willard, L. L. Merrit, J. A. Dean, and F. A. Settle, *Instrumental Methods of Analysis*, 7th ed., p. 500, copyright 1988, Wadsworth Publishing.

the interest of these investigators was the ion–molecule chemistry occurring in the deposition plasma, there are compelling motives for the extension of this work to elemental analysis. The resolution attainable with FT-ICR MS can readily solve the problem of isobaric interference in GDMS. As with QITMS, CID can also be employed to remove polyatomic interferences. Recently, Barshick and Eyler[22] reported on the use of FTMS coupled with an external glow discharge source for elemental analysis. The observed detection was limited by the transport of ions from the external source into the ICR cell of the mass spectrometer. Limits to dynamic range due to space charging effects were minimized by ejection of the major ions prior to data acquisition.

5.2.3. *Mass Spectrometry of Glow Discharge Plasmas*

The operating pressure of glow discharge ion sources is greater than that tolerated by any of these mass analyzers. This necessitates differential pumping between the ion source region and the mass analyzer region. The extent of differential pumping will be a function of the type of mass analyzer employed, the ion trap requiring the least differential pumping, the FT-ICR MS requiring the most. Typically three vacuum regions are employed: the glow discharge source at ≈ 1 Torr, the intermediate vacuum region at $\leq 10^{-4}$ Torr, and the analyzer region at 10^{-6} Torr. In order to maintain analytical signals, ion optics are required to transport ions efficiently through the differential pumping region. Often, additional ion optics that serve to limit the spread of ion kinetic energy at the mass analyzer entrance are employed.[23] The mass analyzer provides temporal or spatial distribution of ions of differing m/z. Except for the case of the FT-ICR MS, mass-selected ions are detected directly with a Faraday cup for major species or with ion conversion and electron multipliers for trace elements. The double-focusing instruments use a Daly type of detector where ions are converted to electrons that are converted to photons and detected with a photomultiplier tube.[24] This detection scheme is less susceptible to degradation caused by vacuum impurities or mishaps. Ion counting can be employed to reduce noise and improve analytical signal. Data acquisition for elemental mass spectrometry differs from that for organic mass spectrometry in two respects: the widest possible dynamic range is desired permitting the observation of major and trace sample constituents in the same mass spectrum and sources of background noise that may be tolerated in organic mass spectrometry but would obscure the signal for trace sample constituents must be reduced significantly.

5.3. Atomization/Ionization and Quantitation

We are treating representation and quantitation together because they are inextricably interwoven by the differential sensitivity among elements resulting from the atomization and ionization processes. Most quantitative protocols in GDMS rely on the use of relative sensitivity factors (RSFs), sometimes referred to as relative ion yields (RIYs). The RSF of an analyte element, A, is the ration of its sensitivity (intensity, I, per unit concentration, C) to the sensitivity of some reference element, R:

$$RSF_{\mathrm{A/R}} = (I_{\mathrm{A}}/C_{\mathrm{A}})/(I_{\mathrm{R}}/C_{\mathrm{R}})$$

These RSFs contain contributions arising from instrumental factors, such as ion transmission and sensitivity, and glow discharge processes, such as differential atomization and differential ionization.[25] Whereas the contribution arising from instrumental factors is generally minimal and uniform for a given system, the contributions arising from glow discharge processes are dominant and vary from sample to sample. Obviously, an understanding of the atomization and ionization processes aids the analyst in evaluation of RSF values. Atomization and ionization will be discussed in the context of their influence on RSFs.

5.3.1. Glow Discharge Processes

As an ion source for elemental mass spectrometry, the glow discharge is characterized by two attractive attributes, cathodic sputtering and Penning ionization, that are inherent to its operation. Cathodic sputtering generates a representative atomic population directly from the solid sample. Penning ionization selectively ionizes these sputtered atoms permitting speciation on the basis of their characteristic mass-to-charge rations by mass spectrometry.

5.3.1.1. Atomization

At the heart of the analytical glow discharge lies the process of cathodic sputtering. Ions accelerated across the cathode dark space impinge on the surface of the sample. These projectiles transfer their kinetic energy to the atoms in the sample lattice through a collision cascade. If the energy imparted to an atom in the lattice exceeds its lattice binding energy, the atom is released. Excess energy above the binding energy can result in the release of the atom in an excited electronic state or as the ion. The electric field in the vicinity of the cathode leads to the redeposition of these secondary cations onto the cathode surface. Electrons or anions are accelerated into the negative glow by this same field to sustain the discharge.

The number of target atoms ejected per incident projectile or sputter yield, S, is a function of the energy deposited in the collision cascade. Under the operating conditions of most analytical glow discharges the sputter yield can be described as a function of the projectile kinetic energy, E, projectile mass, m_p, lattice binding energy, U_0, and target atom mass, m_t:[26]

$$S = 3am_pm_tE/\pi^2(m_p + m_t)^2U_0$$

$$\text{For } m_t/m_p = 0.1,\ a = 0.17 \text{ and for } m_t/m_p = 10,\ a = 1.4$$

(This equation is valid for analytical glow discharges where E is less than 1000 eV; E is typically less than half of the discharge operating potential.[27]) A related value is the sputtering rate, the number of target ions sputtered per unit time. This value is determined by the discharge operating current as well as the factors mentioned for sputter yields.

From this information it is important to identify those GDMS experimental variables that influence atomization and thereby influence RSFs. Obviously, the target matrix material is variable from experiment to experiment. For reasons related to ionization, argon is considered to be the best choice for discharge support gas. Thus, the influence of projectile identity is normalized for all species. For analytical work, glow discharge operation in a constant-power mode is preferred and E is considered to be constant during the time of analysis. The only variable over which the analyst cannot exercise control is the identity of the target atom. When one element dominates the matrix and all other elements are present in trace amounts, it can be assumed that sputtering behaves as for a pure material. Thus, there should be little difference in RSFs determined in a similarly pure matrix as reported by Vieth and Huneke.[28] With alloys containing large percentages of several elements, it is necessary to match the concentrations of the major constituents closely, as determined by a semiquantitative GDMS analysis before determining RSF values. The influence of variation in trace constituents on sputter atomization is minor. It is in ionization where such variations are no longer subtle.

5.3.1.2. Ionization

In general, GDMS does not utilize optical transitions of the analyte atoms; rather, the mass-to-charge ratio of atoms that have been ionized in the plasma affords the speciation of analytes. This fact shifts the emphasis from excitation mechanisms in general to ionization mechanisms specifically. In some ways this simplifies the relationship between analyte signal and analyte concentration in the sample. But if we assume that the atomization does not differ strongly between elements in a given matrix, we cannot assume the same to be true for ionization. RSFs are most likely controlled by differences in the probability of ionization among the elements.

There exist many processes that contribute to ionization in glow discharges[26] (Table 5-2). These processes may be characterized as collisions of the first kind in which kinetic energy is transferred to an atom resulting in ionization or collisions of the second kind in which potential energy is transferred to an atom resulting in ionization.[29] The relative importance of each mechanism will vary with discharge operating conditions, current, voltage, pressure, and geometry. The Penning process is the dominant ionization process occurring in the coaxial cathode-type glow discharge ion sources operating at 0.4–2.0 Torr, 1–1.5 kV, and 0–10 mA.[30] Langmuir probe measurements indicate that the electron population in glow discharges consists of three types of electrons in a non-Maxwell–Boltzmann distribution: fast ($E > 25$ eV), secondary ($E \approx 7$ eV), and ultimate (thermal).[31] The most

Table 5-2. Ionization Processes in Glow Discharge

Collisions of the first kind	
Electron ionization	$A + e^- \rightarrow A^+ + 2e^-$
Collisions of the second kind	
Penning ionization	$Ar^m + X \rightarrow Ar + X^+ + e^-$
Associative ionization	$Ar^m + X \rightarrow ArX^+ + e^-$
Symmetric charge exchange	$A^+ + A \rightarrow A + A^+$
Asymmetric charge exchange	$A^+ + B \rightarrow A + B^+$

energetic of these are localized in the cathode fall region of the glow discharge, whereas ultimate electrons, with energy less than that for optimum electron ionization cross sections, are the majority species in the negative glow. In hollow cathode and Grimm-type glow discharges, charge exchange contributes significantly to sputtered atom ionization.[32] Because Penning ionization in argon glow discharges uniquely results in the ionization of atoms in a manner that is relatively uniform among elements and selective with respect to polyatomic species, ion sources in which this process dominates are preferred.

5.3.2. Quantitation

In the ideal GDMS analysis, the ion signals for all ionized sputtered species are summed and then the ration of the ion signal for individual species is calculated. This ratio then corresponds to the concentration of the species in the bulk. This ion-beam ratio (IBR)[33] method of quantitation would be an absolute, i.e., standardless, method if there were no difference in sensitivity among the elements. Of course, sensitivity is not constant among the elements, and generally varies by less than a factor of ten; therefore, the IBR method provides a straightforward means for reliable semiquantitative analysis. Quantitative analysis is achieved by correcting IBR values employing RSFs determined for a set of standards having the same general matrix as the sample under investigation.[34]

An alternative method of calibration has been reported recently by Klingler and Harrison.[35] They employ a source having a sample cathode and a reference cathode. The discharge potential is alternately applied to these cathodes and mass spectra for both are obtained. The ratio of sample-to-reference signals for a given isotope yields a direct calibration. This method offers the potential to eliminate the need for RSF values.

5.4. Glow Discharge Ion Sources

A wide variety of analytical glow discharge geometries have been investigated as ion sources. In the early 1970s, investigators were interested in monitoring glow discharge plasmas employed in thin-film deposition. For applications in microelectronics, high-purity thin films were required. This, in turn, required that the deposition plasma and sputter target also be of the highest quality. The group at IBM headed by Coburn used mass spectrometry initially to characterize their dc- and rf-powered diode glow discharge deposition plasmas.[36] They soon realized that the mass spectrometric monitoring of the ions produced in these deposition plasmas also provided a means of determining the elemental composition of the sputtering target employed.[37] Their work provided the basis for the development of GDMS as an attractive alternative to spark source mass spectrometry. Hollow cathode glow discharges were coupled with magnetic sector mass analyzers in the preliminary investigations of analytical GDMS.[38] In 1978, the first dedicated elemental analysis system employing a glow discharge sampled by a QMA was developed at the University of Virginia.[39] The ions were sampled through an orifice in the base of the sample cathode. The resulting beam was found to have a wide energy spread due to the sharp potential gradient in the near-cathode region. Investigations employing a variation of the hollow cathode, the hollow cathode plume, exhibited a similar ion energy behavior.[40] In both instances, energy filtering of the ion beam was found to offer optimized performance because of the difference in kinetic energy between sputtered and discharge gas species. The other geometries investigated all employ ion sampling through an orifice in the anode yielding an ion beam characterized by a more narrow ion kinetic energy distribution. Commercial instruments employ a modified coaxial cathode geometry,[41] the so-called "pin-type" glow discharge ion source[42] (Fig. 5-9). This is also the most widely characterized glow discharge ion source. A Grimm type glow discharge was developed as an ion source for use with a quadrupole mass filter by Jakubowski *et al.*[43] A modified version of the Grimm-type glow discharge, the jet-enhanced sputtering cell, has recently been adapted to an ICP-MS instrument to permit GDMS determinations.[44] Comparisons of the operating conditions and the analytical merits of the principal glow discharge ion sources are shown in Tables 5-3 and 5-4, respectively.

Whereas different glow discharge ion sources have not exhibited any significant performance differences, different methods of powering the sources do show specific performance differences. dc-powered sources are

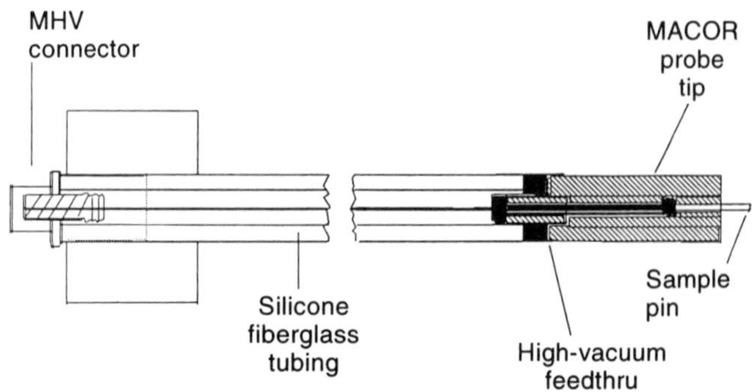

Figure 5-9. Coaxial cathode (pin-type) GD ion source.

Table 5-3. Comparison of Glow Discharge Ion Sources

Source	Voltage (V)	Current (mA)	Pressure (Torr)	Cathode
Hollow cathode	200–500	10–100	0.1–1.0	23-mm-deep cylinder with 5-mm-diameter base
Grimm	500–1000	25–100	1–5	6.5-mm-diameter circle
Jet-enhanced	900	28	2.5	12-mm-diameter circle
Coaxial cathode	800–1500	1–5	0.2–2.0	1.5–2.0-mm-diameter × 4–8-mm-long rod

Table 5-4. Comparison of Glow Discharge Ion Sources

Source	Advantages	Disadvantages
Hollow cathode	>High sputter rate >Intense ion beams >Amenable to powders	>Charge exchange mechanisms important >Complicated sample geometry
Grimm	>Amenable to depth profiling >Easy to use for compacted powders	>Flat samples only >Higher discharge gas flow rates
Jet-enhanced	>High sputter rate >Easy to use for compacted powders	>Flat samples only >Higher discharge gas flow rates
Coaxial cathode	>Amenable to various sample shapes >Ionization dominated by Penning process	>Powders require conversion into solid form

the most common. These are suitable for samples in a conducting matrix. Operation in a constant-power mode has been observed to yield improved signal stability as evidenced by studies of run-to-run and sample-to-sample precision. This results from an inherent compensation for sputter erosion-induced surface area changes. The use of rf-powered glow discharges permits the direct analysis of nonconducting samples that would have to be mixed with a conductive host matrix prior to analysis with a dc-powered ion source.[(45)] Operation of glow discharges with a dc power supply modulated at less than 100 Hz offers two distinct advantages: higher instantaneous signal intensities and temporal differentiation of analyte ions from background gas ions.[(46)]

5.5. GDMS Applications

Newly developed analytical techniques sometimes struggle to justify their existence. Applications are posed that could be accomplished much easier by existing methodologies. Not so for GDMS. It arose out of a need unmet by any available techniques. The driving force in its development was the limited ability at the time to perform trace elemental analysis directly in solids, where so many problems and opportunities were present.

In the 1970s, when GDMS was undergoing its metamorphosis from laboratory curiosity to credible analytical tool, the primary method for survey trace analysis of solids was spark source mass spectrometry (SSMS). This technique was limited in scope by its complexity, expense, and unreliability. GDMS exhibited all of the strengths of SSMS, but also had sufficient critical advantages as to suggest it would serve as the eventual replacement for SSMS, an expectation that has by now generally taken place.

The advantageous features that have led to many GDMS applications include the following:

- Simplicity—The glow discharge source is a simple two-element (cathode/anode) device operated at low power and low gas flows.
- Stability—The source produces a stable supply of low-energy ions characteristic of the sample.
- Sensitivity—The high signal-to-background ratio permits excellent detection limits, often at the sub-ppb levels.
- Low cost—Actually, commercial mass spectrometers are relatively expensive, but for those adapting existing units, inexpensive power supplies are sufficient to drive the source, and the small discharge gas flow volumes make for low-cost operation.
- Uniform response—The RSFs do not differ widely across the periodic table, thus allowing one set of discharge conditions to serve for a broad range of elements.
- Inert environment—The closed argon source yields a generally inert plasma that minimizes the formation of polyatomic species, such as metal oxides.
- Minimal matrix effects—The source can be thought of as a sputter atomization cell, followed by an ionization source; this pseudo-tandem configuration negates chemical memory of the sample to minimize matrix-induced differences.

Limitations of GDMS include the potential presence of interferences, particularly in ultratrace analyses, the inability to handle solutions directly, and the need for differential pumping to handle the gas load.

After a long gestation period, culminating in the first commercial instrument in the early 1980s, GDMS has spread rapidly, finding broad application. The major reason for its success has simply been that it solves problems. GDMS permits the analysis of a broad range of solid sample types, from pure metals to ceramics, from semiconductors to geologicals.

A review of the literature is deceiving in any attempt to determine the popularity of GDMS. Much (probably most) of the day-by-day applications are carried out in commercial or industrial laboratories from whence only technical reports or analytical data sets arise. Some commercial laboratories run several GDMS instruments on multiple shifts in order to meet the demand. GDMS is often the "cash cow" that keeps the business healthy.

We present here only some typical examples of analytical work to which GDMS is applied. Some of the samples are relatively simple; others present extremely difficult problems for the technique and the analyst. In all cases, the need and interest in GDMS is still growing, so that applications that might be reviewed 5 years from now will show increasingly diverse utility of this powerful technique.

5.5.1. Solids Elemental Analysis

5.5.1.1. Bulk Metal Analysis

Of all the sample types, bulk metals are the ones for which GDMS is almost ideal. Little sample preparation is required; simply cutting or machining a pin or disk produces a sample ready for mounting into the source. By use of a presputter period, the plasma cleans the sample surface and readies the material for analysis. Since reasonably large material consumption (mg) occurs during GDMS analysis, small sample inhomogeneities are averaged out.

Sample spectra may be obtained in minutes, and a rapid qualitative analysis can be extracted by examination of the isotopic lines and patterns. Quantitative analysis follows from line intensity measurements compared with the matrix element or some selected internal reference element. Software packages for data treatment can thus very rapidly tell the analyst not only which elements are present but also their concentration, with linear response over nine decades.

a. High-Purity Metals. High-purity metals are held to exacting standards for trace impurities. For example, aluminum used in the electronics industry must be at least 5-9's purity and 6-9's may be desired. The latter would require that the *entire impurity load* be summed to no more than 1 ppm. Within the suite of relevant trace elements, some have relatively relaxed restrictions in the low ppm range, but certain critical elements such as Th and U (alpha emitters) may be unacceptable at anything above 1 ppb. To survey the entire periodic table with the detection limits required for such applications, only GDMS currently meets the need. It is thus not surprising that this type of sample comprises a significant fraction of the work load of GDMS instruments in industry and service laboratories.

Vassamillet[7] describes the role of GDMS in the quality control of high-purity aluminum production. Because trace impurities can be introduced at various treatment and melting steps in the production process, multiple analyses are necessary on a lot-by-lot basis. Before any given shipment of specified purity aluminum is made to customers, a certificate of analysis is necessary, requiring an analytical technique that is not only sensitive and accurate, but also relatively rapid. A VG 9000 GDMS instrument was used to survey aluminum impurities. After a sputter cleaning time of 30 min, an argon discharge was initiated at 1 kV and 2 mA. For each sample, 71 elements were monitored and concentrations calculated by use of RSFs determined from Pechiney reference standards. Most results remained within a 15% range in the aluminum sample, which suffers from inhomogeneity problems in that the trace elements are not present as a homogeneous solid solution, but rather occur as particulate precipitates scattered randomly throughout the sample. Even here, however, with the large sample consumption of GDMS, reasonable results can be obtained by averaging of sequential mass spectral runs. In the high-purity aluminum, magnesium (0.86 ppm) and iron (1.1 ppm) were found as the highest impurities. Gas analyses, such as C, H, and O, are difficult because of the background contributions of each of these species from the glow discharge itself. Gas results vary significantly with time as the source is being cleaned by the discharge sputtering. After about 45 min, the gas impurities reach a minimum. GDMS analyses are not rapid, at least not in comparison with solution measurements. Analysis times of approximately 1 h, which includes sputter cleaning of the sample, are indicated as normal. Although only seven or eight samples per day could be analyzed on that basis, it must be recalled that (1) many elements are analyzed for each sample and (2) very little time is required for sample preparation.

Mykytiuk *et al.*[47] have also analyzed high-purity aluminum with a VG 9000, which they report permits a 100-fold lower detection limit than with SSMS and exhibits fewer spectral interferences. By use of cryogenic cooling, background contributions of C, N, and O were reduced by three orders of magnitude, achieving values in the 10–100 ppb range. Typical analyses of aluminum are shown in Table 5-5, which indicates the reproducibility obtained when seven pins were prepared and run from the same sample. In this paper, the authors also demonstrated the analysis of pure gallium, a metal characterized by a low melting point (28°C). Without sample cooling, the solid metal would quickly become liquid upon initiation of the glow discharge. By use of the cryogenic cell, however, direct analysis of gallium in the solid state can be performed. Interestingly enough, sample preparation is carried out in the liquid state by melting the sample under low heat and then drawing the molten gallium into micropipettes. After cooling the pipettes to resolidify the gallium, sample pins are removed for GDMS analysis.

Table 5-5. Analysis of Aluminum by GDMS:[47] Values Shown in ppm[a]

Element	Run 1	2	3	4	5	6	7	Mean	%S.D.
B	1.1	1.4	1.1	1.8	1.2	1.0	1.2	1.3	22.0
Mg	0.11	0.11	0.11	0.11	0.11	0.12	0.12	0.113	4.3
Si	4.2	4.1	4.1	4.2	4.3	4.0	4.0	4.13	2.7
P	10.1	9.3	8.5	9.2	10.5	8.1	7.9	9.09	11.0
S	0.033	0.018	0.020	0.035	0.015	0.017	0.042	0.026	41.0
Sc	0.054	0.049	0.059	0.059	0.054	0.054	0.058	0.055	6.6
Ti	0.20	0.11	0.10	0.12	0.12	0.16	0.13	0.134	26.0
V	0.029	0.028	0.026	0.030	0.029	0.029	0.027	0.028	4.9
Cr	0.053	0.053	0.045	0.051	0.050	0.040	0.075	0.052	21.0
Mn	0.033	0.038	0.035	0.038	0.033	0.040	0.038	0.036	7.6
Fe	1.8	1.6	1.7	1.8	1.7	1.7	1.8	1.7	4.4
Ni	0.027	0.029	0.025	0.037	0.024	0.023	0.040	0.029	23.0
Cu	0.77	0.58	0.61	0.67	0.85	0.51	0.71	0.67	17.0
Zn	0.12	0.11	0.10	0.12	0.10	0.12	0.11	0.11	8.1
Ce	0.042	0.039	0.038	0.046	0.053	0.043	0.036	0.042	14.0
Th	0.022	0.018	0.020	0.024	0.022	0.021	0.018	0.021	11.0
U	0.017	0.016	0.015	0.017	0.016	0.014	0.014	0.016	8.2

[a]Seven pins were cut from the same sample. Each result is the mean of five replicate firings. Preburn 1 h at 5 mA, analyze at 3 mA, 1.1 kV.

Table 5-6 shows a comparison of GDMS and SSMS for two gallium samples. Because there are no adequate standards for gallium metal, the analyst must rely on comparison data from several analytical techniques to determine the "true" values of the impurities. Sample 1 contains relatively low levels of impurities, with the exception of In, Sn, Hg, and Pb. Sample 2 is a higher-purity sample in which most "analyses" are simply indicated as being less than a certain instrumental detection limit. A comparison of the GDMS estimations relative to those from SSMS shows the lower detection limits available for the former.

b. Alloy Metals. Like pure metals, alloys are easily analyzed by GDMS. Sample preparation is quite straightforward, and data analysis is complicated only by the need to be aware of potential spectral and polyatomic interferences. Even for complex alloys, at least one unaffected isotope line is normally available for each element. Occasional difficulties can be encountered for monoisotopic elements like aluminum and arsenic. A great strength of GDMS is that ion yields are relatively unaffected by alloy composition.[48] Figure 5-10 shows that RSFs collected for seven different materials, ranging from steels to uranium oxides, are grouped rather tightly for each element. This may be contrasted with other multielement analytical techniques (e.g., SIMS, NAA), which exhibit orders of magnitude differences in sensitivity among elements.

Table 5-6. Comparative Analysis of High-Purity Gallium by GDMS Versus SSMS:[47] Values Shown in ppm

	Sample 1		Sample 2	
	GDMS	SSMS	GDMS	SSMS
C	2.0	2.0	0.4	0.2
N	0.2	0.4	0.08	0.2
O	0.6	2.0	0.2	0.2
Na	<0.001	Intf.[a]	<0.0003	Intf.[a]
Mg	0.002	0.001	<0.0007	<0.002
Al	0.005	0.003	<0.0004	<0.002
Si	0.03	0.02	<0.001	<0.08
S	0.03	0.03	0.002	0.002
Cl	0.04	0.03	0.01	0.01
K	0.02	0.05	<0.01	0.005
Ca	<0.03	0.01	<0.03	0.002
Fe	0.02	0.04	0.003	<0.008
Cu	3.0	0.9	<0.001	<0.02
Zn	0.9	0.7	<0.001	<0.02
Se	<0.02	<0.04	<0.01	<0.04
Ge	0.4	0.2	<0.03	<0.2
Cd	0.3	0.2	<0.002	<0.05
In	10.0	16.0	<0.0003	<0.008
Sn	10.0	30.0	<0.003	<0.05
Hg	4.0	9.0	<0.002	<0.09
Pb	13.0	4.0	<0.002	<0.06

[a]Interference.

An example that illustrates typical alloy analysis is found in the work by Vieth *et al.*[49] in which they studied various alloy standards to determine RSFs and then applied these factors to the analysis of nickel-base alloys. An interesting aspect of this study for the purposes of this report was a comparison between two different types of GDMS instruments. The first commercial instrument was of the so-called "high-resolution" type, which incorporated a magnetic sector instrument (reverse Nier Johnson geometry) to permit relatively high ion throughput and the resolution of certain spectral interferences (e.g., ArO^+ on Fe^+). Quadrupole-based instruments have been used for GDMS for many years,[23] but for all their other advantages, such units cannot resolve same-nominal-mass interferences and exhibit a significant decrease in ion throughput at higher masses. Recent access to commercial instruments of both types permitted the comparison shown in Table 5-7, whereby RSFs were determined as an aggregate from five different matrices. What becomes obvious is that the same samples, sputtered and ionized by essentially identical sources, exhibit generally similar RSFs up to a point, but beyond about mass 90 the elemental sensitivities of the quadrupole

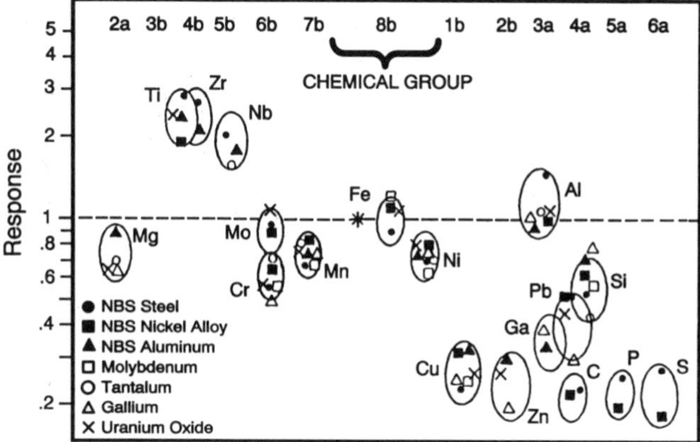

Figure 5-10. Demonstration of the generally uniform relative sensitivity factors characteristic of GDMS.[48]

instrument begin to fall off rapidly, presumably due to lower ion transmission and tuning artifacts. (Note: RSFs calculated by this method are really "insensitivity factors." That is, larger RSFs reflect lower sensitivity.)

Since RSFs are simply correction factors for the entire instrument system (sputtering, ionization, extraction, throughput, and so on), it is feasible to use the different RSFs generated for the sample elements to calculate analytical results and compare two commercial units. This is shown in Table 5-8, in which RSFs compiled from Ni-based alloys were used to analyze a CRM 345 standard IN-100 Ni-base alloy. Several points become clear from an examination of the data. (1) GDMS is able to provide elemental results from the major sample constituents down to the trace level in one analysis. (2) The sector instrument versus quadrupole data are in relatively good agreement; the presence of spectral interferences would make quadrupole results appear high relative to higher-resolution results, and there is little appearance of this in the data. (3) Both GDMS instruments yield results that are reasonably good with respect to the certified value of the alloy standard. Analyses such as these demonstrate the utility of GDMS for qualitative and quantitative analysis of complex alloys.

5.5.1.2. Semiconductors

A third type of bulk sample analysis that is very important in commercial applications is that of semiconductors (or semi-insulators). The electrical properties of these materials are critically dependent on the intrinsic and doped levels of impurities, making it essential that fabricators know not only

Table 5-7. Comparison of RSF Values Obtained in a Steel Sample by Two Commercial GDMS Instruments [49]

Element	RSF (x/Fe)	
	GloQuad	VG-9000
Be	2.0	1.1
B	1.5	1.22 ± 0.17
C	5.7 ± 1.4	4.5 ± 1.1
Mg	1.1	1.3 ± 0.3
Al	0.96 ± 0.30	1.39 ± 0.09
Si	2.27 ± 0.65	2.0 ± 0.3
P	5.1 ± 0.7	3.5 ± 0.1
S	3.5 ± 0.7	3.3 ± 0.1
Ti	0.67 ± 0.11	0.42 ± 0.04
V	0.76 ± 0.06	0.55 ± 0.02
Cr	2.11 ± 0.16	2.2 ± 0.4
Mn	1.20 ± 0.22	1.5 ± 0.2
Fe	≡ 1.00	≡ 1.00
Co	1.07 ± 0.13	1.1 ± 0.4
Ni	1.50 ± 0.08	1.5 ± 0.1
Cu	5.0 ± 0.7	5.0 ± 1.2
Zn	4.5 ± 0.6	5.5 ± 0.9
Ga	6.1	4.5 ± 0.4
Se	3.0	3.1 ± 0.3
Zr	2.2	0.64 ± 0.08
Mo	3.00 ± 0.11	1.3 ± 0.2
Sn	7.8 ± 1.9	2.4 ± 0.2
Ta	7.0	1.1
W	5.5	1.5 ± 0.1
Pb	9.2 ± 2.1	2.2 ± 0.2
Bi	18	4.3 ± 0.7

what is present but also the concentration levels. The problem for the analyst is exacerbated by the fact that even extremely low elemental concentrations of specific elements are known to alter semiconductor properties. Furthermore, the presence of any impurity element can be worrisome enough to require its complete characterization. The presence of C, O, and N is often difficult to measure by elemental techniques (including GDMS), but such information may be crucial. The net result of all this has been a growing need for analytical techniques that will survey semiconductors for all elements and provide at least semiquantitative results. Manufacturers cannot permit surprises to show up later in a batch of substrate material that has been used to prepare many expensive semiconductor components.

GDMS has been particularly valuable in supplying rapid, accurate, and sensitive analyses in this regard. Using GaAs as an example, Evans and co-workers[50] demonstrated that GDMS could achieve excellent detection

Table 5-8. GDMS Analysis of CRM 345 Standard IN-100 Ni-Base Alloy:[49] Values in ppm Except Where Indicated as %

Element	Certified	VG-9000	GloQuad
B	190±10	230	180
C	0.153	0.21%	0.14%
Mg	5±1.2	6.1	6.3
Al	5.58%	5.5%	5.4%
Si	—	520	320
S	—	18	23
Ca	(<1)	≤0.2	0.9
Ti	(4.74%)	5.28%	4.0%
V	1.00%	1.04%	0.85%
Cr	9.95%	10.4%	9.3%
Fe	—	780	670
Co	14.71%	15.3%	14.9%
Zn	<0.5	<0.1	0.49
Ga	8±0.3	8.2	7
As	(2)	2.3	4
Se	<0.5	2	3
Zr	440	425	410
Mo	3.01%	3.05%	3.3%
Ag	<0.2	<2	<0.8
Cd	<0.1	≤0.2	<0.2
In	—	0.04	0.05
Sn	6±1.6	5.2	11
Sb	<2	2.1	3
Te	<0.2	<0.03	<0.3
Tl	<0.2	≤0.007	<0.03
Pb	0.2	≤0.05	0.4
Bi	<0.2	≤0.005	<0.2

limits for many elements, as shown in Table 5-9. However, beyond the 28 elements successfully presented, standing out by their absence are five other elements, namely H, C, N, O, and Cl, which may be of great interest to manufacturers. The authors solved the problem by complementary analyses using SIMS with Cs bombardment. It was estimated that by the use of suitable standards, accuracy within 20% was obtained by GDMS and within a factor of 2 by SIMS.

Mykytiuk *et al.*[47] have also reported the analysis of GaAs, but they have taken a different approach to allow the determination of impurity gases C, N, and O. After GaAs pins are cut, they are rinsed in 5% bromine in methanol, then in high-purity methanol, and then dried in an oven. Their key to obtaining good results for the gases lies in an extended sample sputter burn for 1 h at high-current conditions of 4–5 mA for GDMS. Normal currents lie usually in the 1–2 mA range. The greater bombardment current and energy subject the sample to considerable heating and sputter cleaning.

Table 5-9. Detection Limits for Selected Elements In GaAs Comparing GDMS and SIMS:[(50)] Values Shown in ppm

Element	GDMS	O_2-SIMS	Cs-SIMS
H	—[a]		10,000[a]
Be	20	2	
B	1	2	
C	—[a]		500[a]
N	—[a]		100[a]
O	—[a]		1,000[a]
F	300		20
Na	30	10	
Mg	5	5	
Al	5	500	
Si	10		40
P	40		40
S	200		50
Cl	—[a]		20
Ca	60	6	
Ti	0.7	5	
Cr	4	1	
Mn	1	5	
Fe	3	10	
Cu	10		100
Zn	20		100
Ge	300		100
Br	30		8
Se	400		5
Mo	10	100	
Cd	20	6,000	
In	200	8	
Sn	20		10
Sb	20		20
Te	10		0.5
W	4	5,000	
Hg	9	100,000	
Au	20		1,000

[a]Limited by residual atmospheric gases in the instrument.

Table 5-10 shows the time-dependent concentrations of C, N, and O during the sample treatment step. A question arises as to whether the longer burn time required to conduct GDMS analyses by this approach is preferable to the additional time that would be involved in acquiring SIMS data for these troublesome elements. If no SIMS instrument is available, of course, such conjecture becomes a moot point. Having confirmational data from either SIMS or SSMS is always desirable, but falls into the category of high luxury for most analysts.

Table 5-10. Effect of Preburn Time on GDMS Analysis of GaAs:[47] Values Shown in ppm

Element	5 min	30 min	45 min	60 min
B	0.2	0.08	0.06	0.07
C	0.5	0.03	0.01	0.009
N	0.09	0.02	0.01	0.01
O	0.1	0.09	0.04	0.05
Na	0.02	<0.002	<0.003	<0.003
Mg	0.005	<0.008	<0.003	<0.003
Al	0.05	<0.005	<0.001	<0.001
Si	0.2	0.2	0.004	<0.002
S	0.03	0.01	0.008	0.01
Cl	0.003	0.005	0.004	0.003
Fe	0.002	0.003	0.002	0.001

5.5.1.3. Surface Analysis

In one sense, GDMS is always a surface analysis technique, even though it permits measurement of bulk concentrations. That is, the sputtering process central to the glow discharge acts as an atomic mill that regularly erodes away the surface of the bombarded sample. Whatever atoms are on the surface are sputtered away and measured. Because GDMS consumes significant quantities of material (up to milligrams per minute), these sequential layer analyses combine to yield an averaged composition that is typical of the bulk concentration. By slowing down the ablation process and limiting measurements to a brief time span, data indicative of surface concentrations can be obtained. GDMS and its optical analogue have found considerable application for the analysis of layered samples.

A number of common surface-layered materials benefit from GDMS analysis. For example, metals that are likely to corrode in an oxidizing atmosphere may be coated with a protective layer 1–10 μm in depth. Galvanized steel features a thin layer of Zn on an Fe base material. The emf of Zn allows it to withstand oxidation in the presence of water and oxygen, protecting the Fe, which would readily rust. Typical etch rates of the glow discharge are 1–5 μm/min, permitting mass spectral accumulation for up to several minutes during the ablation of a given thin layer.

It is not realistic to speak of these analyses as thin-film measurements. Given its high sputter rate, GDMS lends itself more to the analysis of layers in the range of micrometers rather than angstroms. In principle, the glow discharge mills itself through a sample, layer by atomic layer. Thus, a layer comprised of element A should provide a large and steady signal of A^+ until that layer is consumed, at which time the A^+ signal should drop sharply to zero, with a simultaneously abrupt appearance of a signal from the next layer, comprised of element B. One reasonable approximation of this behavior is illustrated in Fig. 5-11, in which results from the analysis of a three-component sandwich sample are presented. The thin Cr layer (ca. 1 μm) sustains a Cr ion signal for only a few minutes, at which point it drops dramatically (note the log scale). The next layer, Cu, is represented by a signal that essentially rises vertically to a level typical of a matrix signal. After about 75 min of sputtering, the next interface (Cu/Fe) is reached, at which point the Fe signal becomes prominent while the Cu intensity drops off over an approximately 20 min interval.

It is obvious from Fig. 5-11 that the signal responses do not follow an ideal square wave pattern for their appearance and disappearance. Clearly, all of the metal sample layer does not disappear at the prescribed time, but rather is scattered across a diffuse region that gradually is eroded away. Note that the initiation of a new layer is quite sharp (Cu, Fe), but the termination of that layer lingers for many minutes (Cr, Cu). Part of this is due to the atomic mixing of the sputter process that causes signal from the depleted layer to extend into the next layer. Another factor is the intrinsic redeposition that occurs in the collision-rich (1 Torr) environment of the discharge. It has been estimated[52] that up to 67% of the sputtered atom population is returned to the surface by collisions with argon atoms, to be subsequently resputtered (perhaps several times) before escaping the surface region. In this manner, the analytical signal for a given elemental layer will extend far beyond the nominal demarcation line anticipated.

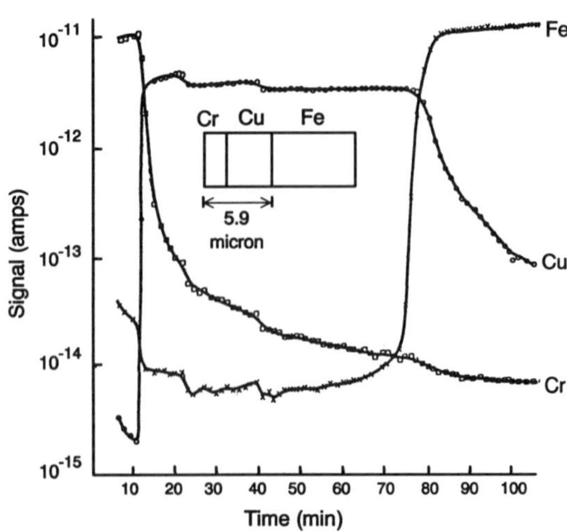

Figure 5-11. Thin-layer analysis by GDMS for a three-element sandwich sample.[51]

It is this thin layer cross talk that prevents GDMS from being used for very thin films. For the micrometer range layers, however, the data obtained from GDMS are quite valuable and capable of providing quantitative analysis and depth profile analysis. Because GDMS permits rapid scans, an entire sweep of the periodic table may be performed in a few tenths of a second. In this way multiple scans may be accumulated and impurity levels in the thin layer detected and their concentrations determined. By calibrating erosion rates using standard thin-layer samples, the thickness of layers may be determined by examination of signal/time profiles. In this way, thin-layer samples can be fully characterized. The increasing popularity of GDMS for thin-layer work arises from the general lack of matrix effects exhibited by this technique, in comparison with the major problems frequently encountered when using SIMS analysis.

5.5.1.4. Compacted Samples

The samples considered to this point have all arisen from bulk solids, permitting the ready formation of an electrode for glow discharge analysis. All solids are not so conveniently prepackaged for the analyst. Many are in a state that requires compaction to form a suitable sample electrode (e.g., metal filings or powder). If, in addition, these powdered samples are non-conductors (e.g., soils, sediments, glasses), then an additional step in the preparation is required to yield a conducting matrix for analysis.

The analysis of metals and alloys calls upon much simpler methodologies than in the case of compacted samples. In addition to the increase in sample preparation time, the need to convert to a conducting matrix introduces the possibility of contamination. Other problems arise because of the trapping of water vapor and atmospheric gases in the sample during the compaction process. Figure 5-12 shows a general schematic of a sample

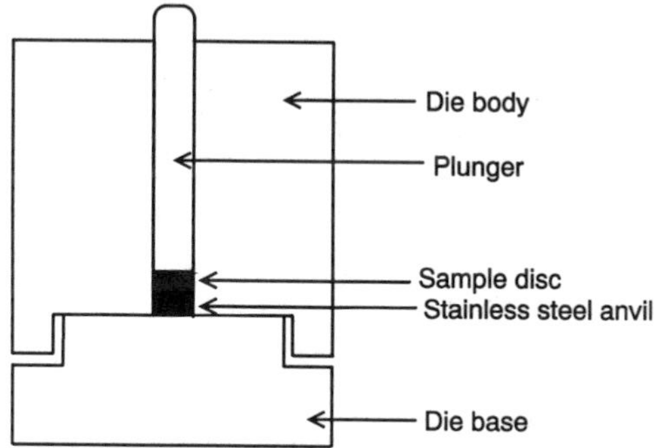

Figure 5-12. Cross section of a custom die used to fabricate GDMS samples from powdered materials.

preparation die. Consider that an ore sample is to be analyzed by GDMS using a dc discharge. A 10–20% by weight sample is prepared by mixing with a suitable high-purity metal powder, usually silver or copper. (Graphite, which is used as a matrix for many optical methods, is less satisfactory for GDMS due to the large amount of hydrocarbon ions produced.) Pressing the matrix/sample mixture at high pressure by means of a suitable die and hydraulic press effects the formation of a conductive electrode disk or pellet for GDMS analysis. If an rf-powered discharge source is available, the sample electrode does not have to be conducting, thus eliminating the need for matrix modification by dilution with a conducting powder. An rf source will also permit the direct analysis of solid nonconductors, such as glasses or ceramics.

The glow discharge is confronted with a more difficult sample matrix in the case of compacted samples. Unlike the case of bulk solid metals and alloys, where the sample surface is initially cleaned by the sputter process, revealing the "true" bulk composition below, a compacted electrode is comprised of an unending series of surfaces. That is, each micrometer-sized metal particle is coated with a contaminant layer, often the metal oxide, through which the plasma must etch to reach the metal sample underneath. The net result is that the discharge is always being presented a combination of bulk material and surface "contaminants." In the case of more complex materials, such as rocks and sediments, the situation is even more difficult. The elemental analytes are chemically bonded (and tenaciously, at that) to other elements that may contaminate the discharge and cause it to be more reactive toward the formation of interferents. For example, examination of a geological sample will reveal the trace metals tied up with oxygen and usually located in a hostile silicate matrix. When the discharge strikes these materials, not only are the analytical elements sputtered, but also the matrix elements, including atomic oxygen, which is then highly reactive in the plasma.

In spite of these impediments, GDMS offers some outstanding opportunities for compacted sample analysis. Figure 5-13 shows the type of spectrum that can result from a compacted sample of a standard reference sample, SCo-1 (USGS). Only the rare earth region is depicted in the spectrum, illustrating the attainable sensitivity and the potential for interferences in this case. Strong oxides of Ce and La are not totally dissociated in the plasma, leading to the isobaric interferences shown in Fig. 5-13. Normally, an uninterfered isotope can be found by which the elemental concentrations can be determined. Table 5-11 shows some calculated concentrations in the standard sample using RSFs previously determined against Ce as an internal standard. These results demonstrate the still untapped potential of using GDMS as a qualitative and quantitative analysis tool for compacted samples.

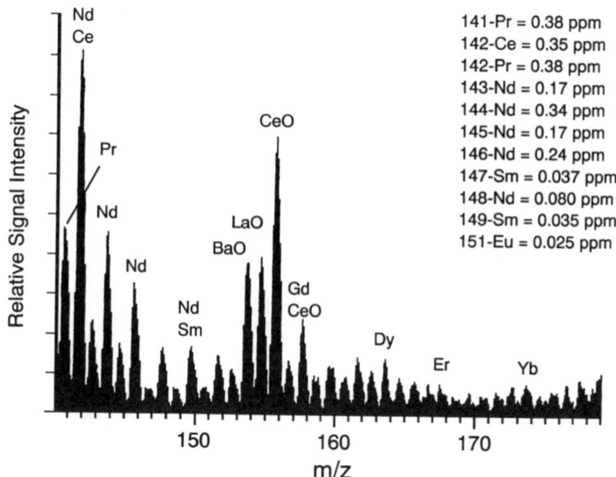

Figure 5-13. Mass spectrum of the rare earth region for an SCo-1 (USGS) standard reference material. Values shown are concentrations in the prepared electrode.

Table 5-11. Rare Earth Concentrations as Determined by GDMS in USGS Standard Reference SCo-1 Values Shown in ppm

	Element concentration	
Ion peaks	Reported	Calculated
La-139	29.7	25.6
Ce-140	63.4	(internal standard)
Pr-141	7.62	6.1
Nd-143, 144, 145, 146	27.9	26.6
Sm-147,149	5.07	5.5
Eu-153	1.03	1.3
Gd-160	4.00	4.8
Dy-162,163,164	3.79	3.8
Ho-165	0.83	0.8
Er-166,167,168	2.39	2.7
Yb-171,172,174	2.25	2.1

5.5.2. Solutions

The point has been strongly made that the major advantage of GDMS is its ability to determine solids directly, without sample dissolution. Thus, the inclusion of solution analysis among GDMS applications seems a contradiction in purpose. Still, there are certain situations in which this approach may make sense. It should be recognized that the direct analysis of solutions by GDMS is not normally attempted. Nebulization techniques that work with atmospheric pressure sources such as flames and ICPs cause quenching of glow discharges. It is necessary to remove the solvent from the sample before introducing it into the plasma. With that caveat, a number of approaches are possible.

The most direct is simply to evaporate a solution sample onto the surface of a conducting electrode, leaving a residue film of the previously dissolved analytes. This thin layer then comprises the sample, which can be sputtered in the glow discharge for analysis. In reality, of course, this is solids analysis, not solutions analysis, as the solution is converted into a solid residue for conventional GDMS. The opportunities and problems are more in keeping with Section 5.5.1.3. On the positive side, the sample is effectively concentrated greatly in that all the analytes are presented to the plasma in the short period of time (seconds) during which the surface layer is sputtered away. This is also a disadvantage in that little time is available to average spectra, or even adjust instrument parameters, creating potential problems of precision.

A somewhat more elegant method of transforming a solution sample into a surface film is by means of electrodepositing selected metals onto a cathode. In this manner, an analyte contained at a trace level in a large solution volume may be isolated and concentrated onto a surface for subsequent sputter atomization into the glow discharge. This approach offers the possibility of both selectivity and enhanced detection limits. A third method serves as a more *in situ* solution/solid transformation approach. Drawing upon well-established electrothermal vaporization methodology, microliter aliquots of solution sample are placed on a filament that will also serve as the GDMS electrode. Currents are passed through the filament to dry the sample to a residue film, then the current is increased to ash away any organic constituents, if necessary, and finally a large step current is applied to atomize the residue rapidly into the glow discharge plasma. Again, all the advantages and disadvantages of microsample techniques are present. Note, however, that in this approach, the primary atomization step is electrothermal, rather than by sputtering. A fourth method of treating solution samples, and one that gets away from creation of a thin surface residue, is the mixing of a solution aliquot with powdered matrix material (e.g., silver or graphite). Drying this mixture under an IR lamp, homogenizing in a Wig-L-Bug vibrator (Crescent Dental Manufacturing Co.), and pressing of the sample by compaction methods (Section 5.1.4) produce a sample electrode containing the solution residue. This amounts basically to a matrix conversion from aqueous solution to metal.

Solution analyses by GDMS have been reported since the mid-1970s,[(41)] but they have rightly remained on the outer fringe of analytical interest. A recent application, however, shows that specialized use of GDMS for solution analysis may be advantageous. Jakubowski *et al.*[(53)] reported the detection of sub-picogram levels for Pt and Ir, elements increasingly of interest in the environment because of their ejection from catalytic converters in automobiles. Samples of 1–10 μl were pipetted onto copper substrates and the solution dried to a residue at room temperature, producing a 4-mm-diameter residue spot for a 10-μl aliquot. Data were acquired by rapid multiple scans over Pt and Ir mass region or by single ion monitoring of selected isotopes. Figure 5-14 shows a time trace of the ion signal from a 1-pg Ir sample. The transient nature of the signal is apparent. The authors show that satisfactory precision was obtainable as indicated by the error bars in Fig. 5-15.

Given the effect of solvent introduction on glow discharge operation, the direct injection of solutions has many obstacles to overcome, but some modified efforts in this direction have been reported. Strange and Marcus[(54)] have used a particle injection system to introduce a solution sample into the glow discharge. For the foreseeable future, GDMS is likely to remain a solids analysis technique, short of some new sample introduction breakthrough.

5.5.3. Nonelemental GDMS Applications

The versatility of the glow discharge relative to elemental applications is generally well recognized in the availability of commercial instrumentation for atomic absorption, atomic emission, and atomic mass spectrometry. Those outside the elemental community have also taken note of the glow discharge and found complementary characteristics that are useful in their field. In general, these emphasize use of the versatile glow discharge plasma processes more than the atomization process. Two examples will be presented here for illustration.

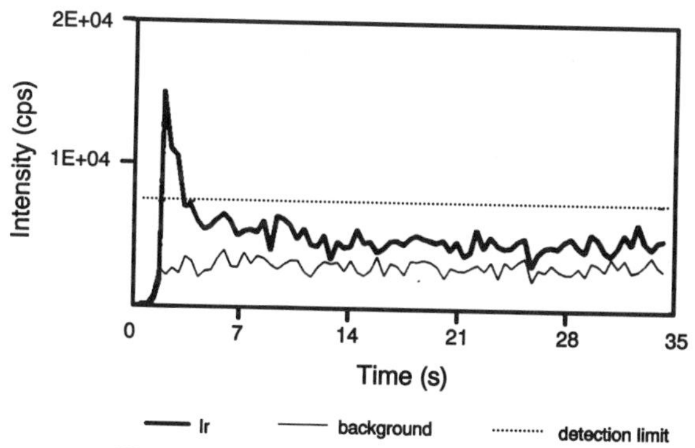

Figure 5-14. Temporal ^{193}Ir ion signal response from 1 pg iridium deposited onto a sample electrode plate.[(53)]

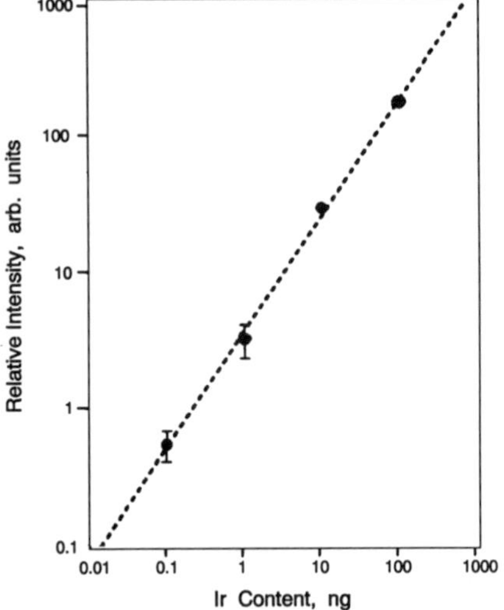

Figure 5-15. Linearity of response for ^{193}Ir ion signal by GDMS.[53]

McLuckey *et al.*[55] took advantage of the inherent ability of the glow discharge to function on almost any readily ionizable gas. While the inert gases are normally used as discharge agents, stable plasmas can be obtained with nitrogen, oxygen, air, and even water vapor. By designing a glow discharge ion source that used ambient air as its discharge gas, the authors were able to analyze trace impurities present in the sampled air. Figure

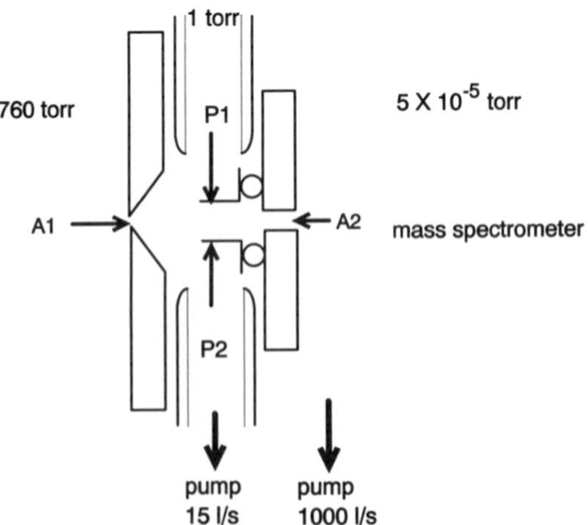

Figure 5-16. Atmospheric sampling GDMS ion source.[55]

5-16 shows a sketch of the atmospheric sampling glow discharge source. By appropriate differential pumping, trace contaminants may be transferred from the sampled air to the glow discharge cell for ionization by the plasma, with the subsequent extraction of the ions into the mass spectrometer for analysis. In this manner a very simple ion source is able to sample directly the ambient air contaminants for sensitive mass spectral analysis. Traces of explosive residues at the ppb level were detected by this arrangement. In this type of application, it is necessary to observe the sample in its molecular form, rather than reducing it to its atomic components. It is the sputter step in the elemental analysis process that produces major bond breaking. By introducing the gaseous sample directly into the low-energy glow discharge plasma, sufficient ionization and residual molecular ions are produced.

All the applications presented have featured the glow discharge as a primary ionization source for analyte determination. An increasing use finds the glow discharge in the role as a detector by ionizing effluents, normally organic materials, from separation processes. Figure 5-17 shows one such use in which the LC effluent is vaporized and directed into a glow discharge plasma. The LC tubing itself serves as the discharge cathode around which the negative glow forms. Sequential chromatographic components pass into the plasma, where they are ionized by collision with electrons and other energetic particles and then extracted into the mass spectrometer for sensitive analysis. The glow discharge serves as a stable, low-maintenance ionizer that creates efficient ionization of the organic effluents from the chromatograph.

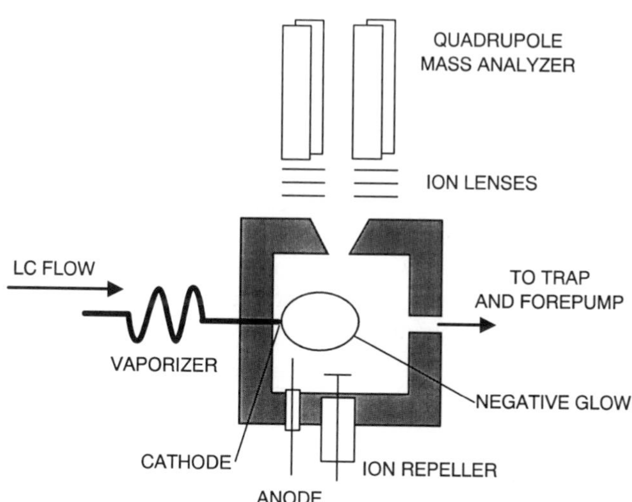

Figure 5-17. Representation of a glow discharge ionization source used as a detector for liquid chromatography.[56]

This success has led others to use the glow discharge as a primary ionization source for the analysis of organic samples.[56]

5.5.4. Future Directions in GDMS Applications

We have reviewed examples of the most common GDMS applications, and these seem likely to show sustained importance in the future. However, it might be worthwhile to suggest a few additional areas that could grow in significance. The dc glow discharge source may find competition from alternative discharge modes. Most promising at this writing is the rf source, which has been demonstrated to exhibit the sensitivity and stability needed for commercial success. At the moment, investigators must build their own rf systems, limiting the extent of usage. The anticipated appearance of a commercial rf discharge system and associated direct insertion probe will stimulate more interest in the analysis of nonconductors such as soils, sediments, glasses, insulators, and ceramics. Careful shielding of the rf generator is critical to prevent serious noise problems, and the spectra generated by rf sources must also prove to have the atomic simplicity characteristic of dc plasmas. Pulsed glow discharges also offer advantages to the analyst for time-resolved measurements, although the additional gating electronics required will be a disadvantage for routine work, not so much in terms of cost, but in the greater complication of data acquisition.

Other types of spectrometers may find application with GDMS in the future, particularly if they are able to assist in solving the problem of isobaric interferences. A magnetic sector instrument (resolution ca. 8000) currently controls the greatest market share, while a second sector unit has only recently appeared that claims advantages based on 20,000 resolution. Two other spectrometer styles that are receiving attention include the FT-ICR and TOF. Preliminary data[22] from the interfacing of a glow discharge to an FT-ICR are quite promising based on attainment of up to 40,000 resolution, which would permit separation of potentially interfering species that currently are unresolvable. Finally, recent improvements in TOF spectrometer designs suggest that, given the powerful intrinsic advantages of this technique, it would be worthwhile to couple a pulsed glow discharge to a state-of-the-art TOF. We can likely expect this in the near future, based on the growing interest in glow discharges.

References

1. C. R. Aita, *J. Vac. Sci. Technol. A* 3 (1985) 625.
2. J. W. Coburn, *Thin Solid Films* 171 (1989) 65.
3. G. D. Patterson, in: *Treatise on Analytical Chemistry* (I. M. Kolthoff, P. J. Elving, and E. B. Sandell, eds.), Vol. 10, Wiley–Interscience, New York, 1972.

4. VG 9000, VG MicroTrace, Cheshire, UK. VG Glow Quad, UK. TS SOLA, Turner Scientific, JEOL, Kratos.
5. F. W. Aston, *Isotopes*, 2nd ed., Longmans, Green, New York, 1924.
6. J. Roboz, *Introduction to Mass Spectrometry: Instrumentation and Techniques*, Interscience, New York, 1968.
7. L. F. Vassamillet, *J. Anal. At. Spectrom.* 4 (1989) 451.
8. P. E. Miller and M. B. Denton, *J. Chem. Educ.* 63 (1986) 617.
9. T. Niehuis, H. F. Heller, and A. Benninghoven, *J. Vac. Sci. Technol. A* 5 (1987) 1243.
10. P. H. Dawson and J. E. Fulford, *Int. J. Mass Spectrom. Ion Phys.* 42 (1982) 195.
11. R. S. Houk, *Anal. Chem.* 58 (1986) 97A.
12. R. E. March and R. J. Hughes, *Quadrupole Storage Mass Spectrometry*, p. 43, Wiley, New York, 1989.
13. P. H. Dawson, in: *Quadrupole Mass Spectrometry and Its Applications* (P. H. Dawson, ed.), Elsevier Scientific, New York, 1976.
14. J. E. Campana, *Int. J. Mass Spectrom. Ion Phys.* 33 (1980) 101.
15. F. L. King, A. L. McCormack, and W. W. Harrison, *J. Anal. At. Spectrom.* 3 (1988) 883. F. L. King and W. W. Harrison, *Int. J. Mass Spectrom. Ion Proc.* 89 (1989) 171.
16. J. F. J. Todd, in: *Quadrupole Storage Mass Spectrometry* (R. E. March and R. J. Hughes, eds.), Wiley, New York, 1989.
17. G. C. Stafford, P. E. Kelley, J. E. P. Syka, W. E. Reynolds, and J. F. J. Todd, *Int. J. Mass Spectrom. Ion Proc.* 60 (1984) 85.
18. S. A. McLuckey, G. L. Glish, and K. G. Asano, *Anal. Chim. Acta* 225 (1989) 25.
19. P. H. Hemberger, J. S. Crain, and J. Williams, Data presented at ASMS Sannibel Island Conference on Ion Trapping in Mass Spectrometry, 1990.
20. N. M. M. Nibbering, *Acc. Chem. Res.* 23 (1990) 279.
21. J. L. Shohet, W. L. Phillips, A. R. T. Lefkow, J. W. Taylor, C. Bonham, and J. T. Brenna, *Plasma Chem. Plasma Process.* 9 (1989) 207.
22. C. M. Barshick and J. R. Eyler, *J. Am. Soc. Mass Spectrom.* 3 (1992) 122.
23. B. L. Bentz, C. G. Bruhn, and W. W. Harrison, *Int. J. Mass Spectrom. Ion Phys.* 28 (1978) 409.
24. N. R. Daly, *Rev. Sci. Instrum.* 31 (1960) 264.
25. W. Vieth and J. C. Huneke, *Spectrochim. Acta* 46B (1991) 137.
26. B. Chapman, *Glow Discharge Processes*, p. 181, Wiley–Interscience, New York, 1980.
27. W. D. Davis and T. A. Vanderslice, *Phys. Rev.* 131 (1963) 219.
28. W. Vieth and J. C. Huneke, *Spectrochim. Acta* 45B (1990) 941.
29. E. W. McDaniel, *Collision Phenomena in Ionized Gases*, Wiley, New York, 1980.
30. M. K. Levy, D. Serxner, A. D. Angstadt, R. L. Smith, and K. R. Hess, *Spectrochim. Acta* 46B (1991) 253.
31. F. Howorka and M. Pahl, *Z. Naturforsch.* 27A (1972) 1425.
32. B. E. Warner, K. B. Persson, and G. J. Collins, *J. Appl. Phys.* 50 (1979) 5694. E. B. M. Steers and R. J. Fielding, *J. Anal. At. Spectrom.* 2 (1987) 239.
33. R. W. Smithwick, *J. Am. Soc. Mass. Spectrom.* 3 (1992) 79.
34. F. L. King and W. W. Harrison, *Mass Spectrom. Rev.* 9 (1990) 285.
35. J. A. Klingler and W. W. Harrison, *Anal. Chem.* 63 (1991) 2982.
36. J. W. Coburn, *Rev. Sci. Instrum.* 41 (1970) 1219.
37. J. W. Coburn and E. Kay, *Appl. Phys. Lett.* 19 (1971) 350.
38. W. W. Harrison and C. W. Magee, *Anal. Chem.* 46 (1974) 461. B. N. Colby and C. A. Evans, *Anal. Chem.* 46 (1974) 1236.
39. C. G. Bruhn, B. L. Bentz, and W. W. Harrison, *Anal. Chem.* 50 (1978) 373.
40. R. K. Marcus, F. L. King, and W. W. Harrison, *Anal. Chem.* 58 (1986) 972.
41. W. A. Mattson, B. L. Bentz, and W. W. Harrison, *Anal. Chem.* 48 (1976) 489.

42. W. W. Harrison, C. M. Barshick, J. A. Klingler, P. H. Ratliff, and Y. Mei, *Anal. Chem.* 62 (1990) 943A.
43. N. Jakubowski, D. Stuewer, and G. Toelg, *Int. J. Mass Spectrom. Ion Proc.* 71 (1986) 183.
44. H. J. Kim, E. H. Piepmeier, G. L. Beck, G. G. Brumbaugh, and O. T. Farmer III, *Anal. Chem.* 62 (1990) 639, 1366.
45. D. L. Donohue and W. W. Harrison, *Anal. Chem.* 47 (1975) 1528. D. C. Duckworth and R. K. Marcus, *Anal. Chem.* 61 (1989) 1879.
46. W. A. Mattson, Ph.D. dissertation, University of Virginia, 1976. J. A. Klingler, P. J. Savickas, and W. W. Harrison, *J. Am. Soc. Mass Spectrom.* 1 (1990) 138. J. A. Klingler, C. M. Barshick, and W. W. Harrison, *Anal. Chem.* 63 (1991) 2571.
47. A. P. Mykytiuk, P. Semeniuk, and S. Berman, *Spectrochim. Acta Rev.* 13 (1990) 1.
48. N. E. Sanderson, E. Hall, J. Clark, P. Charlambous, and D. Hall, *Mikrochim. Acta* I (1987) 275.
49. W. Vieth, A. Raith, and J. C. Huneke, Charles Evans & Associates, unpublished data.
50. Application Note, GDMS-AN-002-01, Charles Evans & Associates, Redwood City, Calif., 1988.
51. GDMS Technical Bulletin, APP/GQ/001, Fisons/VG Instruments.
52. C. M. Barshick and W. W. Harrison, *Mikrochim. Acta* III (1989) 169.
53. N. Jakubowski, D. Stuewer, and G. Toelg, *Spectrochim. Acta* 46B (1991) 155.
54. C. M. Strange and R. K. Marcus, *Spectrochim. Acta* 46B (1991) 517.
55. S. A. McLuckey, G. L. Glish, and K. G. Asano, *Anal. Chim. Acta* 225 (1989) 25.
56. R. Mason and D. Milton, *Int. J. Mass Spectrom. Ion Proc.* 91 (1989) 209.

6

Hollow Cathode Discharges

Sergio Caroli and Oreste Senofonte

6.1. Introduction

6.1.1. Historical Background

Pioneering observations on the luminescent phenomena generated in evacuated tubes belong to the history of spectroscopy and date as far back as the mid-1800s.[1,2] Within this field of research, the first description of a hollow cathode discharge (HCD) published in the scientific literature can be found in the early years of this century, when a German physicist, Friedrich Paschen, reported on the quite unique features of this radiation source.[3] At the time he was mainly engaged in the investigation of the spectral series of H_2 in the IR region and in the distribution of energy in the spectra emitted by glowing gases. Together with Back in 1913 he discovered the effect (named for both scientists) that gives rise to the splitting of emission lines when the source is subjected to a very strong magnetic field. This phenomenon is actually a modification of the Zeeman effect that requires less intense fields.

Thus, interested as Paschen was in the processes occurring when electric currents are passed in a gas at reduced pressure, he arrived almost directly at the special electrode arrangement to which this chapter is devoted. The specific reason that prompted him to develop the HCD prototype was the fact that only a couple of years earlier H. Bartels and co-workers had noted

Sergio Caroli and Oreste Senofonte • Department of Applied Toxicology, Istituto Superiore di Sanità, 00161 Rome, Italy.

Glow Discharge Spectroscopies, edited by R. Kenneth Marcus. Plenum Press, New York, 1993.

how intense the Fowler lines in the He spectrum were in the luminescent layer within the cylindrical cathode of a Geissler tube filled with the said rare gas operated in the dc mode. Experimental evidence available at that time showed that a progressive decrease in the He pressure led to the confinement of the glow within the cathode bore and that an enhancement in current intensity had as a consequence an increase in the radiation output, while the outer surface of the electrode remained entirely unaffected.

On the basis of these previous findings, Paschen devised a glass tube whose main axis coincided with that of the cathode. The anode, instead, was placed perpendicularly to it in a side arm. It may seem surprising that the cathode, worked out by bending a thin Al foil, was shaped so as to form a box open at both ends with a pronounced rectangular cross section, in order that the luminescence in its cavity was ellipsoidal. Paschen noted that at the relatively low applied direct current of 100 mA the expected lifetime of the tube did not exceed 70–100 h, but in spite of this, and in obvious conflict with the long time necessary to photographically record the emitted lines in higher orders (several hours), he was able to fully reach his goal of elucidating the hyperfine structure of the spectrum. The HCD was thus expediently used for the first time in recognition of one of its more striking properties, i.e., the ability to generate spectra with exceptionally sharp lines, and Paschen should be given credit for identifying this potential and exploiting it to the benefit of his studies.

Later on, further applications of this discharge type were reported by Paschen in investigations of the spectrum emitted by a hollow cathode of Al.[(4)] This research should be regarded in its proper light in that what is quite clear to the experimentalist today was relatively unknown in Paschen's time. In other words, it became apparent that not only the filler gas, but also the material forming the negative electrode could issue spectral lines with the same basic features as the former. A detailed description of the various zones distinguishable within the cathode cavity was also given, whereby the presence of the cathode layer of the Crookes dark space and of the negative glow were highlighted. It was noted in particular that in the HCD, contrary to what happens with a flat-cathode He discharge, the more intense radiation is found at greater distances from the dark space boundary, with a maximum of intensity along the axis at a given (and relatively low) value of the gas pressure. The occurrence of Al lines emitted by atoms ionized once and twice was also stressed as a characteristic of the spectrum. The outstanding and exhaustive information thus obtained by Paschen was a further proof of the innovative character of the HCD and to a certain extent marked a turning point in its history as it was brought to the attention of a much larger number of researchers. In two subsequent papers, a complete interpretation of the spectra emitted by an Al HCD in an atmosphere of He was achieved, with particular regard to the lines of singly ionized Al atoms.[(5,6)]

During this same period, other groups started devoting themselves to this branch of spectroscopy. Substantial contributions in fact were made by Schüler and his associates for over 35 years. In one of his preliminary communications, the potential fall in an HCD was thoroughly investigated employing Al or Fe as the cathode material and H_2 as the carrier gas (also He in the case of Al).[7] The major conclusions reached at the time established unequivocally that virtually the entire potential drop resides within the cathode cavity with minimal anodic fall (which is in no case mandatory to sustain the discharge). Simple relations were also established between the ionization energy of a given gas, the ratio of electron free path to that of gas atoms, and the potential fall at the electrodes. It is worth mentioning that the experimental configuration adopted by Schüler for his discharge tube was based on a coaxial mounting of two hollow cylinders, the much smaller inner one being the cathode and the outer one the anode. Since this brilliant beginning, Schüler and co-workers successfully achieved in the subsequent three decades comprehension and applicability of the HCD.[8–12] His originality was also demonstrated in the use of glow discharge (GD) to obtain emission spectra from organic substances through electron impact excitation processes.[13–15]

In addition, among the founders of the technique one should certainly include the name of Sawyer. After some time spent at Paschen's laboratory, he soon developed his own approach to the study of HCD. Extensive investigations were carried out to assess the discharge behavior of numerous elements, such as Al, Ca, Cu, Gd, Hg, Mg, T1, and Zn, and its dependence on different noble gases (Ar, He, and Ne).[16] The parameters governing the excitation of metals in the HCD were thoroughly examined taking into particular account the ability of cathodic material to sputter. It became undeniably clear that the complexity and variety of possible situations inside the glow would not always allow results to be predicted. Sawyer is additionally credited with "exporting" the HCD technique from Europe to the United States while it was still relatively unknown and in doing so set the premises for its widespread dissemination.

In the 1930s the concoction was thus reaching the right cooking temperature as some preliminary steps were being taken toward quantitative analysis. In 1934, Konovalov and Frish had already demonstrated the suitability of the HCD for detecting Ar and N_2.[17] It was only 13 years later, however, when it was possible for McNally and co-workers to establish that quantitative relationships could be derived between spectral emission and concentration of analytes present in the cathodic sample.[18] These authors were able, by means of a water-cooled HCD source, to determine F in metal oxides in the visible region at concentrations as low as 1 $\mu g\ g^{-1}$ with a sample weight of only 10 ng.

The facts that ensued after this breakthrough belong more to the recent chronicle than to history and will therefore be dealt with, to the extent that is necessary to better evaluate the present state of the art, in the following sections.

6.1.2. Present Impact on the Progress of Spectroscopy

Undoubtedly, the first real success of the HCD should be ascribed to its use in atomic absorption spectrometry (AAS) as a device capable of emitting particularly sharp and stable spectral lines.[19,20] The birth and sudden widespread diffusion in the 1960s of this technique were in fact made possible by the availability of a source with such unique characteristics. One might wonder why the fortune of HCD should have been bound to the development of another methodology where the spectral properties of the discharge simply played an ancillary role and were not directly exploited for analytical purposes. This was certainly related to the profound crisis undergone by emission techniques in general in that period, mainly because there was increasing awareness of their inadequacy in the face of the emerging analytical challenges. This distrustful attitude has in practice favored the search for more promising alternatives while discouraging innovative investigation in the field of atomic emission spectroscopy (AES).

The enormous surge forward given by AAS to modern analytical chemistry is, however, out of the scope of this survey, which instead focuses essentially on more specific applications of HCD in AES. The last decade witnessed a genuine renewal in this branch of spectroscopy, mostly thanks to the maturity reached by the inductively coupled plasma (ICP) methodology, paired to a sort of rediscovery of HCD. The host of element-related problems that today must be faced and that require accurate and precise analyses to be carried out in an endless variety of matrices has decidedly contributed to its renaissance, not only because the HCD–AES is an inherently multielemental technique, but also because it offers benefits not easily achieved through other sources of excitation.[21]

From this standpoint one striking feature is the ability of HCD to be employed for the analysis of solid samples as a direct consequence of its basic sputtering mechanism. This property makes the discharge at least complementary to what is afforded by the much more popular ICP source, in its turn particularly suited to assay liquid specimens. This is, however, only one of the aspects characterizing the wide area of coverage of the HCD, as gases and solutions (mainly, but not only, in the form of dry residues) fall within the realm of this atomization/excitation system.[22–27] In turn, it is precisely the possibility of separating these two fundamental processes typical of any low-pressure discharge (LPD) that has recently opened a few major new avenues. Both mechanisms, in fact, display unequaled merits of their own, exploitable to the benefit of other investigative approaches. The facets of this analytical potential will be treated in detail in Section 6.4; here it suffices to recall that sputter atomization by means of LPDs is an expedient means to bring to the vapor state any kind of sample for subsequent introduction into other spectroscopic sources,[28] whereas an alternative strategy has led to the excitation within the glow of the HCD of thermally volatilized materials, as in furnace atomic nonthermal excitation spectroscopy (FANES)[29] (see Chapter 7).

The renewed interest of the experimentalist in LPDs is moreover well documented by the increasing frequency with which ad hoc meetings or special symposia hosted by broader-coverage conferences have been successfully organized in the last few years. Among these, mention should be made, for the time being at least, of the Winter Conference on Plasma Spectrochemistry series, of the traditional FACSS congress, and of the periodic Anwendertreffen held in Germany. All this rapidly developing background presumably will open a brilliant future for the HCD source in many respects. The properties with which this discharge is endowed make this prospect more than likely and the technical evidence provided in the sections to follow will give it the necessary credibility.

6.2. Fundamental Aspects

6.2.1. Working Characteristics of a Hollow Cathode Discharge

Most of the basic characteristics inherent in the functioning mechanism of LPDs are obviously shared by the HCD and are, therefore, exhaustively dealt with elsewhere in this volume. Consequently, the present discussion will focus chiefly on those peculiarities that are unique to the HCD. As already briefly touched upon, the inner configuration of the cathode can be thought of as a series of planar longitudinal narrow cathodic elements placed in parallel fashion along a main axis so that each lamina has its specular counterpart with respect to the said axis. The prototypical hollow cathode set up by Paschen[3] can thus be regarded as two pairs of opposite flat electrodes forming a box open on both sides. Generally speaking, the macroscopic outcome of such an arrangement is that the characteristic luminescence of the GD arising from that side of the plane cathode facing the anode coalesces, in the case of the HCD, with that produced by the opposite segment, provided that the gap between these two is small enough and the rare gas pressure is within a suitable interval. In other words, the brightly emitting zone occupies the space contained within the set of cathodic laminas. By extrapolation to an infinite number of these laminas, each with an infinitesimal width and a given length, it is easy to understand that the overall structure will evolve toward a hollow cylinder, which, if closed at

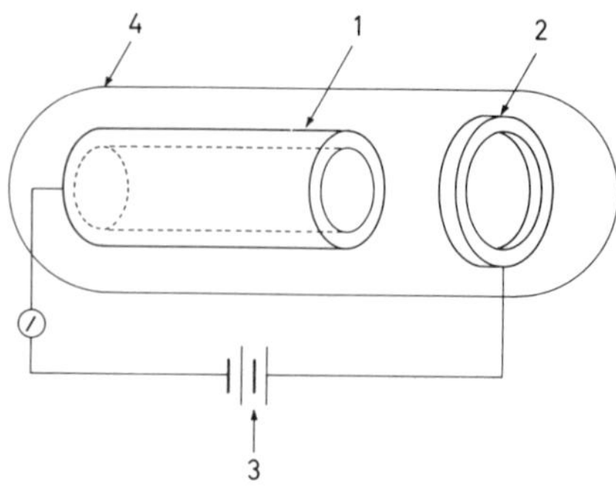

Figure 6-1. Schematic diagram of an HCD setup. 1, hollow cathode; 2, anode; 3, power supply; 4, housing.

one end, gives rise to the configuration known as hollow cathode. The basic scheme of an HCD lamp is given in Fig. 6-1.

The advantages of the cuplike shape of the negative electrode are remarkable: (1) the previously mentioned formation of the negative glow takes place exclusively inside the cavity, and is accompanied by an increase in current density of no less than three orders of magnitude;[30] (2) closely related to the former property, the cathodic fall is contained almost entirely within the bore, where virtually all the relevant processes occur (atomization, ionization, excitation, emission). All this implies that the sample, be it the cathode itself or a specimen placed within the hollow, is confined to the inner volume when it volatilizes and is capable of leaving the cavity only with difficulty. The practical consequences of such a pattern are very expedient from many points of view, since consumption of the sample loaded into the cathode as well as loss of cathode material are minimized.

In contrast to the behavior shown by planar GDs, the mass ablated by the impinging ions and neutrals undergoes repeated cycles of volatilization and redeposition before eventually escaping the bore. This characteristic can be extremely beneficial for a variety of purposes, primarily including analytical determinations. In this context it is useful to recall what was stated in a memorable review published in 1984 by Mavrodineanu.[31] According to this author and in full agreement with present knowledge in the field of LPDs, the features of such devices can be traced back to:

- The production of narrow spectral lines for atoms, ions, and radicals of all known chemical species
- The limited extent of self-absorption
- The nonthermal and practically nonselective mechanism of volatilization induced by the impact of particles from the carrier gas

- The use of a noble gas at low pressure to sustain the discharge, this giving straight access to the vacuum ultraviolet (VUV) region
- The noticeable stability of the process, as the current–voltage characteristics clearly allow one to assume
- The relatively low working temperature even when no cooling of the system by means of a circulating fluid is provided
- The silent performance of the discharge

The HCD shares these properties to the highest degree, particularly the subtle natural width of the emitted spectral lines, which is a major consequence of the virtual absence of both the Doppler effect (linked to the low temperature and pressure of the plasma) and the Stark effect (due to the very weak electric field existing in the negative glow). Together with the state of nonlocal thermal equilibrium (non-LTE, or NLTE), the above peculiarities confer to the HCD unique features, as reflected in the nature of its basic processes.

6.2.2. Plasma Formation

The abundant literature available on the general mechanism of plasma generation in LPDs provides, on the one hand, evidence of the fact that the fundamental principles are well understood, while on the other hand, it testifies to the complexity of the whole phenomenon and its dependence on a large number of physical and electrical parameters, which to a certain extent make some minor facets not predictable.[19,32–34] This also holds true for the HCD, where the geometrical confinement of the glow makes matters less simple. In brief, filling the sealed tube that houses the HCD circuitry with a noble gas at pressures from tenths to thousands of pascals and applying a direct current from a few milliamperes upward, triggers the onset of the discharge and the attendant luminescence. From the standpoint of voltage–current characteristics, as a special case of LPDs, the HCD places itself in the region pertaining to GDs with abnormal cathode fall.[35] In other words, there is an increase in voltage paralleling that of current intensity, less pronounced, however, than the one typical of the abnormal planar model. In this connection it is worth recalling that the three basic quantities presiding over the HCD operation are gas pressure, voltage, and current intensity. These are so interrelated that only two can be freely chosen, while the third is unequivocally fixed by them.

Under the effect of the applied electric field the gas ionizes as the few electrons and ions always present start heading toward the respective electrodes and, in so doing, collide with other, mostly neutral, species. In a cascade process, further charged particles are thus generated until equilibrium is attained between the two contrasting actions of ionization and recombination. A possibly significant contribution to this overall balance is given by electrons ejected from the cathode after the impact of energetic species, as they may play an important role in sustaining the discharge. This is not so for the thermionic electrons, whose aliquot is normally negligible unless the temperature of the cathode is allowed to rise, as may be desirable under circumstances and for preestablished sets of working conditions. Ultraviolet photons also significantly contribute to the generation of electrons and are obviously able to travel in any direction in the negative glow.

Apparently, in an HCD configuration the momentum-gaining particles and photons have a very small chance of not falling onto the cavity surface and the electron-production efficiency is significantly higher than for the planar discharge. A number of physically different zones can be perceived in the plasma moving away from the cathode toward the axis, the sequence of which coincides with that displayed by flat cathode arrangements.

The first of these zones, spreading over the inner surface, is called the Aston primary (or dark) space. It is followed by a weakly luminous region and then by a second dark space (named after Crookes or Hittorf). The bulk of the discharge is, however, filled by the intensely luminescent negative glow. Subsequent to these zones, LPDs are generally characterized by the presence of the Faraday dark space, the positive column, the thin anodic dark space, and the anode glow. A more detailed treatment of their function and relative importance can be found in other chapters of this volume, even though in this context it is important to stress that some of the mentioned regions may actually disappear (e.g., the Faraday dark space), due among other reasons to physical constraints.

What certainly cannot be eliminated (being essential to the discharge functioning) are the Crookes dark space and the glow region. Both zones interact in the case of an HCD in a very particular way, as the electrons are accelerated across the former, where virtually all the cathodic fall concentrates. Some of them gain sufficient kinetic energy to travel through the latter and enter the opposite portion of the dark space, where they experience a reversal of direction. The same electron can thus accomplish several passages across the glow and exert an ionization action far larger than that of the planar system before eventually being quenched. The cathode dark space acts therefore as an electron gun whose function is to transfer energy to the equipotential plasma in the glow region. The yield of secondary electrons generated in the dark space increases steadily with the atomic weight of the inert gas. For He, it is almost negligible and, consequently, a large part of the primary electrons can cover more than once the whole distance between the cathode walls. On the contrary, the ionization cross section of Ar makes this yield of the same order of magnitude as the primary electron current.

Quantitative predictions are rather complex; surely, type and pressure of filler gas play a decisive role in that they determine the yield of secondary electrons from the cathode, this quantity also depending on the nature of the material. Photoelectric emission is estimated to be the main cause of the HCD phenomenon [(36)] and increases in the order Xe (9.57), Kr (10.03), Ar (11.83), Ne (16.85), He (21.22) (values in parentheses are the photon energy in electron volts at selected wavelengths).[(29)] For He, photoelectric emission can reach even 70% of the total. Depending on the type of rare gas and experimental parameters, a different proportion of metastable atoms is also expected, whose role in the excitation process may be of primary importance. Summarizing, there are plenty of events likely to actively participate in the generation and sustenance of the HCD plasma, as set forth in Table 6-1.

The plasma that ensues from this multiplicity of either competing or reinforcing processes is highly energetic and provides an ideal environment for the excitation of foreign species. The remaining paragraphs in this section will elucidate some of the aspects implied in the above capability. Further and exhaustive information on this subject matter can be found in the relevant literature.[(e.g.,37)]

6.2.3. Energy Distribution

Whatever the HCD variant used, one common feature emerges, namely, considerable variations of the electron energy distribution from the one predicted by the Maxwellian model. Experimental observations made by

Table 6-1 Main Ionization Processes in HCD

Ionization type	Event[a]
Primary	
Electron impact	$X + e^- \rightarrow X^+ + 2e^-$
	$X^* + e^- \rightarrow X^{*+} + 2e^-$
Secondary	
Penning type	$G^m + e^- \rightarrow G^+ + 2e^-$
	$G^* + E^* \rightarrow G + E^+ + e^-$
	$G^m + E \rightarrow G + E^{*+} + e^-$
	$G^m + E^* \rightarrow G + E^+ + e^-$
Charge transfer	$G^+ + E \rightarrow G + E^+$
	$G_a^+(\text{fast}) + G_b(\text{slow}) \rightarrow G_a^+(\text{slow}) + G_b(\text{fast})$
Associative	$G^m + E \rightarrow GE^+ + e^-$
Photoinduced	$X^* + h\nu \rightarrow X^+ + e^-$
Cumulative	$E + e^- \rightarrow E^+ + 2e^-$

[a]E, element sputtered from the cathode; G, gas atom; X, either E or G; *, excited state; m, metastable state.

various authors[38,39] using probes and electrostatic analyzers lead to a self-consistent interpretation of the electron energy distribution as expressed by the equation

$$f^*(E_e) = \left[\int_0^{E_e} f(E'_e)\, dE'_e\right] \Big/ \left[\int_0^{\infty} f(E'_e)\, dE'_e\right] \tag{6-1}$$

where E_e is the electron energy. In the case of He and for settings of 270–400 Pa and 30–60 mA, the function $f^*(E)$ was found to change very little and to predict that most electrons possess energy below 20 eV.[38–40] The findings obtained at the time by the same team pointed to the presence of a number of fast electrons (i.e., with energy >20 eV) higher in the negative glow than in the vicinity of the anode and, surprisingly enough, even a large number of slow electrons. The noticeable fraction of electrons possessing energy close to the cathodic fall accounts for the previous statement about the similarity to the planar abnormal discharge.

At lower gas pressures (10–100 Pa), the fast electrons were found to split into two groups, containing very fast (not undergoing any collision) and fast electrons (with little residual energy of about 4 eV available for further ionization and excitation), while the slow electrons were relatively thermalized at approximately 0.5 eV.[41] It is only the first group that is responsible for the high current intensities characteristic of an HCD with electrons repeatedly oscillating from point to point of the inner wall.

All these observations lead to the conclusion that for a given set of electrical parameters there is a particular pressure of the carrier gas at which the portion of electrons with higher energy reaches the maximum. Gas pressure, therefore, exerts a strong influence on the excitation efficiency, the latter quantity being also a function of the atomic energy level. Typically, number densities for fast and slow electrons fall in the range of $2\text{–}20 \times 10^8$ and $2\text{–}20 \times 10^9\ \text{cm}^{-3}$, respectively, and their ratio can be as high as 1:50.[42] The radial electron number density is not uniform, but shifts from a distribution with a maximum along the axis for pressures as low as 10–30 Pa to an annular one with the minimum on the axis for higher pressures (50–100 Pa).[43] This pattern is also shown in relation to increasing current intensities.

Positive ions, in turn, float in the almost field-free glow until they enter the Crookes dark space and are then accelerated toward the cathode surface. In the planar, abnormal GD, ions are prevented from acquiring the full energy of the voltage drop by the occurrence of the so-called symmetrical charge exchange, i.e., the capture of an electron by the fast ion colliding with an atom of the same species. The net balance will be an ion with thermal speed, whereby the overall loss of kinetic energy clearly depends on the total number of collisions, directly related to the gas pressure and the length of the dark space.

This phenomenon is, however, much less serious for an HCD, where a considerable number of ions can display energy up to 80% of the entire cathodic drop.(44) A justification for this difference lies in the very thin dark space peculiar to this type of discharge, which allows only a limited number of collisions to occur whatever the pressure is. Moreover, similarly to what has been observed for electrons, the doughnut configuration is also present for the ion density, in good conformity with the experimentally ascertained practical neutrality of the plasma. Besides this, an annular distribution of brightness of light emitted by the HCD is visible, as would be suggested by the higher density of energy-rich charged species toward the edge of the inner space and the consequently more suitable local conditions for excitation and emission.

6.2.4. Sputtering Process

The sea of corpuscles hitting the cathode surface exert a major influence in that single atoms and clusters are released in a physical state resembling that of a gaseous phase, albeit the temperature is much lower than that in true volatilization.(45) The temperature of a working cathode rarely reaches 800–1000 K. Heat production, however, is always concomitant with the erosion process and may be intentionally augmented through adoption of suitable working parameters to facilitate the extraction of volatile species and hence the separation from a refractory matrix.

This kind of vaporization is termed *sputtering*, as suggested by J. J. Thompson in the early 1920s, and is mainly of a physical rather than chemical nature, since an interaction occurs between the impinging particle and the lattice structure of the cathodic material without giving rise to significant formation of volatile compounds which can evaporate on their own.(46) The ablated species are, as a general rule, neutral as ions are inevitably attracted back to the surface (the probability of ion ejection can be as low as 1/100). For impacting ions of 100–300 eV, the kinetic energy of the sputtered material is relatively small, with part of it immediately diffusing back to the solid phase. Under such conditions, the released species may travel at a speed of roughly 3×10^5 cm s^{-1}.

In the pseudovapor formed in this way, molecular aggregates are also detectable, whose origin can be ascribed to the presence of both exogenous impurities in the cavity and the sample loaded on the cathode bottom, if any. While it is obvious that the crystalline structure is important in the establishment of the overall sputtering rate, the atomic mass of the gas employed is of equal worth, as the closer it is to that of the material to be sputtered, the more efficient is the erosion process. High-purity He is from this point of view of poor consequence and the reported use of this gas together with appreciable ablation may well have been due to traces of other gases like Ne.

Although target atoms to a certain extent are ejected randomly, it has been ascertained that the preferential direction of departure is specular to that of the incident projectile. The duration of the impact may be as short as 10^{-12} s; during this time the colliding particle can penetrate the target surface to a depth of even 0.5 nm. The thickness crossed by the incoming particle is approximately four times the escape depth (the layer from which the target atoms are released). In spite of the internal collisions suffered, enough energy is available to eject atoms from the solid phase (normally, from 12 to 35 eV is necessary for this purpose). The ablation rate can reach up to 200 monolayers min^{-1} and is greatly affected by the presence of quenching species like oxygen and water vapor.

So far, numerous published studies have aimed at elucidating the diffusion mechanism of the sputtered sample cloud and its density in the cathode bore.[47–50] Nonetheless, the intricacy of the factors involved in the global phenomenon allows for few generalizations, such as enhancement of the vapor density of cathode material with increasing tube current, often linked by a linear correlation.[51] Similarly, an increase in gas pressure may cause the maximum density of sputtered material to shift from the cathode axis to the walls. If heat is efficiently dissipated and high-vacuum conditions are attained so as to minimize backscattering phenomena, a general formula for the sputtering yield γ has been found to hold:[34]

$$\gamma = (k\pi R_0^2 n_0 mME)/(m - M) \tag{6-2}$$

where k is a constant, R_0 the closest distance of approach under hard-sphere conditions, n_0 the density of target atoms, E the energy of the impinging body, and m and M the masses of the impacting and leaving particles, respectively. On the other hand, more empirical expressions have been worked out to better account for the experimentally observed behavior in real-world cases.[52]

Thermal equilibrium of the discharge gas with the incoming pseudo-vapor is attained after very few collisions and hence can be considered complete at a distance of only a few tenths of a millimeter from the cathode surface. It is likely that this process ends when the sputtered material is still in the transition region at the border between the cathode dark space and the negative glow. In the long term, the cathode inner cavity will progressively modify its shape under the effect of sputtering, approaching ever more that of an hourglass because of both redeposition phenomena and definitive loss of material through the cathode mouth.

6.2.5. Excitation Mechanism

Since the early work on the HCD, there has been wide consensus that excitation and emission phenomena normally do not obey classical, blackbody-based spectral distributions. In connection with this, one aspect is of great relevance for a correct understanding of the correlations existing between the dynamic atomic–molecular interactions and spectral emission, namely, the nonthermal energy distribution with the ensuing different temperatures for the various microscopic states. In more detail, the temperature of the filler gas atoms can be up to 1200 K, as determined for wide intervals of current intensity,[(53)] while several thousand degrees Kelvin marks the possible electron temperature, with a manifest lack of LTE. The average path length of electrons in the negative glow is a function, among others, of carrier gas type and pressure and can range from a few millimeters to several decades as the gas atomic mass increases.

Considering the multiplicity of parameters that are active in the HCD phenomenon, the overall excitation process will necessarily be the outcome of several concomitant actions, a few of which may prevail over others according to the operational conditions adopted. Collisions of the second type, in the broadest meaning of the term, are generally acknowledged as one of the major contributors to this effect, while a less essential role can be ascribed to collisions of the first kind, i.e., with electrons. As regards the former, the assumption behind it is that conversion of excitation energy into kinetic is minimal and that the total spin of colliding atoms is conserved. On the basis of present knowledge, a number of reactions in the gaseous phase are deemed to occur, as listed in Table 6-2.

Attention is drawn to the important function that in many instances metastables from the carrier gas may have in producing excited states. Just to mention the most frequently employed gases, Ar has important metastable states at 11.49 and 11.83 eV, Ne at 16.85 eV, and He at 19.77 and 20.55 eV. Whether excitation is principally due to electron or metastable–atom collisions is a matter of debate and the response largely depends on the particular situation at hand.[(47,54,55)] The nonthermal character of the interactions *per se* is a definitive advantage that ultimately leads to considerably favor particular emission transitions, with the possibility of reaching high spectral levels

Table 6-2 Main Excitation Processes in HCD[a]

$X + e_f^- \rightarrow X^* + e_s^-$	$E + G^m \rightarrow E^* + G$
$X^+ + e_f^- \rightarrow X^{*+} + e_s^-$	$G + e_f^- \rightarrow G^m + e_s^*$
$X + h\nu \rightarrow X^*$	$G^+ + E + e^- \rightarrow G + E^*$

[a]E, element sputtered from the cathode; G, gas atom; X, either E or G; e_f^-, fast electron; e_s^-, slow electron; *, excited state; m, metastable state.

provided suitable parameters and operation modes are chosen. Strong upper-level first- and second-spectrum emissions are, however, more the rule than the exception, wherefore nonmetals become wholly accessible to the HCD. Concerning the radial distribution of radiation emission within the cathode bore, conclusions reached so far have not been unambiguous, even though a general correspondence can be foreshadowed with the pattern followed by electron and ion density touched on in Section 6.2.3. Intensity of the emitted radiation and electron density are linked by the relationship worked out by Falk and Lucht:[56]

$$dI_e = h\nu_0 n_e n \int_0^\infty Q_a(E_e) v_e(E_e)\, dE_e\, dV \tag{6-3}$$

where dI_e is the whole of the radiation emitted by the infinitesimal volume dV, h Planck's constant, ν_0 the central frequency of the spectral line, n_e the number of electrons, n the number of atoms in the base state, $Q_a(E_e)$ the cross section of excitation, and $v_e(E_e)$ the velocity of the electrons.

Of all the possible uses of the HCD, analytical applications in AES are of consolidated importance. Basic to them is the assumption that the vapor obtained through sputtering from the cathode or sample contained therein and to be excited in the plasma has a stable composition over the entire discharge duration such as to reflect that of the solid phase if this is multi-elemental. Normally, this is achieved after a very short predischarge time, during which the elements with higher sputtering yields are progressively depleted from the cathode surface until an equilibrium is established.

6.3. Main Variants of Hollow Cathode Tubes

6.3.1. Preliminary Comments

The proliferation of HCD lamp configurations for a variety of purposes has encouraged, from the very beginning, research in this field and has been limited only by the inventiveness of the experimentalist in facing a given analytical or technological challenge. In categorizing such devices different criteria can be followed, each being equally valid in describing the various models so far developed. Consequent with the multipurpose-oriented approach chiefly adopted here, it seems more expedient, however, to survey the principal types of tubes according to their general design and operation mode rather than to classify them according to their specific uses, since a given instrumental version is liable to be employed in different sectors.

With regard to the analytical applications (undoubtedly a major portion of the whole scope), description is focused exclusively on exploitation of the

HCD in AES, deliberately ignoring, for the reasons given in Section 6.1.2, its use as a spectral line source for AAS. It goes without saying, at any rate, that every technical improvement in stability, emission intensity, and lifetime of HCD lamps is obviously a potential benefit for this technique.

6.3.2. Selected Models

One fundamental distinction when referring to models is whether the cathodic block of the lamp has been designed to be cooled by a circulating fluid, be it plain water, frigorific solution, or liquid nitrogen. The final decision on the use of a cooled or hot HCD lamp clearly depends on the desired effect: heat dissipation allows the purest features of the discharge to emerge in terms of spectral line sharpness and sampling of the cathode (or specimen) surface through real sputtering, which is most frequently the case.[27] This may not be so when other requirements become a priority, such as the need for selective volatilization of certain components to separate them from an analytically disturbing matrix,[57–59] or for a complete and instantaneous vaporization and atomization of the cavity-hosted sample to further excite it in the HCD plasma, as in the FANES approach.[29,60,61]

A solution intermediate between the cooled and hot HCD is that reported by Eichhoff[62] with the so-called transitional version. In it, thermal effects prevail on the cathode bottom and progressively disappear as efficiency of heat removal increases along the cathode mantle. The sample vapor thus experiences transport into a cooler, more HCD-genuine environment and the spectra emitted are exempt from the thermal influence. Both this artifact and the FANES system coincide (in spite of the would-be important differences) in that production of the sample cloud and its excitation are clearly separate (spatially in the former and temporally in the latter), with the ensuing possibility of optimizing the two processes independently from one another.

To try to list all of the types of HCD lamps reported to date in the literature on the basis of their different geometrical arrangements would be an overwhelming task. Therefore, it seems much more advisable to spotlight those tubes that deserve particular attention owing to their innovative characteristics at the time of their introduction, referring the reader to the appropriate bibliographic sources for a more methodical coverage.[21,28,31,63,64] Starting with the already quoted prototype by Paschen,[3] an almost endless series of lamps have been proposed, among which mention should be made of the one devised by Schüler and Gollnow in 1935.[10] This displays, in fact, the basic features of a demountable device, which has since been adopted by others.

Representative of the class of deep-cooled lamps is the twin hollow cathode set conceived by Török and Záray,[55] which employs liquid nitrogen

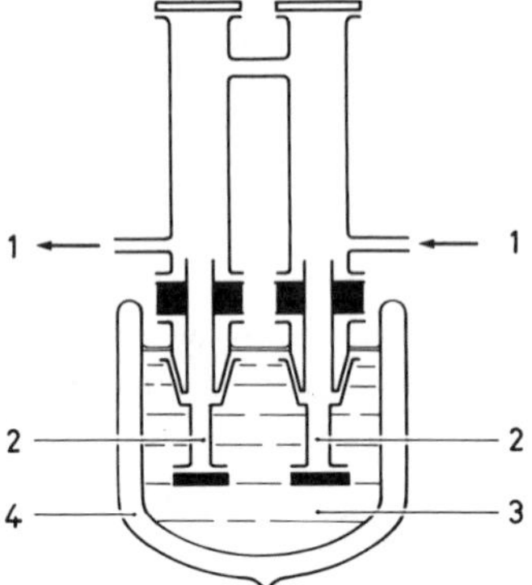

Figure 6-2. Sketch of the deep-cooled, twin hollow cathode assembly. 1, carrier gas pathway; 2, twin hollow cathode arrangement; 3, liquid nitrogen; 4, Dewar. Reproduced by permission of Pergamon Press, with full acknowledgment to the authors of Ref. 55.

as a coolant. The system, depicted in Fig. 6-2, was coupled to an interferometer–spectrometer assembly to thoroughly investigate the excitation process in an HCD in view of its exploitation in trace element analysis. In turn, the equipment shown in Fig. 6-3 was set up to ascertain whether the spectrum of singly ionized Cu was convenient for wavelength calibration in the VUV region.[65] Several Cu(II) lines among the 30 tested were found to be sufficiently narrow and symmetrical for use as wavelength standards. In fact, provided that current intensity and He carrier gas pressure were in the 200–300 mA and 400–800 Pa ranges, respectively, accuracy was better than ±0.01 pm, whereas no significant shifts were observed with the gas

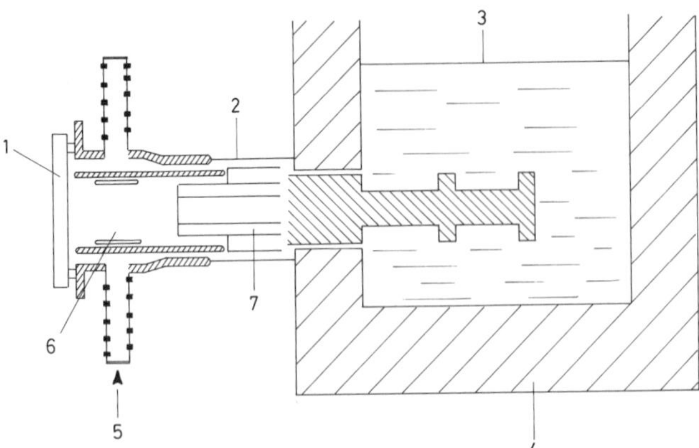

Figure 6-3. Scheme of the liquid nitrogen-cooled HCD tube according to Freeman and King. 1, MgF_2 window; 2, glass sleeve; 3, liquid nitrogen; 4, expanded polystyrene Dewar; 5, carrier gas inlet; 6, aluminum anode cylinder; 7, hollow cathode. Reproduced by permission of IOP Publishing Ltd., with full acknowledgment to the authors of Ref. 65.

flow. Another interesting variant of the deep-cooled HCD tube was launched by Appleblad and Schmidt to study the emission spectra of diatomic molecules .[66] In particular, this apparatus (Fig. 6-4) was devoted to investigation of the CuO excitation under various operating conditions in order to optimize the emission of the molecular spectrum.

Among the many attempts made to better understand the plasma behavior in an HCD, it is worth mentioning the stationary hollow cathode arc (HCA) developed by Bessenrodt-Weberpals *et al.*,[67] a detailed scheme of which is given in Fig. 6-5. Attachments like this are becoming more and more popular for diagnostic development, spacecraft charge control, plasma wake-field accelerators, and others, and allow for the extraction of high-intensity, low-emittance, steady-state electron beams. The one described here achieves discharge stabilization by means of a longitudinal magnetic field and permits the influence of the anode on the spatial distribution of densities, temperatures, velocities, and excitation of plasma particles to be ascertained. For this purpose the authors resorted to powerful investigative techniques such as Thomson scattering, laser-induced fluorescence, and collective scattering diagnostics. The hot and dense plasma that forms in such cases instantaneously expands outside the cathode cavity and hence brings about a dramatic drop in density and temperature of the charge carriers, this phenomenon being accompanied by an equally impetuous outward radial diffusion of particles. Proceeding toward the anode, electron density was found to decrease by roughly one order of magnitude (from 3×10^{13} cm^{-3} to 0.4×10^{13} cm^{-3}). In turn, acceleration of electrons in front of the anode causes their temperature to considerably increase and consequently thermal-

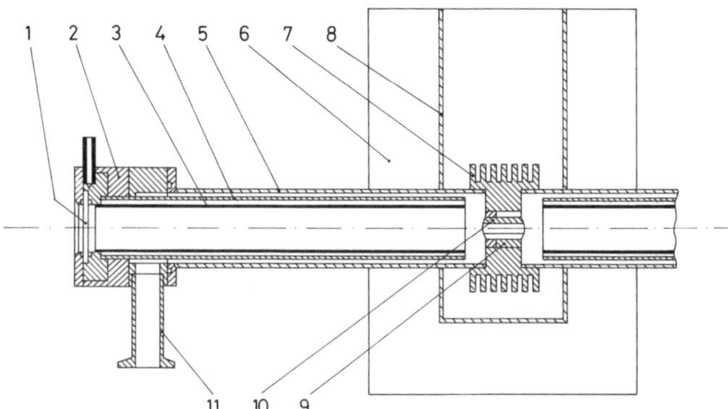

Figure 6-4. Cross-sectional view of the composite wall HCD tube. 1, window; 2, nylon insulation; 3, anodic brass cylinder; 4, glass tube; 5, stainless steel cylinder; 6, polystyrene thermal insulation; 7, copper cooling flange; 8, stainless steel cooling bottle; 9, pump connection; 10, hollow cathode; 11, conical copper socket. Reproduced by permission of the University of Stockholm, Department of Physics, with full acknowledgment to the authors of Ref. 66.

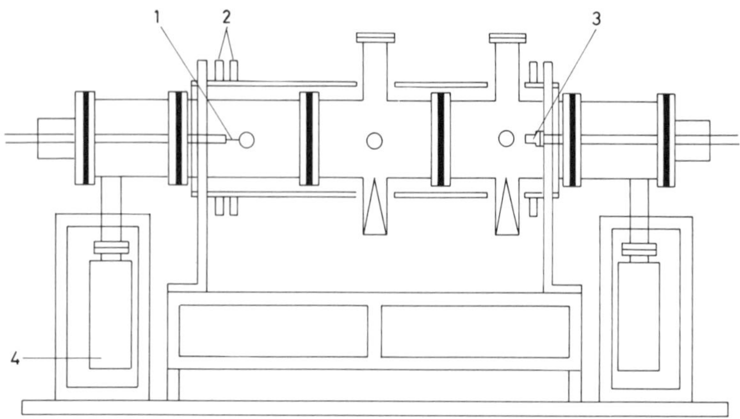

Figure 6-5. Hollow cathode arc device. 1, hollow cathode; 2, coils; 3, anode; 4, pumping system. Reproduced by permission of the Institute of Electrical and Electronics Engineers, with full acknowledgment to the authors of Ref. 67.

ization to be attained. The conditions that favor the establishment of the HCA (e.g., overall voltage 40 V, current intensity 60 A, magnetic field 0.074 T, Ar working gas pressure 1.3 Pa) are quite different from those adopted for the conventional HCD and open new paths for yet unheard-of technological applications.

The analytical constraints set by the very limited availability of certain classes of samples for clinical testing, such as spinal cord or renal tubule fluids, have stimulated the miniaturization of HCD sources, such as the version described by Williams and co-workers.[(68)] With the lamp shown in Fig. 6-6, they were able to quantify Ca, K, Li, and Na at the sub-picogram level (volumes from microliters to nanoliters were requested). The method thus seems able to compete with electrothermal AAS, not considering its multielemental capability.

Atomic fluorescence spectrometry (AFS), once a technique considered a hybrid of AES and AAS, now enjoys a dignity of its own, profiting from the HCD properties, such as the brightness of the spectra emitted and the small width of its lines, which are fundamental requisites for good analytical performance. This warrants the efforts by many research groups to develop HCD models apt to best fit the AFS requirements. Of particular value in this context is the demountable, water-cooled lamp conceived by Rossi and Omenetto,[(69)] illustrated in Fig. 6-7. The very favorable fluorescence-to-scattering ratios and rapidity in interchanging the cathodes and regaining discharge stability were the major advantages of this device, by means of which detection limits of between 0.03 and 2 $\mu g\ ml^{-1}$ could be reached for various elements (Ag, Co, Cr, Cu, Fe, Ga, In, Mg, Mn, Ni, Pb, and Tl).

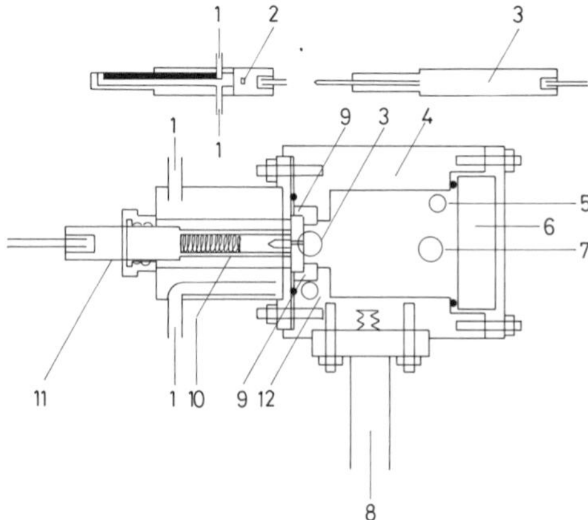

Figure 6-6. Mini-HCD source. 1, cooling water inlet/outlet; 2, auxiliary cathode; 3, anode; 4, hollow cathode housing made of molybdate nylon; 5, carrier gas inlet; 6, quartz window; 7, water-cooled secondary electrode; 8, lamp holder; 9, quartz insulator with a hole; 10, spring-held hollow cathode; 11, hollow cathode holder; 12, carrier gas outlet. The two components in the upper part of the figure are seen as a cross section in the lower part. Reproduced by permission of the Society for Applied Spectroscopy, with full acknowledgment to the authors of Ref. 68.

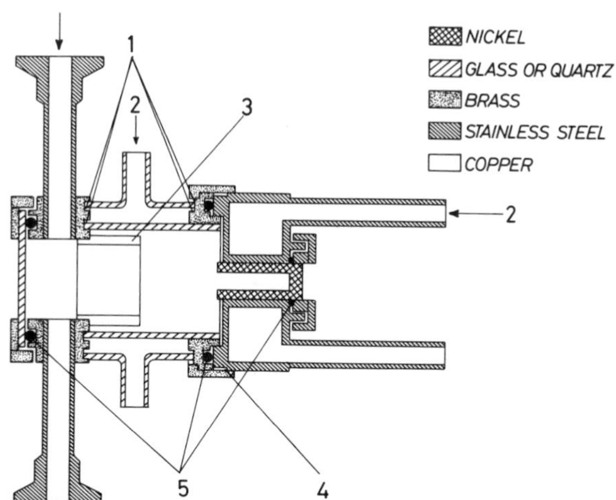

Figure 6-7. Layout of the water-cooled HCD tube for AFS. 1, glass-to-metal seal; 2, water cooling; 3, anode; 4, mica shield; 5, O-ring. Reproduced by permission of Pergamon Press, with full acknowledgment to the authors of Ref. 69.

Sampling of both positive and negative ions for further studies of gas-phase reactions is another area of investigation in which the role of HCDs has recently become ever more important. One of the first effective solutions to this problem was presented by Howorka *et al.*[70] who devised the movable apparatus of Fig. 6-8. The system, able to sample the entire diameter of the discharge, was bakeable at least up to 200°C (400°C was still considered achievable). Residual water may in fact seriously limit the type of processes that can be investigated. Reactions of positive ions of rare gases with neutral species, neutral–radical distributions, radial spectral emission of molecular and atomic states, and measurement of metastable density are among the applications possible with this device. Similarly, the generation of an atomic beam for facilitating spectroscopic investigations of Mo, Th, U, Zr, and other refractory elements was accomplished by means of the appliance presented in Fig. 6-9.[71] This attachment was specifically employed to evaluate the characteristics of U atomic beams by laser-induced fluorescence spectroscopy to determine the beam divergence and the most probable axial velocity.

Metastable population of Li ($1s2s2p^4\ P^0$ level at 57.4 eV) and Na ($2p^5 3s 3p^4\ D^{7/2}$ level at 33.1 eV) were measured using the HCD tube shown in Fig. 6-10.[72] The design includes a 5-cm-diameter stainless-steel tube serving as the discharge anode and container of the metal vapor, and a 1.9-cm-diameter, 30-cm-long stainless-steel tube functioning as the cathode and lying inside the former. The assembly is heated to several hundred degrees centigrade depending on the working conditions adopted. By using a tunable

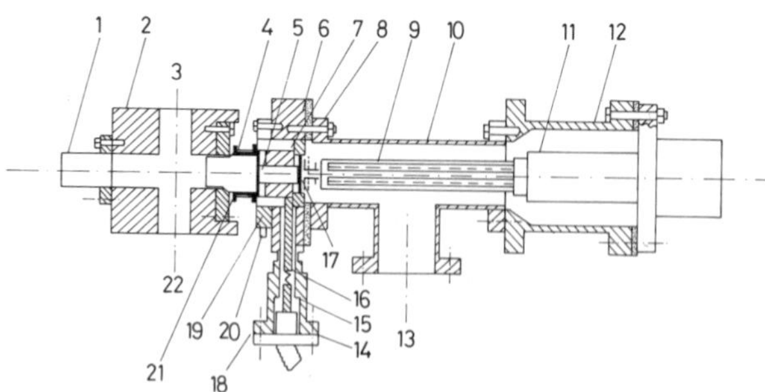

Figure 6-8. Scheme of the bakeable hollow cathode/mass spectrometer system. 1, quartz window; 2, anode flange; 3, connection to pump; 4, metal-to-ceramic fitting; 5, cathode; 6, cathode flange; 7, cathode body; 8, Teflon washer; 9, quadrupole mass spectrometer; 10, mass spectrometer housing; 11, multiplier; 12, Teflon washer; 13, mass spectrometer flange; 14–16, 18, shift mechanism; 17, hole probe assembly; 19, shutter; 20, gas inlet; 21, anode; 22, connection to pressure sensor. Reproduced by permission of Elsevier Science Publishers, with full acknowledgment to the authors of Ref. 70.

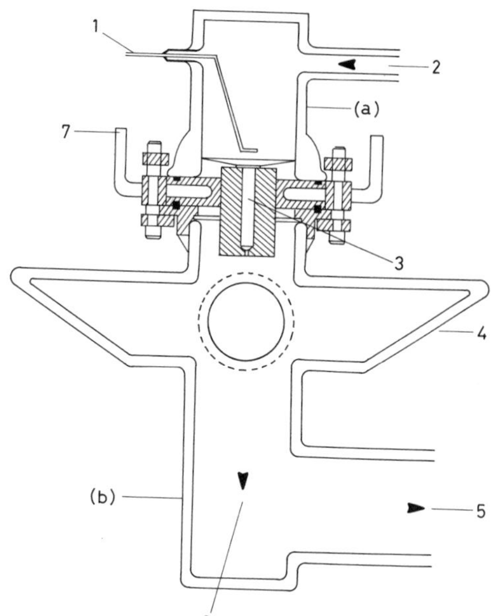

Figure 6-9. Diagram of the hollow cathode observation chamber apparatus. (a), hollow cathode assembly: 1, anode; 2, buffer gas; 3, cathode; 7, cooling water; (b), observation chamber: 4, laser; 5, connection to pump; 6, atomic beam. Reproduced by permission of the American Institute of Physics, with full acknowledgment to the authors of Ref. 71.

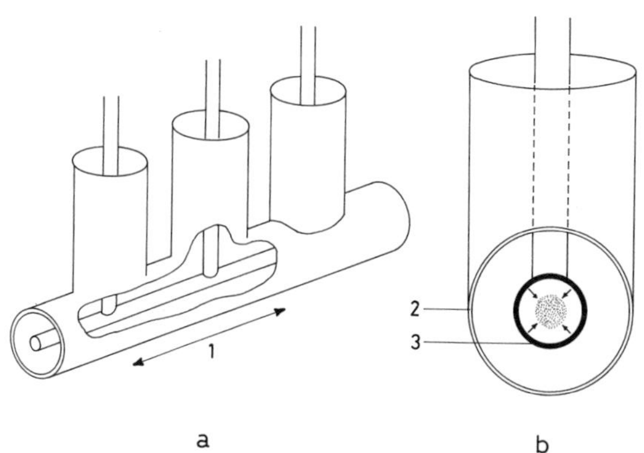

Figure 6-10. Sketch of the pipelike HCD apparatus. (a): 1, Li heat pipe zone; (b): 2, anode; 3, cathode (the end view shows the plasma and arrows represent high-energy electrons). Reproduced by permission of the Optical Society of America, with full acknowledgment to the authors of Ref. 72.

probe laser, densities of 3×10^{10} and 10^{11} atoms cm^{-3} were determined for Li and Na, respectively.

Approaches of this kind enrich present knowledge of the atomic levels likely to be used for energy storage for extreme-UV lasers. An exhaustive survey of the potential possessed by the HCD for the development of a new class of metal ion lasers reports the principal transverse type available at the time, as shown in Fig. 6-11.[(73)] These devices are comparable in UV output power to noble gas ion lasers although they require no more than 1/20th of the normally employed threshold current intensities. Another configuration was developed by Iijima[(74)] to improve the oscillation characteristics of the He–Zn^+ ion laser. The structure, shown in Fig. 6-12, permitted high voltage to be obtained through restriction of the cathode inner wall by an insulator. The importance of the HCD in this technological area should not be underestimated as attested by the multiplicity of configurations continuously being elaborated.

Further possibilities offered by the HCD extend to deposition of coatings or erosion of surfaces. Figure 6-13 shows, as an example, an assembly constructed to deposit Al layers on a substrate, elucidating the influence of cathode separation, bias voltage, and working gas pressure on film thickness and structure.[(75)]

This hasty, largely incomplete scan witnesses the ingenuity of experimentalists active in this field of research. Whatever the utilization of the

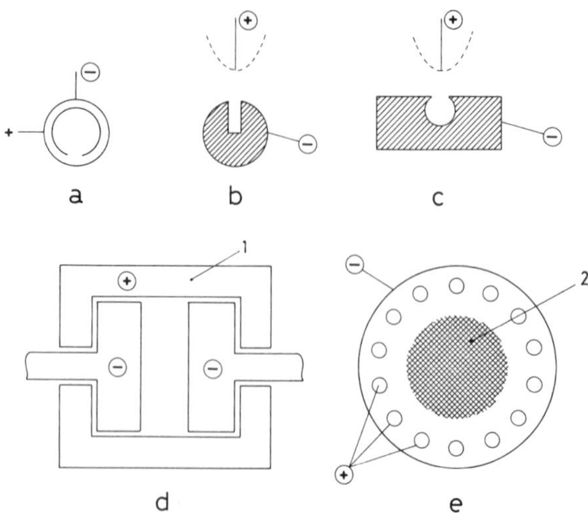

Figure 6-11. Hollow cathode laser types with transverse excitation. (a) Schuebel model; (b) rectangular slotted HCD; (c) circular slot HCD; (d) waveguide structure: 1, envelope; (e) hollow anode or obstructed glow: 2, discharge. Reproduced by permission of the Institute of Electrical and Electronics Engineering, with full acknowledgment to the authors of Ref. 73.

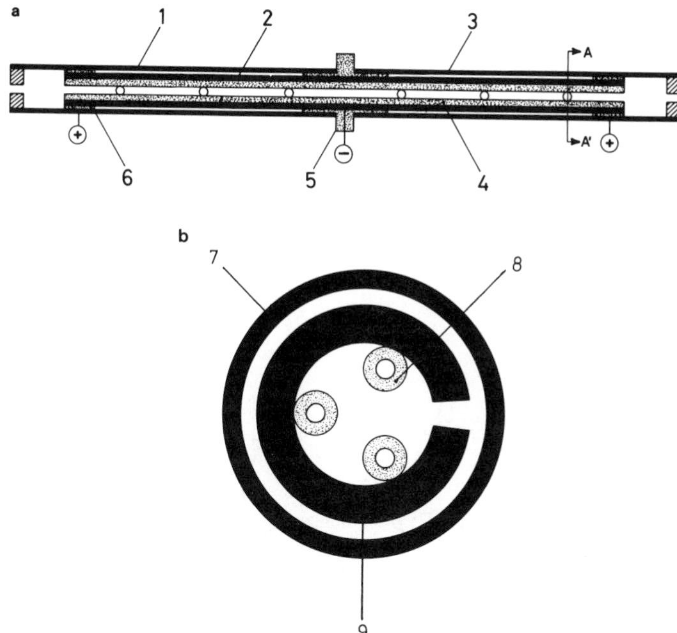

Figure 6-12. Cylindrical-type hollow cathode discharge after Iijima. (a) longitudinal section: 1, anode (I); 2, perforated hollow cathode; 3, anode (II); 4, steatite pipe; 5, macor; 6, macor space; (b) cross section at A–A′: 7, anode; 8, steatite pipe; 9, perforated hollow cathode. Reproduced by permission of the *Japanese Journal of Applied Physics*, with full acknowledgment to the authors of Ref. 74.

HCD's various versions, the overall trend is more than ever toward an increase of research and development.

6.3.3. Boosted Models

The need to enhance the emission output of HCD lamps can be considered a common feature of all application sectors. From an analytical point of view this is all the more true, because one of the major drawbacks of the HCD as an excitation source lies in the fact that, in spite of the very favorable intensity-to-background ratio, the net emission is not sufficient, under normal working parameters, to detect analytes at ultratrace levels. Therefore, great effort has been and still is being devoted to the design of discharge tubes capable of decidedly higher emission power without compromising the valuable HCD features. Generally, this goal can be achieved either by expediently modifying the geometrical configuration of the cathode or by

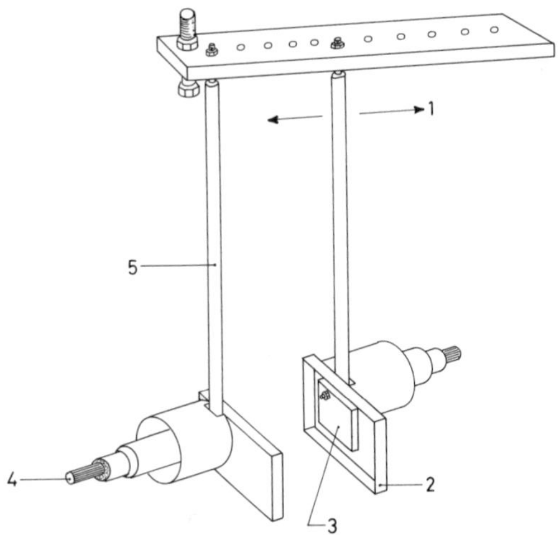

Figure 6-13. Schematic view of an HCD arrangement for ion plating. 1, variable position insulated support; 2, mask; 3, cathode block; 4, high-voltage feedthrough; 5, insulated support. Reproduced by permission of Elsevier Sequoia Science Publishers, with full acknowledgment to the authors of Ref. 75.

enhancing the discharge by superposing an external form of energy compatible with the one inherent to the HCD, without unacceptably altering its essential positive features.

Examples of the first category include inserting a cone or, better still, a rod inside the cathode cavity and coaxial with it. The sample to be assayed is placed on its free end. This version was proposed by Papp and his associates as illustrated in Fig. 6-14.[76] The advantage of this configuration is that

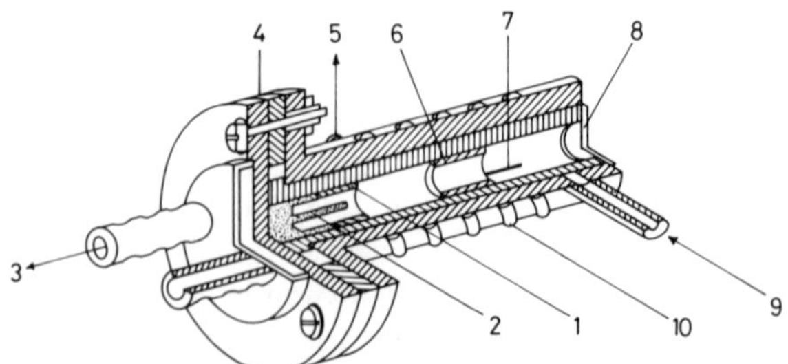

Figure 6-14. Structure of the three-electrode HCD lamp. 1, hollow cathode; 2, coaxial rod; 3, cooling water; 4, insulator; 5, carrier gas outlet; 6, anode; 7, anode connection; 8, quartz window; 9, carrier gas inlet; 10, cooling coil. Reproduced by permission of the Royal Society of Chemistry, with full acknowledgment to the authors of Ref. 76.

it increases the sputtering action on the surface of the specimen by staying in the very core of the plasma, and also because the bar, adequately insulated from the cathode, can be kept at a more negative potential than the latter and may be varied independently. The net result is a strong enhancement of the ablation rate and the ensuing reinforced emission.

Further options are the microcavity HCD described in Section 6.3.2,[68] which has a prominent predecessor in the model described by Czakow.[77] The desired effect of intensifying the emitted radiation has its rationale in the relation found by Novoselov and Znamenskii[78] linking in an inverse fashion the intensity I of a spectral line with the cathode inner diameter:

$$I = \text{const}/d \tag{6-4}$$

A greater diversification of design is encountered in the case of boosted versions of the HCD. One of the first approaches simply consisted in coupling a secondary discharge to the hollow cathode to increase the electron density within the cavity.[79,80] This effect can be obtained either by a thermionic source (a heated W filament or plate) or, more satisfactorily and elegantly, through the lateral insertion into the main bore of an auxiliary hollow cathode operated in the hot mode.

Superposition of transversal or longitudinal magnetic fields was also deemed particularly promising and was primarily described by Popovici and Someşan[81] and Rudnevsky and Maksimov.[82] Under this external influence, charged particles are greatly deflected from their expected pathways with the consequence that the density of slow electrons tends to rise. The process favors an enhancement in the intensity of the spectral lines with low excitation potential. Quite recently, another step ahead in this area of investigation has been done by Pavlović and Dobrosavljević[83] who applied a rotating magnetic field to the HCD (Fig. 6-15). Although clearly

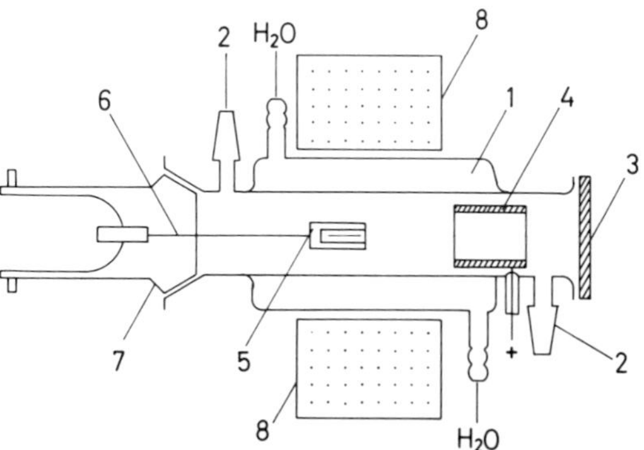

Figure 6-15. Diagram of the rotating magnetic field HCD. 1, Pyrex tube; 2, glass joints; 3, quartz window; 4, aluminum anode; 5, graphite cathode; 6, tungsten electrical connection; 7, spherical ground joint; 8, solenoids. Reproduced by permission of Pergamon Press, with acknowledgment to the authors of Ref. 83.

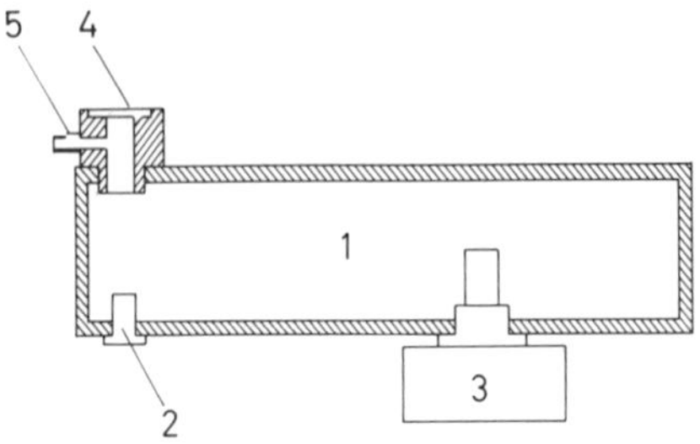

Figure 6-16.
Sectional scheme of the microwave-boosted HCD. 1, lamp body; 2, hollow cathode; 3, magnetron; 4, quartz window; 5, pump and gas ports.

a function of carrier gas type and discharge settings, the improvement of emission intensity for analytes like Cs, Pb, Sb, and Zn did not in the best case exceed 50–70% of the conventional source.

Another strategy consists in pulsing the discharge, which leads to higher spectral intensity, lower noise, and often a reduction of linewidth.[84] The current pulse may be superimposed to the normal dc continuous operation mode or even replace it. The end application of the pulsed lamp (in AES, AFS, or AAS) foreordains the optimization of the working parameters. Also of value in this connection is the rf-boosted HCD proposed by Farnsworth and Walters,[85] where the current pulse is used to sample the cathodic material and the rf burst reexcites the atomic vapor thus generated. The main benefit is due to the noticeable reduction in self-absorption of resonance lines, a phenomenon not at all negligible with a conventional pulse discharge.

Finally, coupling with microwave (MW) irradiation is particularly enticing as the increase in the signal-to-background ratio for the analytes can be of one order of magnitude or more, while emission from the filler gas is reduced and self-absorption almost vanishes.[86] The positive effect is thought to be related—especially for Ar—to the larger production of metastables from the rare gas consequent to the reciprocal reinforcement of the two congenial types of plasma. A new model of the MW-coupled HCD tube has been designed, as illustrated in Fig. 6-16, in which the magnetron is placed within the discharge housing, thus eliminating any need for coaxial cable and the attendant loss of MW energy due to transmission.[87]

6.4. Recent Applications

6.4.1. Sample Introduction

In these and the following sections, attention will principally focus on the developments, trends, and uses disclosed in the scientific literature over the last decade with the purpose of citing those more representative. From what has been said so far, it follows that the hollow cathode source, as in general all GDs, is of considerable importance in the light of volatilization of solid materials whose vapors can act as atom reservoirs for an array of other instrumental techniques.[28] Today this is ever more true given the astonishing development of modern analytical methodologies, which are plagued by the need for regretfully time-consuming and tedious procedures to dissolve the samples in a reproducible and straightforward way, when liquid specimens are required.

For solid conducting materials, this is easily achieved with the HCD, as all that has to be done is to machine the material into a suitable form (e.g., an entire hollow cathode or a disc to be placed on the bottom of a supporting electrode) and clean its surface of residual contaminants. Generally, a short period of presputtering can remove even the slightest traces of exogenous chemicals. If any powders or ashes are to be assayed, they should preliminarily be pelletized with pulverized graphite, copper, or silver at ratios from 1:3 to 1:9 in order to make them conductive and, therefore, liable to attack from the bombarding particles of the noble gas. The use of these binding agents is also beneficial from the standpoint of smoothing off inhomogeneities which might seriously affect signal stability as well as of minimizing matrix effects thanks to the presence of a dispersing medium.

For the above reasons, if solutions are amenable to the technique at hand it would apparently be preposterous to convert them into dry residues for the subsequent atomization, were it not for the fact that chemical interferences from other constituents (in the first place the major and the minor ones) can be noticeably reduced, while the signal is enhanced owing to the analyte concentration in the resulting thin film of salts. In consideration of this very small thickness, emission from the sample will be limited in time, but its transient nature is less than a drawback with multielemental approaches. Alternatively, solutions can be absorbed on porous supports (graphite or pellet disks of pressed copper powder) for a longer duration of the sputtering step.

Obviously, these considerations are particularly valid when the discharge is operated to supply the vaporized sample to the very HCD plasma, in other words, when the source simultaneously performs the two tasks of vaporizing and exciting the material under test without any practical distinction between the two processes, or when the two functions are temporally separated as in the FANES system.[29] Beyond AES, however, the techniques that at the moment greatly benefit from the availability of HCD devices are AAS, AFS, plasma etching and deposition, mass spectrometry, laser resonance ionization, and laser technology at large.[88–93]

6.4.2. Analysis of Solids

One of the best explored application areas of the HCD, the determination of trace, minor, and major elements in massive or pelletized specimens still records numerous and often innovative uses.

Resorting to the procedure of machining massive cathodes from the specimens to be tested, brass and stainless steel samples were analyzed for their Cu and Zn, and Cr, Fe, and Ni contents, respectively, with accuracy comparable to that of ICP–AES.[94] Analogously, the quantification of Au in Au alloys containing from 79.91 to 97.87% of the element was carried out employing the internal standardization method.[95] Both accuracy and precision were about 0.03% (w/w), thus comparing favorably with fire assays.

A comparison of spectral patterns obtained from dc and high-frequency (27.2 MHz) HCD source was carried out for solid copper, brass, and steel using Ar as the carrier gas.[96] Results showed that the electronic energy distributions are similar for Ar II ions, while they may be quite different for cathode elements. Spectral lines from Cd, Cu, and Zn showed increases in emission intensity of more than one order of magnitude for the high-frequency HCD. Mixture of fluoride salts (NaF, PbF_2, LaF_3, ZrF_4, Na_3AlF_6, and K_2NbF_7) with SiO_2 were analyzed in a hot HCD tube through quantification of the InF bands.[97] The method was able to measure F in the $1–10^4$ μg g^{-1} concentration range, the lowest amount detectable being 10 ng.

After introduction of metal chips of Pb, Sb, and Zn or metal salt of Cs and T1 into the cathode cavity, the behaviour of these elements in the rotating magnetic field HCD was reported by Pavlović and Dobrosavljević.[83] The MW–HCD was also employed to analyze Cr, Cu, Mn, Mo, and Ni in stainless steel with considerable improvement over the conventional dc mode.[98,99] The briquetting procedure, in turn, was used to discharge nonconductive powders containing trace elements like Ag, Al, Cr, Ga, Mn, Ni, Pb, Sn, Ti, and Zn with detection limits between 1 and 10 μg g^{-1} [100] Cu, Fe, Mg, Mn, Ti, and Zn could be quantified in Al alloys with the MW–HCD operation mode at concentrations from 0.01 to 6%.[101] Finally, the differences in sputtering of solid samples between the conventional HCD and the MW-boosted version were examined in view of the benefits of the latter in the analysis of trace elements.[102]

On the basis of these examples, one can certainly infer how broadly the HCD is being exploited in this specific field.

6.4.3. Analysis of Liquids

Apart from a few attempts to introduce solutions as such into the discharge by spraying them,[24] the most widely adopted mode resorts to the sputtering of the thin layer of salts left by the liquid sample after drying within the cathode bore.

In Section 6.3.2 we discussed the ability of a mini-HCD source to quantify Ca, K, Li, and Na.[68] Absolute detection limits of 0.70, 0.38, 0.88, and 0.72 pg, respectively, could be obtained. Several rare earth elements (Er, Ho, Nd, Tm, and Y) were determined after drying of solutions within a Cu hollow cathode operated in the hot mode.[103] Absolute detection limits ranged from 0.02 (Y) to 0.25 (Nd) μg, with an improvement of about one order of magnitude with respect to those attained in graphite cathodes.[104]

Given the occurrence of serious spectral interferences in steel matrices, the determination of P in this material has to be performed after dissolution of the solid and drying of the solution within the cathode cavity.[105] Using He as the carrier gas in the MW-coupled discharge, amounts of P as low as 20 μg liter $^{-1}$ could be detected. The ability of the HCD as such or combined with an MW field to analyze trace elements in solutions obtained after digestion of biological materials was extensively verified by Caroli and his team.[106–110]

Cr, Mn, Ni, and Pb were analyzed in solution dry residues using both a conventional dc and an MW-boosted HCD lamp.[99] The data obtained with the two operation modes using Ar or He as the filler gas point to a significantly better detection power and sensitivity for the MW–HCD. Desiccated solutions of fluoride salts were analyzed with an HCD setup using graphite supporting cathodes.[97] Quantities from 10 to 1000 μg g^{-1} F were amenable to determination.

An elegant way to circumvent the problems posed by liquid samples in an HCD was proposed by Foss *et al.*,[27] who held the solutions at liquid nitrogen temperature within the cathode and thus could sputter them in the solid form. Seventy elements, variously grouped, could be analyzed in volumes from 20 to 50 μl extracting the ions from a miniaturized hollow cathode and sending them to a double-focusing mass spectrometer. Detection power ranged from submicrogram to milligrams per liter.

A number of elements, such as Ag, Al, Au, B, Cd, Co, Cr, Cu, Fe, Mn, Mo, Ni, Pb, Se, Sn, W, and Zn, were analyzed by means of the HCD-excited ICP–AFS.[111] To achieve acceptable compromise conditions, rf power, carrier-gas flowrate, and nebulization type were thoroughly examined and optimized. The use of an ultrasonic nebulizer afforded detection limits in the low microgram per liter range, with the exception of refractory elements, which showed significantly higher levels (up to 200 μg liter^{-1} for W). Further improvements in ICP–AFS were gained by using an HCD pulsed at peak current higher than 1 A.[112] Detection limits for Ag, Al, Cu, Cr, Mo, Sr, and Zn were in the interval from 0.02 μg liter^{-1} for Cu to 5.0 ng liter^{-1} for Mo. Molecular species were also investigated by probing the vapors sputtered from the Cu–Pb cathode in the GD source using an N_2 pump dye laser to excite the fluorescence.[113]

Again, the suitability of boosted HCD sources for use in ICP–AFS was scrutinized and compared with that of the ICP source.[114] The test was performed for Cu, Ni, and Pb. It was concluded that the HCD is less suitable than the ICP torch in terms of detection ability, although other advantages are obvious, e.g., low cost of the device. In an attempt to overcome the rather poor detection power of HCD-excited ICP–AFS (one to two orders of magnitude lower than in the case of ICP–AES), some technical improvements were made such as the use of ultrasonic nebulization and of potentiated HCD tubes.[115] The potential of this analytical approach was tested for numerous trace elements in water, blood, and fuel oils.

Although not as straightforward as direct nebulization, the procedure generally adopted to analyze liquid specimens is reasonably reliable and rapid, and the preconcentration undergone by analytes on the cathode surface during the desiccation step enhances the intensity of the emission signal resulting in better detection power.

6.4.4. Analysis of Gases

Although inherently suited to quantify components in a gaseous mixture, very few such instances have been reported. This is all the more surprising if one considers that most studies have been, and are being, conducted with gases. Nor should it be overlooked that often analysis of gases with HCD means their quantification in a solid host matrix, as was reported for oxygen in iron[116] or for nitrogen in refractory alloys.[117] One significant example of application of this kind of analysis, however, can be found in the monitoring of artificial atmospheres in spacecraft.[22] Again not really a determination of gases, the system devised by Iida[118] permitted the analysis of Al alloys for the content of Cu, Fe, Mg, Mn, and Ni (besides Al) after either single or multiple pulse laser vaporization of the solid sample. Detection power was in the range of a few micrograms per gram.

Perhaps the most significant example in this field relates to the qualitative and quantitative assay of Br, C1, and F in a He gas stream.[119] The system, based on the use of a microcavity HCD, permitted the determination of amounts in the low nanogram range and was seen as a more than likely candidate for an element-specific detector in vacuum-outlet gas chromatography.

No further significant analytical applications of this kind have been reported recently.

6.4.5. Special Uses

6.4.5.1. Optogalvanic Spectroscopy

An exhaustive study of the optogalvanic effect in the HCD plasma was made by van Veldhuizen *et al.*[120] The time-dependence of the effect was analyzed in terms of ambipolar diffusion. Further, the relative enhancement of optogalvanic signal related to the presence of the ^{235}U species was described for the U atomic lines at 436.3 and 437.2 nm, the cause of which was ascribed to saturation and radiation trapping effects.[121]

Of the two possible concomitant processes leading to the optogalvanic effect (increased cross section for electron impact ionization and increased electron temperature), a study conducted in an Ne HCD yielded results favoring temperature coupling as the dominant event under the chosen experimental conditions.[122] Agreement with the predictions of a mathematical model was more than satisfactory.

Two-step optogalvanic spectroscopy was applied to the measurement of the radial distribution of the electric field in the cathode fall region of an Ne HCD. Observations of the linear Stark effect of the nd' ($n =$ 10–12)–$3p'[1/2]1$ transitions were made. The depth of the cathode fall in the HCD is one-fourth to one-fifth (ca. 0.8 mm) that of a flat cathode GD under similar experimental conditions (3–4 mm), whereas the electric field at the cathode surface is approximately 5.2 and 3–4 kV/cm in the two cases.[123]

A rather complicated relationship could be arrived at through laser optogalvanic measurements linking the spatial distribution of the electric field in an HCD with the spatial variation of the spectral width of the Ar $7d[5/2]3$–$4p[3/2]2$ transition.[124] No inverted population between the levels $^2P_{1,2,3,4,5}$ and $^1P_1$ was detected in an Ne HCD by using the two-step optogalvanic method, in full agreement with the observed very large enhancement of the optogalvanic signals.[125] Further optogalvanic observations revealed a linear decrease in electric field with radial distance from the cathode surface in the case of a Kr HCD laser through study of the Stark shift of the Kr $8d[3/2]^0_2$–$5p[3/2]_2$ atomic line.[126]

The inherent ability of the HCD to lend itself to such investigations favors both the comprehension of its basic working mechanism and the development of other technological applications, such as those dealt with in the next subsection.

6.4.5.2. Laser Technology

A fascinating area of investigation, laser development through the use of the HCD process has gained tremendous momentum over the last decade.

The problems of efficiency, instability, cataphoresis, and sputtering underlying the development of a proper discharge tube for cw laser operation were faced by Apai *et al.*[127] They designed a dc He–Kr^+ laser affording an output power of 5 mW at 469.4 nm which could adequately control the sputtering rate by increasing the total gas pressure, with a considerably longer lifetime of the discharge tube.

To overcome the drawbacks inherent in the excessive length of flutelike or hybrid-sized HCD tubes for laser operation, a compact (46 cm long) device of cylindrical structure was fabricated for white light laser operation, whose performance is equivalent to that of the other systems mentioned above.[(128)] Transverse and longitudinal HCDs for laser operation were compared with regard to the 469.4 nm Kr transition. The first type affords better axial uniformity, considered to be superior for laser media excitation.[(129)] Longitudinal HCD is affected by plasma inhomogeneity seriously affecting the laser tube operation.[(130)]

A new model of HCD He–Zn laser at 492.4 nm was developed with a grid as the auxiliary electrode to influence the gain of the laser line through variations in the charge flow distribution.[(131)] The Cu HCD He–Ne laser at 632.8 nm ($3s_2$–$2p_4$ transition) produces a higher efficiency than an Al one.[(132)] This could be ascribed to the fact that the upper laser and $^4S_0$ He metastable levels are more populated in the former case. Cd and Zn neutral atom densities were measured by Arai *et al.*[(133)] in the HCD He–Cd^+ and He–Zn^+ laser discharge by monitoring the decays of the Cd(II) 636.0-nm and Zn(II) 610.2-nm line intensities. Density values of 6.2×10^{14} and 7.1×10^{14} cm^{-3} were obtained, respectively. Although these values are lower by a factor of 0.3 than expected, they still are higher by one order of magnitude than those found in the positive column-type laser discharges.

A number of elements (Mo, Sm, Ta, V, Y, Zr) were studied after machining them in the form of massive hollow cathodes to demonstrate the suitability of the HCD for very high resolution laser saturation spectroscopy and consequently to ascertain the isotope shifts and hyperfine structure.[(134)]

As some amount of ions in the interelectrode space is necessary before every pumping pulse for regular and stable pulse laser operation, a high-voltage electrodeless discharge was combined with a pulsed HCD Cu^+ laser, resulting in a roughly 30% enhancement in laser output power and lower likelihood of arc formation.[(135)] For the first time four laser transitions [Te(II) at 629.4, 564.9, and 548.8 nm and Ge(II) at 589.3 nm] were reported in HCD-operated laser systems.[(136)] With these elements it was possible to verify that laser action can take place via charge exchange, Penning ionization, and radiation cascade processes. A lifetime of ca. 360 h was achieved in a dc He–Kr ion HCD laser with an output power of 5 mW at 469.4 nm.[(137)]

Reportedly, laser light noise may be a major problem in writing applications and means to abate it are constantly being sought.[(138)] An He–Zn HCD laser tube, used to conduct an investigation of this issue, revealed that fluctuations in He^+ are responsible for the instability of the power output from which noise originates. The addition of tiny amounts of Ne suppressed noise down to less than 0.2% (rms).

Additional information on the use of HCD systems in laser operation can be found in an excellent Hungarian review.[(139)]

6.4.5.3. Deposition and Etching

The ability of the HCD to both erode and coat surfaces by appropriately guiding the sputtering mechanism and the chemical reactions occurring in the plasma can be expediently used to achieve results in the treatment of surfaces that could not otherwise be obtained.

How deposition processes can be advantageously combined in a single HCD chamber was brilliantly dealt with by Horwitz *et al.*[(140)] The system was reported to behave excellently in mixed metal–polymer deposition and SiO_2 etching on commercial complementary metal oxide semiconductor wafers, especially with regard to process kinetics and uniformity. High deposition rates together with obtaining abrasion-resistant films were the points of merit in a new procedure developed by Horwitz and McKenzie[(141)] to deposit amorphous silicon with an rf-operated HCD.

The effect of the metal cathode on methane dissociation was elucidated in order to generate precursors for the polymerization of methane.[(142)] Mixed with Ar, methane could be polymerized at a deposition rate of 70 nm s^{-1} in a 19% Pt–81% W cathode and 42 nm s^{-1} for a W cathode, both rates being much higher than more conventional procedures based on rf. The same authors had previously obtained very encouraging results in similar undertakings using He, which suppressed the phenomenon of carbon formation.[(143)]

Ion beam sputter deposition of metallic films was achieved by means of a superdense HCD plasma.[(144)] Moreover, a triode device made up of a magnetron cathode coupled with an HCD source as a second cathode to enhance the plasma density was devised.[(145)] The coupling efficiency was $\leq 60\%$. Compared with conventional magnetrons, the combined system allows for a tenfold increase in discharge current and deposition rate.

Although still in their early stage, such applications are definitely promising and open entirely new avenues to GDs in general.

6.4.5.4. Beam Generation

HCD properties allow for the production of ionic and atomic beams that can be extracted for many other scopes. The combination of this source with mass spectrometry is the focus of the studies carried out by Harrison and his team[(50,146)] and is certainly among the most exciting promises of this technique. A small hole on the bottom of the cathode acts as a minielectrode cavity giving rise to an increased HCD effect with ejection of a "plume," whose ions are then assayed on the basis of their masses.

A neutral beam injector ion source consisting of a combination of HCD and ion accelerator system was described by Aston.[(147)] In it, the cathode works at high stagnation pressure and plasma density, which ensures field-enhanced thermionic emission. The ion accelerator system amplifies the extracted ion current density by a factor of four through a set of multiple hole and slit apertures.

The suitability of the He HCD as a neutral beam injector was further investigated using a BaO-impregnated porous W tube in an attempt to overcome the drawbacks involved in directly heated pure W or oxide-coated filaments.[148] The combination of Mo buttons set in front of the orifice and antiparallel magnetic fields was found effective in reducing the minimal gas flowrate down to ca. 320 Pa liter s^{-1}

The production of a U atomic beam by Babin and Gagné[71] was mentioned in Section 6.3.2. The two-dimensional emittance and brightness for a given high ion current of an HCD ion source were quantified by Saad and Holk.[149] The use of the HCD as an electron beam is also compatible with the technique and has been exhaustively described in the literature.[150,151]

6.4.5.5. Plasma Diagnostics

Not to be forgotten are the impressive number of recent contributions attempting to shed further light on unclear aspects of the HCD processes.

The use of a steady-state low-pressure HCD arc in Ar for diagnostic studies was described in Section 6.3.2.[67] In continuation of previous works, Kułakowska and Żyrnicki[152] studied the chemiluminescent reaction $SnCl_4+[O_2+Ar(He)]^*$ and also compared the characteristics of this process with those of excitation in an HCD. The 0–0 and 0–1 bands of the $^1\pi$–$X^1\Sigma^+$ system and the $^3\pi_{0,1}$–$X^1\Sigma^+$ 0–0 subbands of the InCl spectrum were subjected to further examination by Borkowska-Burnecka and Żyrnicki.[153] Rotational constants of the upper states could thus be recalculated. Similarly, the rotational constants of $^{115}In^{79}Br$ and $^{115}In^{81}Br$ were calculated by a high-resolution study of the $A^3\pi_0$–$X^1\Sigma^+$ and $B^3\pi_1$–$X^1\Sigma^+$ subsystems. As already reported, the two-component electron population was investigated by Hershcovitch *et al.*[154]

The occurrence of plasma satellites was detected by an assembly of two pulse-operated hollow cathodes and its dependence on voltage modulation was clarified.[155] Preliminary observations by Hildebrandt[156] evidenced the advantages offered by the HCD in the study of such plasma satellites. The temporal dependence of these results on the increase of the forbidden component of He(I) at 447 nm showed that their intensities can exceed that of the said forbidden component while, for a short period, the increase in the near-satellite intensity is inversely proportional to that of the far-satellite.

Reflection of 8-mm microwaves was used to determine both the radius and speed of the cylindric ionization front in an HCD under the expansion conditions that occur when applying high-voltage pulses.[157] In an attempt to model the radial dependence of electron concentration in an He HCD it was found that it becomes nonuniform at pressures higher than ca.

130 Pa.[158] A simple one-dimensional equation was deduced. Later, it was pointed out that for electrons with energy of hundreds of electron volts, their distribution relaxes faster than the angular one.

Spectral lines of weak intensity emitted by Ne atoms and ions were examined by a microcomputer-controlled photon-counter system, which proved to be particularly advantageous.[159] By measuring variations in plasma conductivity, internal elemental processes in the plasma could be ascertained for Al, Cd, Cr, Cu, and Fe.[160] Along the same line, the utility of the HCD for high-resolution, Doppler-free saturation experiments was further demonstrated by ascertaining the hyperfine structure of V atoms and the relevant constants for the five $3d^3 4s(^5F)4p\,^6D^0$ levels and the four $3d^4(^5D)4s\,^6D$ metastable levels.[161]

The absolute values of the electric field, measured by Stark optogalvanic spectroscopy, were found not to decrease linearly from the cathode surface in an Ar HCD, but to follow a radial pattern mirroring the linear enhancement of space charge density along the current flow.[162] Further experiments revealed that this behavior is described by a function of the $\varrho(r) = -Ar + B$ type, where ϱ is the space charge distribution and r the radial distance from the cathode axis in the zone between the cathode surface and the boundary with the negative glow.[163] Finally, the same team ascertained the pressure broadening and frequency shift of the Ar $7d[5/2]3$–$4p[3/2]2$ transition.[164]

How discharge temperature and hydrogen density can influence the intensity of the H_3^* emission was elucidated by Miderski and Gellene by monitoring the $3s^2A_1' \rightarrow 2p^2A_2''$ transition.[165] The effect was found to be accurately described by assuming that the primary process leading to H_3^* is the H_5^+/e^- dissociative recombination.

A good agreement between experimental findings and theoretical predictions was obtained in the Stark width measurements of the 589.6 nm neutral Na hyperfine lines in the presence of charged perturbers.[166] Another team strove to elucidate the processes involved in high-current HCD devices for metal ion lasers by examining the case of an Ne–Cu discharge. A set of expressions was derived that accounted for the dependence of current density on the spatially averaged densities of noble gas ions and metal ions and atoms.

In a series of precise measurements of saturation spectroscopy in the HCD, the isotope shifts and hyperfine splitting were determined for the 270.24- and 270.59-nm transitions of Pt(I).[168] Notwithstanding relatively large specific mass variations, the isotope-shift ratios turned out to be transition independent.

The incorporation of a confined positive column in front of the hollow cathode could feed additional electrons into the HCD plasma without increasing sputtering or broadening self-absorption.[169] The net enhancement of the hyperfine doublet of the Cu(I) 324.8 nm transition was measured as a function of current intensity and gas type and pressure. One of the more annoying problems with spacecraft operation is reported to be electrical charging, especially for satellites in geosynchronous orbits, with spacecraft frame potential of thousands of volts negative with respect to the ambient space plasma.[170] To compensate for this effect, an HCD neutral plasma source was designed to serve as a plasma emitter, capable of operating with either Ar or Kr and to supply ion currents of up to 325 μA and electron currents of 0.02 to 6.0 A. Another exotic use of the HCD consisted in the precise determination of the radiance output of a sealed Pt/Cr–Ne tube to calibrate the faint object spectrograph on the Hubble space telescope.[171]

The applicability of fluorescence techniques with a repetitively pulsed tunable dye laser to investigate electron densities and neutral-particle temperatures was thoroughly examined.[172] A power-broadening relationship was developed to accommodate rather large saturation parameters. To study the mechanism and kinetics of HCD plasma chemical reactions, the decomposition of $TiCl_4$ under conditions of non-LTE was also clarified.[173] While the interaction of Ar metastables with $TiCl_4$ molecules appears to play only a minor role, the overall reaction is assumed to be $TiCl_4 + ne^{-*} \rightarrow Ti^* + Cl_2 + 2Cl + ne^-(n = 4–6)$.

The remarkable improvement in the signal-to-background ratio in the MW-boosted HCD lamp, especially when Ar is used as the carrier gas, was examined in the spectral range 200–400 nm.[174] Moreover, Pataky-Szabó and her associates gained further data on the functioning of the three-electrode HCD source.[175] The crater that formed on top of the central rod turned out to be quite independent of the physical and chemical inhomogeneities.

A number of other contributions in this field deserve mentioning: for example, those studies devoted to the mechanism of Th atomization,[176] the measurement of the density of U ions,[177] the effects of a magnetic field parallel to the cathode axis on the radial emission profiles[178] (also investigated for commercial AAS lamps[179]), the achievement of optical phase conjugation in an HCD tube through degenerate four-wave mixing,[180] the performance of the Cu–Ne HCD in the current density interval of 10–100 mA cm^{-2},[181] the formation of metastable Ar atoms in an Ne HCD arc,[182] the ablation pattern of Al and Cd under normal dc and MW-coupled HCD discharge conditions,[183] the generation of Cu vapor[184] and its emission and absorption characteristics in a pulsed HCD,[185] the lifetime of some P(II) levels,[186] the appearance of incoherent Hanle signals caused by mutual orthogonalization of the entrance slit height,[187] the determination of the excitation temperature of Ar(I), Ar(II), Cr(I), Cr(II), Cu(I), Si(I), V(I), and V(II) species[188] and of Al(I), Mg(I), Mg(II), Pb(I), Ti(I), and Ti(II) species,[189] and the study of HCD tubes operating at both dc and high-frequency currents.[190]

For those interested in obtaining further information, updated reviews on the various analytical, technological, and diagnostic uses of the HCD are available.[e.g., 21,31,37,88]

6.5. Future Trends

Although divination is an unreliable art, the only chance of succeeding lies in the accurate and sound analysis of past and present events. Trying to apply this general rule to the disclosure of the most probable pathways the HCD will take, one must scrutinize foregone facts and current tendencies. Of the many possible ways to achieve this, the approach adopted herein consists in searching the available literature on this subject (more than 1300 documents) from the early years up to 1980 and comparing the data thus obtained with the data for the last decade, in order to highlight any significant changes.

The issues being considered (selected on the basis of their role as trend markers) are the topic of published document, the nationality of the authors, the medium used to report the results, and finally the language in which the document is written. The information obtained is set forth in Tables 6-3 through 6-6. With regard to subject matter, the highest effort has been devoted to classifying each document as systematically as possible according to its predominant content. Despite this effort a certain amount of arbitrariness in the categorization process cannot be avoided, although this would not affect the overall breakdown and the hierarchy that descends from it.

The data allow some main conclusions to be drawn, namely, that works centered on the investigation of fundamental aspects of the HCD are still about a third of the total, and more than half the volume of the literature up to 1980, but still clearly highlighting the need for a better understanding

Table 6-3. Percentage Distribution of Documents on HCD with Reference to the Topic

Topic	Fraction %	
	Up to 1980	From 1980
Analysis	24.6	11.1
Basic mechanism	65.1	33.7
Beam generation	0.1	4.2
Deposition/etching	0.1	13.0
Instrumentation	9.5	8.3
Laser technology	0.1	22.3
Optogalvanic spectroscopy	0	5.8
Review	0.5	1.6

Table 6-4. Percentage Distribution of Documents on HCD with Reference to the Nationality of the Authors

Country	Fraction %	
	Up to 1980	From 1980
Australia	0.7	3.0
Canada	0.7	4.1
China	0	6.2
Europe	45.5	38.7
Japan	5.3	18.1
Soviet Union	33.2	13.2
United States	14.3	14.7
Others	0.3	2.0

Table 6-5. Percentage Distribution of Documents on HCD with Reference to the Language Used

Language	Fraction %	
	Up to 1980	From 1980
Chinese	0	5.1
English	39.5	65.4
French	5.6	1.5
German	14.3	3.3
Japanese	2.3	8.2
Russian	34.8	13.4
Others	3.5	3.1

of some facets of the discharge processes. At the same time, the percentage of analytical papers has more than halved, reflecting the difficulty in coping with the present requirements of trace determinations. While studies reporting on the development of instrumentation do not show appreciable changes, other previously unrecorded areas have become dramatically important, as is the case for laser research and coating and etching technology. This trend clearly points to the ongoing diversification of HCD applications.

With regard to nationality of the authors, there is a marked decline in the number of contributions from Europe and the USSR (once clearly the major sources of activities in the field) and an extraordinary increase in the number of Japanese and Chinese teams. The leading role of Europe and the USSR seems still indisputable, accounting for more than half of the total publications. It should be noted that the most active countries in Europe have been Germany (both West and East), France, and Italy and that most of the above decrease can be ascribed to the presently reduced amount of research with HCDs by the first two.

Table 6-6. Percentage Distribution of Documents on HCD with Reference to Publication Type

Publication type[a]	Fraction %	
	Up to 1980	From 1980
Journals		
Canada	0.7	1.3
China	0	5.7
Europe	34.3	18.6
Japan	3.1	6.4
USA	17.4	31.2
USSR	32.6	12.5
Conference proceedings	5.5	8.2
Technical reports	4.3	1.8
Dissertations	1.0	0.9
Patents		
Europe	0.2	2.7
Japan	0	9.1
USA	0.9	1.6

[a]Journals also include chapters in books or monographs.

Quite understandably, such changes are reflected in the languages used with the additional point that the enhanced diffusion of English in the scientific literature is now documented by approximately two-thirds of publications being written in this language.

An analogous pattern is shown in the breakdown of diffusion media, as journals printed in North America are now in the lead, replacing Europe as the most productive region. Among journals preferred for disseminating results, until 1980 an apparent predominance was shown for *Spectrochimica Acta* (7.8%), followed by *Applied Spectroscopy* (2.7%) and *Analytical Chemistry* (1.6%) (values in parentheses refer to the frequency of appearance in journals only). An appreciable number of papers on HCD also used to be published in *Fresenius' Zeitschrift für physikalische Chemie*. Over the last decade, the situation has moved toward a better balance, with *Spectrochimica Acta* still prevailing (4.5%), but with a smaller gap to *Applied Spectroscopy* (3.6%) and *Analytical Chemistry* (1.7%). A significant number of papers on HCD have appeared in a periodical first released in the mid-1980s, the *Journal of Analytical Atomic Spectrometry*, totaling 1.7% of publications on HCD. In both periods (before and after 1980) the preferred media for publishing studies on HCD have been two European and two North American journals. Another phenomenon that should not be overlooked is certainly the quickly increasing number of patents pivoting on HCD developments and applications. The economic implications of this spectroscopic branch seem to be well understood by Japan, followed by Europe and the USA.

Therefore, a crystal ball is not needed to forecast that the above-outlined trends will be all the more valid in the future and that technological exploitation of HCD in laser development and surface treatment procedures will play a major role. Regarding analytical impact, current premises do justify the belief in an imminent adequacy of the source to the challenging requirements of trace determinations as posed by today's problems. Whether these avenues will be explored and accomplished is a question for the year 2000.

Acknowledgment

The expertise and patience of Dr. Isabel A. Robinson in revising the English style of the various drafts of this chapter are gratefully acknowledged.

References

1. J. Plücker, Continued observation on the electrical discharge through rarefied gas tubes, *Ann. Phys.* 104 (1858) 113.
2. J. Plücker and J. W. Hittorf, On the spectra of ignited gases and vapors, with especial regard to the different spectra of the same elementary gaseous substance, *Philos. Trans.* 115(IV) (1865) 1.
3. F. Paschen, Bohrs Heliumlinien, *Ann. Phys.* 50 (IV) (1916) 901.
4. F. Paschen, Die Funkenspektren des Aluminiums. I Teil, *Ann. Phys.* 71 (IV)(1923) 142
5. F. Paschen, Die Funkenspektren des Aluminiums. II Teil, *Ann. Phys.* 71(IV) (1923) 538.
6. R. A. Sawyer and F. Paschen, Das erste Funkenspektrum des Aluminiums Al II, *Ann. Phys.* 84 (IV) (1927) 1.
7. H Schüler, Ueber Potentialgefälle an Elektroden in Gasentladungsröhren, *Phys. Z.* 22 (1961) 264.
8. H Schüler, Ueber eine neue Lichtquelle und ihre Anwendungsmöglichkeiten, *Z. Phys.* 35 (1926) 323.
9. H. Schüler and J. E. Keystone, Hyperfeinstrukturen und Kernmomente des Quecksilbers, *Z. Phys.* 72 (1931) 423.
10. H. Schüler and H. Gollnow, Ueber eine lichtstarke Glimmentladungsröhre zur spektroskopischen Untersuchung geringer Substanzmengen, *Z. Phys.* 93 (1935) 611
11. H. Schüler, Possibility of applying the hollow cathode discharge to spectroanalytical investigations, *Proceedings of the I Colloquium Spectroscopicum Internationale*, Strasbourg, 1950, pp. 169–171.
12. H. Schüler and A. Michel, Ueber zwei neue Hohlkathodenentladungsröhren, *Spectrochim. Acta* 5 (1952) 322.
13. H. Schüler, Ueber die Emissionsspektroskopie organischer Substanzen mit Hilfe der Elektronenstossanregung in der Glimmentladung. I, *Spectrochim. Acta* 4 (1950) 85.
14. H. Schüler and L. Reinebeck, Ueber die Emissionsspektroskopie organischer Substanzen mit Hilfe der Elektronenstossanregung in der Glimmentladung. II, *Spectrochim. Acta* 6 (1954) 288.
15. H. Schüler and L. Reinebeck, Ueber eine Methode der Variation der Anregungsbedingungen organischer Substanzen, *Z. Naturforsch.* 5a (1950) 657.
16. R. A. Sawyer, Excitation processes in the hollow discharge, *Phys. Rev* 36 (1930) 44.

17. V. A. Konovalov and E. S. Frish, Illumination of the mixture of argon and nitrogen, *Zh. Tekh. Fiz.* 4 (1934) 523.
18. J. R. McNally, Jr., G. R. Harrison, and E. Rowe, A hollow cathode source applicable to spectrographic analysis for the halogens and gases, *J. Opt. Soc. Am.* 37 (1947) 93.
19. A. Walsh, The application of atomic absorption spectra to chemical analysis, *Spectrochim. Acta* 7 (1955) 108.
20. A. Walsh, Atomic absorption spectroscopy and its applications—Old and new, *Pure Appl. Chem.* 49 (1977) 1621.
21. S. Caroli, Low-pressure discharges: Fundamental and applicative aspects, *J. Anal. At. Spectrom.* 2 (1987) 661.
22. E. L. Grove and W. A. Loseke, Hollow cathode excitation of air-type atmospheres, *Can. J. Spectrosc.* 18 (1973) 33.
23. J. A. C. Broekaert, Emission spectrographic determination of all rare earths in solutions by hollow cathode excitation, *Bull. Soc. Chim. Belg.* 85 (1976) 261.
24. P. Zanzucchi, A study of direct solution analysis with a new hollow cathode discharge system, Dissertation, University of Illinois, Urbana, University Microfilms Ltd., Ann Arbor, Mich., 1967.
25. S. Caroli, A. Alimonti, P. Delle Femmine, and S. K. Shukla, Determination of gallium in tumor-affected tissues by means of spectroscopic techniques, *Anal. Chim. Acta* 136 (1982) 225.
26. S. Caroli, O. Senofonte, and P. Delle Femmine, Determination of trace elements in biological materials using a hollow-cathode discharge: Comparative study of matrix effects, *Analyst (London)* 108 (1983) 196.
27. J. O. Foss, H. J. Svec, and R. J. Conzemius, The determination of trace elements in aqueous media without preconcentration using a cryogenic hollow-cathode ion source, *Anal. Chim. Acta* 147 (1983) 151.
28. S. Caroli, Low-pressure discharges, in: *Sample Introduction in Atomic Spectroscopy* (J. Sneddon, ed.), pp. 225–253, Elsevier, Amsterdam, 1990.
29. H. Falk, Hollow cathode discharge within a graphite furnace: Furnace atomic non-thermal excitation spectrometry (FANES), in: *Improved Hollow Cathode Lamps for Atomic Spectroscopy* (S. Caroli, ed.), pp. 74–118, Ellis Horwood Ltd., Chichester, UK, 1985.
30. A. Günther-Schulze, Glow discharge in a hollow cathode, *Z. Techn. Phys.* (1930) 49.
31. R. Mavrodineanu, Hollow cathode discharges: Analytical applications, *J. Res. Natl. Bur. Stand.* 89 (1984) 143.
32. G. Francis, The glow discharge at low pressure, in: *Handbuch der Physik* (S. Flügge, ed.), Vol. 22, pp. 53–208, Springer-Verlag, Berlin, 1956.
33. R. Papoular, *The Glow Discharge, Electrical Phenomena in Gases*, pp. 123–140, Iliffe Books, London, 1965.
34. M. Kaminsky, *Atomic and Ionic Impact Phenomena on Metal Surfaces*, Springer-Verlag, Berlin, and Academic Press, New York, 1965.
35. F. M. Penning, *Electrical Discharges in Gases*, Philips Technical Library, Eindhoven, Servire B.V., Katwijk, The Netherlands, 1957.
36. P. F. Little and A. V. Engel, The hollow cathode effect and the theory of glow discharges, *Proc. R. Soc. London Ser. A* 224 (1954) 209.
37. M. E. Pillow, A critical review of spectral and related physical properties of the hollow cathode discharge, *Spectrochim. Acta* 36B (1981) 821.
38. V. S. Borodin and Yu. M. Kagan, The investigation of a hollow cathode discharge, *Zh. Tekhn. Fiz.* 36 (1966) 181.
39. H. Falk, Einige theoretische Ueberlegungen zum Vergleich der physikalischen Grenzen thermischer und nicht-thermischer spectroskopischer Strahlungsquellen, *Spectrochim. Acta* 32B (1977) 437.

40. V. S. Borodin, Yu. M. Kagan, and L. I. Lyagushchenko, Investigation of a hollow cathode discharge. II. *Zh. Tekh. Fiz.* 11 (1967) 887.
41. F. Howorka and M. Pahl, Experimental determination of internal and external parameters of the negative glow plasma of a cylindrical hollow cathode discharge in argon, *Z. Naturforsch.* A 27 (1972) 1425.
42. P. A. Büger and W. Fink, Analysis of solutions in a hollow cathode, *Fresenius Z. Anal. Chem.* 244 (1969) 121.
43. C. Howard, M. E. Pillow, E. B. M. Steers, and D. W. Ward, Intensities of some spectral lines from hollow-cathode lamps, *Analyst* (*London*) 108 (1983) 145.
44. F. Howorka, W. Lindinger, and M. Pahl, Ion sampling from the negative glow plasma in a cylindrical hollow cathode, *Int. J. Mass Spectrom. Ion Phys.* 12 (1973) 67.
45. H. Patterson and D. H. Tomlin, Experiments by radioactive tracer methods on sputtering by rare gas ions, *Proc. R. Soc. London Ser. A* 265 (1962) 474.
46. R. E. Honing, Sputtering of surfaces by positive ion beams of low energy, *J. Appl. Phys.* 29 (1958) 549.
47. K. B. Mitchell, Spectroscopic studies of ionization in a hollow cathode discharge, *J. Opt. Soc. Am.* 51, (1961) 846.
48. T. Musha, Cathodic sputtering in a hollow cathode discharge, *J. Phys. Soc.* (*Jpn.*) 17 (1962) 1440.
49. A. D. White, New hollow cathode glow discharge, *J. Appl. Phys.* 30 (1959) 711.
50. E. H. Daughtrey, D. L. Donohue, P. J. Slevin, and W. W. Harrison, Surface sputter effects in a hollow cathode discharge, *Anal. Chem.* 47 (1975) 683.
51. B. E. Warner, Investigation of the hollow cathode discharge at high current density, Dissertation, University of Colorado, Boulder, 1979.
52. G. Knerr, J. Maierhofer, and A. Reis, Application of high-current hollow cathode for quantitative analysis of conductors and glasses, *Fresenius' Z. Anal. Chem.* 229 (1967) 241.
53. H. Falk, The limiting factors for intensity and line profile of radiation sources for atomic absorption spectrometry, *Prog. Anal. At. Spectrosc.* 5 (1982) 205.
54. J. G. Hirschberg, E. Hinnov, and F. W. Hofmann, Spectroscopic investigations of a weakly ionized plasma in a helium hollow cathode discharge, *MATI* (Plasma Physics Laboratory, Princeton University), 236, 1963.
55. T. Török and G. Záray, Versuche mit einem tiefgekühlten Zwillingshohlkathoden Interferometer-Spektrometer. I. *Spectrochim. Acta* 30B (1975) 157.
56. H. Falk and H. Lucht, Investigation of excitation processes in a hollow-cathode discharge by time-resolved measurements in the vacuum UV, *J. Quant. Spectrosc. Radiat. Transfer* 16 (1976) 909.
57. T. Lee, S. Katz, and S. A. MacIntyre, The spectrographic determination of uranium 235. V. Routine application of a multiple hollow grating spectrograph, *Appl. Spectrosc.* 16 (1962) 92.
58. I. A. Berezin and K. V. Aleksandrovich, Determination of sulfur, chlorine and fluorine in beryllium oxide by a spectrographic method, *Zh. Anal. Khim.* 16 (1961) 613.
59. M. P. Chaika, Analysis of low-volatile oxides for halogens, *Opt. Spektrosk.* 2 (1957) 421.
60. H. Falk, E. Hoffmann, and C. Lüdke, FANES (furnace atomic nonthermal excitation spectrometry). A new emission technique with high detection power, *Spectrochim. Acta* 36B (1981) 767.
61. H. Falk, E. Hoffmann, and Ch. Lüdke, A comparison of furnace atomic nonthermal excitation spectrometry (FANES) with other atomic spectroscopic technique, *Spectrochim. Acta* 39B (1984) 283.
62. H. J. Eichhoff and R. Voigt, Use of a metallic hollow cathode for spectrochemical analyses, in: *Proceedings of the IX Colloquium Spectroscopicum Internationale Lyon*, 1961, Vol. 3, pp. 309–317 (1962).

63. S. Caroli (ed.), *Improved Hollow Cathode Lamps for Atomic Spectroscopy*, Ellis Horwood Ltd., Chichester, UK, 1985.
64. S. Caroli, Hollow cathode lamps as excitation sources for analytical atomic spectroscopy, *Fresenius' Z. Anal. Chem.* 324 (1986) 442.
65. G. H. C. Freeman and W. H. King, Cu II spectral lines and their suitability as wavelength standards in the vacuum ultraviolet, *J. Phys. E* 10 (1977) 984.
66. O. Appelblad and K. Schmidt, A liquid nitrogen cooled composite wall hollow cathode, USIP Report (University of Stockholm), 85–07, 1985.
67. M. Bessenrodt-Weberpals, A. Brockhaus, P. Jaunernik, H. Kempkens, C. Nieswand, and J. Uhlenbusch, Diagnostics of a steady-state low-pressure hollow cathode arc in argon, *IEEE Trans. Plasma Sci.* PS-14 (1986) 492.
68. Y. R. Jong, R. L. Davis, J. C. Williams, and J. C. Williams, Jr., Evaluation of the mini-hollow cathode emission source for the analysis of microsamples, *Appl. Spectrosc.* 42 (1988) 1379.
69. G. Rossi and N. Omenetto, Application of a demountable water-cooled hollow cathode lamp to atomic fluorescence spectrometry, *Talanta* 16 (1969) 263.
70. F. Howorka, A. Scherleitner, V. Gieseke, and I. Kuen, Bakeable hollow cathode for the study of ion–molecule reactions in discharge in gaseous mixtures, *Int. J. Mass Spectrom. Ion Phys.* 32 (1980) 321.
71. F. Babin and J.-M. Gagné, Characterization of an atom beam produced with the help of a hollow cathode discharge, *Rev. Sci. Instrum.* 57 (1986) 1536.
72. D. E. Holmgren, R. W. Falcone, D. J. Walker, and S. E. Harris, Measurement of lithium and sodium metastable quartet atoms in a hollow cathode discharge, *Opt. Lett.* 9 (1984) 85.
73. D. C. Gerstenberger, R. Solanki, and G. J. Collins, Hollow cathode metal ion lasers, *IEEE J. Quantum Electron.* 8 (1980) 820.
74. T. Iijima, He-Zn^+ ion laser using hollow cathode discharge with high-voltage operation, *Jpn. J. Appl. Phys.* 21 (1982) 1732.
75. N. A. G. Ahmed and D. G. Teer, Characterization of aluminium coatings deposited in a hollow cathode discharge, *Thin Solid Films* 80 (1981) 49.
76. L. Papp, Development of hollow cathode radiation sources. Part 2. Study of the effect of a cylinder placed in the cathode cavity on the emitted light intensity, *J. Anal. At. Spectrom.* 2 (1987) 407.
77. J. Czakow, The microcavity hollow cathode and its analytical potential, in: *Improved Hollow Cathode Lamps for Atomic Spectroscopy.* (S. Caroli, ed.), Ellis Horwood Ltd., Chichester, UK, 1985, pp. 35–51.
78. V. A. Novoselov and V. B. Znamenskii, Correlations between the intensity of spectral lines, discharge parameters in a hollow cathode and its diameter, *Spektrosk. Tr. Sib. Soveshch.*, 4th, 1965, pp. 273–278.
79. Z. Szilvássy, Principles and use of a boosted hollow cathode discharge source for atomic spectroscopy, in: *Improved Hollow Cathode Lamps for Atomic Spectroscopy.* (S. Caroli, ed.), Ellis Horwood Ltd., Chichester, UK, 1985, pp. 178–202.
80. Z. Szilvássy, A. Buzási, and E. Házi, Effect of cathode material and sample composition on the plasma characteristics of hollow cathode discharge, *Acta Chim. Hung.* 126 (1989) 353.
81. C. Popovici and M. Someşan, On the emission spectrum of the negative glow plasma of a hollow cathode discharge in magnetic field, *Appl. Phys. Lett.* 8 (1966) 103.
82. N. K. Rudnevsky and D. E. Maksimov, Hollow cathode discharge in a magnetic field, in: *Improved Hollow Cathode Lamps for Atomic Spectroscopy.* (S. Caroli, ed.), Ellis Horwood Ltd., Chichester, UK, 1985, pp. 148–177.
83. B. Pavlović and J. Dobrosavljević, Influence of a rotating magnetic field on a hollow cathode discharge, *Spectrochim. Acta* 44B (1989) 1191.

84. R. B. Djulgerova, The pulsed hollow cathode discharge. New spectroanalytical possibilities, in: *Improved Hollow Cathode Lamps for Atomic Spectroscopy* (S. Caroli, ed.), Ellis Horwood Ltd., Chichester, UK, 1985, pp. 55–73.
85. P. B. Farnsworth and J. P. Walters, The radiofrequency-boosted, pulsed hollow cathode lamp, in: *Improved Hollow Cathode Lamps for Atomic Spectroscopy* (S. Caroli, ed.), Ellis Horwood Ltd., Chichester, UK, 1985, pp. 119–147.
86. S. Caroli, A. Alimonti, and F. Petrucci, Analytical capabilities of the microwave-coupled hollow cathode discharge, *Anal. Chim. Acta* 136 (1982) 269.
87. R. Tomellini, M. Cilia, O. Senofonte, G. Guantera, M. G. Del Monte Tamba, and S. Caroli, A newly devised microwave-boosted low pressure source in atomic emission spectrometry. *Berichte des* 3. *Anwendertreffen Analytische Glimmentladungs-Spektroskopie*, Jülich (FRG), April, 1990.
88. S. Caroli, The hollow cathode emission source: A survey of the past and a look into the future, *Prog. Anal. At. Spectrosc.* 6 (1983) 253.
89. Atomsource Direct Solids Atomizer, Varian Report, 1988.
90. W. Pekruhn, L. K. Thomas, I. Broser, A. Schroder, and U. Wenning, Chromium/silicon monoxide on copper solar selective absorbers, *Sol. Energy Mater.* 12 (1985) 199.
91. N. Jakubowski, D. Stuewer, and W. Vieth, Performance of a glow discharge mass spectrometry for simultaneous multielement analysis of steel, *Anal. Chem.* 59 (1987) 1825.
92. W. A. Mattson, B. L. Bentz, and W. W. Harrison, Coaxial cathode ion source for solids mass spectrometry, *Anal. Chem.* 48 (1976) 489.
93. K. Rózsa, Hollow cathode discharges for gas lasers, *Z. Naturforsch.* 35a (1980) 649.
94. R. De Marco, D. Kew, and J. V. Sullivan, Determination of major constituents in metal samples by emission spectrometry using a demountable hollow cathode source and internal standardization, *Spectrochim. Acta* 41B (1986) 591.
95. R. De Marco, D. J. Kew, C. Chadjilazarod, D. W. Owen, and J. V. Sullivan, Precision and accuracy of quantitative emission spectrometry with particular reference to gold alloys, *Anal. Chim. Acta* 94 (1987) 189.
96. J. Borkowska-Burnecka and W. Żyrnicki, Comparison of spectra excited in HF and dc hollow cathode discharges, *Spectrosc. Lett.* 20 (1987) 795.
97. J. Borkowska-Burnecka and W. Żyrnicki, Fluorine determination in a hollow cathode discharge by monofluoride emission spectra, *Anal. Lett.* 18 (1985) 1539.
98. A. Buzási Győrfiné, S. Caroli, A. Alimonti, and O. Senofonte, Comparative study of the hollow cathode and glow discharge radiation sources for spectrographic determination of alloying elements and impurities in steel, *Acta Chim. Hung.* 113 (1983) 295.
99. S. Caroli, F. Petrucci, and A. Alimonti, The microwave-coupled hollow cathode discharge and its analytical potential for the determination of trace elements in steel, *Can. J. Spectrosc.* 28 (1983) 156.
100. S. Caroli, O. Senofonte, N. Violante, F. Petrucci, and A. Alimonti, An investigation of the power of detection of the hollow cathode source in emission spectroscopy, *Spectrochim. Acta* 39B (1984) 1425.
101. S. Caroli, F. Petrucci, A. Alimonti, and G. Záray, Analysis of minor elements in metals by microwave-coupled hollow cathode discharge, *Spectrosc. Lett.* 18 (1985) 609.
102. O. Senofonte, N. Violante, O. Falasca, and S. Caroli, Solid sample investigation by means of a novel version of the microwave-coupled hollow cathode discharge (MW-HCD), *Acta Chim. Hung.* 126 (1989) 317.
103. J. Mierzwa and W. Żyrnicki, Determination of Nd, Ho, Er, Tm, and Y in solutions by hollow cathode discharge with copper cathodes, *Anal. Lett.* 21 (1988) 115.
104. J. A. C. Broekaert, The investigation of two sample preparation techniques applied to the determination of rare earths in solutions with the aid of hollow cathode excitation, *Spectrochim. Acta* 35B (1980) 225.
105. S. Caroli, O. Falasca, O. Senofonte, and N. Violante, The hollow cathode emission source and its analytical potential for the determination of major, minor and trace elements. II. Phosphorus, *Can. J. Spectrosc.* 30 (1985) 79.

106. S. Caroli, Development of a hollow cathode method for the spectroanalytical determination of trace elements in biological materials, *Ann. Ist. Super. Sanità* 19 (1983) 495.
107. S. Caroli, Katódporlasztasos sugárforrások jelenlegi alkalmazásai és várható fejlesztési irányzatai az emissziós spektroskópiában, *Kém. Közl.* 62 (1984) 57.
108. A. Alimonti, S. Caroli, F. Petrucci, and C. Alvarez Herrero, Determination of arsenic by hollow-cathode emission spectrometry, *Anal. Chim. Acta* 156 (1984) 121.
109. S. Caroli, A. Alimonti, and K. Zimmer, Applicability of a hollow cathode emission source for determining trace elements in electrically non-conducting powders, *Spectrochim. Acta* 38B (1983) 626.
110. S. Caroli, A. Alimonti, P. Delle Femmine, and S. K. Shukla, Determination of gallium in tumor-affected tissues by means of spectroscopic techniques, *Anal. Chim. Acta* 136 (1982) 225.
111. E. B. M. Jansen and D. R. Demers, Hollow-cathode lamp-excited inductively coupled plasma atomic-fluorescence spectrometry: Performance under compromise conditions for simultaneous multi-element analysis, *Analyst (London)* 110 (1985) 541.
112. W. R. Masamba, B. W. Smith, R. J. Krupa, and J. D. Winefordner, Atomic and ionic fluorescence in an inductively coupled plasma using hollow cathode lamps pulsed at high currents as excitation sources, *Appl. Spectrosc.* 42 (1988) 872.
113. B. M. Patel and J. D. Winefordner, Laser-excited fluorescence of diatomic molecules of copper and lead in glow discharge sputtering, *Appl. Spectrosc.* 40 (1986) 667.
114. S. Greenfield, T. M. Durrani, and J. F. Tyson, A comparison of boosted-discharge hollow cathode lamps and an inductively coupled plasma (ICP) as excitation sources in ICP atomic fluorescence spectrometry, *Spectrochim. Acta* 45B (1990) 341.
115. D. R. Demers and C. D. Allemand, Atomic fluorescence spectrometry with an inductively coupled plasma as atomization cell and pulsed hollow cathode lamp for excitation, *Anal. Chem.* 53 (1981) 1915.
116. B. Rosen, New developments in the application of a hollow cathode discharge tube designed for the quantitative determination of oxygen in metals, *Appl. Spectrosc.* 5 (1951) 20.
117. N. A. Zakorina, I. S. Lindstrem, and A. A. Petrov, Spectral-isotopic determination of nitrogen in refractory alloys by using a discharge with a hot hollow cathode, *Zavod. Lab.* 50 (1984) 30.
118. Y. Iida, Laser vaporization of solid samples into a hollow-cathode discharge for atomic emission spectrometry, *Spectrochim. Acta* 45B (1990) 427.
119. L. Puig and R. Sacks, Hollow cathode plasma emission determination of F, Cl and Br in gas streams, *Appl. Spectrosc.* 43 (1989) 801.
120. E. M. van Veldhuizen, F. J. de Hoog, and D. C. Schram, Optogalvanic effects in a hollow cathode glow discharge plasma, *J. Appl. Phys.* 56 (1984) 2047.
121. B. M. Suri, R. Kapoor, G. D. Saksena, and P. R. K. Rao, Relative enhancement of optogalvanic signal of less abundant isotope in uranium hollow cathode discharge, *Opt. Commun.* 49 (1984) 29.
122. R. A. Keller, B. E. Warner, E. P. Zalewski, P. Dyer, R. Engleman, Jr., and B. A. Palmer, The mechanism of the optogalvanic effect in a hollow-cathode discharge, *J. Phys.* 44 (1983) C7/23.
123. N. Ami, A. Wada, Y. Adachi, and C. Hirose, Optogalvanic measurement of the electric field inside the cathode fall region of neon hollow cathode discharge, *Appl. Spectrosc.* 43 (1989) 245.
124. C. Hirose, T. Masaki, A. Wada, and Y. Adachi, Radial profile of the spectral width of the Ar $7d[5/2]3$–$4p[3/2]2$ line inside the cathode fall region of Ar hollow cathode discharges, *Appl. Spectrosc.* 43 (1989) 87.
125. T. Caesar and J. L. Heully, Experimental evidence of non-inverted population in a neon hollow cathode, *Opt. Commun.* 45 (1983) 258.

126. S. Fujimaki, Y. Adachi, and C. Hirose, Optogalvanic measurement of the cathode-fall region of Kr hollow cathode discharge, *Appl. Spectrosc.* 41 (1987) 567.
127. P. Apai, M. Jánossy, I. Pálmai, K. Rózsa, and G. Rubin, D.c. hollow cathode He-Kr discharge, *SPIE* 473 (1984) 198.
128. I. Ebina, W. Sasaki, and T. Ohta, A 46 cm-discharge-length white light He-Cd II laser with coaxial type hollow cathode, *Trans. IECE Jpn.* 69 (1986) 367.
129. J. Mizeraczyk, J. Wasilewski, J. Konieczka, W. Urbanik, M. Grozeva, and J. Pavlik, Comparison of He-Kr^+ laser oscillations in transverse and longitudinal hollow-cathode discharges, *Proceedings of the International Conference on Lasers*, 1981, pp. 877–881.
130. J. Mizeraczyck, J. Konieczka, J. Wasilewski, and K. Rózsa, High-voltage hollow-cathode He-Cd^+ laser, *Proceedings of the International Conference on Lasers*, 1980, pp. 177–181.
131. M. Cilea, C. P. Cristescu, I. M. Popescu, and A. M. Preda, Hollow cathode He-Zn laser with an additional command electrode, *Rev. Roum. Phys.* 27 (1982) 357.
132. S. S. Cartaleva, S. V. Gateva, and V. J. Stefanov, Investigation of spontaneous lines from upper and lower levels of He-Ne laser line 632.8 nm excited in hollow cathode, *SPIE* 473 (1984) 192.
133. T. Arai, K. Nihira, T. Iijima, and T. Goto, Excitation mechanism in the hollow cathode He-Zn^+ laser, *Ik. Kog. Doi. Ken. Hok.* B-9 (1985) 67.
134. D. S. Gough and P. Hannaford, Very high resolution laser saturation spectroscopy in hollow-cathode and glow discharges, *Opt. Commun.* 55 (1985) 91.
135. A. Baczyński, M. Dzwonkowski, and P. Targowski, The influence of ignition method on laser output of pulse hollow cathode copper ion laser, *Opt. Appl.* XIII (1983) 231.
136. J.-b. Liu, Investigation of hollow cathode Ge II and Te II lasers, *Appl. Phys. B* 32 (1983) 211.
137. M. Jánossy, K. Rózsa, P. Apai, and L. Csillag, D.c. hollow cathode He-Kr ion laser, *SPIE* 473 (1984) 177.
138. T. Iijima, Noise measurement in a He-Zn hollow cathode laser tube, *Physica* 115C (1983) 257.
139. K. Rózsa, Gáz- és fémgözlézerek céljára alkalmazott üreges katódu kisülések, *Magy. Fiz. Foly.* XXXIV (1986).
140. C. M. Horwitz, S. Boronkay, M. Gross, and F. Davies, Hollow cathode etching and deposition, *J. Vac. Sci. Technol.* A6 (1988) 1837.
141. C. M. Horwitz and D. R. McKenzie, High-rate hollow-cathode amorphous silicon deposition, *Appl. Surf. Sci.* 22/23 (1985) 925.
142. P. Meubus, H. Lange, and G. Jean, Methane polymerization with a hollow cathode: Influence of the cathode metal, *Plasma Chem. Plasma Process.* 9 (1989) 527.
143. P. Meubus and G. Jean, Methane polymerization using a hollow cathode, *Prepr. Pap. Am. Chem. Soc. Div. Fuel Chem.* 32 (1987) 260.
144. H. Nagasaka, K. Yanagida, T. Tanabe, M. Takeuchi, and H. Mase, Compact ion source using superdense hollow cathode discharge and its application to thin film formation, *Pur. Kot.* 5 (1985) 155.
145. J. J. Cuomo and S. M. Rossnagel, Hollow-cathode enhanced magnetron sputtering, *J. Vac. Sci. Technol.* A4 (1986) 393.
146. W. W. Harrison and B. L. Bentz, Glow-discharge mass spectrometry, *Prog. Anal. Spectrosc.* 11 (1988) 53.
147. G. Aston, Hollow cathode startup using a microplasma discharge, *Rev. Sci. Instrum.* 52 (1981) 1259.
148. S. Tanaka, M. Akiba, Y. Arakawa, H. Horiike, and J. Sakuraba, Reduction of gas flow into a hollow cathode ion source for a neutral beam injector, *Rev. Sci. Instrum.* 53 (1982) 1038.
149. H. M. Saad and O. Holk, Investigation of beam emittance of a hollow cathode ion source, *Egypt. J. Phys.* 15 (1984) 45.

150. R. J. Barker, S. A. Goldstein, and R. E. Lee, Computer simulation of intense electron beam generation in a hollow cathode diode, *NRL Memorandum Report* 4279, 1980.
151. J. J. Rocca, J. Meyer, and G. J. Collins, Hollow cathode electron gun for the excitation of cw lasers, *Phys. Lett.* 87A (1982) 237.
152. B. Kułakowska and W. Żyrnicki, Spectroscopic studies of low pressure plasma. Tin and its compounds, *Phys. Scr.* 33 (1986) 424.
153. J. Borkowska-Burnecka and W. Żyrnicki, High resolution study of the $A^3\pi_0$–$X^1\Sigma^+$ and $B^3\pi_1$–$X^1\Sigma^+$ subsystems of the $^{115}In^{79}Br$ and $^{115}In^{81}Br$ molecules, *Phys. Scr.* 35 (1987) 141.
154. A. I. Hershcovitch, V. J. Kovarik, and K. Prelec, Observation of a two-component electron population in a hollow cathode discharge, *J. Appl. Phys.* 67 (1990) 671.
155. J. Hildebrandt, Voltage modulation in a pulsed hollow-cathode discharge and its relation to the occurrence of plasma satellites, *Phys. Lett.* 95A (1983) 365.
156. J. Hildebrandt, Temporally and spatially resolved plasma satellites in a hollow-cathode source, *J. Phys. B* 16 (1983) 149.
157. J. Hildebrandt, Microwave diagnostic of the pulsed generation in the hollow-cathode glow discharge, *Z. Naturforsch.* 38a (1983) 1088.
158. Yu. M. Kagan, Rate of ionisation and density of electrons in a hollow cathode, *J. Phys. D* 18 (1985) 1113.
159. M. Kimura, Time-resolved spectral measurement in hollow cathode discharge tube by photon-counting method, *Jpn. J. Appl. Phys.* 23 (1984) 105.
160. Z. Szilvássy, A. Buzási, and E. Házi, Effect of cathode material and sample composition on the plasma characteristics of hollow cathode discharge, *Acta Chim. Hung.* 126 (1989) 353.
161. D. S. Gough, P. Hannaford, M. Lowe, and A. P. Willis, Hyperfine structures in ^{51}V using saturation spectroscopy in a hollow-cathode discharge, *J. Phys. B* 18 (1985) 3895.
162. T. Masaki, A. Wada, Y. Adachi, and C. Hirose, Space charge distribution in the cathode fall region of an Ar hollow cathode discharge, Appl. Spectrosc. 42 (1988) 49.
163. T. Masaki, A. Wada, Y. Adachi, and C. Hirose, Pressure dependence of space charge distribution in the cathode fall of Ar hollow cathode discharges, *Appl. Spectrosc.* 42 (1988) 51.
164. T. Masaki, Y. Adachi, and C. Hirose, Application of optogalvanic spectroscopy to the measurement of pressure broadening and shift in the negative glow region of an Ar hollow cathode discharge, *Appl. Spectrosc.* 42 (1988) 54.
165. C. A. Miderski and G. I. Gellene, Experimental evidence for the formation of H_3* by $H^+{}_5/e^-$ dissociative recombination, *J. Chem. Phys.* 88 (1988) 5331.
166. F. Moreno, J. M. Alvarez, J.C. Amaré, and E. Bernabeu, Stark effect of atomic sodium measured in a hollow cathode plasma by Doppler-free spectroscopy, *J. Appl. Phys.* 56 (1984) 1939.
167. H. Koch and H. J. Eichler, Particle densities in high current hollow discharges, *J. Appl. Phys.* 54 (1983) 4939.
168. R. D. LaBelle, W. M. Fairbank, Jr., R. Engleman, Jr., and R. A. Keller, Isotope shift and hyperfine structure of Pt I transitions in a hollow-cathode discharge, *J. Opt. Soc. Am. B* 6 (1989) 137.
169. H. A. Phillips, H. L. Lancaster, M. Bonner Denton, K. Rózsa, and P. Apai, Self-absorption in copper hollow cathode discharges: Effects on spectral line shape and absorption sensitivity, *Appl. Spectrosc.* 42 (1988) 572.
170. W. D. Deininger, G. Aston, and L. C. Pless, Hollow cathode plasma source for active spacecraft charge control, *Rev. Sci. Instrum.* 58 (1987) 1053.
171. J. Z. Klose and J. M. Bridges, Radiance of a Pt/Cr-Ne hollow cathode spectral line source, *Appl. Opt.* 26 (1987) 5202.
172. P. Chall, E. K. Souw, and J. Uhlenbusch, Laser diagnostics of a low-pressure hollow-cathode arc, *J. Quant. Spectrosc. Radiat. Transfer* 34 (1985) 309.

173. J. F. Behnke, H. Lange, and H.-E. Wagner, Dissociation of titanium tetrachloride to titanium in hollow cathode discharges. Part I. Determination of the dissociation degree, *Pol. J. Chem.* 56 (1982) 1175.
174. S. Caroli, O. Senofonte, N. Violante, and R. Astrologo, Analytical capabilities of a microwave-coupled hollow-cathode discharge, *J. Anal. At. Spectrom.* 3 (1988) 887.
175. M. Pataky-Szabó, L. Papp, and B. Derecskei, Investigations of processes in the three-electrode hollow cathode emission source, *Acta Chim. Hung.* 126 (1989) 359.
176. P. Pianarosa, Y. Demers, and J. M. Gagné, Atomization of thorium in a hollow-cathode type discharge, *Spectrochim. Acta* 39B (1984) 761.
177. P. Pianarosa, P. Bouchard, J. P. Saint-Dizier, and J. M. Gagné, Density of uranium ions in the $^4I^0_{9/2}$ ground state in a hollow-cathode type discharge, *Appl. Opt.* 22 (1983) 1568.
178. R. Simonneau and R. Sacks, Modulation of commercial hollow cathode lamps by a magnetic field in a magnetron configuration, *Appl. Spectrosc.* 42 (1988) 1032.
179. S. Tanaka, M. Akiba, H. Horiike, Y. Okumura, and Y. Ohara, Effect of magnetic field on the characteristics of hollow cathode ion source, *Rev. Sci. Instrum.* 54 (1983) 1104.
180. W. G. Tong and D. A. Chen, Doppler-free spectroscopy based on phase conjugation by degenerate four-wave mixing in hollow cathode discharge, *Appl. Spectrosc.* 41 (1987) 586.
181. E. M. van Veldhuizen and F. J. de Hoog, Analysis of a Cu-Ne hollow cathode glow discharge at intermediate currents, *J. Phys. D* 17 (1984) 953.
182. M. J. Verheijen, H. C. W. Beijerinck, P. W. E. Berkers, D. C. Schram, and N. F. Verster, A hollow cathode arc in neon: Simultaneous laser probing and molecular-beam sampling of metastable atoms as a plasma diagnostic, *J. Appl. Phys.* 56 (1984) 3141.
183. N. Violante, O. Senofonte, A. Marconi, O. Falasca, and S. Caroli, An investigation of the sputtering process in the microwave-coupled hollow cathode discharge, *Can. J. Spectrosc.* 33 (1988) 49.
184. W. Winiarczyk and L. Krause, Production of copper vapour in a pulsed hollow cathode discharge, *J. Quant. Spectrosc. Radiat. Transfer* 33 (1985) 581.
185. W. Winiarczyk and L. Krause, Emission and absorption studies of copper vapor ejected from a pulsed hollow cathode discharge, *J. Quant. Spectrosc. Radiat. Transfer* 34 (1985) 163.
186. D. Z. Zhechev and I. T. Koleva, Self-alignment and radiative lifetime measurements of some P II levels in a hollow-cathode discharge, *Phys. Scr.* 34 (1986) 221.
187. D. Z. Zhechev, Incoherent signal at Hanle experiments in a hollow cathode discharge, *Spectrosc. Lett.* 20 (1987) 111.
188. W. Żyrnicki and A. B. Basily, Atom and ion temperatures in a hollow cathode discharge, *Spectrosc. Lett.* 19 (1985) 713.
189. W. Żyrnicki, Excitation temperatures of atoms and ions in a hollow cathode discharge, *Spectrochim. Acta* 40B (1985) 995.
190. W. Żyrnicki, Z. Tomasik, and I. Nowicka, Spectroscopic measurement of plasma temperatures in a hollow cathode discharge, *Spectrosc. Lett.* 17 (1984) 207.

7

Analysis of Nonconducting Sample Types

Michael R. Winchester, Douglas C. Duckworth, and R. Kenneth Marcus

7.1. Introduction

New developments in the field of specialized materials continually present new challenges to the analytical chemist. The improved performance of novel glasses and advanced ceramics demands improved analytical techniques because trace contaminants in such materials critically affect their performance. Advances in the material sciences are thus dependent on improved analytical chemistry and more sensitive methods for the analysis of nonconducting materials. Broekaert *et al.* have reviewed state-of-the-art techniques for the analysis of advanced ceramics, comparing and contrasting the performance of each.[1]

Traditional methods for the elemental analysis of bulk materials, such as atomic absorption and inductively coupled plasma mass spectrometry and atomic emission spectroscopy, are typically solution-based techniques. These are tried-and-proven successful analytical techniques; however, they require dissolution procedures that can be difficult and time-consuming. Dilution of the analyte and concomitant contamination, resulting in lowered sensitivities, quantitative errors, and spectral interferences, are obviously a concern.

Michael R. Winchester • Inorganic Chemistry Research Division, National Institute for Standards and Technology, Gaithersburg, Maryland 20899. **Douglas C. Duckworth** • Analytical Chemistry Division, Oak Ridge National Laboratory, Oak Ridge, Tennessee 37831. **R. Kenneth Marcus** • Department of Chemistry, Howard L. Hunter Chemical Laboratories, Clemson University, Clemson, South Carolina 29634

Glow Discharge Spectroscopies, edited by R. Kenneth Marcus. Plenum Press, New York, 1993.

Direct solids techniques for nonconducting material analysis are quite powerful in some respects and do not require dissolution procedures; however, they are not free of analytically undesirable effects. Techniques such as secondary ion mass spectrometry provide very low detection limits and lateral spatial resolution, though matrix and chemical sensitivities in the ion formation process can be severe. X-ray and electron spectroscopies offer lateral spatial resolution but typically lack the sensitivity for trace component detection. Erratic vaporization and ionization processes in spark sources require long integration times, although relative and absolute sensitivities are very good.

Much research effort has been devoted to the analysis of nonconducting species by glow discharge techniques. The methods to be presented in this chapter include mixing the nonconducting material with a conducting host matrix (much like spark source methodology) and direct analysis of solids by radio frequency-powered glow discharges. Both techniques are solids techniques and though dilution of the analyte is necessary in the first, none is necessary in the latter. As mentioned in the preceding chapters, glow discharges provide a steady-state source of excitation and ionization that is relatively free of matrix effects due to separate atomization and excitation/ionization steps. Also, the simplicity of the glow discharge makes it an attractive and a relatively inexpensive source for nonconducting materials analysis. In light of the specialized analytical techniques that are necessary for the development of materials with improved performance, the glow discharge may be an attractive alternative to more traditional approaches.

7.2. Application of the Direct Current Glow Discharge

7.2.1. Introduction

In the direct current (dc)-powered glow discharge, the solid sample serves as the cathode, and so must be electrically conductive. In order to atomize nonconductive solids in the dc glow discharge, the sample must first be rendered conductive, generally by compaction into a suitable host matrix material. This methodology is similar to the briquetting technique used with arc and spark discharge sources and was first introduced for use with the dc glow discharge by Dogan *et al.* in 1972.[2]

The advantages that characterize the application of the dc glow discharge to the direct analysis of nonconductive solids are basically those that characterize the glow discharge itself. In particular, in comparison with solution-based analytical methodologies, the need for dissolution of the sample is avoided, often resulting in a savings of time, effort, and expense. This is particularly evident in the case of nonconductive materials that are usually difficult to dissolve. Also, since the cathodic sputtering step is nonthermal in nature, sample atomization rates for various species are quite uniform, varying typically by less than an order of magnitude. Because of this fact, and aided by the process of steady-state sputtering (see Chapter 2), the gas-phase concentrations of sputtered species in the negative glow region of the discharge are thought to be directly representative of the bulk solid concentrations. Direct solids techniques that rely on thermal atomization mechanisms are often characterized by atomization rates that vary by several orders of magnitude from element to element, making representative gas-phase concentrations an impossibility. Uniformity of sputter rates also results in the fact that analyte speciation effects usually prove to be insignificant.

An additional advantage of the glow discharge is the relative absence of matrix effects. This matrix insensitivity results from the fact that the atomization, excitation, and ionization steps are segregated in time and space. This characteristic of the glow discharge is particularly relevant in comparison with x-ray spectroscopies, which are inherently limited by severe matrix effects. Another advantage is that the discharge is operated in a low-pressure, inert atmosphere, which effectively inhibits efficient formation of interfering molecular species often found in atmospheric flames and plasmas. Furthermore, relatively low degrees of collisional and Doppler broadenings result in narrow optical transitions, allowing optical isotopic analysis of constituents with suitably large isotopic shifts. The narrow atomic transitions in the glow discharge also lead to enhanced sensitivity and linear dynamic range in atomic absorption analysis because of the close match of line profiles in the glow discharge atomizer and the hollow cathode light source.

There are, however, some disadvantages inherent to the sample preparation process. For example, mixing with an appropriate host matrix material is equivalent to analyte dilution and loss of sensitivity. Furthermore, problems of contamination, loss of analyte, and inhomogeneity may also result. Despite these pitfalls, the application of the dc glow discharge to the direct solids analysis of electrically nonconductive samples has been shown by a number of researchers to be an attractive alternative to competing methods of analysis. In the following sections, the factors and analytical practices important to this methodology are discussed in detail. Also, example applications are present. It is hoped that the discussion will provide the reader with an understanding of the advantages and limitations of the dc glow discharge for such applications.

7.2.2. Sample Preparation

Conventional dc glow discharge devices can be applied to the direct analysis of electrically nonconductive solids if the sample is first rendered conductive by compaction into a suitable conductive host matrix material.[(2)] This involves intimately mixing the sample with a conductive host matrix material, both in powder form, and pressing a portion of the resulting mixture to form an overall conductive solid sample that can serve as the cathode in the glow discharge. In order to perform this type of sample preparation, and obtain acceptable analytical data, a number of factors that affect analytical performance must be considered.

Handling of sample materials prior to mixing is an important consideration. This is because atmospheric gases trapped inside the interstices of the sample mixture during pressing can degrade the analytical characteristics of the glow discharge. Additionally, water, which may exist as adsorbed or hydrate waters on the surface of the sample particles can be particularly damaging. Particle size control may be somewhat effective at reducing interstitial trapped contaminants. Heating procedures are often utilized to reduce the amounts of water and other atmospheric gases on the surfaces of the sample particles prior to pressing. However, if strongly hygroscopic materials and volatile analyte species are simultaneously present in the sample, it may not be possible to remove the majority of the chemisorbed water without volatilizing analyte-containing compounds. Thus, heating procedures may be inappropriate. All sample materials should be stored in a desiccator prior to use. Likewise, pressed samples should be isolated from the atmosphere if immediate analysis is not anticipated.

Winchester *et al.*,[(3)] investigating the implementation of glow discharge devices for the direct analysis of automotive catalysts, have reported the observation of deleterious effects of water trapped in compacted samples. Automotive catalysts are composed principally of γ-alumina particles into which noble metals are chemisorbed. In this work, chemisorbed Pt and Rh were determined by means of atomic absorption spectrophotometry. Alumina is a highly hygroscopic material. Overnight drying of the samples at 110°C proved unsuccessful at removing a sufficient amount of the water present in the samples, but more rigorous drying procedures were not attempted in order to avoid altering the chemical composition of the samples. Therefore, significant amounts of water were incorporated into the compacted samples. Conveniently, sputtering of a sample in a glow discharge results in sample heating due to the transfer of kinetic energy from the impinging ions to surface atoms. In this way, water may be removed from the compacted sample through vaporization. Unfortunately, if the rate of vaporization is too fast, damage to the sample may result. In this particular case, for example, as the compacted catalyst samples were heated by this mechanism, the trapped water expanded too rapidly as it was vaporized, resulting in sample cracking. In some instances, the samples were even rendered unusable by this process. At the least, the precision of analysis was worsened, and presputter routines designed to slowly eliminate trapped water, as well as any other adsorbed gases, were undertaken. Even with this difficulty, however, the glow discharge analysis of these catalyst materials was successful. As will be discussed in a subsequent section, the presence of water in the glow discharge environment is harmful for other reasons as well.

Mixing of sample powders with host matrix materials is usually accomplished by means of a mechanical shaker or grinder. These vibrating mills usually produce changes in the particle sizes through attrition. This is not normally seen as detrimental and may in fact be viewed as advantageous. Mixing times of approximately one-half to several minutes are necessary to produce mixtures of sufficient homogeneity. Lomdahl *et al.* utilized the glass mixing vessel illustrated in Fig. 7-1 to homogenize sample powders.[4] The vessel is rotated about a horizontal axis at 20 rpm, such that each rotation results in an approximately 50% division and recombination. After a 30-min mixing period, mixture homogeneities superior to that obtainable in vibrating mills are claimed. Additionally, little or no particle attrition occurs. Such a procedure would be advantageous if studies of the effects of particle size distributions on analytical results are to be performed.

Once the sample/host mixture has been satisfactorily homogenized, an appropriate portion is compacted in a die press assembly to form the final

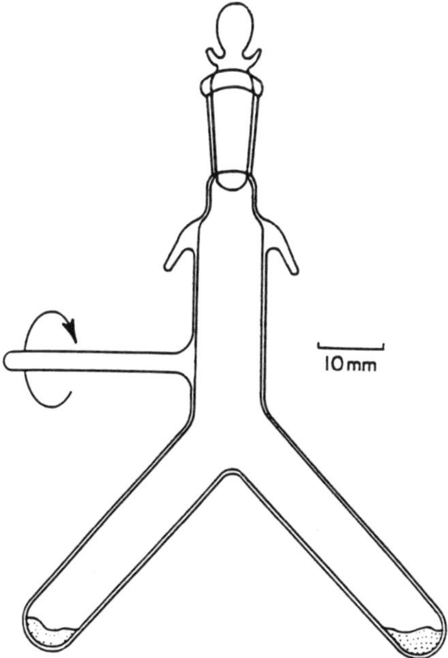

Figure 7-1. Glass mixing vessel (sample divider) for mixing powders.[4]

electrically conductive solid sample used in the analysis. Figure7-2 illustrates a typical die press assembly used to produce sample cathodes. Normally, a pressure of at least several hundred psi is necessary to produce cathodes of suitable strength. Pressed samples may take the form of disks,[4–10] pins,[11,12] or hollow cathodes.[13,14] The pressed cathode may constitute the entire cathode, or only part of the cathode. A multiple-section hollow cathode assembly is illustrated in Fig. 7-3a.[13] With this assembly, the compacted sample mixture may comprise the entire inner surface of the hollow. However, sputtering in the closed-end hollow cathode has been shown to be concentrated toward the closed end.[15–17] Therefore, pressing an entire hollow cathode may be considered unnecessary. The multiple-section assembly offers the analyst the flexibility to utilize the compacted sample mixture as only the base of the hollow cathode if so desired. The remainder of the cathode inner surface may consist of a blank material that does not contribute analyte species to the negative glow. As shown in Fig. 7-3b, a compacted pellet of sample mixture may alternatively be placed into a base plate, which acts as a sample holder and forms the closed end of a hollow cathode assembly.[14]

Compacted samples may not be capable of withstanding a great deal of mechanical stress, such that a specialized sample holder may be required. This is usually the case in the application of the Grimm-type or similar glow discharge lamps, in which the sample is mounted on the outside of the atomizer by pressing it firmly against an O-ring by means of a compression bolt assembly and thereby forming a vacuum seal. For example, using a sputtering cell similar in construction to the Grimm emission source, Winchester and Marcus employed a brass sample holder with a 1.25-cm-diameter, 0.1-cm-deep, recess in one face.[5] This sample holder is diagrammatically illustrated in Fig. 7-4. The inside walls of the recess are tapered

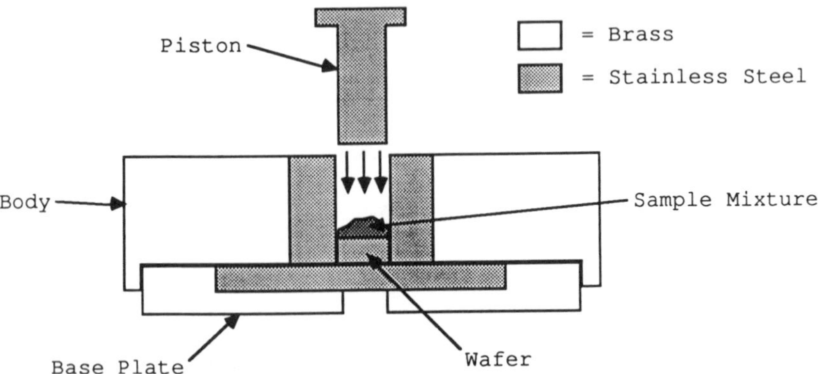

Figure 7-2. Typical die press assembly used for pressing powder sample mixtures.

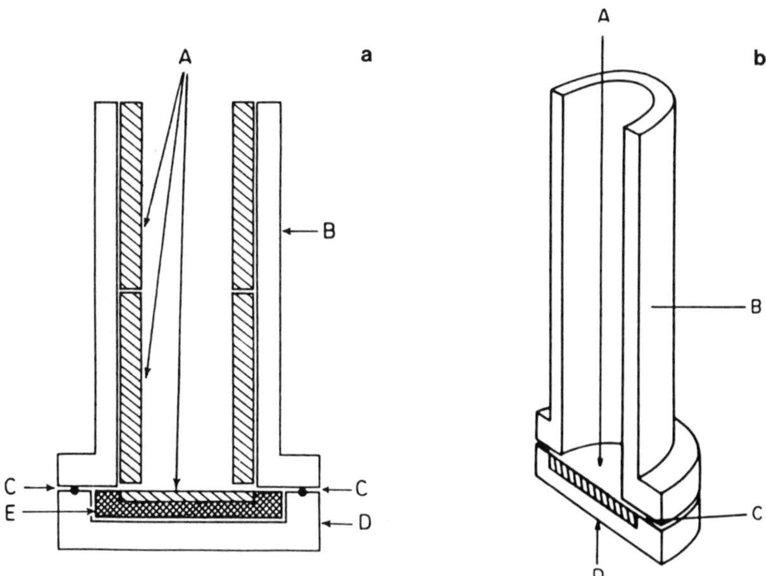

Figure 7-3. Hollow cathode assemblies used for the analysis of compacted powder sample mixtures. In panel a, the compacted sample mixture may constitute any or all of the inner surface sections,[13] while in panel b, it constitutes only the end.[14] A, compacted sample; B, copper or graphite supporting cylinder; C, sealing O-ring; D, supporting plate; E, copper pellet incorporating sample disk.

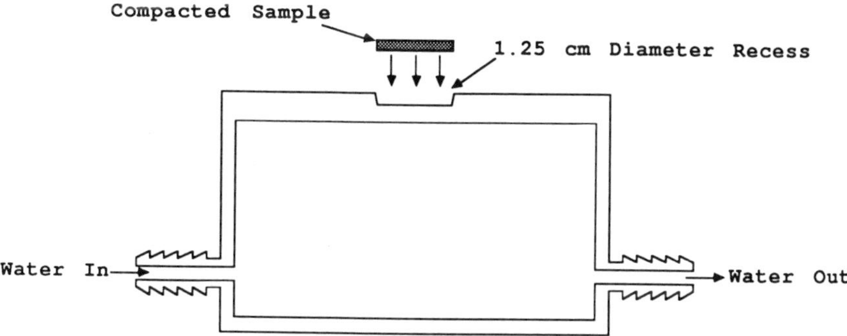

Figure 7-4. Brass, water-cooled, sample holder for the analysis of powder mixtures compacted into the shape of disks.

such that the diameter of the recess decreases slightly with increasing depth. In this way, compacted sample disks of 1.25-cm diameter and approximately 1-mm thickness can be press-fit into the recess against a clean, flat surface until flush with the holder surface. The holder containing the sample may then be mounted for analysis. Only the holder is in physical contact with the O-ring and the compression bolt, such that no stress is placed on the sample. Additionally, since only the compacted sample is exposed to the discharge environment, only the sample is sputtered. The conductive holder serves to couple the electrical power to the sample and is water-cooled in an attempt to maintain ambient sample temperatures. Various solid-state processes, such as diffusion, which can be detrimental from an analytical standpoint, are known to occur in compacted samples (and alloys to a lesser extent) at rates that are a function of sample temperature. These types of processes will be discussed at length in a subsequent section. An additional advantage of such a holder is that relatively little sample is needed for an analysis, with each disk containing only about 200 mg of actual sample material.

7.2.3. Analytical Sample Composition

The composition of the sample is an important consideration for the successful implementation of the dc glow discharge to the analysis of electrically nonconductive solids. Compacted sample composition directly affects the performance of the methodology, principally by determining the sputter rate and structural integrity of the compacted sample and the amounts of water and trapped gases in the compacted solid. These unwanted constituents are harmful to glow discharge analysis in terms of their effects both on the structural integrity of the compacted sample and on gas-phase processes within the discharge. Sample composition factors to consider include the choice of host matrix material, the relative percentages of sample and host in the sample/host mixture, and the particle sizes of the sample constituents. Each of these considerations is discussed in the sections that follow.

7.2.3.1. Choice of Host Matrix Material

The choice of conductive host matrix material is paramount to the success of the analysis. Boumans has shown that the sputter rate of an alloy is determined by the sputter rates and concentrations of the individual components comprising the alloy.[18] This should also be true of compacted samples. It follows that for good sensitivity, host materials with high sputter rates are desirable. Marcus and Harrison compared the use of graphite, silver, and copper as conductive host matrix materials for the emission spectrometric analysis of flint clay added in equal mass portions with the host (see Table 7-1).[19] Emission intensities for the analyte elements were significantly higher when the clay sample was mixed with copper than when the sample was mixed with either silver or graphite, which provided similar intensities. The reduction in copper sputtering rate is similar to that found by

Table 7-1. Comparison of Emission Intensities (Counts) and Net Sputter Weight Losses for 1:1 Mixtures of NBS 97 Flint Clay in Graphite, Silver, and Copper Matrices[19]

		Relative intensities in:		
Element	% in sample disk	Graphite	Silver	Copper
Al	10.25	2507	3183	6354
Si	10.28	641	810	1133
Fe	0.16	265	288	325
Mg	0.05	454	534	796
30-min weight loss (mg)				
Sample and matrix		1.71	1.62	2.21
Pure matrix		2.31	18.52	9.02

Winchester and Marcus who observed that 1:4 (w:w) sample mixtures of bauxite (NBS SRM 69a) in copper sputter only 65% as fast as pressed pure copper powder under similar discharge conditions.[20]

Silver is intrinsically a rapid sputterer, as shown in Table 7-1. However, the presence of the clay in the mixture degraded the sputter rate by a factor of 11.4. This effect was attributed to the large clay/silver volume ratio in the mixture, owing to the relatively large density of silver, which caused the conductance of the mixture to be seriously depleted relative to the conductance of the silver host itself. The large suppression of sputter rate with the silver host as compared with the copper host has also been observed by El Alfy *et al.*[7] The low sputter rate of the graphite/clay mixture was attributed to the intrinsically low sputter rate of graphite.

It is also desirable that the host matrix material not interfere with the determination of the analytes of interest. In the implementation of optical spectroscopies, such an interference would most likely take the form of overlapping absorption or emission lines, but because of the inherently narrow linewidths of atomic transitions in the glow discharge, such spectroscopic interferences are rare. Isobaric interferences in low- to medium-resolution mass spectrometry, on the other hand, are not rare, and so the choice of host matrix material is somewhat more important in such applications. Therefore, it may be necessary to judiciously choose a host matrix material that does not interfere isobarically. Other desirable characteristics of the host include good mechanical strength upon pressing, high thermal conductivity, low cost, and availability in high purity and a variety of particle sizes. Copper,[4,6,8–10,14] silver,[8] graphite,[11,12] tantalum,[21,22] iron,[7] nickel,[7] and a 4:1 (w:w) mixture of copper and graphite[13] have all been used successfully. It is interesting to note that Ta possesses the added advantage of behaving as a getter, meaning that it tends to form strong oxide bonds, effectively "sucking" oxygen from the discharge environment. Thus, if it is used as the host matrix material, isobaric molecular oxide interferences in glow discharge mass spectrometry experiments may be substantially reduced.[22]

7.2.3.2. Relative Percentages of Sample and Host in Mixture

When choosing an appropriate sample/host composition, a trade-off between sputtering rate and analyte concentration must be considered. As the percentage of nonconductive sample material in the sample/host mixture is increased, the concentrations of analyte species in the compacted sample also increase. However, sputter rates are seen to decrease with increasing percentages of nonconductive materials.[22] The reasons for this are somewhat complex and not well understood. Because of the larger impedance associated with the incorporation of the nonconductive material, the discharge voltage is larger at constant current and pressure, meaning that ions of the support gas impinge on the surface of the cathode with a higher average kinetic energy. This should result in enhanced sputtering, but due to the usually stronger molecular bonding in the nonconductive sample (typically oxides, carbides, and the like) as compared with the metallic bond strengths of the conductive host matrix material, lower sputter rates are expected.

The choice of sample composition may be complicated if high sputter rates are desired and the sputtering source utilized is characterized by spatially nonuniform sputter patterns across the surface of the sample, such as with the so-called "jet-enhanced" discharges (see Chapter 3). The nonuniform sputtering results in spatially nonuniform heating of the sample surface, which imposes considerable stress on the compacted sample, particularly at the rapid heating rates induced by high discharge powers. Because of the presence of temperature gradients across the sample surface, the compacted sample may fracture, producing an erratic discharge. The thermal conductivity and mechanical strength of compacted samples usually decrease with increasing nonconductive material content, meaning that higher percentage compacted samples cannot necessarily be sputtered at discharge powers as large as lower percentage compacted samples. The effects of sample heating may be somewhat alleviated by use of sample cooling. Even using water cooling, however, near-surface sample temperatures of 800–900 K have been measured.[23] As a result, the choice of nonconductive material percentage must incorporate considerations of this effect.

Discharge current and pressure (voltage) influence not only sample sputter and heating rates, but also excitation and ionization conditions in the discharge. Therefore, in the case of spatially nonuniform sputtering, the most effective method of optimization involves monitoring the analytical signal of interest while varying discharge current and pressure, for a range of sample makeups. Although somewhat time-consuming, this methodology is the only way to deduce the parameters necessary for optimum analytical operation. In those systems where spatially nonuniform sputtering is not problematic, the nonconductive percentage may generally be optimized independent of other discharge conditions. In such cases, maximum analytical signals are usually obtained at nonconductive material percentages of 20% and most researchers agree on this.[(7,13,16,17)] However, other percentages may be used successfully. In cases involving spatially nonuniform sputtering, lower percentages may be analytically optimum. For example, 10% samples were found to be optimal in one study involving the analysis of alumina-based automotive catalyst materials.[(3)]

7.2.3.3. *Effect of Particle Size*

Another important factor influencing the success of the technique is the range of particle sizes present in the sample mixture. In order to obtain a discharge whose gas-phase concentrations are representative of the bulk sample, the particle sizes present in the pressed sample must be small compared with the rate of removal of surface layers. Furthermore, discharge stability appears to be inversely related to particle size. Owing to these facts, small particle sizes are preferable in terms of accuracy and precision. It should be kept in mind, however, that the use of smaller particles increases the specific analyte surface area, allowing for more adsorbed or chemisorbed water in the final pressed solid. Nevertheless, homogeneity and discharge stability are usually the dominant concerns, and many researchers agree that particle sizes should be kept below approximately 30–40 μm in diameter in order to obtain acceptable accuracy and precision of analysis.[(4,7,13)] However, so long as thorough mixing procedures are adopted and sputter rates are relatively high, successful analyses may be performed with little or no attention paid to particle size.[(5)]

7.2.3.4. *Effect of Trapped Gases and Water*

Pressing of powders inevitably results in the incorporation of trapped gases, including water vapor, into the pressed sample. The presence of such unwanted constituents can effectively degrade the analytical characteristics of the glow discharge environment. Completely ridding the sample of these types of components is not possible. In fact, even the sputtering of alloys, rather than compacted solids, does not totally eradicate the problems associated with these constituents, because the materials commonly used to construct glow discharge sources possess some propensity to maintain adsorbed water and gases on their surfaces. These gases will outgas during discharge operation at a rate determined by the temperature of the glow discharge source and the nature of the material from which it is constructed.[24]

Nitrogen is a common contaminant in compacted samples and in the glow discharge source since it composes 78% of the Earth's atmosphere. The effects of molecular nitrogen on argon and neon discharges have been studied.[25] In these studies, N_2 was found to effectively quench argon metastable atoms through collisions in which the nitrogen molecule is excited into the $C\,^3P_u$ level. The $C\,^3P_u$ level of N_2 lies 11.05 eV above the ground state while the argon metastable levels reside 11.55 and 11.72 eV above the ground state. The resulting small energy defect of 0.50 or 0.67 eV accounts for the successful quenching of argon metastables by N_2. Since argon ions are most rapidly produced in the discharge by electron impact ionization of argon metastables, quenching is observed to lower the discharge current under constant voltage conditions. Sputter rates and emission intensities of sputtered species also decrease compared with pure argon conditions, as a result of the lower current density at the cathode surface. Many other gases have been found to have similar quenching effects on the metastables of argon, krypton, and xenon.[26] In fact, molecular oxygen has been found to quench argon metastable atoms at a rate that is a factor of seven faster than molecular nitrogen. Other common gases on the list of perpetrators include H_2, CO, NO, N_2O, CO_2, HCl, CCl_4, H_2O, and CH_4.

Because of the fact that waters of hydration may often accompany sample components and the fact that water vapor may be the atmospheric constituent most prevalently adsorbed onto the inside surfaces of the discharge source, water may be the most analytically damaging of the commonly encountered contaminants. As noted above, water is known to collisionally quench metastable species. Through various excitation and ionization processes, water vapor yields several contaminant ion species in the discharge, including H^+, H_2^+, O^+, OH^+, H_2O^+, H_3O^+, and $H_3O^+ \cdot nH_2O$, where $n = 1$ to 5.[11] Mass spectra are often hampered by isobaric interferences from such species. Water, however, is particularly detrimental for another reason as well. Stern and Caswell approximated the mobilities of H^+ and H_2^+, the primary dissociated ionic species of water, as being between 20 and 40 times that of Ar^+.[27] These ions may thus carry a proportionately large percentage of the total discharge current and produce very little sputtering. This is because the ratio of the masses of the ions to the atoms present in the typical sample is such that energy transfer cross sections are quite low, resulting in very low sputter yields. Gough *et al.* observed that absorbance values of sputtered species, which should reflect relative sputter rates, were indeed lower when "moist" argon with only 17 ppm water was used as opposed to "dry" argon with only 3 ppm water.[24] Similar correlations between absorbance values (sputter rates) and approximate hydration water concentrations of compacted samples have been observed by Winchester and Marcus.[20]

Although sputtering and other discharge processes may be altered by the presence of contaminants, analyses should still be relatively successful, provided the concentrations of the contaminant species remain constant. Unfortunately, this is never the case, and so methodologies for reducing the effects of contaminants are essential to the successful analysis. Proper cleaning and outgassing of the glow discharge source prior to analysis can be quite effective. As mentioned previously, control of the particle sizes of sample components, heating of sample materials before mixing, and storage of samples in a desiccator can help prevent the trapping of gases, especially water, into the compacted sample. As a further precaution, a mild vacuum may be applied to the sample mixture during pressing.[3] Additionally, the compacted sample may be outgassed under vacuum and/or mild heat prior to analysis. Perhaps the most effective means of reducing the effects of contaminants, however, is the use of a suitable presputtering period prior to data acquisition. Loving and Harrison demonstrated mass spectrometrically the erratic behavior of glow discharge-produced ions immediately after discharge initiation as a result of changing concentrations of water vapor.[11] After the discharge had time to "dry out," ion signals became temporally stable, intense, and reproducible.

Presputter procedures are generally undertaken simply by allowing the analytical signals to stabilize at the discharge conditions desired for analysis before acquiring data. However, the use of a sputtering cell with spatially nonuniform sputtering may necessitate more complicated procedures. If large discharge powers are desired, such that sputter-induced heating is fast, the compacted sample may not be able to withstand the mechanical stress caused by the rapidly expanding trapped gases. An erratic discharge is usually the result. For this reason, presputter methods in such cases should involve attaining the desired discharge conditions through several steps, each successive step employing a somewhat higher discharge power than the preceding step. This methodology has been demonstrated to be quite appropriate, though possibly time-consuming.[3]

7.2.4. Atomic Transport in the Compacted Solid

Atomic transport processes have been suggested to occur in compacted samples under glow discharge sputtering. These are important to be aware of, since their presence contributes to the approach to steady-state sputtering conditions. At first glance, it would appear that these processes are so severe that the acquisition of accurate quantitative data would be impossible. However, numerous experiments have demonstrated the feasibility, and appropriateness, of acquiring such data with the glow discharge. The following discussion is offered simply to provide insight into the processes *possibly* occurring in the sample during glow discharge operation.

The approach to a steady state of sputtering necessarily results in the formation of a near-surface altered layer that possesses a composition somewhat different than the bulk material deeper in the sample. A number of factors contribute to the formation of the altered layer and thus the development of and approach to the steady-state sputtering condition. Differing sputter rates of the various components at the surface, as well as solid-state diffusion and chemical reactions occurring in the near-surface layers, will affect sputter-equilibration times.

The role of surface temperature in determining the composition and thickness of the altered layer has been observed in a number of studies involving alloys. Swartzfager *et al.*, sputtering a 50:50 Cu:Ni alloy with 2.0-keV Ne^+ (SIMS-type sputtering), found an exponential increase in the altered layer thickness with sample temperature, particularly above 200 K.[(28)] Similar results have been reported by Lam *et al.*[(29)] In another study, the surface of a Cu–Ni alloy was found by Auger electron spectrometry to be enriched in Ni after sputtering at near room temperature, while sputtering at 250°C effectively enriched the surface with Cu atoms.[(30)] The Cu enhancement was attributed to solid-state diffusion of the Cu from the bulk to the surface, accelerated by the higher temperature. Anderson reported that the relative sputter rates of Ag and Cu in Ag:Cu eutectic (60:40) targets bombarded by 100-eV Ar^+ are significantly different at low sample temperatures than they are at high sample temperatures.[(31)] Since sputter yields of pure metals are *thought* to be relatively independent of temperature,[(30)] this implies that the differing surface compositions as a function of temperature are the result of the presence of temperature-dependent solid-phase processes.

Solid-state diffusion and chemical phenomena are likely present in compacted samples. However, solid-state processes may be somewhat different in compacted samples than they are in alloys and pure metals, because of the presence of the individual particles that compose the sample. Mai and Scholze have developed a model of the formation of the altered layer in compacted samples (see Fig. 7-5).[(23,32)] Similar to alloy sputtering, bulk diffusion (within individual particles), surface diffusion, and radiation-enhanced diffusion are present. Contact boundaries (the boundaries between individual particles) and dislocations within individual particles formed by the compaction process provide regions of enhanced solid-phase diffusivity relative to the more dense regions of a bulk solid. Therefore, "short-circuit" diffusion along contact boundaries or dislocations will dominate in such samples. In fact, diffusion of large groups of atoms may also be possible

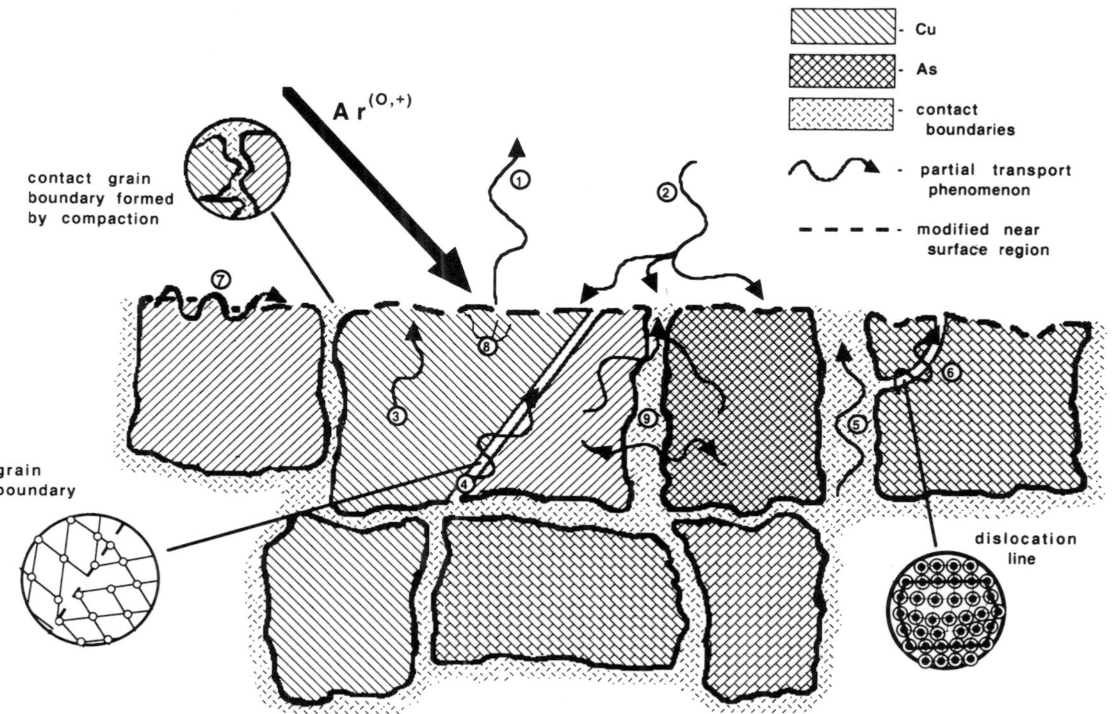

Figure 7-5. Schematic representation of partial transport phenomena supposed to occur during glow discharge excitation. 1, sputtering; 2, redeposition; 3, bulk diffusion; 4, grain boundary diffusion; 5, contact grain boundary diffusion; 6, diffusion along dislocations; 7, surface diffusion; 8, radiation-enhanced diffusion; 9, solid-state reactions.[32]

along boundaries. It has been estimated that for a wide range of sample temperatures the rates of surface and "short-circuit" diffusions may be quite similar, while bulk diffusion may be orders of magnitude slower. The presence of the boundaries also results in enhanced chemical reaction rates in these areas as well. The rates of any such reactions will of course also depend on temperature and the reactive characteristics of the components.

Small-diameter component particles were shown previously to be an important factor in terms of accuracy and precision of analysis. Particle size also partially determines the rates of the various solid-state processes being discussed here. For example, as the particle sizes of the components of a compacted sample decrease, the density of contact boundaries increases. This could result in a further enhancement of the rate of short-circuit diffusion along contact boundaries, as well as the overall rates of chemical reactions. Simultaneously, small particles may decrease these rates by increasing the thermal conductivity of the sample and thereby effectively decreasing sample surface temperatures.

Although the precise effect of particle size on the rates of solid-state processes is not known, the rates of these processes will determine the formation of the altered layer and the time necessary to reach the condition of steady-state sputtering, presputter times necessary for the establishment of equilibrium sputtering conditions may in fact be shortened by the use of highly homogenized samples of small particle size. Mai and Scholze observed this type of effect by emission spectroscopy for samples of $ZnFe_2O_4$ and ZnO/Fe_2O_3 mixed in Cu powder.[23] As illustrated in Fig. 7-6, emission intensities from the coarse mixtures needed nearly 9 min for stabilization, while those from the highly homogenized mixtures required only about 4 min. Additionally, the behavior of the two coarse sample mixtures containing identical overall stoichiometries but different speciations was significantly different before the establishment of equilibrium, while the behavior of the two isostoichiometric homogenized sample mixtures was quite comparable both before and after the establishment of equilibrium. These data are strong support for the importance of solid-state processes in establishing steady-state sputter conditions. Furthermore, the data indicate the need for homogenized sample mixtures of small particle sizes in terms of reducing necessary presputter times. Thus, from all viewpoints, small particle sizes may be considered advantageous to the analysis.

The necessary presputter time required for attainment of steady-state sputtering has also been observed to depend on the discharge power used.[23] As illustrated in Fig. 7-7, the emission signals of W from compacted WC (both W and C are electrically conductive) and from a mixture of W in Cu powder (50% w/w) do not coincide even after 30 min of sputtering at 14 W. Sputtering at 30 W attains equilibrium only after approximately 11 min, while sputtering at 100 W produces the altered layer and steady-state sput-

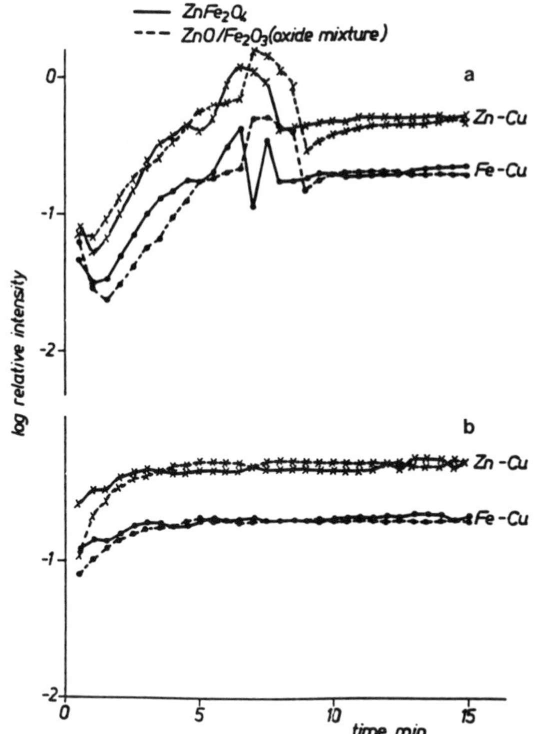

Figure 7-6. Temporal behavior of emission intensities of Zn (crosses) and Fe (solid dots) in different compounds compacted into Cu host matrix for (a) coarse mixture and (b) sufficiently homogenized mixture.[23]

tering in less than 3 min. The use of large discharge powers during presputtering, however, may not necessarily be advantageous. This is due to the fact that the altered layer composition and thickness may be quite different from one set of discharge conditions to the next, owing to the differing current densities and heating rates. The most expedient procedure may be to presputter the sample at whatever discharge conditions are necessary for analysis. Additionally, as mentioned previously, the use of stepped presputtering routines may be needed in some cases.

In the experience of the authors, the sputtering rates of even compacted samples are so high as to make the influence of any local atomic transport within the sample negligible in the attainment of steady-state sputtering conditions. Studies of the sputtering of various inorganic compounds of the same metal ions show little influence on sputtering characteristics,[5,19] suggesting minimal amounts of "chemistry" within the sample matrix. In

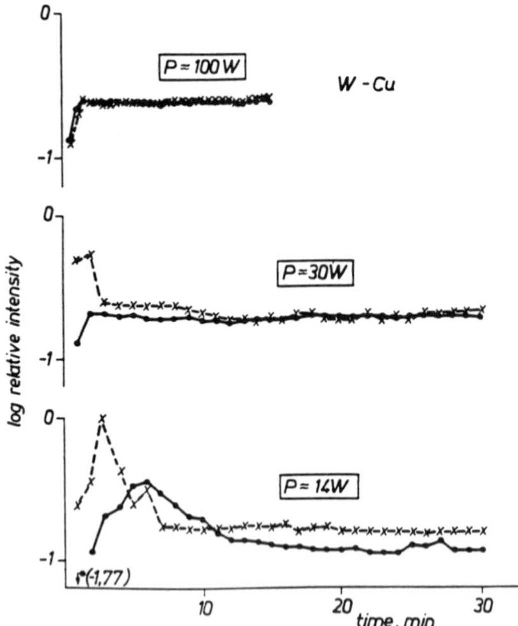

Figure 7-7. Temporal behavior of W emission intensity for WC single phase (solid dots) and W/C mixture (crosses) compacted into Cu host matrix for three different discharge powers.[23]

terms of plasma stabilization and the like, the effects of outgassing and thermal breakdown seem to be far more important than transport phenomena.

7.2.5. Methods for Improved Quantitation

Qualitative, semiquantitative, and quantitative analytical data may be obtained from the glow discharge through optical and mass spectroscopies. In the simplest experiments, data are collected as raw signals using dc glow discharges that are continuously operational, i.e., nonpulsed. Quantitative analyses done in this way rely on the reproducibility of the glow discharge environment and conditions, and the lack of systematic errors, to obtain acceptable accuracy. Owing to the fact that glow discharge devices are in fact quite reproducible and appear to suffer from few systematic errors, analyses can often be performed in this way with accuracies of better than 5% relative. However, electrically nonconductive sample materials compacted into conductive host matrices introduce other variables that may be somewhat irreproducible. These include the amount of trapped water vapor and atmospheric gases contained within the compacted sample, the sample structural integrity, and the like. Therefore, experimental conditions may be somewhat less reproducible when analysis of electrically nonconductive materials is to be compared with that of alloys. As a result, the accuracies of such simple analyses may be affected.

The analytical performance of the glow discharge source for the analysis of electrically nonconductive materials may be improved by the use of internal standardization, as well as by the use of a pulsed discharge. In fact, those methodologies may in some cases be essential to the successful implementation of the discharge. Therefore, each is discussed in some detail in the following sections.

7.2.5.1. Internal Standardization

Determinations of analytes in nonconductive sample types compacted into host matrix materials have been done most often by comparison with suitably prepared external standards (i.e., through calibration curves). Because of the relative lack of matrix and speciation effects in the glow discharge, matrix matching of external standards to samples is not a necessity, although it may be advantageous. However, the reproducibility and stability of analytical signals obtained from the glow discharge are affected by a number of parameters. Some of these have already been discussed, such as the presence of contaminant species. This is most likely a problem if the discharge source must be vented to atmosphere to change samples, such as with Grimm-type sources. The placement of the sample into the discharge source may also affect the analytical signal attained. This may be particularly true with the use of sample holders, where the efficiency of sample cooling could be changed by irreproducible placement of the sample in the holder.

If the assumption is made that factors affecting a given analyte will similarly affect every analyte in the sample, then utilization of internal standardization is suggested. This approach assumes that the ratios of the concentrations of analyte species in the vapor phase are the same as those in the solid sample. This is likely a valid assumption, provided that a steady state has been attained and the only physical sputtering is involved. This is also true of chemical sputtering, which is species specific. Nevertheless, McDonald has presented evidence to support the validity of this assumption for five types of alloyed samples analyzed by atomic absorption spectrophotometry, as shown in Table 7-2.[8] Although the ratio of Ni to Cu concentrations (R_C) in the alloy samples varied over four orders of magnitude, the ratio of their absorbances (R_A) in the negative glow were observed to behave similarly, such that R_C/R_A remained relatively constant.

Because of the relative absence of matrix effects in the glow discharge, practically any atom present in the sample, including the host matrix material, a constituent in the original sample material, or a deliberately

Table 7-2. Proportionality of Concentration Ratios to Absorbance Ratios for Solid Alloy Samples[8][a]

Sample[b]	$R_C = C_{Ni}/C_{Cu}$	$R_A = A_{Ni}/A_{Cu}$	R_C/R_A
SS406	5.28	0.851	6.2
H1168	3.96	0.572	6.9
SAC1576	0.625	0.0952	6.6
C1100	0.000771	0.000116	6.6
C1115	0.000841	0.000136	6.2

[a]Cu measured at 324.7 nm and Ni measured at 352.5 nm.
[b]See Ref. 8 for sample descriptions.

added component, may be employed as the internal standard. This is in contrast to the use of internal standardization in other spectrochemical sources prone to gas-phase matrix effects, where the internal standard must be quite similar to the analyte chemically and spectroscopically in order to adequately compensate for interferences that might be present. However, even in the glow discharge several characteristics of the internal standard are desirable. In particular, the added internal standard should be chemically inert, nonhygroscopic, available in high purity, and easily handled. For use of a constituent other than the host matrix material, it should be an element with reasonable sensitivity in whatever analytical methodology is to be employed, such that a small but easily measured quantity can be added reproducibly to the sample mixture. In this context, use of the host matrix material requires the use of spectroscopic transitions with low sensitivity for optical spectrometries. For mass spectrometry, a minor isotope of the host matrix may be used.

According to synthesis of variance,[33] the use of internal standardization necessarily reduces analytical precision by virtue of the variance associated with the added measurement of the reference signal from the internal standard. Limits of detection may be worsened by this phenomenon, but the improvement in accuracy of the analysis usually makes internal standardization quite beneficial.

Because only a specific volume of the discharge is monitored analytically during analysis, accuracy of determinations involving internal standardization may be affected by changing concentration gradients of sputtered species in the negative glow if various species cannot be expected to behave similarly.[34] Analyte species of widely differing masses are known to diffuse at differing rates in the gas phase, producing concentration gradients.[34] Other studies have revealed the presence of significant concentration gradients of sputtered species in the discharge,[22,35,36] due not only to differing diffusion rates, but also to spatially differing atomization[20] and ionization[36] rates. Provided that concentration gradients remain constant throughout the analysis, no problems should result. Nevertheless, the presence of gradients should be kept in mind in case of any problems that might ensue.

Use of the host matrix material as the internal standard has met with mixed success. Determination of Ni (in samples of nickel salts compacted in Cu) using Cu as the internal standard was quite successful in one atomic absorption study.[8] In another study involving the analysis of automotive catalysts, the host matrix material, again Cu, was used successfully in AAS.[3] However, Lomdahl *et al.* reported an unsuccessful attempt to determine nickel and iron, as oxides, compacted in Cu, utilizing the Cu as the internal standard for emission spectroscopy.[4] The reason for this apparent discrepancy is unknown. However, the authors suggested that the high concentration of oxygen in the discharge environment may have contributed to environmental instabilities.

Use of an additive (spike) as the internal standard, on the other hand, has been shown to be quite appropriate in several investigations.[3,8,13] In particular, McDonald reported accurate determinations of Cu (in Cu slag samples incorporated into Ag) using Ni as the internal standard, even though sputter rates for the compressed samples varied by a factor of five at constant discharge current and pressure.[8] Ni was added to the sample mixtures to yield a final concentration of 1%. For five samples, relative quantitative errors were less than 4.5%. Without the use of the internal standard, errors of up to 500% would have been expected, based on the widely varying sputter rates. Additionally, Marcus and co-workers have demonstrated the successful use of Ba as an added internal standard in the determination of Pt and Rh in γ-alumina-based automotive catalyst materials by AAS.[3] Average accuracy was improved by a factor of 6 for Pt and a factor of 9 for Rh by using the internal standard. Ba was added as Ba_2CO_3 to obtain a final Ba concentration in the compacted solid of 0.27%. It is interesting to note that the Ba(II) 455.4 nm line was utilized to measure the reference absorbance signal. Although likely to be affected by changing ionization rates in the negative glow, this ionic transition was still useful for the internal standardization. Quantitation in this study used both calibration plots and relative sensitivity factors. These quantitation methodologies will be outlined further in a subsequent section.

In short, although glow discharge analyses of compacted samples can be performed quite successfully with direct external standardization (calibration curves), it would appear that internal standardization might be a generally more attractive means of improving analytical determinations. Certainly, this form of quantitation should be investigated further.

7.2.5.2. Pulsed Glow Discharges

An electrically nonconductive material placed directly in contact with an electrically conductive cathode being sputtered in a dc glow discharge will eventually sputter. However, the onset of sputtering of the nonconductive sample will begin only after the material has been reduced to the metallic state,[37] possibly by thermal decomposition or random bombardment with energetic particles near the cathode, or by redeposition. If the discharge is pulsed rapidly (>ca. 1 kHz), the nonconductive sample may be sputtered directly immediately after discharge initiation, in the same way that alloys may be sputtered in the dc glow discharge.[38] High-frequency sputtering of this type will be discussed in detail in the second half of this chapter.

Even if electrically nonconductive samples compacted into host matrix materials are to be sputtered in the dc glow discharge, use of a pulsed (power modulated) power supply also holds analytical benefits. Pulsed glow discharge operation has proven to be quite beneficial in the analyses of electrically nonconductive materials compacted into host matrix materials.[3] For example, high discharge powers which result in arcing in the continuous dc mode may be applied while avoiding arcing. This is possible because the pulse length is kept short enough that any electric arc does not have sufficient time to be established before the end of the pulse. Duty cycles may be kept low to maintain a low average power, such that the problems associated with excessive sample heating may be avoided. Shorter presputter times may result from the higher discharge powers, because of the more rapid but still gentle release of trapped gases from the compacted solid.

The use of the large discharge powers during the pulse, coupled with synchronous detection, enables higher sensitivities and signal-to-noise ratios to be obtained relative to those attainable in the continuous dc mode, because of higher sputter rates during the pulse and a more energetic gas-phase environment.[39] Higher sputter rates are produced by virtue of the fact that sputter rates are determined primarily by discharge power.[40] The sensitivity improvement associated with enhanced sputtering is especially advantageous when determining analytes in samples with intrinsically low sputter rates (e.g. compacted samples). The more energetic gas-phase environment is a result of increased electron number density and electron temperature during the high-power pulse, promoting more efficient electron impact collisional processes in the gas phase.

When using the pulsed glow discharge, accuracy and precision can possibly be degraded if pulse lengths utilized are too short. As species are sputtered from the cathode, different analytes may diffuse into the sampling volume of the negative glow at differing rates, because of dissimilar masses and gas-kinetic cross sections. If the pulse length is short relative to the time necessary to establish an equilibrium concentration of the various analytes in the sampled volume, then the gas-phase concentrations of sputtered species sampled will not necessarily accurately reflect their concentrations in the original bulk solid,[39] even in the presence of steady-state sputtering. Provided that the behavior is reproducible, no inaccuracies should ensue with short pulse durations. However, reduced accuracy cannot necessarily be ruled out, and worsened precision is likely to result. This fact should be considered when choosing a pulse duration.

Pulse power, duration, and repetition frequency are mutually dependent in terms of their effects on discharge behavior, and optimization of these pulse parameters may be somewhat time-consuming. Typical pulse parameters might include pulse powers in hundreds of watts, pulse durations on the order of milliseconds, and pulse repetition frequencies less than 100 Hz.

7.2.6. Illustrative Analytical Applications

Employing the sampling methodologies described in preceding sections, the dc glow discharge source may be applied to the analysis of electrically nonconductive materials. The remainder of this section focuses on practical applications of dc glow discharge devices for the analyses of nonconductive sample types. Since atomic absorption, emission, and mass spectrometries are the most commonly utilized detection methodologies, one example of each is presented. These examples are given to illustrate the range of sample types amenable to dc glow discharge analyses, as well as the sample preparation considerations, quantitation methods, and representative figures of merit. The reader is referred to the original publications for further details.

7.2.6.1. Atomic Absorption Spectrophotometry: Analysis of Automotive Catalyst Materials

Winchester and colleagues have described the determination of Pt and Rh in automotive catalyst materials using atomic absorption spectrophotometry.[(3)] Such samples consist of the noble metals chemisorbed onto γ-alumina particles. The Atomsource glow discharge atomizer (Analyte Corp., Medford, Oreg.) was used to atomize the compacted solid samples. The Atomsource is described in some detail in Chapter 3. Of interest here is the ability of the Atomsource to operate in a pulsed mode. As discussed in a preceding section, such atomizer operation holds significant analytical advantages. The Atomsource's mate spectrometer, the Analyte 16, atomic absorption spectrophotometer, was used to detect the sputtered atoms. Internal standardization, in conjunction with calibration curve standardization, and the use of relative sensitivity factors were employed as possible means of quantitation.

Compacted calibration standards consisting of Ba_2CO_3, $PtO_2 \cdot H_2O$, Rh_2O_3, and diluent in a Cu host matrix (total nonconductive content = 5%) were utilized. The raw support material onto which active platinum and rhodium were deposited in the actual catalysts, consisting primarily of

γ-alumina, was employed as the diluent. Since alumina is considerably hygroscopic, it was dried overnight at approximately 110°C. Ba_2CO_3 was added to the samples, Ba being used as an internal standard to compensate for sample-to-sample sputter rate differences. The Cu host was also investigated as a possible internal standard. Samples and standards were presputtered before analysis in order to allow water and trapped gases to escape slowly without rupturing the sample disks and to allow absorbance values to stabilize. Analyses were performed under discharge conditions of 10 Torr argon and 300 mA for platinum and 10 Torr argon and 255 mA for rhodium, barium, and copper. During both presputtering and data collection the Atomsource was operated in a 7.75-Hz pulsed mode, with either a 26- or 52-msec pulse length, under current regulation.

The results of the analyses employing calibration curves are presented in Table 7-3. Those data in the "corrected by . . ." rows consist of the ratio of analyte absorbance to internal standard absorbance plotted against their relative concentrations. The data in the "raw absorbance" rows are simply the analyte absorbance versus concentration. As is illustrated by the R^2 (goodness of fit) values, linearity was acceptable in all cases. The determined Pt and Rh concentrations are presented, along with "accepted" values. The first three samples were produced gravimetrically from the metal oxides, in order to determine the accuracy of the methodology in the absence of certified standards, since no such standards are commercially available. Therefore, the accepted values for these samples are simply calculated from the amounts of the constituents known to have been added to the sample mixtures. The fourth sample was an actual catalyst obtained from a commercial producer. The accepted values listed for this sample were determined by competing methods of analysis as indicated in Table 7-3. Of particular interest, the value for the platinum in Sample 4 was determined by the stannous chloride method (differential spectrophotometry), which is currently considered to be the method of choice.

As indicated in Table 7-3, the glow discharge analyses using raw (unweighted) absorbances were somewhat successful, with average relative errors of 13.5% for Pt and 8.3% for Rh. This degree of accuracy may be acceptable for less stringent analyses. The employment of the internal standard elements greatly improved the performance of the method. Using the Cu internal reference, the average relative errors were 2.59% for Pt and 3.27% for Rh. Using the Ba reference, the average relative errors were 1.91% for Pt and 1.85% for Rh.

Relative sensitivity factors (RSFs) were also used in this study as a possible means of quantitation. An RSF is an empirically determined quantifier, based on the assumption that variations in analytical signal intensity caused by variations in glow discharge conditions are not analyte specific (see Section 7.2.5.1). An RSF is defined mathematically as the second

Table 7-3. Pt and Rh Determinations via Calibration Plots[(3)]

	Equation	R^2	Sample No. 1	2	3[a]	4
Platinum						
Raw absorbances	$(1.11 \times 10^{-5})x - 7.07 \times 10^{-5}$	0.998	0.578[b] (±0.040%)	2.124 (±0.078%)	1.305%	0.463 (±0.049%)
Corrected by Cu(I) 324.7 nm	$(13.0)x + 3.58 \times 10^{-4}$	0.997	0.470 (±0.023%)	1.938 (±0.140%)	1.164%	0.418 (±0.024%)
Corrected by Ba(II) 455.4 nm	$(9.93 \times 10^{-2})x + 4.75 \times 10^{-5}$	0.992	0.487 (±0.040%)	1.873 (±0.163%)	1.197%	0.432 (±0.052%)
Accepted value			0.477%	1.877%	1.149%	0.437%[c] (0.450%)[d]
Rhodium						
Raw absorbances	$(1.91 \times 10^{-4})x + 8.49 \times 10^{-5}$	0.998	0.467 (±0.036%)	0.186 (±0.003%)	0.400%	0.0860 (±0.0023%)
Corrected by Cu(I) 324.7 nm	$(225)x + 1.30 \times 10^{-3}$	0.995	0.449 (±0.043%)	0.164 (±0.007%)	0.364%	0.0798 (±0.0068%)
Corrected by Ba(II) 455.4 nm	$(1.70)x + 2.50 \times 10^{-4}$	0.996	0.431 (±0.012%)	0.167 (±0.014%)	0.373%	0.0838 (±0.0025%)
Accepted value			0.427%	0.163%	0.366%	0.0856%[e]

[a]Only one sample disk analyzed.
[b]All values represent concentrations in original solid sample. Values in parentheses are one standard deviation.
[c]Value obtained by stannous chloride method (current method of choice).
[d]Value obtained by ICP-AES.
[e]Average of values obtained by flame AAS, DCP-AES, and ICP-AES.

bracketed term in the equation

$$C_a^s = [C_i^s(A_a^s/A_i^s)][(C_a^r/C_i^r)(A_i^r/A_a^r)] \tag{7-1}$$

where C_a and C_i represent the concentrations of the analyte and internal standard, A_a and A_i their respective absorbances, and the subscripts r and s represent the values obtained for reference and analytical samples, respectively. Therefore, RSF values can be determined from a single external standard, or multiple standards, containing known concentrations of analyte and internal standard. The analyte concentration in the unknown sample may then be determined directly using the RSF, the absorbances of the analyte and internal standard, and the known concentration of internal standard in the unknown. Results using RSFs in this study proved to be quite similar to those obtained with the calibration plots, and so will not be presented here. The advantage of using RSFs as opposed to calibration plots is that a single standard could conceivably be utilized for the analysis of a large number of samples of widely varying composition.

In conclusion, the glow discharge methodology was shown to compete quite well with more established methods of analysis, and both the Cu host and the added Ba are suitable as internal standards. In general, analytical precisions were found to be better than 7.5% (relative) when internal standardization was employed. No limits of detection were reported in the paper, but were in the single-digit ppm range for Rh and in the double-digit ppm range for Pt.[(20)]

7.2.6.2. *Atomic Emission Spectrometry: Analysis of Geological Samples*

Marcus and Harrison have described the analysis of geological materials by means of glow discharge atomic emission spectrometry.[(19)] In that work, the hollow cathode plume (HCP) discharge source[(41)] was used to atomize compacted solids and excite the sputtered analyte atoms. The hollow cathode plume, depicted in Fig. 7-8, is a modified hollow cathode source incorporating a gas flow through the center and an exit orifice in the base, which comprises the compacted sample disk. Under certain conditions of pressure, current, and orifice diameter, two hollow cathode discharges exist, one in the main body and the second inside the orifice, which acts as a microcavity hollow cathode under such conditions. Microcavity hollow cathode discharge devices are known to produce intense analyte emission.[(42)] The gas flowing through the plume device causes the negative glow region of the dual discharge to extend out of the orifice as depicted in Fig. 7-8. Intense analyte emission from this extended glow region was monitored.

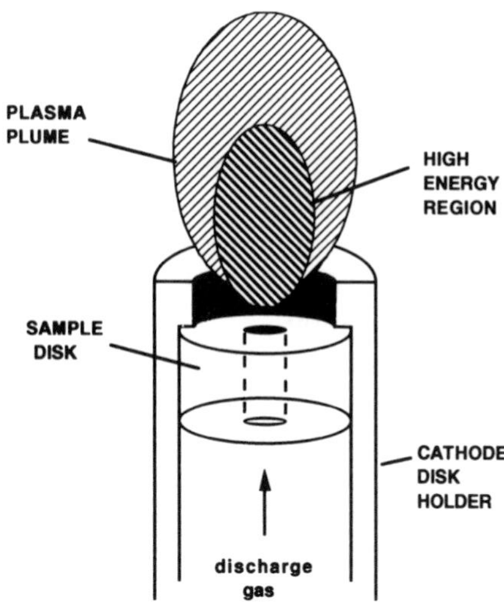

Figure 7-8. Schematic representation of the hollow cathode plume source.[41]

Graphite, silver, and copper were investigated as possible host matrix materials for geological specimens. These investigations showed that, on the basis of analyte emission intensities (see Table 7-1), Cu was the best choice. Studies of the emission intensities of various analytes (from a flint clay sample compacted into Cu) as a function of the percentage of sample in the sample/host mixture are shown in Table 7-4. These results show that 10–25% sample mixtures may be considered analytically optimal. The authors concluded that 10% sample mixtures could be considered better, however, on the basis of the smaller required sample sizes, the consequential lowering of the amounts of trapped gases and water incorporated into the compacted samples, and the normalization of the sample matrix material's effect on sputtering as a result of the larger host fraction.

Table 7-4. Elemental Emission Intensities as a Function of Flint Clay Concentration in a Copper Host Matrix[19]

	Relative intensities from single scan				
Element	5%	10%	25%	50%	75%
Cu	1220730	1124420	711744	242160	22720
Al	7204	11984	11964	6710	1435
Si	1356	2245	2119	1121	522
Fe	277	321	337	284	250
Mg	803	1007	1026	600	297
Na	482	538	435	269	390

As was noted in earlier discussions, presputtering is often necessary with compacted samples in order to allow trapped gases and water to escape slowly from the samples and to allow analytical signals to stabilize. The authors found this to be the case in these studies as well. Figure 7-9 illustrates the temporal response of Al(I) 309.2-nm emission upon initiation of the discharge. A 5-min presputter (preburn) at a lower current (50 mA, 2 Torr argon) was followed by an increase in current to optimal analytical conditions (150 mA, 2 Torr argon). With this routine, about 10 min was required to attain temporally stable emission.

In order to estimate the feasibility of using direct, single-standard analysis of geological samples with this technique, the authors employed NBS 97a flint clay as the single external standard (i.e., no calibration plots were generated) for the analysis of NBS 98 plastic clay. Their results are given in Table 7-5. The calculated concentrations of the plastic clay constituents demonstrate reasonable agreement with the assayed values. The deviations ranged from 3.4% for Al to 14% for silicon, with an average of 8.1%. As pointed out by the authors, the determinations were quite successful even in the cases of Fe and Mg, which were present at a factor of three difference in concentration between the two clays. Reproducibilities were reported to be in the range of 5% relative. Detection limits for five of the elements are also given in Table 7-5. These values were reported to be approximately twice those found for the same elements by the HCP atomic emission technique applied to conductive alloys, the lower sputter rate of the compacted geologicals accounting for the lower sensitivity.

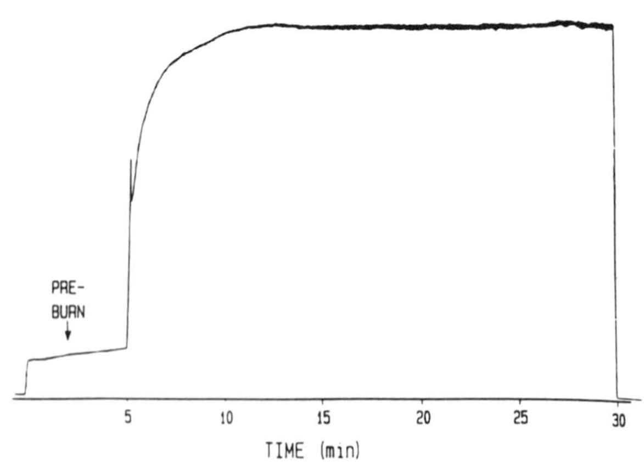

Figure 7-9. Temporal behavior of Al(I) 309.2-nm emission from a 10% flint clay/Cu matrix sample (Al concentration 2.05%): presputtered (preburned) at 50 mA, final current of 150 mA; 2 Torr.[19]

Table 7-5. Quantitative Analysis of NBS 98 Plastic Clay Using NBS 97a Flint Clay as an External Standard[19]

Element	I_{plas}/I_{flint}	Plastic clay		Flint clay	LOD
		Calcd	Assay	(assay)	
Al	261516/296188[a]	1.8%	1.74%	2.05%	5 ppm
Si	111966/87698	2.61%	2.28%	2.04%	10 ppm
Fe	13715/4823	903 ppm	945 ppm	317 ppm	2 ppm
Na	21549/10793	54.6 ppm	60 ppm	27 ppm	0.1 ppm
Mg	86077/32678	241 ppm	256 ppm	92 ppm	0.2 ppm
Cr	3818/4267	18.7 ppm	21 ppm	21 ppm	—

[a]Average of three 3-s integrations.

7.2.6.3. Atomic Mass Spectrometry: Analysis of Nuclear Materials

Robinson and Hall have described the analysis of uranium oxide (U_3O_8) reactor fuels by glow discharge mass spectrometry (GDMS).[43] They employed the VG9000 glow discharge mass spectrometer (VG Isotopes, Limited, Cheshire, UK) for these analyses. The VG9000 is a high-resolution magnetic sector instrument, somewhat unusual for "atomic" mass spectrometers, which usually employ quadrupole mass filters. The instrument uses pin cathodes, typically 15 mm long by 1.5 mm wide, mounted in a removable sample holder. The U_3O_8 samples were mixed with graphite (1:3 by weight) and compacted into the shape of pin cathodes.

The graphite matrix was used as an internal reference element, and ion beam ratios, analyte versus carbon, were utilized as the means of quantitation. Thus, no external standards were employed, and "no standards" analysis was achieved. The results of the analysis for 30 elements in four standard reference U_3O_8 nuclear fuels, along with certified values, are presented in Table 7-6. As is evident, the glow discharge analyses were quite successful. Analytical precisions, as indicated in Table 7-6, were generally in the range of 5 to 10% relative. Although no limits of detection were reported in this study, they are normally in the sub ppm range for dc GDMS applied to electrically nonconductive samples.

7.3. Application of rf-Powered Glow Discharges

7.3.1. Introduction

As discussed in the previous section, the mixing of nonconductors with a conducting host matrix is a viable method for nonconductor analysis. However, this method of nonconductor analysis may be undesirable due to loss of homogeneity (or inhomogeneity), entrained gases, sample contami-

Table 7-6. Comparison of VG9000 Results with Certified Values for Analysis of NBL98 Series Standard U_3O_8 [(43)a]

Element	NBL98/7 (ppm)			NBL98/5 (ppm)			NBL98/3 (ppm)			NBL98/1 (ppm)		
	Cert	VG	1σ	Cert	VG	1σ	Cert	VG	1σ	Cert	VG	1σ
Li	0	<0.04	0.01	1.2	1.6	0.22	5.0	8.4	1.0	26.2	30.3	1.0
Be	0	<0.04	0.01	0.8	1.0	0.16	5.3	6.1	0.8	25.7	23.8	4.0
B	0.1	0.81	0.51	0.4	0.3	0.08	1.2	1.8	0.3	5.5	6.8	0.3
Na	4.0	9.3	4.8	16.0	15.7	2.7	88.0	55.0	2.0	455.0	230.0	12.6
Mg	1.0	<0.4	—	4.0	2.5	0.45	17.0	13.5	2.2	91.0	42.0	3.9
Al	5.0	0.8	0.05	25.0	20.0	4.9	115.0	120.0	7.7	522.0	429.0	13.4
Si	2.0	—	—	10.0	10.2	1.9	65.0	410.0	5.3	315.0	—	—
P	3.7	0.7	0.04	23.0	7.2	0.82	99.0	39.5	2.1	505.0	175.0	9.8
K	2.3	4.8	2.1	28.0	22.2	2.6	138.0	102.0	7.3	725.0	582.0	40.0
Ca	—	—	—	4.5	9.6	5.0	19.0	40.0	5.1	100.0	167.0	19.0
Ti	0.3	0.44	—	2.1	3.6	0.6	11.0	14.0	0.5	50.0	78.0	2.3
V	0	<0.05	0.01	10.0	12.0	1.4	50.0	55.0	3.3	250.0	328.0	11.7
Cr	2.0	<1.4	—	9.0	6.7	1.3	22.0	16.1	3.4	101.0	43.0	4.2
Mn	0.8	—	—	2.9	1.3	0.11	10.6	7.5	0.9	49.0	28.0	0.5
Fe	13.0	9.3	0.9	32.0	26.6	4.0	110.0	87.3	6.4	515.0	394.0	17.0

Table 7-6. Continued

	NBL98/7 (ppm)			NBL98/5 (ppm)			NBL98/3 (ppm)			NBL98/1 (ppm)		
Element	Cert	VG	1σ	Cert	VG	1σ	Cert	VG	1σ	Cert	VG	1σ
Ni	2.0	1.0	0.2	5.6	3.5	0.66	22.0	12.8	1.7	103.0	61.0	2.1
Co	0.06	<0.04	0.01	1.0	0.84	0.11	5.0	3.5	0.1	25.0	22.0	0.6
Cu	0.4	<0.06	—	2.4	0.38	0.10	10.0	2.6	0.1	51.0	9.6	0.4
Zn	1.5	<0.2	—	19.0	3.1	0.77	96.0	24.8	3.5	480.0	96.0	2.70
Sr	0	—	—	2.6	1.9	0.21	10.0	9.0	0.9	55.0	52.0	2.0
Mo	<0.1	<0.28	0.07	2.0	2.3	0.62	10.0	9.5	2.7	51.0	45.0	0.8
Ag	0.1	<0.08	—	0.3	0.25	0.1	0.8	0.43	0.1	6.0	2.7	0.5
Cd	0.3	<0.33	—	0.6	—	—	1.9	<1.0	—	5.6	1.6	0.65
In	0.2	<0.04	—	0.3	—	—	1.4	0.65	0.13	8.0	1.45	0.7
Sn	<1.0	<0.5	—	2.5	—	—	10.0	5.7	1.5	50	18.3	0.2
Sb	0	0.14	1.0	1.0	—	—	5.0	1.8	0.3	25.0	6.4	0.5
Ba	0	5.2	1.6	2.0	1.6	0.91	10.0	8.2	0.5	50.0	28.8	4.2
W	0.1	<0.13	0.04	2.0	1.8	0.38	9.9	7.1	1.0	48.0	42.5	0.54
Pb	0.8	<0.08	—	2.5	0.53	0.16	9.0	3.2	0.4	46.0	18.0	1.0
Bi	<0.2	<0.04	0.01	1.0	0.7	0.26	7.0	3.2	0.7	46.0	10.0	0.6

[a]NBL values: provisional certificate concentrations by weight. VG9000 values: ion beam ratio concentrations.

nation, and dilution. As described briefly in Chapter 2, rf-powered glow discharges allow one to avoid most of these sample preparation-related problems because no mixing with a conducting host is required. The sample need only be shaped, or pressed in the case of a powder, into the appropriate form. An attractive benefit of the rf-powered glow discharge is its applicability to most sample types. Because the fundamental dc bias may be sustained on both conducting and nonconducting surfaces, both conductors and nonconductors can be analyzed without altering the experimental arrangement. Generally, any glow discharge source designed for rf operation is amenable to dc operation, although the reverse is not usually true because of fundamental differences in the discharge operation. The potential of such a source is clear, yet its analytical applications have been surprisingly limited. Before looking at the analytical applications of the rf-powered glow discharge, a brief review of its history, theory, and other fundamental considerations is merited.

7.3.2. Historical

Jackson has speculated that rf-powered plasma sputtering may have been observed as early as 1920 during investigations of rf plasmas.[44] Robertson and Clapp[45] were perhaps the first to record (1933) the observation that material was removed from the walls of a glass tube when a high-frequency discharge was initiated between two external electrodes. Hay[46] observed that material removal, which he attributed to sputtering, occurred only at high frequencies, although the reason for this was not understood. In 1948, Lodge and Stewart[47] compared the nature of dc and high-frequency discharges along with the spatially dependent removal of material from the walls of the discharge and attributed the removal to sputtering induced by a high negative wall charge and a high potential field at the surface of the insulator. Levitskii[48] investigated ion energies, performed probe measurements, and sputter studies of high-frequency discharges maintained between internal metal electrodes. In 1955, Wehner[49] proposed a rationale for sputtering nonconducting materials, which Wehner and co-workers[50] demonstrated in 1962 with the application of an external high-frequency potential to a glass wall of a thermionically supported sputtering system. The discharge served to clean the glass, and therefore, it was proposed that because of the high sputtering rates observed, the method could be used for relative yield measurements, etch studies, and sputtering of insulating layers. Indeed, in subsequent studies by Davidse and Maissel,[51,52] the technique of depositing insulating thin films was developed into a practical method involving a relatively simple two-electrode geometry, the grounded diode arrangement.

7.3.3. Theory of Operation

A thorough discussion of the theory of rf glow discharges has been presented elsewhere,[53] and will therefore be presented here in limited detail. The consideration of a dc potential applied to a sample (electrode) made of a nonconducting material provides a simplistic explanation of rf plasma operation. Obviously, the current necessary to sustain a glow discharge cannot flow through the target. The resulting response of the insulator surface to the applied potential is analogous to the charging of a capacitor as demonstrated in Fig. 7-10. When a high negative voltage ($-V_s$) is applied to the insulator, the surface potential drops initially to $-V_s$ and decays with time to more positive potentials because of ion neutralization reactions at the cathode surface.[53] A short-lived discharge will exist until a minimum threshold voltage is reached and the plasma is extinguished. Thus, the solution to the problem of sputtering insulators is to provide a source of electrons for ion neutralization while maintaining a controlled negative potential at the cathode. As previously mentioned, Wehner[49] correctly proposed that the application of a high-frequency potential to a metal base at the back of an insulator would cause the positive charges, which accumulate on the surface of the insulator, to be neutralized with plasma electrons during the positive portion of each cycle, maintaining a negative potential on the surface of the nonconductor. Based on the rate of positive charge accumulation in reduced pressure environments, a high-voltage pulse of 1 MHz or greater will sustain a pseudocontinuous discharge.[50]

The key fundamental phenomenon occurring from the application of a high-frequency ac potential is the "self-biasing" of the electrodes. It is this self-biasing that supports cathodic sputtering in the plasma. In order to demonstrate this phenomenon, consider the 2-kV peak-to-peak square-wave potential (V_a) illustrated in Fig. 7-11 and the resultant potential on the cathode surface (V_b). As the potential is applied during the first half-cycle, the surface charges to −1 kV followed by a decay to approximately −0.7 kV as the surface is bombarded by positive ions. As the second half-cycle begins,

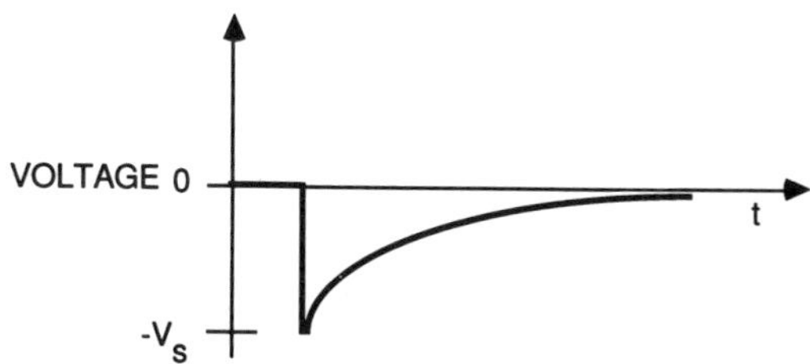

Figure 7-10. Temporal response of an insulating material to an applied negative voltage pulse.[114]

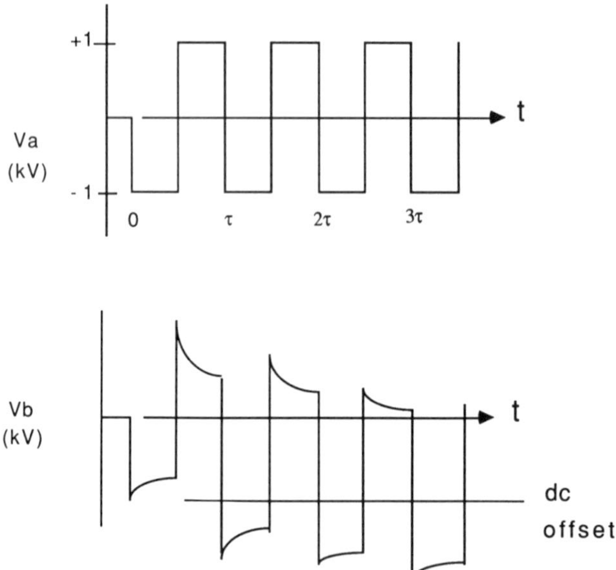

Figure 7-11. Electrode response to an applied square-wave potential.[114]

the applied +1 kV (+2 kV, relative) potential results in a +1.3 kV potential on the surface. During this half-cycle, electrons are accelerated to the insulator surface and charging results analogous to that of the prior half-cycle. However, because of the greater mobility of the electrons relative to the heavier positive ions, the surface potential decays toward zero at a faster rate than during the previous half-cycle, thus reaching a value of +0.5 kV. As the second full cycle is initiated and the polarity of the electrode is switched, the resulting potential is −1.5 kV (+0.5–2 kV). After several such cycles, the waveform of V_b will reach a constant dc offset. This dc offset is the self-bias potential that maintains a time-averaged cathode and sustains the sputtering ion current. The dc offset potential is approximately one-half the applied peak-to-peak voltage for analytical applications, depending on discharge parameters and source geometry. The insulating target material is alternately bombarded by high-energy ions and low-energy electrons; however, the discharge is, for all intents and purposes, continuous, and can be considered a dc discharge with a superimposed ac potential.

7.3.4. Fundamental Considerations

Anyone wishing to use an rf-powered glow discharge as an analytical source will soon come to appreciate the wealth of prior investigations that arose from the industrial application of these sources in deposition and etching processes. The electronics industry was the first to apply rf-powered glow discharge sputtering sources to the development of thin films. Since then, the rf-powered glow discharge has proven itself useful in the preparation of various thin films including semiconductors,[54–57] dielectrics,[58–61] and superconducting films.[62,63] Radio-frequency-powered discharges have also been used for generating polymer films by plasma-induced polymerization.[64–68]

With the development of deposition and etching processes arose the need for a better understanding of these discharges both in terms of monitoring the fabrication processes as well as obtaining a more fundamental understanding of the collisional processes occurring in the negative glow. Successful methods of monitoring these deposition and etching systems include optical emission,[69–72] fluorescence,[73,74] and luminescence from the growing film.[75] Various researchers have also successfully demonstrated mass spectrometric monitoring of the deposition process by sampling pertinent discharge ions.[76,77]

Various diagnostic techniques have been utilized in studying the processes occurring in the plasma. Gottscho and Miller[78] have reviewed the optical diagnostic techniques that have been applied to glow discharge plasmas. (Techniques for plasma diagnostics are described in Chapter 11.) Diagnostic techniques applied to rf-powered glow discharges include emission,[79–81] atomic absorption,[82] laser-induced fluorescence,[83,84] and actinometry.[81] Electrostatic probes[85–87] and mass spectrometers[82,88–93] have been used to study both the energy of charged species in the plasma and factors affecting their formation. The various techniques utilized in applications of monitoring and diagnostics suggest a range of useful analytical applications of the rf-powered glow discharge.

Such fundamental studies of rf deposition processes have generated a wealth of information and have suggested typical operating considerations for analytical devices. While a detailed discussion of such findings is beyond the scope of this work, a presentation of the most fundamental and relevant considerations necessary to the successful application of rf glow discharges to analytical techniques is merited. The critical parameters in rf glow discharges—discharge geometry, frequency, and electrical coupling—have been the focus of much of the diagnostic investigations to date.[79,89–91,94–96] The discussion that follows merely highlights these critical parameters. For a more detailed consideration, the reader is referred to Chapman[53] and other references included therein.

7.3.4.1. Source Geometry

The geometry of the rf glow discharge source is critical in terms of isolating the bombarding species to the sample and in determining the energy of the bombarding ions as related to the dc bias. In highly asymmetric rf discharges (those with large difference in electrode surface areas), the ion bombardment is generally isolated to the smaller electrode and the negative bias approaches half of the rf peak-to-peak voltage. As presented by Kohler *et al.*[89] in a thorough investigation of rf plasma potentials, Fig. 7.12 demonstrates the effect of geometry and mode of rf coupling, either directly or capacitively coupled, on the dc bias voltage (V_{dc}) the excitation electrode voltage (V_t, dashed curve), and the plasma potential [$V_p(t)$, solid curve], for

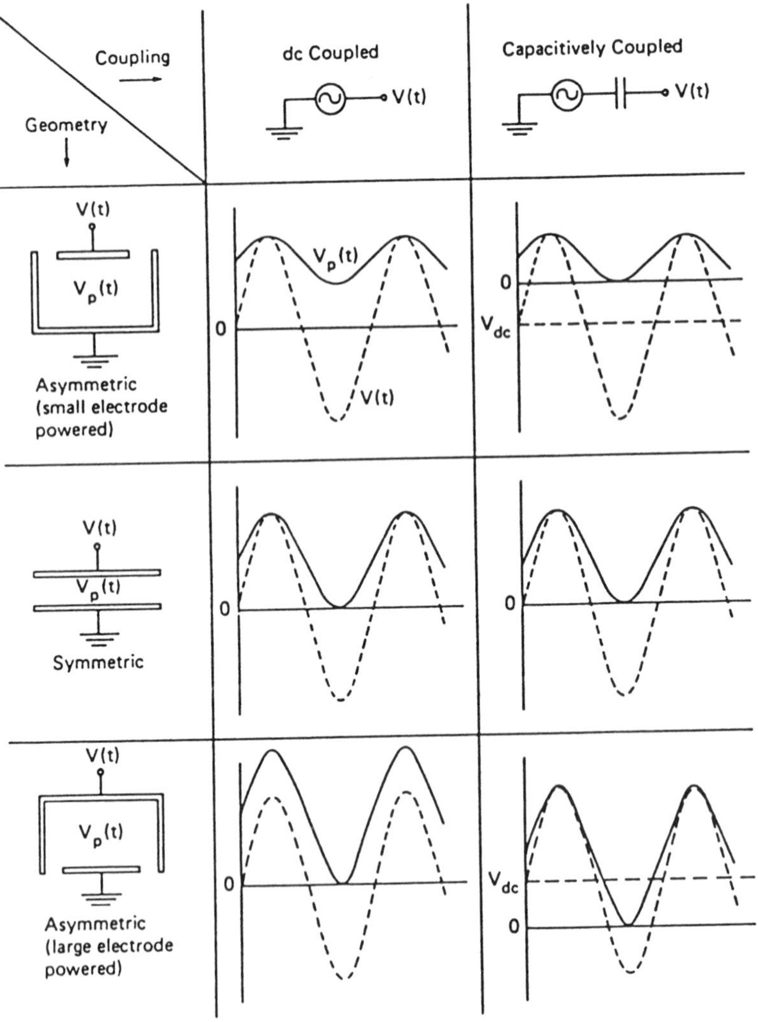

Figure 7-12. Plasma potentials [$V_p(t)$, solid curves], excitation electrode voltages [$V(t)$, dashed curves], and dc bias voltages (V_{dc}) as a function of discharge coupling and geometry.[89]

a sinusoidal waveform (assuming a purely capacitive sheath). As illustrated for the capacitively coupled excitation electrode, decreasing the cathode-to-anode ration causes the dc bias voltage potential on the target electrode to become more negative with the plasma potential following suite. This distribution of voltage between the electrodes is caused by differences in the capacitances of the dark space sheaths of the two electrodes. At sufficiently high frequencies, a few megahertz, the sheaths can be considered to behave capacitively[(97)] because the rf frequency is below the electron resonant frequency and above the resonant frequency for most ions. Kohler *et al.* concluded that the capacitive sheath model was in good agreement with experimental results in most applications.[(89)] The relative capacitances of the dark spaces are proportional to the relative electrode surface area and determine the distribution of the dc bias developed across the electrodes. The larger electrode has a larger capacitance and thus a smaller capacitive reactance. Therefore, the impedance of its sheath should be smaller than the impedance of the sheath of the smaller electrode, and its bias potential is closer to zero than that of the smaller electrode. Koenig and Maissel[(94)] developed (to a first approximation) the relationship $V_1/V_2 = (A_2/A_1)^4$, where V_1/V_2 and A_2/A_1 are the ratios of the dc potential difference between the glow space and the electrodes and the electrode surface areas, respectively. More recent studies indicate that this relationship may be closer to the first power than the fourth in cases of higher power densities and for cathode-to-anode area ratios less than 0.6.[(91,98)] In analytical applications, sputtering of the chamber or resputtering of the deposited films is obviously undesirable and thus the cathode-to-anode surface area ratio should be as small as practical.

It is not the absolute relative sizes of the electrodes that are critical, but rather it is the area of the electrode surfaces in contact with the discharge species that determines the relative electrode areas. In some instances the nature of the discharge gas can affect the symmetry of the discharge. For example, Roth[(79)] observed spatially resolved emission profiles between the electrodes of a closely spaced parallel plate rf discharge positioned in a larger vacuum chamber. Emission profiles of nonreactive gases indicate that the symmetry of the discharge can be dramatically affected by the nature of the gas without physical changes in the areas of the parallel plate electrodes. This is explained in terms of the degree to which the plasma diffuses into the chamber. As the plasma diffuses from the grounded electrode (anode) into the chamber, which is also at ground, the discharge becomes asymmetric with respect to the powered electrode because of the differences in surface areas. The diffusion is dependent on both the pressure and the electron temperature of the gas. In molecular gases the electron temperature is lower than that of atomic gases because of vibrational and rotational energy losses not available in the latter. Thus, diffusion in the atomic discharges results in increased asymmetry of the discharge as demonstrated by a change in the spatially resolved emission profiles of the discharge gases at each electrode. This study demonstrates the need for asymmetry between the electrodes. Generally, analytical applications of rf glow discharges use argon or other atomic gases and are asymmetric in design.

7.3.4.2. rf Power Coupling

The electrical coupling of the rf power to the target electrode is also critical. Without capacitive coupling in the circuitry, any dc potential difference that would otherwise exist between the electrodes cannot be maintained and the dc bias potential would be divided equally between the electrodes via the external circuitry. Note that in Fig. 7-12 the bias potential is zero for every geometry where the rf power is directly coupled. With direct coupling, little sputtering of the target electrode is likely because of the absence of a steady potential bias between the electrodes. Koenig and Maissel[(94)] described an equivalent electrical circuit of an rf glow discharge for sputter applications where the effects of geometry and other parameters such as pressure and magnetic fields can be evaluated. As adapted from their circuit, an equivalent circuit for analytical applications is shown in Fig. 7-13. The blocking capacitor serves to maintain the dc bias voltage (capacitive

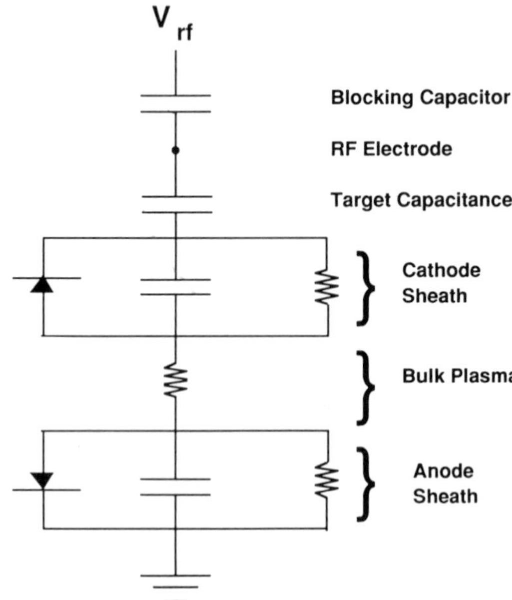

Figure 7-13. Equivalent circuit for an rf glow discharge plasma.

coupling). When analyzing insulating materials, an external capacitor is not necessary because of the capacitance of the nonconducting surface. In this case, the sample provides the required target capacitance. For conductors, the matching network provides the capacitance required to sustain the bias potential. The magnitude of the dc bias is largely responsible for the energies of incident ions on the sample surface and subsequent sputter rates. The division of the bias voltage between electrodes determines the degree of sputtering at each electrode. Various methods, such as substrate tuning,[(99)] have been developed to control the division of the dc bias between the electrodes. Division of the bias potential between electrodes promotes higher film quality in deposition systems as a result of resputtering of impurities from the surface of the growing films by energetic ion bombardment.[(100,101)] However, for analytical purposes, sputtering from surfaces other than the target is undesirable because of resulting spectral interferences. Resputtering would also degrade any depth profiling analyses. Therefore, discharge systems that maximize the dc bias at the target surface are desirable for most analytical purposes. (Jackson has reviewed rf sputtering with respect to the equipment and process mechanisms related to deposition and etching systems.[(44)]) Also illustrated in Fig. 7-13 are the electrical properties of the bulk plasma and the anode and cathode sheaths. The glow region of the plasma is resistive in nature while the sheaths have both resistive and capacitive properties. Resistances in the sheath are typically very large at low pressures and therefore the rf discharge is primarily capacitive in nature. Only at higher pressures, and at lower operating frequencies do the sheaths exhibit resistive behavior. The diodes in the circuit represent the high electron mobility.

7.3.4.3. Operating Frequency

The operating frequency of the discharge generator is another parameter that should be considered. Typically, a 13.56-MHz potential is chosen for rf glow discharge applications simply because this frequency has been designated by the Federal Communications Commission as a "free frequency." This frequency is generally suitable for all analytical applications, with detrimental effects occurring only at lower frequencies where neutralization of ions on the sample surface is less efficient, resulting in a lower dc bias. For optimum results, low megahertz frequencies are required. The effect of frequency has been investigated primarily in terms of the energy of ions bombarding the substrates in deposition systems. Such studies have provided important information for the mass spectrometrist while little emphasis has been placed on optical phenomena occurring as a function of frequency.

Analytically, phenomena occurring at lower frequencies have implications for mass spectrometric analyses in terms of ion energy distributions observed in the discharge. If the frequency chosen for the discharge is too low, rf modulation may occur. Radio frequency modulation is an ion transit time-related broadening of the ion energy distribution. Kohler *et al.*[90] have investigated the frequency dependence of ions bombarding grounded surfaces in a low-pressure planar rf argon discharge. In order to minimize the spread of the ion energy distribution, the frequency chosen should be high enough that ion transit across the plasma sheath between the bulk plasma and the grounded surface occurs over several cycles of the applied potential. The energy of the ions arriving at a grounded surface is determined primarily by the plasma potential, $V_p(t)$, and collisional processes occurring in the sheath. If the frequency is greater than the ion resonant frequency the energy of the ions will be determined by the time-averaged plasma potential, V_p. This is demonstrated in Fig. 7-14 where the ion energy distributions are illustrated for both 100-kHz and 13.56-MHz plasmas. The ion energy distribution at 13.56 MHz is relatively monoenergetic as determined by V_p because the ion transit time across the sheath occurs over several rf periods. In comparison, the 100-kHz ion energies are a function of the temporal

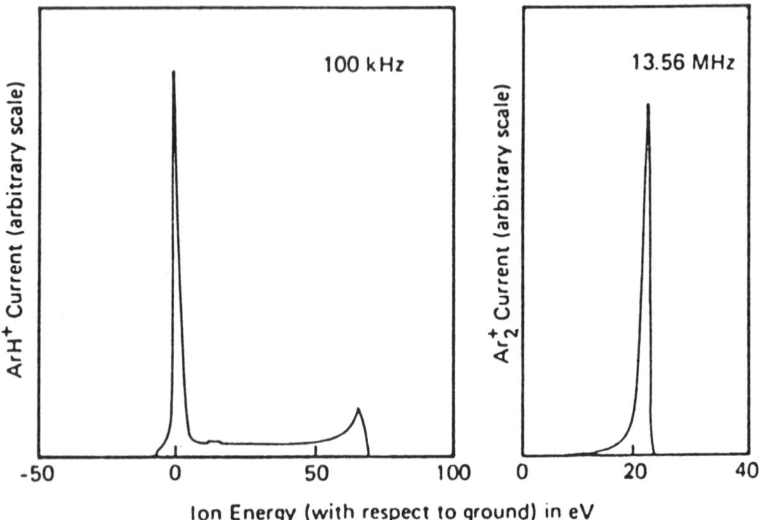

Figure 7-14. Ion energy distribution of discharge species sampled through the grounded anode of a planar diode rf glow discharge with 100-kHz and 13.56-MHz excitation potentials.[90]

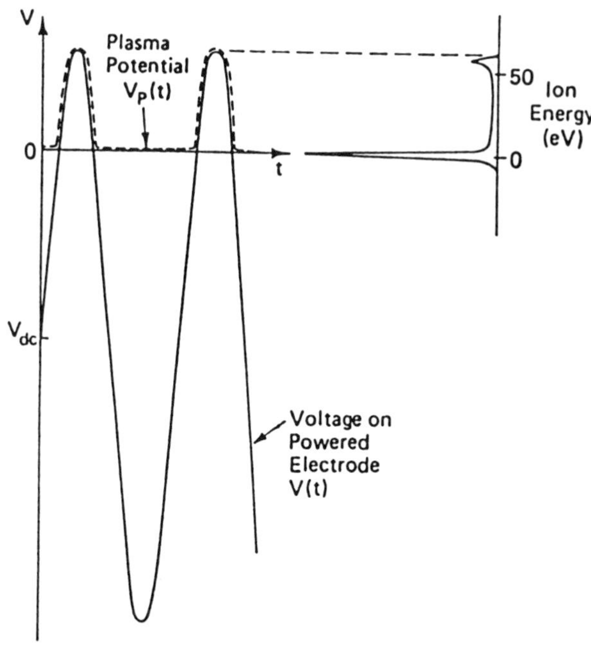

Figure 7-15. Illustrative figure of the relationship between the 100-kHz energy distribution in Fig. 7-14 and the plasma potential, $V_p(t)$.[90]

plasma potential $V_p(t)$ because ion transit times are less than one rf period. Figure 7-15 illustrates the ion energy distribution dependence on $V_p(t)$ at the 100-kHz excitation frequency. At low frequencies, the resistive nature of the plasma sheath is demonstrated as ion energies are dictated by the positive excursions of the sinusoidal excitation electrode potential $V(t)$.

Even at higher excitation frequencies transit effects may still be observed. Figure 7-16 shows the ion energy distributions of ions traversing the ground sheath at the substrate plane in a 13.56-MHz discharge.[91] The discharge was confined by a cylinder which increased the plasma density. Increasing the plasma density decreased the sheath thickness. Subsequently, the high mass Eu^+ required several rf periods to traverse the sheath while the much lighter H_3^+ ion traversed the sheath within one rf period or less as evidenced by the relatively large spread in the energy distributions.

7.3.5. Analytical Applications

Sputter deposition and etching of nonconducting materials, as well as the fundamental studies mentioned, suggest the potential of the rf-powered glow discharge for the analysis of nonconducting materials. Radio-frequency-powered glow discharges are useful as both primary and second-

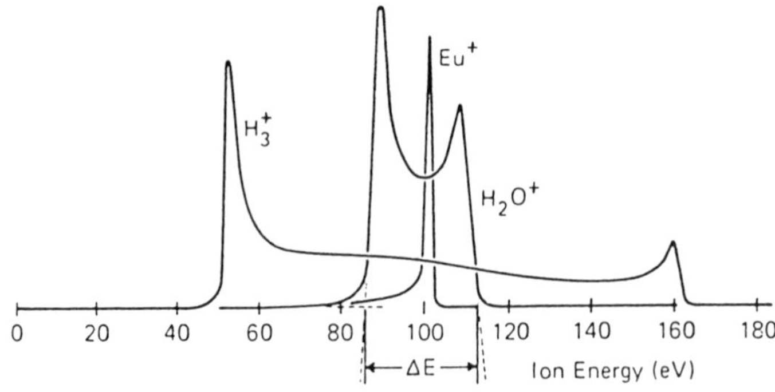

Figure 7-16. Energy distribution of H_3^+, H_2O^+, and Eu^+ at the substrate plane in a confined discharge (13.56 MHz, 100 W, 75 mTorr).[91]

ary sources of excitation and ionization. This section first describes the analytical applications of rf-powered sources as secondary excitation/ionization sources and concludes in describing their use as primary atomization/excitation/ionization sources.

7.3.5.1. rf Discharges as Secondary Excitation/Ionization Sources

A variety of applications have utilized rf glow discharges as secondary excitation or ionization sources. As mentioned in Chapter 4, secondary discharges within hollow cathode lamps can boost the emission intensity by enhancing electron impact in the gas phase while minimizing resonance broadening by controlling the sputtering current. Unfortunately, any attempts to increase excitation intensity by increasing the current or decreasing the pressure of the discharge gas in the hollow cathode discharge are accompanied by an increase in the vapor pressure of the analyte, and subsequently resonance broadening. Secondary sources of excitation allow increased excitation without increasing the vapor pressure of the analyte, minimizing self-absorption. Direct current,[102–106] and microwave[107–109]-boosted discharges have demonstrated increased emission intensity of spectral lines and lower background levels relative to nonboosted lamps. Likewise, Walters and Human demonstrated that an inductively coupled, rf-boosted glow discharge lamp resulted in higher emission intensities and more linear calibration curves over a wider dynamic range as a result of decreased self-absorption.[110]

The technique of sputtered neutral mass spectrometry (SNMS) developed by Oeschner[111] incorporates an inductively coupled high-frequency (27 MHz) discharge as either the primary sputter/ionization or secondary ionization source. In the direct bombardment mode a small dc potential (a few hundred electron volts), which is applied to the cathode surface, extracts ions formed in the plasma to the surface. Neutrals sputtered from the surface are then ionized in the rf plasma, and sampled mass spectrometrically. As in the glow discharge, separation of the atomization step from the ionization process results in minimized matrix effects and improved relative elemental sensitivities.

7.3.5.2. The rf Glow Discharge as a Primary Spectrochemical Source

The rf discharge has served as the primary atomization/excitation/ionization source for various analytical applications. The sputtering process that occurs in the rf discharge is analogous to that of the dc discharge. High-energy ions, under the influence of the bias potential, bombard the sample (cathode) surface. Sputtered neutral atoms diffuse into the negative glow where subsequent excitation and ionization occur. Analytical applications of the rf-powered glow discharge source have been limited primarily to mass spectrometry with a few applications in atomic absorption, emission, and laser-induced fluorescence spectroscopies.

Sample preparation is typically similar for any of the rf glow discharge techniques. Because of the nature of the rf discharge (i.e., its ability to sputter nonconductors), sample preparation is quite simple. For solid samples, no grinding or mixing of the samples is necessary, and samples need only conform to the appropriate shape and dimensions of the holder. Powdered samples can often be pressed in an appropriate die assembly without the need for binders which would dilute and possibly contaminate the sample. In the case of solids, no additional gases are entrained in the preparation step; thus, gas analysis of nonconducting materials may be possible. As will be presented, depth profiling of elemental compositions in nonconducting solids is possible because sample homogeneity is preserved during sample preparation.

Source designs for analytical applications vary but obviously share common features. The discharge is maintained between two electrodes, either external or internal with respect to the vacuum system employed. Donohue and Harrison(112) employed an rf-powered hollow cathode for the analysis of conductors, nonconductors, and solution residues. Their rf-powered glow discharge cavity ion source is shown in Fig. 7-17. The hollow cathode, B, served as the sample electrode and was also used for analyzing deposited solutions. The excitation potential (30 kV, 500 kHz) was supplied from a spark source generator to electrode C, externally positioned around the discharge tube as shown. The resulting plasma could be sampled optically through the quartz window (D) or mass spectrometrically through a small hole in the base of the hollow electrode. Typical operating pressure for this

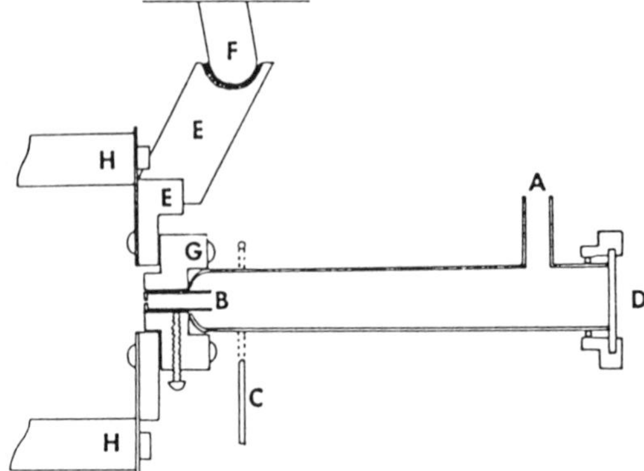

Figure 7-17. Radio-frequency-powered hollow cathode source. A, gas inlet; B, hollow electrode; C, rf counter electrode; D, quartz optical window; E, heat sink; F, liquid N_2 cold finger; G, electrode block; H, insulators.[112]

source was approximately 300 mTorr. Acquired mass spectra of Zn, Sn, and glass (Pyrex) electrodes, as well as an analysis of elements deposited on glass from solution, demonstrate the analytical capabilities of this source.

The grounded, diode rf discharge source design is most common. In this configuration the sample is mounted on the powered electrode while the vacuum chamber typically serves as the opposite (grounded) electrode for experimental simplicity. This geometry also serves to minimize the cathode-to-anode ratio, which, as previously discussed, distributes the dc bias on the smaller electrode and thus isolates sputtering to the sample surface. A practical concern in such a design is proper shielding of the sample holder in order to restrict sputtering to the sample surface. A typical rf source design is illustrated in Fig. 7-18. The sample is mounted on the rf excitation electrode,

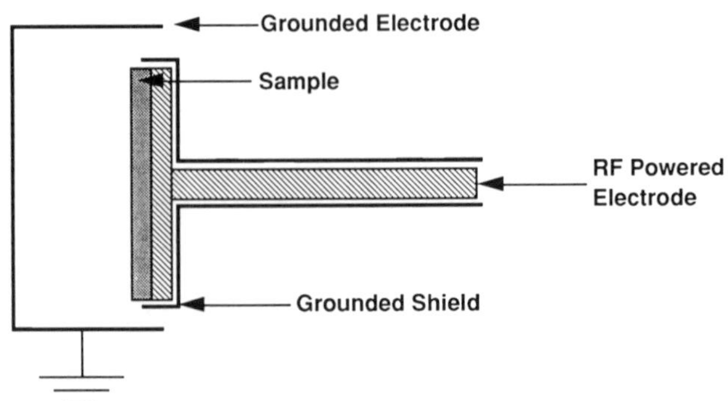

Figure 7-18. Typical rf glow discharge source design.

which may take many forms. Typically, the sample is deposited on, press-fit into, or otherwise mounted onto the rf electrode. The grounded shield is positioned within the cathode dark space (typically about 0.5–1.0 mm spacing) and prevents a discharge from forming on the rf backing electrode, thereby limiting sputtering to the sample surface.

Coburn *et al.*[113] utilized a six-position, rotary, water-cooled sample holder for depth profile analyses. In this configuration, shown in Fig. 7-19, up to six samples could be mounted and analyzed separately. Only one sample is exposed to the discharge at a time while the others are protected within a grounded shield. Such a design maintains vacuum integrity between analyses and increases sample throughput. The exposed sample holder is

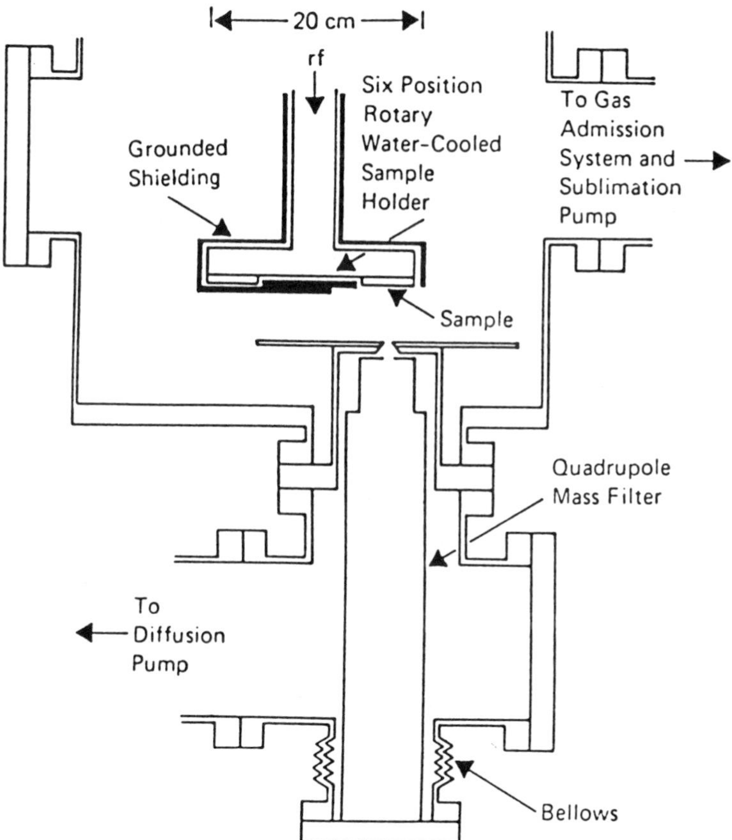

Figure 7-19. Experimental design incorporating a six-position, rotary, water-cooled rf discharge source.[113]

approximately 5 cm in diameter, though sample sizes were somewhat reduced in this application.

Duckworth and Marcus developed an rf glow discharge source (Fig. 7-20) that is more amenable to typical sample sizes found in bulk analytical applications.[114] In this design sample sizes are reduced to approximately 0.5 inch in diameter. Again the grounded steel sleeve is positioned to restrict sputtering to the sample surface and the source is water-cooled in order to minimize volatilization of atoms and/or changes in relative sputter yields. Samples are machined to the appropriate height (~0.5 inch) for analysis. Conducting sample mounts of various heights are also used to position nonconducting or small samples relative to the grounded steel sleeve. A Macor insulating sleeve is used to isolate the rf-driven sample holder from the conflat flange (2.75 inch)-mounted source housing. This source is analytically sound, but sample throughput is low because of loss of vacuum integrity upon changing samples and the time required to remove and reassemble the conflat flange source mounting hardware.

A direct insertion probe (Fig. 7-21) has been designed that increases sample throughput, maintains vacuum integrity, and is more amenable to small sample sizes.[115] The probe accommodates both pins and disk samples of up to 3/16 inch in diameter. In this design the rf is supplied through a high-voltage feedthrough that is vacuum-welded to the end of the probe. The samples are press-fit or silver-painted onto a copper sample support.

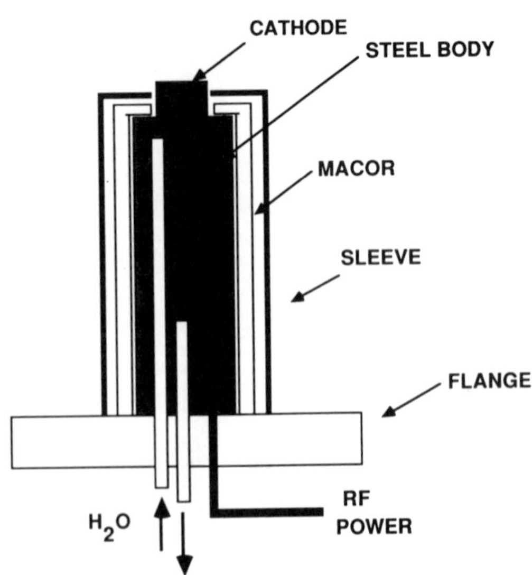

Figure 7-20. Flange-mounted ($2\frac{3}{4}$ inch) rf-powered glow discharge source for analytical applications.[114]

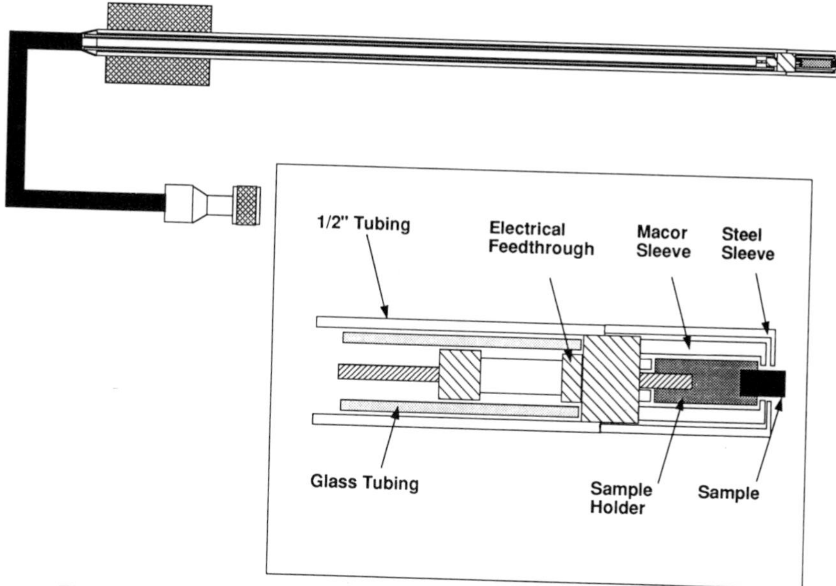

Figure 7-21. Radio-frequency-powered glow discharge direct insertion probe.[115]

This support is further insulated from ground by a Macor cap and a final stainless steel cap that restricts sputtering to the sample surface. Shielding on the coaxial cable is pulled back to allow the insulated (Teflon) rf lead to fit through the half-inch stainless-steel probe body. Coaxial shielding integrity is maintained by the probe body over the length of the probe to reduce radiated power losses. The probe is inserted through a ball valve assembly, thus maintaining vacuum integrity and allowing rapid sample interchange. The addition of an expandable baffle to the ball valve assembly allows the axial positioning of the discharge relative to the sampling orifice and subsequent optimization of the analyte signal. As will be discussed in Section 7.3.5.3, Duckworth and Marcus[114] have observed a strong spatial dependence of ion densities in rf GDMS. Movable source assemblies, which allow the sampling position to be optimized, were critical for optimum analytical performance.

The limiting feature of the direct insertion probe is that solids must be machined into a compatible size (e.g., <3/16 inch in diameter) and shape. This is often difficult and time-consuming. Nonconducting samples are often nonmachinable and must be ground and pressed into a solid of appropriate dimensions. Such sample preparation is restrictive and in the latter case will result in a loss of depth profiling capabilities. The latest design in rf discharge sources incorporates an external sample mount that circumvents such

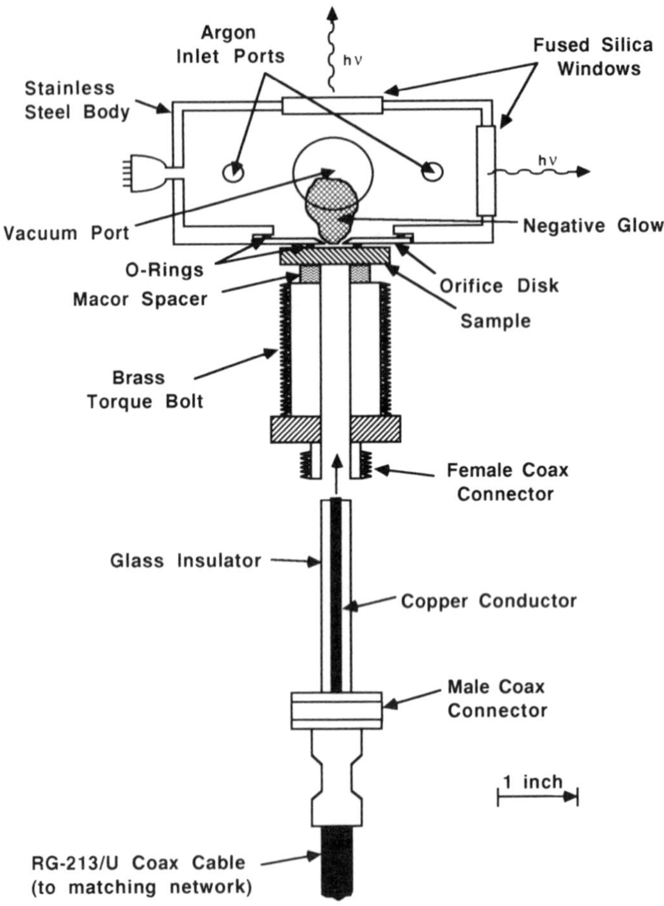

Figure 7-22. Diagrammatic representation of the rf glow discharge atomization/excitation source.[116]

problems.[116,117] The source, shown in Fig. 7-22, is amenable to various sample sizes—the only requirement is that the sample surface be flat enough to form a vacuum seal on an O-ring (typical for many nonconductors such as glasses and ceramics). Powder samples may also be analyzed with this source by compacting them into a brass disk that is pressed against the O-ring (Fig. 7-3). The sample is pressed against a 10-mm-diameter Teflon O-ring on an orifice disk by a brass torque bolt, forming a vacuum seal. A Macor spacer electrically isolates the sample from the torque bolt. A copper conductor, insulated in glass tubing, couples the rf power from the HN coaxial connector to the sample. In atomic emission applications, the source is equipped with two optical ports, one on and one perpendicular to the sample axis. With these optical configurations, the authors were able to compare analyte emission intensities, signal-to-noise ratios, and emission spectra from both optical geometries.[116] These comparisons will be summarized in Section 7.3.5.4. A similar source, shown in Fig. 7-23, has been designed for mass spectrometry.[117] This source is comparable in design in all respects except for sample size and an ability to position the sample relative to the mass spectrometer sampling orifice. A torque screw, acting in concert with the rf feedthrough bolt, serves to move the sample (discharge) with respect to the sampling orifice in order to optimize the mass spectrometric sampling of the plasma ions. Because of the sample positioner, sample sizes were restricted to dimensions varying from 1.3 to 2.2 cm, though larger versions of the same design are feasible.

Initial mass spectrometric studies using the external sample mount design indicate that the total preparation and plasma stabilization times for analysis are on the order of a few minutes, consisting of 1 min of pumpdown time and 1–2 min presputtering. Figure 7-24 indicates the temporal stability of the $^{63}Cu^{+}$ ion signal. In less than 10 s the ion current stabilizes and remains stable throughout 18 min of monitoring. In order to minimize pumping

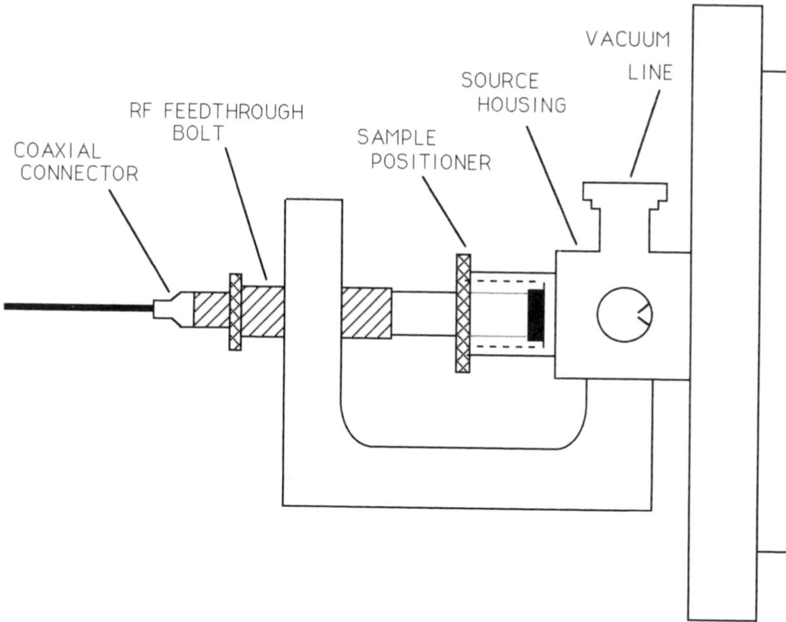

Figure 7-23. Diagrammatic representation of the external sample mount geometry for rf glow discharge mass spectrometry.[117]

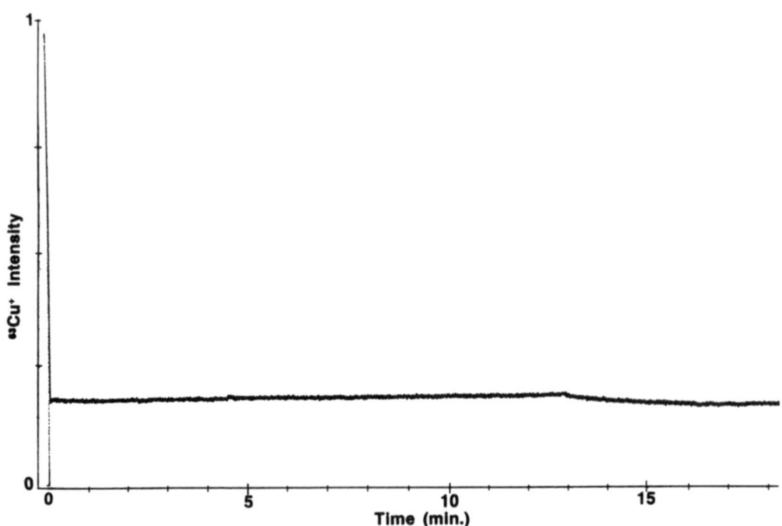

Figure 7-24. Temporal stability of $^{63}Cu^+$ intensity utilizing the external sample mount, rf-powered glow discharge (13.56 MHz, 15 W, 0.11 Torr).[117]

times, the discharge volume is reduced as much as possible while maintaining asymmetry. Large vacuum lines were utilized to increase the conductance and thus minimize initial rough pumping time. Short pumpdown and pre-sputter times, as well as source amenability to sample shape and size, should allow for convenient routine analyses.

It must be stressed that in any rf-powered glow discharge source design, radiated noise must be minimized. Complete coaxial shielding of all rf circuitry and proper grounding are critical requirements. Obviously, any break in coaxial shielding will result in rf power losses. Such noise is known to interfere with associated electronic instrumentation. For further consideration of proper grounding and shielding, the reader is referred to Morrison.[118]

7.3.5.3. rf Glow Discharge Mass Spectrometry

Mass spectrometric sampling of ions formed within the discharge comprises most of the analytical applications of rf glow discharge to date. Most of the rf GDMS has been limited to quadrupole instruments although some attempts to interface rf glow discharges to high-resolution mass spectrometers have been made with limited success. Ion energies in the rf glow discharge are typically 20–30 eV with energy spreads of a few electron volts. These energies, as well as the lower operating pressures relative to dc sources, are quite amenable to quadrupole mass spectrometry. Coburn[93] has reviewed quadrupole mass spectrometric sampling of ions formed in rf glow discharges and concluded that the most promising application of the technique appears to be elemental analysis of solids.

Ionization in the rf-powered glow discharges is known to be quite efficient, attributed in part to the oscillating nature of the rf discharge. Electron impact is thought to be more efficient at high frequencies and other mechanisms such as "multipacting," "electric field amplification," and "surf riding" have been proposed as means of increased ionization in high-frequency discharges.(53) As in the case of dc glow discharges, Penning ionization is thought to be prominent in rf discharges. Eckstein *et al.*(82) have demonstrated quite clearly that Penning ionization was the predominant mechanism for ions sampled in their discharge. As shown in Fig. 7-25, they found a linear correlation between the observed ion signal and the product of Cu neutral and Ne metastable atom densities as determined by atomic absorption spectrophotometry. The dependence on metastable concentrations indicates that inelastic collisions with the neon metastables is the primary source of analyte ionization.

Some studies suggest that Penning ionization may not be the dominant ionization mechanism in other rf discharges. In a similar set of experiments

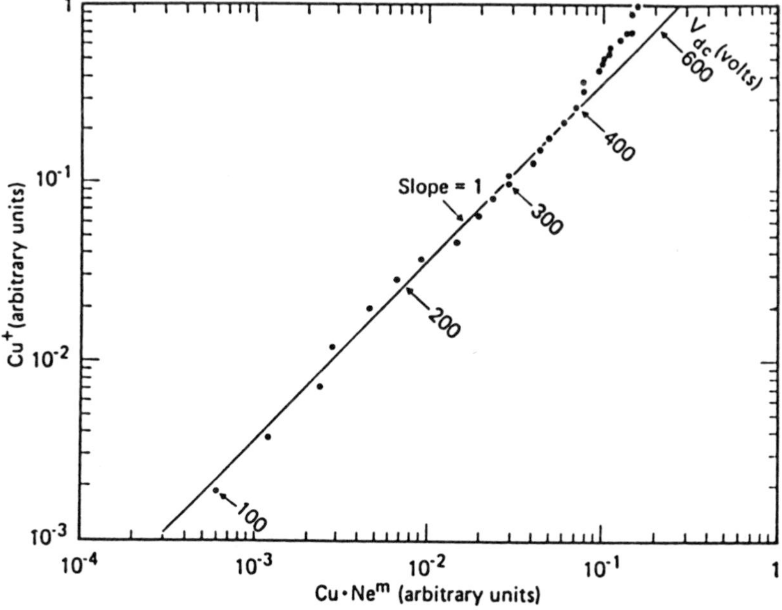

Figure 7-25. First-order relationship between Cu^+ intensity and Cu · Ne^m products (20 mTorr Ne).(82)

by Duckworth and Marcus, no correlation was observed between the analyte ion signal and discharge gas metastable atom densities.[119] The lack of correlation has been attributed to differences in the experimental parameters including source design, discharge parameters, and, perhaps most importantly, the sampling position in the discharge. Demonstrated in Fig. 7-26 is a diagrammatic representation of fundamental sampling considerations within an rf discharge as a function of pressure.[114] The authors observed the greatest analyte signals when sampling from the cathode dark space–negative glow interface, shown in the middle panel at a typical operating power and pressure (25 W and 0.2 Torr). As pressure is decreased (0.1 Torr) and mean electron free paths are increased, the interface expands and only discharge gas ions, predominantly formed by electron impact and charge exchange, are detected. At higher pressure (0.35 Torr), analyte intensities are increased though not as high as when sampling from the interface. Such spatial phenomena have not been mentioned by other authors and it is believed that ions sampled by Eckstein *et al.*[82] were extracted from the negative glow region where electron energies are lowered because of multiple collisions. An increased contribution by electron impact ionization at the negative glow–cathode dark space interface may explain the lack of correlation between the mass spectrometric and atomic absorption measurements. It should also be noted that average electron energies in the rf plasma (4–6 eV) are much higher than those in dc plasmas (~1 eV) because of the rf oscillating fields superimposed on the dc bias potential.[120] Increased contributions to ionization from high-energy electrons would not be surprising, particularly at the interface of the cathode fall region.

Mass spectral characteristics of rf-powered glow discharges are very similar to those of dc glow discharges, with some exceptions. In Fig. 7-27 a mass spectrum of a stock copper sample is shown. Typical matrix and discharge gas species are observed. The most notable difference between the dc and rf spectra is the increased intensity of hydrides and molecular species in the rf mass spectrum. Water and contaminant species in the discharge are efficiently ionized by the enhanced electron impact in the rf plasma and promote the formation of hydride species. Most notable are the argon hydride species, which are typically at a factor of ten less for dc-powered

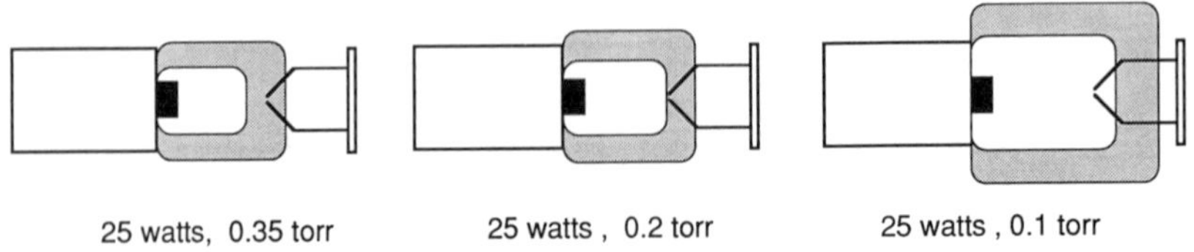

Figure 7-26. Illustration of the influence of discharge pressure and power on the cathode dark space/negative glow interface location.[114]

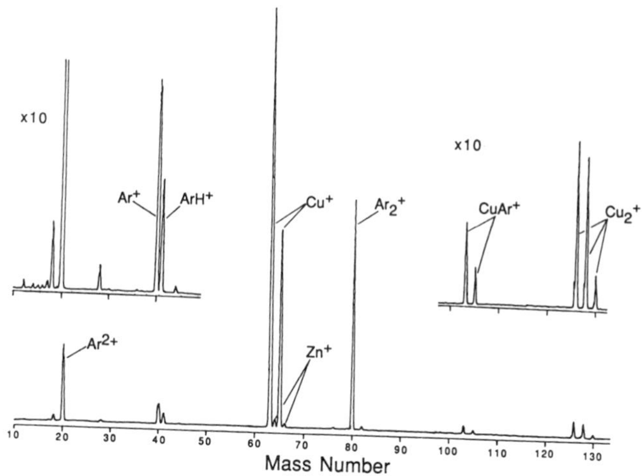

Figure 7-27. Radio frequency glow discharge mass spectrum of a stock copper alloy (0.20 Torr, 10 W).[114]

discharges. Copper dimers as well as argides, thought to be formed through associative collisions,[53] are quite prominent in the rf discharge. This is likely the result of the lower operating pressures in the rf glow discharge, which should reduce the efficiency of collisional dissociation within the plasma. King *et al.* have demonstrated that such polyatomic species are reduced at higher discharge pressures because of the increased collision frequency within the plasma.[121]

Isobaric interferences in the rf glow discharge are similar to those of its dc counterpart; however, the intensities of polyatomic species may be greater. The preferential removal of polyatomic isobaric interferences, generated in the glow discharge, by collision-induced dissociation (CID) in the collision cell of a multiquadrupole mass spectrometer is discussed in Chapter 5. Duckworth and Marcus have demonstrated the utility of performing CID on polyatomic species in the collision cell of a double-quadrupole mass spectrometer.[122] Polyatomic species are preferentially lost relative to the analyte as are atomic species which are lost through symmetric and asymmetric charge exchange. Though effective in the preferential removal of interfering species, analyte intensities are reduced by approximately 80% because of scattering within the collision cell.

The interfacing of an rf glow discharge with a high-resolution mass spectrometer is another alternative to the quadrupole mass filters typically employed. However, because of the required high accelerating potentials of sector mass spectrometry and the required electronic configurations of the rf discharges, attempts have met with only limited success.[123] Efforts typically have employed an isolation transformer to float the rf components to the mass spectrometer's accelerating potential (~8 kV). Since probe inlets and vacuum chambers are at ground potential, this high reference voltage makes complete shielding of the rf cables and source particularly difficult. Consequently, appropriate shielding requirements are not met and rf noise adversely affects detectors, Hall probes, and other electronic systems. Recent modifications in electrical design have allowed complete shielding of the rf components.[124] Radio frequency power is efficiently coupled to the sample through a direct insertion probe similar to that of Fig. 7-21 and is isolated from the high voltage power supply. Likewise, the high voltage is coupled to the sample and source housing while isolated from the rf-generating electronics. A typical mass spectrum of Sn in NIST SRM 1103 brass is shown

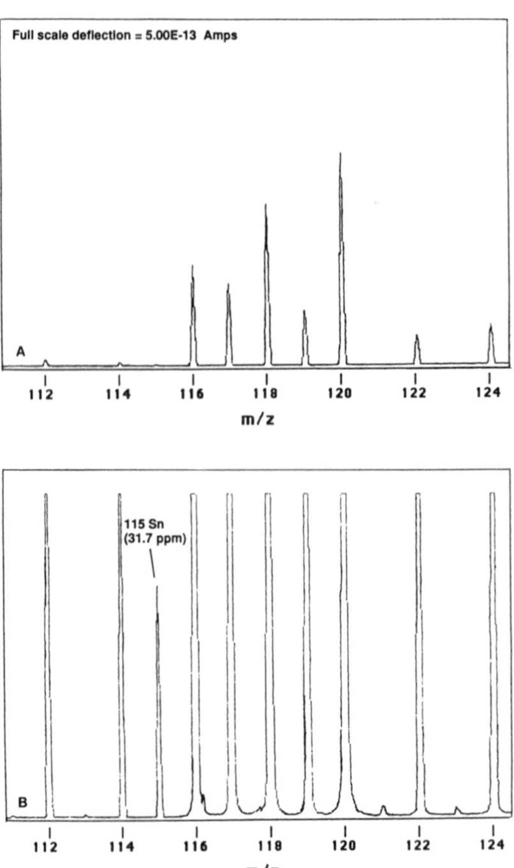

Figure 7-28. Radio frequency glow discharge mass spectrum of NIST SRM 1103 brass. (A) ×1, (B) ×100.[124]

in Fig. 7-28a. The Sn isotopes (0.88% by weight) are clearly detected. This spectrum was acquired utilizing an rf-powered source coupled to the VG9000 double-focusing glow discharge mass spectrometer (VG Isotopes, Ltd., Cheshire, UK).(124) The rf glow discharge provides a very stable ion signal in this application. The electrical design limits the noise level to only a few counts per second. As noted in Fig. 7-28b, ^{115}Sn, present at 31.7 ppm, is detected quite easily under moderate mass resolution conditions ($M/\Delta M \sim 1200$). Limits of detection under these conditions are calculated to be ~11 ppb (2σ).

Comparisons of the analytical characteristics of rf and dc glow discharges as ion sources are difficult because of the large parametric differences—pressure and power regimes in particular. However, the optimum performances of each are similar. Ion currents for the rf sources are comparable to dc discharge ion currents. Ion currents are typically about 10^{-11} for both conducting and nonconducting matrices.(114) Conducting samples which tend to form oxides, such as aluminum, can be quite difficult to analyze with a dc source because of nonconductive coatings that develop on the surface of the sample. Such dielectric layers are efficiently sputtered and dissociated in an rf glow discharge. Duckworth and Marcus have reported increases in ion currents of a factor of 50 over dc glow discharges for Al matrix samples.(125) In the same study, relative sensitivity factors for rf and dc glow discharges were investigated for a variety of aluminum samples, and the relative ion yields were found to be similar. It has also been suggested that matrix effects may be small even between conducting and nonconducting samples.(114)

In the course of the development of rf GDMS, much of the analytical studies and development have been performed on conducting materials such as copper, brass, or steel. This is primarily related to the ease of sample preparation, high sputter yields for some metals, and the need to compare sources with similar materials (i.e., dc versus rf). However, spectra such as that shown in Fig. 7-29 demonstrate the analytical potential of rf-powered glow discharge sources in the analysis of nonconducting materials.(114) Figure 7-29 is a mass spectrum of a mixture of transition metal oxides (As, Cr, Ga, Ni) pressed, without binder, into a pellet form. Important features to note are that the oxides are efficiently sputtered, dissociated, and ionized in what is reported to be a very stable plasma. Analyte ion currents, approximately 10^{-11} A, are comparable to that observed for conducting matrices. The spectrum indicates that dissociation of oxides is rather efficient in this plasma (0.3 Torr, 25 W). No enhancement in oxygen species such as ArO^+ and $ArOH^+$ (as a result of the oxide dissociation) is evident, although hydrides such as AsH^+ are observed. Such hydrides result from water species which are commonly entrained in the preparation of powder samples.

All oxide species may not be dissociated equally. In fact, Coburn and colleagues(126) found that in certain cases the observed atomic signal may be

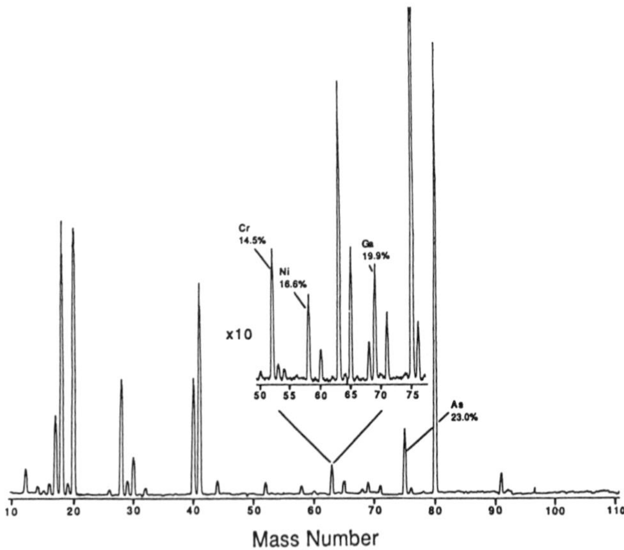

Figure 7-29. Radio frequency glow discharge mass spectrum of a mixture of transition metal (Cr, Ni, Ga, As) oxides (0.30 Torr, 25 W).[114]

less than molecular ion signals. Figure 7-30 indicates the relative amounts of molecular-to-atomic ions as a function of M–O bond energies where η is the fraction of molecular species reaching the sampling orifice. The dissociation efficiency of metal oxides was found to be related to the metal–oxygen bond strength to a large extent, although contributions from other factors are evident. Of note is the arsenic oxide fraction, approximately 0.5, indicated here. Arsenic appears to be one of the more difficult oxides to dissociate although its binding energy is not as high as some of the more easily dissociated oxides indicated. In fact, arsenic oxide is present at a similar fraction in the spectrum shown in Fig. 7-29. One factor proposed that may account for its unlikely tenacity is a mass-dependent bias in the degree of dissociation during the sputtering event. As argon ions impinge on oxygen species at the surface of the sample, the energy of the sputtering atom is transferred to the oxygen atom. The ability of oxygen atoms to elastically exchange the kinetic energy to the metal atoms is a function of the mass of the metal, with more efficient energy exchanges occurring with atoms of like mass. Subsequently, oxygen atoms bonded to light metal atoms are likely to be sputtered as molecules while oxygen atoms bonded to higher-mass metals would leave the surface as atoms. This phenomenon is also illustrated qualitatively in Fig. 7-30 where oxides with lower-mass ions have higher η values (i.e., above the dashed line) and oxides with higher-mass cations have lower η values

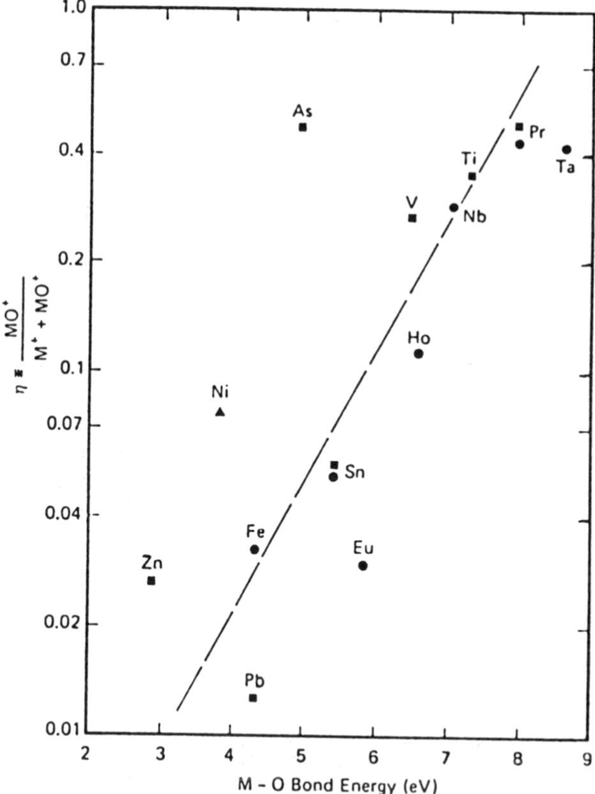

Figure 7-30. The dependence of η on the M–O bond energy (60 mTorr Ar, 13.56 MHz, 100 W).[126]

(i.e., below the dashed line). The authors also concluded that oxygen-bearing contaminant species tend to increase the amount of molecular species observed because of chemisorption of the contaminant onto the sample surface. Good vacuum practice tended to reduce the relative amounts of molecular species.

Other exemplary applications include various vitrified glass matrices. An rf glow discharge mass spectrum of NIST SRM 1412 multicomponent glass is shown in Fig. 7-31a. These ions were generated from the flange-mounted source shown in Fig. 7-20 and were analyzed by a quadrupole mass spectrometer.[127] The sample is a silicon-based vitrified glass. The low signal intensity for silicon is related primarily to its low concentration (19.8%), even though it is the "matrix" element. Higher-mass analyte ions, indicated in the expansions, represent various minor elements present at concentrations of 3–4% atomic. These species are readily atomized, including barium,

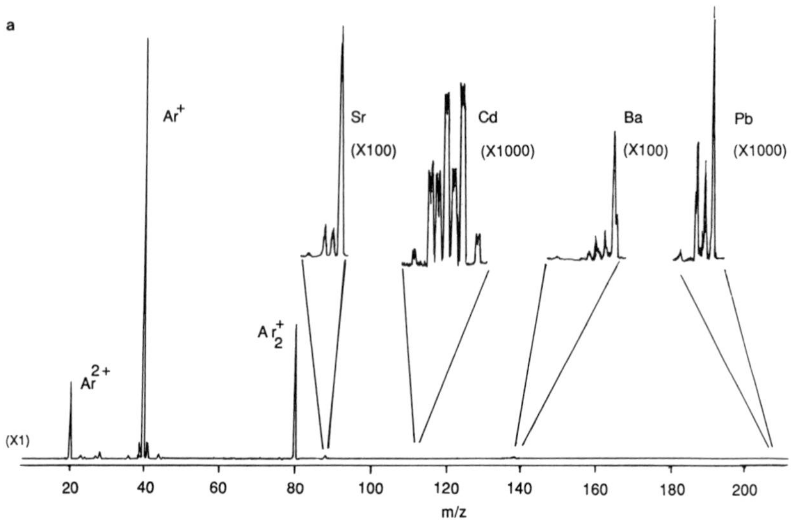

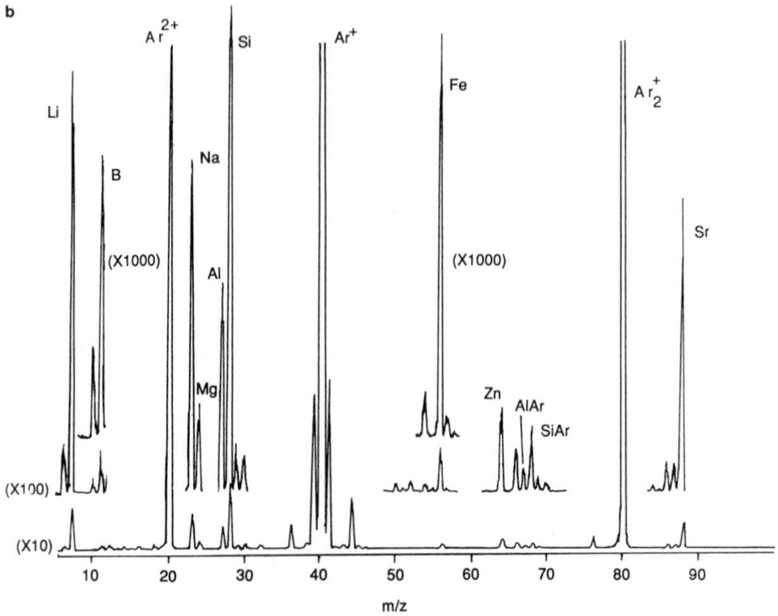

Figure 7-31. Radio frequency glow discharge mass spectrum of (a) high- and (b) low-mass elements in NIST SRM 1412 multicomponent glass.[127]

which tends to form strong metal oxides. Various low-mass and transition elements (1–4% atomic) of interest are seen in Fig. 7-31b. Few molecular interferences are noted at masses of interest. Argides of low-mass elements, such as the labeled silicon and aluminum argide species, can present problems in the transition element region. Such interferences often do not pose a problem in that other isotopes of the element of interest may be chosen, but care must be exercised.

Radio frequency glow discharges also find applications in the area of depth profiling. Lower operating pressures of rf-powered glow discharges should reduce redeposition of sputtered materials and thus should increase depth resolution relative to dc glow discharges. Coburn *et al.* demonstrated and characterized the use of rf glow discharges for the determination of elemental composition profiles in solids.[113,128,129] The samples were mounted on target holders that were much larger in diameter (typically 10–50 times larger) in order to provide a uniform ion current density across the sample surface. Uniform sputtering ion current densities help to maintain uniform etching rates across the sample. Using relatively small samples also tends to normalize discharge characteristics between samples, as well as to minimize sample heating that could lead to vaporization and unacceptably high etch rates. One disadvantage of the technique is that the holder is also sputtered during the analysis, and all constituents of the holder material contributed to the observed ion signal. In such applications care must be exercised in choosing an appropriate backing electrode. Analyses indicated a lack of matrix effects on ion signals and the possibility of performing standardless quantitations. In characterizing the technique, the authors found that sensitivity increased with pressure from 10 to 160 mTorr, and was attributed to increased ionization efficiency resulting from an increased metastable density and collision frequency. Sensitivity was found to decrease with increasing power, although the reason for this was not clear. As expected, depth resolution was found to decrease with increasing pressure because of collisional redeposition. Obviously, a trade-off between sensitivity and resolution exists in terms of optimum discharge pressure. Depth resolution was also found to decrease with increasing power as well. Parametric control of the rf discharge allowed sputter etch rates to be varied from 1 Å/min for low-power He discharges to 1000 Å/min for high-power Ar discharges. Single ppm detection limits were observed in the course of these investigations.

An example of the depth profiling potential of rf-powered glow discharge mass spectrometric analysis of thin films is shown in Fig. 7-32.[129] The sample, which is illustrated in the inset, consists of a 1000-Å copper layer on a fused quartz substrate. A 3000-Å nickel film was evaporated on the copper film. Resolution, as defined by $[t(0.1) - t(0.9)]/t(0.5)$, is 3.3% for Ni^+ and 2.95% for Cu^+ where $t(0.5)$ is the time where the ion signal is 0.5 times the steady-state ion current; $t(0.1)$ and $t(0.9)$ are defined similarly.

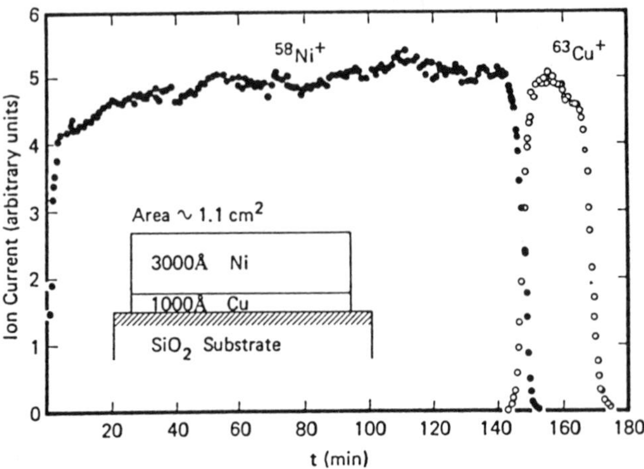

Figure 7-32. Radio frequency glow discharge mass spectrometry-determined depth profile of the thin-film structure shown (13.56 MHz, 50 W, 30 mTorr Ar).[129]

Differences in the resolution reflect the differences in relative ion yields. (Figure 7-32 also indicates a higher sputter yield for copper, in that sputtering times are not directly proportional to the thickness of the layers.) The authors present possible mechanisms that influence interface broadening including depth-dependent mechanisms such as nonuniform sputter etching and depth-independent mechanisms such as surface mixing and diffusion. Depth resolution is proposed to be determined by a 30-Å depth-independent interface broadening plus a broadening that varies between 2 and 8% of the total depth of analysis.

7.3.5.4. *rf Glow Discharge Optical Spectroscopies*

Analytical applications of rf-powered glow discharge are only beginning to attract the interest of optical spectroscopy practitioners. Publications of analytical applications of the rf glow discharge source for spectroscopy have been limited; however, the recent renaissance of rf GDMS is beginning to spawn interests in optical applications.

Winchester and Marcus demonstrated the potential applicability of rf glow discharge atomic emission spectroscopy.[130] Initial work demonstrated the potential of the rf glow discharge source for atomization and subsequent excitation of both electrically conducting and nonconducting materials. These studies were performed using the six-way cross mounted direct insertion probe of Fig. 7-21. Subsequent studies[116] employed the external sample mount shown in Fig. 7-22. As mentioned in that section (7.3.5.2), this source incorporates two optical ports, one viewing down the sample axis (glow + cathode) and one perpendicular to the sample axis (glow only). This geometry allowed the comparison of the contribution of the cathode blackbody radiation and spurious electrical breakdowns to the observed signal. It was found that the "glow only" and the "glow + cathode" optical geometries had signal-to-noise ratios that were statistically equivalent, while analyte emission intensities were a factor of 2–5 higher for the latter configuration. Optimization of source parameters, discharge pressure and power, demonstrated a trade-off between the atomization rate and the extent of excitation. Atomization rates increase with increasing discharge power at constant pressure and increasing pressure at constant discharge power. Excitation temperatures (though not at local thermodynamic equilibrium conditions) tend to increase with increasing power but decrease with increasing pressure. Conditions of high pressure (6–10 Torr) and high power resulted in strong analyte emission intensities. These conditions are radically different from those in the rf GDMS case and point to another difference between the rf and dc, where AES and MS conditions are more comparable. Exemplary emission spectra of both nonconducting and conducting samples were presented. The emission spectrum of Macor, a nonconducting glass ceramic, is shown in Fig. 7-33. Discharge pressures were decreased in order to minimize the sputter rate and therefore the subsequent rapid release of trapped water which contributed a large amount of band structure to the spectrum. At these compromised discharge conditions the discharge emission is still atomic in nature, even though oxides are being sputtered. Several emission lines of Al and Si, sputtered from the Macor, were identified. Excited argon emission lines are also observed in the 350–500 nm range.

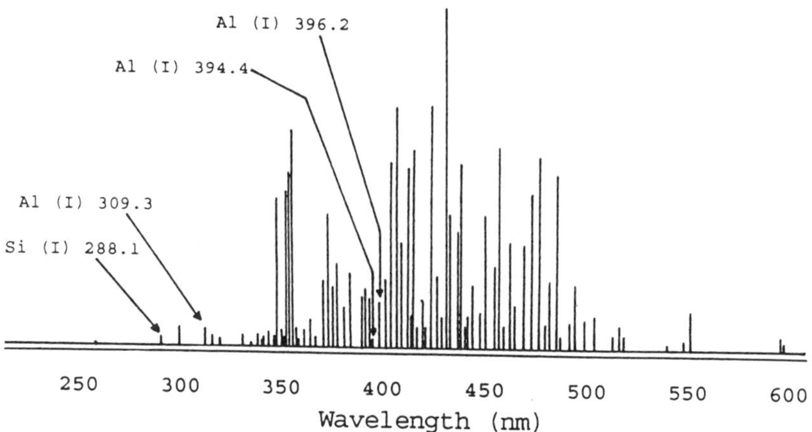

Figure 7-33. Radio frequency glow discharge emission spectrum of Macor collected with the "glow + cathode" optical geometry (13.56 MHz, 2 Torr Ar, 60 W).[116]

Page *et al.*[131] used an rf sputter source to produce atoms and clusters for study by absorption, emission, and laser-induced fluorescence spectroscopies. In comparison with more exotic and expensive sources such as lasers, the rf discharge serves as an economical and efficient source of atoms and small clusters. The rf source used was of a typical deposition (diode) design, such as that of Fig. 7-18, and is mounted via a 6-inch conflat flange to a six-way cross. The source is water-cooled and utilizes 3-inch-diameter targets. The advantages of such a source are its simplicity, low cost, production of large fluxes of both atomic and small cluster species (i.e., dimers), and low blackbody radiation because of the near room temperature of the target and the discharge chamber. The atom number density of an iron target was found by atomic absorption to be 2×10^{12} cm^{-3}, with a translational temperature of 800 K. Copper dimer number densities were estimated to be 10^{10}–10^{11} cm^{-3} for copper targets with a 500 K rotational temperature calculated. As an example of the source utility for the study of small cluster species, an LIF spectrum of the (0,0) band of the Cu_2 C $\leftrightarrow$ X transition is shown in Fig. 7-34. This is the first rotationally resolved spectrum of this transition, with the Q branch head shown at $\approx$21842 cm^{-1} and the R branch at $\approx$21845 cm^{-1}. High signal levels allowed the delayed detection (1 μs) of the A state while ignoring interference from the short-lived B state. With rotational temperatures only slightly above room temperature, rotationally

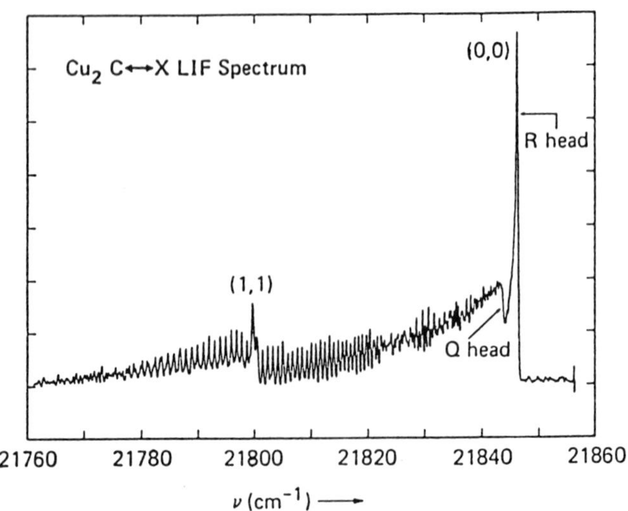

Figure 7-34. LIF spectrum of the Cu_2 C $\leftrightarrow$ X transition, obtained from the sputtering of a copper target in an rf-powered glow discharge.[131]

resolved spectra of small clusters should be possible for other metals. As in this case, assignments of the excited-state symmetry and rotational constants may be determined.

7.4. Conclusions

While bulk metal and semiconductor analyses have been, and will continue to be, the major applications of glow discharge spectroscopies, the needs presented in those areas where oxide/refractory materials are the raw and finished products are so great that future investigations into glow discharge sampling methodologies are guaranteed. The two basic approaches outlined in this chapter demonstrate the basic characteristics and great potential that glow discharge spectroscopies hold for this very important set of sample matrices.

Early in the development of glow discharge atomic emission spectroscopy, the application of the sample briquetting technique showed promise. The combination of analytical methodology development and fundamental plasma studies have allowed for the more straightforward and precise application of the compaction technique. Problematic inclusion of water and gases in the compaction process has been greatly minimized by more careful sample handling and the use of conducting host matrices which have gettering properties. The ability to easily add an internal standard element, or in the best case, an isotopic spike in mass spectrometry applications is an important asset to this method. The compaction technique will see increased usage as it is applicable to any GD system without modification of the base instrument or source, permitting optimum system flexibility.

The inherent capability of radio-frequency-powered glow discharges to operate with either conductive or insulating sample types is probably the single most important development in this field over the last decade. Sputter atomization of bulk nonconductors such as glasses and ceramics is direct and straightforward. Analysis of powder samples still requires compaction to form the appropriate sample shape. While many powders may be compressed to sufficient strength, addition of some amount of metal binder may be necessary from a practical standpoint. Conversion of an existing GD system to rf operation may require some capital investment, but it is expected that the long term benefits for total solids sample coverage may be well worthwhile.

The analysis of nonconductive sample types will always pose unique challenges not faced in traditional GD applications. It is these applications, however, that may generate the most new interest in the use of GD techniques.

References

1. J. A. C. Broekaert, T. Graule, H. Jenett, G. Tölg, and P. Tschopel, *Fresenius Z. Anal. Chem.* 332 (1989) 825.
2. M. Dogan, K. Laqua, and H. Massmann, *Spectrochim. Acta* 27B (1972) 65.
3. M. R. Winchester, S. M. Hayes, and R. K. Marcus, *Spectrochim. Acta* 46B (1991) 615.
4. G. S. Lomdahl, R. McPherson, and J. V. Sullivan, *Anal. Chim. Acta* 148 (1983) 171.
5. M. R. Winchester and R. K. Marcus, *Appl. Spectrosc.* 42 (1988) 941.
6. G. S. Lomdahl and J. V. Sullivan, *Spectrochim. Acta* 39B (1984) 1395.
7. S. El Alfy, K. Laqua, and H. Massmann, *Z. Anal. Chem.* 263 (1973) 1.
8. D. C. McDonald, *Anal. Chem.* 49 (1977) 1336.
9. H. Bubert, W. -D. Hagenah, and K. Laqua, *Spectrochim. Acta* 34B (1979) 19.
10. H. Bubert and W. -D. Hagenah, *Spectrochim. Acta* 36B (1981) 489.
11. T. J. Loving and W. W. Harrison, *Anal. Chem.* 55 (1983) 1526.
12. R. B. Keefe, Ph.D. dissertation, University of Virginia, Charlottesville, August, 1983.
13. S. Caroli, A. Alimonti, and K. Zimmer, *Spectrochim. Acta* 38B (1983) 625.
14. S. Caroli, O. Senofonte, N. Violante, F. Petrucci, and A. Alimonti, *Spectrochim. Acta* 39B (1984) 1425.
15. S. Caroli, A. Alimonti, P. D. Femmine, and S. K. Shukla, *Anal. Chim. Acta* 136 (1982) 225.
16. S. Caroli, O. Senofonte, A. Alimonti, and K. Zimmer, *Spectrosc. Lett.* 14 (1981) 575.
17. P. J. Slevin and W. W. Harrison, *Appl. Spectrosc. Rev.* 10 (1975) 201.
18. P. W. J. M. Boumans, *Anal. Chem.* 44 (1972) 1219.
19. R. K. Marcus and W. W. Harrison, *Anal. Chem.* 59 (1987) 2369.
20. M. R. Winchester and R. K. Marcus, unpublished work.
21. F. L. King, A. L. McCormack, and W. W. Harrison, *J. Anal. At. Spectrom.* 3 (1988) 883.
22. W. J. Kolar, M. S. thesis, Clemson University, Clemson, S. C., May, 1989.
23. H. Mai and H. Scholze, *Spectrochim. Acta* 41B (1986) 797.
24. D. S. Gough, P. Hannaford, and R. M. Lowe, *Anal. Chem.* 61 (1989) 1652.
25. K. Wagatsuma and K. Hirokawa, *Anal. Chem.* 61 (1989) 326.
26. J. E. Velazco, J. H. Kolts, and D. W. Setser, *J. Chem. Phys.* 69 (1978) 4357.
27. E. Stern and H. L. Caswell, *J. Vac. Sci. Technol.* 4 (1967) 128.
28. D. G. Swartzfager, S. B. Ziemecki, and M. J. Kelley, *J. Vac. Sci. Technol.* 19 (1981) 185.
29. N. Q. Lam, H. A. Hoff, H. Wiedersich, and L. E. Rehn, *Surf. Sci.* 149 (1985) 517.
30. H. Shimizu, M. Ono, and K. Nakayama, *J. Appl. Phys.* 46 (1975) 460.
31. G. S. Anderson, *J. Appl. Phys.* 40 (1969) 2884.
32. H. Mai and H. Scholze, *Spectrochim. Acta* 42B (1987) 1187.
33. J. Workman, Jr., and H. Mark, *Spectroscopy* 3(3) (1988) 40.
34. D. C. McDonald, *Anal. Chem.* 54 (1982) 1057.
35. N. P. Ferreira and H. G. C. Human, *Spectrochim. Acta* 36B (1981) 215.
36. M. R. Winchester and R. K. Marcus, *J. Anal. At. Spectrom.* 5 (1990) 9.
37. H. Schuler and H. Gollnow, *Z. Phys.* 93 (1935) 611.
38. A. I. Drobyshev and Y. I. Turkin, *Spectrochim. Acta* 36B (1981) 1153.
39. R. B. Djulgerova, in: *Improved Hollow Cathode Lamps for Atomic Spectroscopy* (S. Caroli, ed.), Ellis Horwood Ltd., Chichester, England, 1987.
40. D. Fang and R. K. Marcus, *Spectrochim. Acta* 43B (1988) 1451.
41. R. K. Marcus and W. W. Harrison, *Anal. Chem.* 58 (1986) 797.
42. J. Czakow, in: *Improved Hollow Cathode Lamps for Atomic Spectroscopy* (S. Caroli, ed.), Ellis Horwood Limited, Chichester, England, 1987.
43. K. Robinson and E. F. H. Hall, *J. Met.* 39 (1987) 14.
44. G. N. Jackson, *Thin Solid Films* 5 (1970) 209.

45. J. K. Robertson and C. W. Clapp, *Nature* 132 (1933) 479.
46. R. H. Hay, *Can. J. Res.* A16 (1938) 191.
47. J. I. Lodge and R. W. Stewart, *Can. J. Res.* A26, (1948) 205.
48. S. M. Levitskii, *Sov. Phys. Tech. Phys.* 27 (1957) 913.
49. G. K. Wehner, *Adv. Electron. Electron Phys.* 7 (1955) 239.
50. G. S. Anderson, W. N. Mayer, and G. K. Wehner, *J. Appl. Phys.* 33 (1962) 2991.
51. P. D. Davidse and L. I. Maissel, *3rd International Vacuum Congress*, Stuttgart, 1965, 651.
52. P. D. Davidse and L. I. Maissel, *J. Appl. Phys.* 37 (1966) 574.
53. B. N. Chapman, *Glow Discharge Processes*, Wiley–Interscience, New York, 1980.
54. J. E. Greene and J. M. Whelan, *J. Appl. Phys.* 44 (1973) 2509.
55. S. B. Hyder, *J. Vac. Sci. Technol.* 8 (1971) 228.
56. J. L. Vossen and E. S. Poliniak, *Thin Solid Films* 13 (1972) 281.
57. K. Wasa and S. Hayakawa, *Jpn. J. Appl. Phys.* 12 (1973) 408.
58. R. M. Goldstein and S. C. Wigginton, *Thin Solid Films* 3 (1969) R41.
59. J. G. Titchmarsh and P. A. B. Toombs, *J. Vac. Sci. Technol.* 7 (1970) 103.
60. S. Iida and S. Kataoka, *Appl. Phys. Lett.* 18 (1971) 391.
61. I. H. Pratt, *Thin Solid Films* 3 (1969) R23.
62. R. Frerichs, *J. Appl. Phys.* 33 (1962) 1898.
63. R. Frerichs and C. J. Kircher, *J. Appl. Phys.* 34 (1963) 3541.
64. P. K. Tien, G. Smolinsky, and R. J. Martin, *Appl. Opt.* 11 (1972) 637.
65. M. J. Vasile and G. Smolinsky, *Int. J. Mass Spectrom. Ion Phys.* 12 (1973) 133.
66. G. Smolinsky and M. J. Vasile, *Int. J. Mass Spectrom. Ion Phys.* 12 (1973) 147.
67. M. J. Vasile and G. Smolinsky, *Int. J. Mass Spectrom. Ion Phys.* 13 (1974) 381.
68. G. Smolinsky and M. J. Vasile, *Int. J. Mass Spectrom. Ion Phys.* 16 (1975) 137.
69. H. Biederman and L. Martinu, *Acta Phys. Slovaca* 35 (1985) 207.
70. N. Mutsukura, K. Kobayashi, and Y. Machi, *J. Appl. Phys.* 66 (1989) 4688.
71. H. Chatham and P. K. Bhat, *Mat. Res. Soc. Symp. Proc.* 118 (1988) 31.
72. J. W. Coburn and M. Chen, *J. Appl. Phys.* 51 (1980) 3134.
73. H. Ratinen, *Appl. Phys. Lett.* 21 (1972) 473.
74. H. Ratinen, *J. Appl. Phys.* 44 (1973) 2730.
75. H. Ratinen, *J. Appl. Phys.* 44 (1973) 3817.
76. T. C. Tisone, B. F. T. Bolker, and T. S. Latos, *J. Vac. Sci. Technol.* 17 (1980) 415.
77. C. R. Aita, T. A. Myers, and W. J. LaRocca, *J. Vac. Sci. Technol.* 18 (1981) 324.
78. R. A. Gottscho and T. A. Miller, *Pure Appl. Chem.* 56 (1984) 189.
79. R. M. Roth, *Mat. Res. Soc. Symp. Proc.* 98 (1987) 209.
80. C. R. Aita and M. E. Marhic, *J. Vac. Sci. Technol. A* 1 (1983) 69.
81. A. D. Kuypers, A. Koch, and H. J. Hopman, *J. Vac. Sci. Technol. A* 8 (1990) 3736.
82. E. W. Eckstein, J. W. Coburn, and E. Kay, *Int. J. Mass Spectrom. Ion Phys.* 17 (1975) 129.
83. R. A. Gottscho and M. L. Mandich, *J. Vac. Sci. Technol. A* 3 (1985) 617.
84. V. M. Donnelly, D. L. Flamm, and G. Collins, *J. Vac. Sci. Technol.* 21 (1982) 817.
85. M. B. Hopkins, C. A. Anderson, and W. G. Graham, *Europhys. Lett.* 8 (1989) 141.
86. T. I. Cox, V. G. I. Deshmukh, D. A. O. Hope, A. J. Hydes, N. St. J. Braithwaite, and N. M. P. Benjamin, *J. Phys. D* 20 (1987) 820.
87. D. N. Ruzic and J. L. Wilson, *J. Vac. Sci. Technol. A* 8 (1990) 3746.
88. J. W. Coburn and E. Kay, *Appl. Phys. Lett.* 18 (1971) 435.
89. K. Kohler, J. W. Coburn, D. E. Horne, E. Kay, and J. H. Keller, *J. Appl. Phys.* 57 (1985) 59.
90. K. Kohler, D. E. Horne, and J. W. Coburn, *J. Appl. Phys.* 58 (1985) 3350.
91. J. W. Coburn and E. Kay, *J. Appl. Phys.* 43 (1972) 4965.
92. F. Shinoki and A. Itoh, *Jpn. J. Appl. Phys. Suppl.* 2, Pt. 1 (1974).

93. J. W. Coburn, *Thin Solid Films* 171 (1989) 65.
94. H. R. Koenig and L. I. Maissel, *IBM J. Res. Dev.* 14 (1970) 168.
95. R. A. Gottscho, G. R. Scheller, D. Stoneback, and T. Intrator, *J. Appl. Phys.* 66 (1989) 492.
96. J. W. Coburn and K. Kohler, *Proc. Electrochem. Soc.* 87 (1986) 13 (Proc. Symp. Plasma Process., 6th, 1986).
97. R. W. Gould, *Phys. Lett.* 11 (1964) 236.
98. C. M. Horwitz, *J. Vac. Sci. Technol. A* 1 (1983) 60.
99. J. S. Logan, *IBM J. Res. Dev.* 14 (1970) 172.
100. L. I. Maissel and P. M. Schaible, *J. Appl. Phys.* 36 (1965) 237.
101. J. E. Greene, *CRC Crit. Rev. Solid State Mater. Sci.* 11 (1983) 47.
102. J. V. Sullivan and A. Walsh, *Spectrochim. Acta* 21 (1965) 721.
103. R. M. Lowe, *Spectrochim. Acta* 31B (1976) 257.
104. D. S. Gough and J. V. Sullivan, *Analyst* 103 (1978) 887.
105. J. V. Sullivan and J. C. Van Loon, *Anal. Chim. Acta* 102 (1978) 25.
106. G. S. Lomdahl and J. V. Sullivan, *Spectrochim. Acta* 39B (1984) 1395.
107. F. Leis, J. A. C. Broekaert, and K. Laqua, *Spectrochim. Acta* 42B (1987) 1169.
108. N. Violante, O. Senofonte, A. Marconi, O. Falasca, and S. Caroli, *Can. J. Spectrosc.* 33 (1988) 49.
109. F. Leis, J. A. C. Broekaert, and E. B. M. Steers, *Spectrochim. Acta* 46B (1991) 243.
110. P. E. Walters and H. G. C. Human, *Spectrochim. Acta* 36B (1981) 585.
111. J. F. Geiger, M. Lopnarski, H. Oeschner, and H. Paulus, *Mikrochim. Acta* (*Wien*) 1 (1987) 497.
112. D. L. Donohue and W. W. Harrison, *Anal. Chem.* 47 (1975) 1528.
113. J. W. Coburn, E. Taglauer, and E. Kay, *J. Appl. Phys.* 45 (1974) 1779.
114. D. C. Duckworth and R. K. Marcus, *Anal. Chem.* 61 (1989) 1879.
115. D. C. Duckworth and R. K. Marcus, 16th Annu. Meet. Fed. Anal. Chem. Appl. Spectrosc. Soc. 1989, Paper No. 138.
116. M. R. Winchester, C. Lazik, and R. K. Marcus, *Spectrochim. Acta* 46B (1991) 483.
117. P. R. Cable and R. K. Marcus, 1991 European Winter Conference on Plasma Spectrochemistry.
118. R. Morrison, *Grounding and Shielding Techniques in Instrumentation*, Wiley–Interscience, New York, 1986.
119. D. C. Duckworth, P. R. Cable, and R. K. Marcus, 38th ASMS Conference on Mass Spectrometry and Allied Topics, 1990.
120. D. Fang and R. K. Marcus, 17th Annu. Meet. Fed. Anal. Chem. Appl. Spectrosc. Soc. 1989, Paper No. 339.
121. F. L. King, A. L. McCormack, and W. W. Harrison, *J. Anal. At. Spectrom.* 3 (1988) 883.
122. D. C. Duckworth and R. K. Marcus, *Appl. Spectrosc.* 44 (1990) 649.
123. N. Ketchell, Ph.D. dissertation, University of Manchester, 1989.
124. D. L. Donohue, W. H. Christie, D. C. Duckworth, D. H. Smith, and R. K. Marcus, 39th ASMS Conference on Mass Spectrometry and Allied Topics, 1991.
125. R. K. Marcus and D. C. Duckworth, 1989 Pittsburgh Conference and Exposition, 1989, Paper No. 657.
126. J. W. Coburn, E. Taglauer, and E. Kay, *Jpn. J. Appl. Phys. Suppl.* 2, Pt. 1 (1974) 501.
127. D. C. Duckworth and R. K. Marcus, 1989 Pittsburgh Conference and Exposition (1989), Paper No. 548.
128. J. W. Coburn and E. Kay, *Appl. Phys. Lett.* 19 (1971) 350.
129. J. W. Coburn, E. W. Eckstein, and E. Kay, *J. Appl. Phys.* 46 (1975) 2828.
130. M. R. Winchester and R. K. Marcus, *J. Anal. At. Spectrom.* 5 (1990) 575.
131. R. H. Page, C. S. Gudeman, and M. V. Mitchell, *Chem. Phys.* 140 (1990) 65.

8

Thin Film Analysis

Hubert Hocquaux

8.1. Historical Introduction

8.1.1. Bulk Analysis

Grimm (1968) was the first to demonstrate the principle of using a glow discharge lamp for the analysis of flat samples.[1] Since the Grimm lamp appeared, low-pressure gas discharges have found many applications. Several authors have investigated the potentialities of such a discharge.[2–9] Principal applications are in the bulk analysis of metals or of nonconducting materials pressed into pellets with a conducting binder. Several other configurations have been described for bulk analysis, mostly to improve the sensitivity, e.g., hollow cathodes,[10,11] boosted lamps,[12,13] and magnetic field-enhanced glow discharges.[14]

8.1.2. Surface Analysis

Berneron and co-workers (1971) introduced the use of the glow discharge lamp for depth profiling.[15] They plotted the variation of the light intensity of aluminum and manganese versus sputtering time in order to study defects on the surface of low-carbon steels. In 1972, Belle and Johnson published some practical applications of the Grimm lamp for depth profiling.[16] They measured the composition of various elements in steel within the depth range of 0.1–40 μm. In the same year, Greene and Whelan proposed the use of glow discharge optical spectrometry as a method for

Hubert Hocquaux • IRSID UNIEUX, BP 50, 42702 Firminy, Cedex, France.

Glow Discharge Spectroscopies, edited by R. Kenneth Marcus. Plenum Press, New York, 1993.

analyzing thin films.[17] Concentration profiles of boron and phosphorus implanted into silicon are described in several publications of Greene *et al.*[18,19] The lamp had a flat cathode but the geometry differed from that of the original Grimm tube. The stability of the discharge was poor during the first 50 s and the authors estimated the depth resolution with such a system to be approximately 80 nm. Schneider and Shuman described in 1974 one of the first examples of surface analysis by GD-OES,[20] but the recording of intensities on photographic medium resulted in a rather poor depth resolution.

Following these pioneering studies, Berneron and Moreau carried out extensive work exploring the use of GD-OES in the determination of gases in metals and depth profiling analysis.[21,23] As a result of this work, a prototype was made commercially available by RSV in 1978. This system was bought by Unirec and designed specially for depth profiling analysis. Between 1978 and 1985, GD-OES emerged as a technique among the most widely used for surface analysis, first in France and thereafter predominately in Germany and Japan. Today, systems for depth profiling are available from several manufacturers of OES equipment.

8.2. Principle of Surface Analysis by Glow Discharge Sputtering

The glow discharge lamp operates by removing material from the sample surface by cathodic sputtering in an atmosphere of a carrier gas at low pressure, generally argon. The material is removed layer by layer, more or less parallel to the original surface. Therefore, this technique is suitable for depth profiling analysis. The ablation can be easily controlled by adjustment of operating parameters: burning voltage, current, and gas pressure. The sputtering process produces a crater the size of which is determined by the anode geometry. The sputtering rate is defined as the sample erosion per unit time.

Atoms released by the impact of positive ions and neutral atoms are raised to an excited state by further collisions, producing a cathodic glow. The resulting plasma emits radiation consisting of characteristic lines of the elements eroded from the sample and the lines of carrier gas. The emission from the plasma is analyzed by an optical spectrometer. By integrating the element-specific signals at a high rate, information on the in-depth variation of elemental composition is obtained.[23] The high sputtering rate, combined with very high speed intensity measurement of analytical signals, led to a very attractive technique capable of in-depth profiling of surface layers with thicknesses ranging from a few nanometers to several tens of micrometers.

8.3. Equipment

8.3.1. Glow Discharge Source

Various modes of glow discharge operation have been used as radiation sources in atomic spectroscopy. For analytical applications the lamp consists of a low-pressure electrical discharge in a noble gas at pressures from 10 to 1000 Pa. The analytical sample acts as the cathode. For surface analysis only plane cathodes have been considered. The principle of glow discharge lamps and their electrical characteristics are described in several publications[8,24–26] and summarized in previous chapters.

For surface analysis the original Grimm Lamp[1] is the most frequently used (Fig. 8-1). This tube consists of a particular shape discharge chamber. The sample to be analyzed acts as the cathode: it seals and separates the chamber from the external atmosphere. In order to eliminate the heat generated, the cathode must be cooled. The anode is tubular in shape and its frontal distance to the sample is 0.2 mm. Because of the shape of the tube, only the negative glow is present. A first pump maintains the gas pressure in the lamp. In order to avoid a second discharge being created between the sample and the face of the anode tube, an additional pump is used to evacuate this interspace. A further development is the glow discharge lamp with floating anode, which has been introduced by Siemens (Fig. 8-2).[27] This system is now used by the firm Leco. Anode, floating anode, and cathode are separated from each other by an isolator. The distance from the cathode

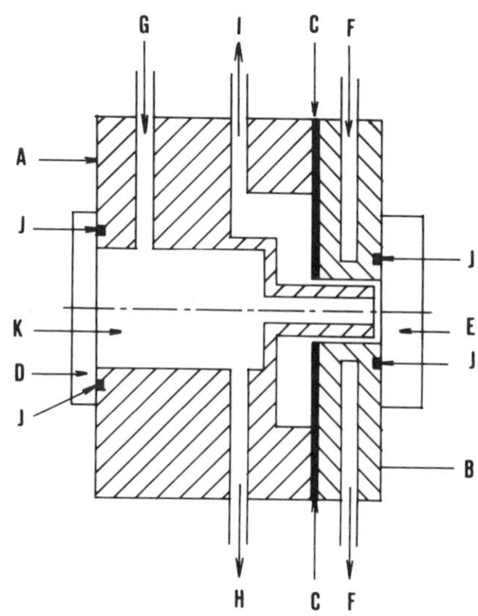

Figure 8-1. Schematic diagram of a Grimm glow discharge lamp. A, anode; B, cathode; C, anode/cathode insulator; D, MgF_2 window; E, sample; F, cooling; G, argon inlet; H, P_1 vacuum pump; I, P_2 vacuum pump; J, O-ring; K, anodic chamber.

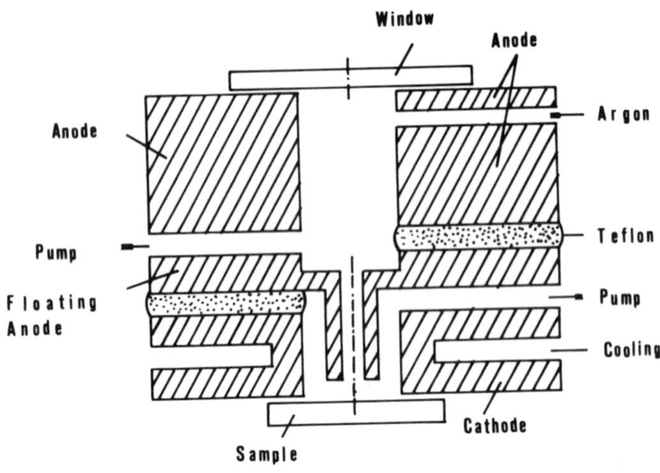

Figure 8-2. Schematic diagram of the glow discharge lamp introduced by Siemens.

to the anode is longer and the positive column is present contributing to the radiation output.

8.3.2. Spectrometer

The glow discharge lamp is placed in front of the slit of an optical spectrometer. A few studies have been performed with a monochromator or a fast sequential spectrometer,[28] but for surface analysis it is necessary to use a multichannel spectrometer. The sinoptic diagram of the more sophisticated commercial system is presented in Fig. 8-3. The light emitted by the plasma enters the spectrometer via a window and a primary slit. The optical components are made with magnesium fluoride (MgF_2) in order to extend the useful spectral range to 110 nm. The light is spectrally resolved by a diffraction grating and led through a secondary slit to a photomultiplier where it is transformed into an electric signal. The principal characteristics of the first prototype used by IRSID are summarized in Table 8-1. Some complementary information on the theory of optical spectrometers can be found in various papers.[22,29,30]

Most frequently a wavelength range from 110 nm to 460 nm is chosen by a suitable selection of the individual optical components. For the detection of spectral lines with wavelengths lower than 150 nm, special photomultipliers are needed. But another solution is commonly used, consisting of covering the envelope of the photomultipliers with a fine coat of sodium salicylate. Thus, the analysis of H, N, O, Cl, and C is possible in the UV region. Some studies have been made in the visible spectral region for

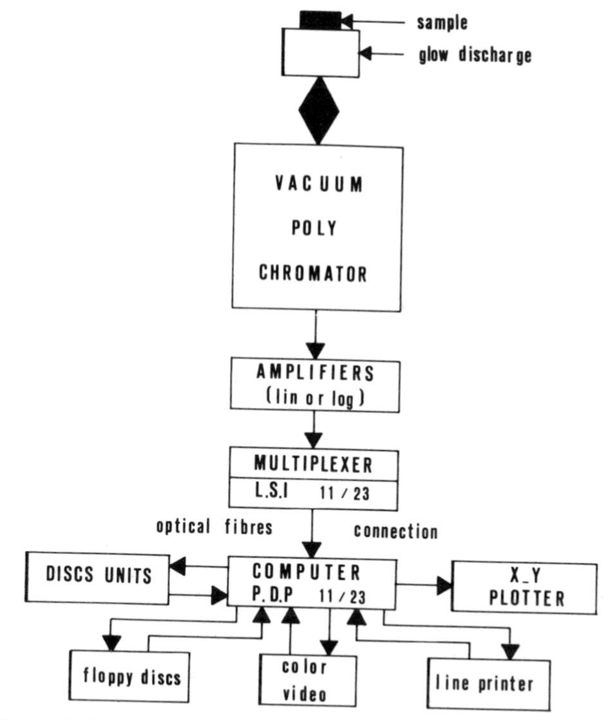

Figure 8-3. Schematic diagram of the IRSID surface analysis system.

Table 8-1. Optical System of the RSV Analymat 2504

Paschen–Runge configuration
Focal length: 2 m
Jobin–Yvon holographic concave grating with a surface covered with MgF_2
Reciprocal dispersion: 0.22 nm/mm

the determination of these elements.[31] It is expected that, with neon or mixtures of argon and helium as carrier gas, spectral lines of high excitation energy (F, Cl, O, N) can be excited, which are not possible in an Ar atmosphere, but at present better detection limits are obtained in the UV region (Cl, O, H, N). In Table 8-2, a selection of analytical lines usable for surface analysis is presented. As can be seen, there are a number of unusual lines that may not be incorporated in commercially available multichannel emission spectrometers. An interesting optical device to increase the spectral range has been proposed recently by Jobin–Yvon (Longjumeau, France) (Fig. 8-4). In the near future, such developments will be necessary to improve the

Table 8-2. Analytical Lines for Surface Analysis

Element	λ	Element	λ	Element	λ
Ag	338.28	Cd	228.80	H	121.56
			346.62		486.13
Al	237.84				656.30
	256.80	Ce	413.76		
	394.40			Hf	286.63
	396.15	Cl	118.87		
			133.57	Hg	253.65
Ar	137.72		134.72		
	157.49		479.50	I	145.79
					183.04
As	189.04	Co	340.51		
	200.33		345.35	In	410.17
Au	242.79	Cr	267.71	Ir	203.35
			298.92		322.07
B	208.95		425.43		
	249.60			K	404.72
		Cu	219.22		766.49
Ba	230.42		327.39		
	455.40			La	408.67
		Fe	259.90		
Be	313.04		271.40	Li	323.26
	332.12		273.95		610.36
			371.99		670.77
Bi	306.77				
		Ga	403.14	Mg	277.66
Br	148.84				280.27
		Gd	376.83		383.82
C	156.14				
	165.70	Ge	303.90	Mn	257.61
					403.14
Ca	393.36				403.44

performance of GD-OES for research purposes. The systems for routine control are less sophisticated.

8.3.3. Electronics and Computer

Various systems are proposed by the manufacturers of optical spectrometers for GD-OES. The equipment shown in Fig. 8-3 is not up-to-date, but for the moment it exhibits the best performance among the commercially available systems. It was first introduced by Siemens. In the near future other systems with microcomputers will become commercially available. Some design criteria should be mentioned:

Table 8-2. *(Continued)*

Element	λ	Element	λ	Element	λ
Mo	317.03	Pd	360.95	Te	200.20
	386.41				238.57
		Pr	433.39		
N	149.26			Ti	337.20
	174.20	Pt	265.94		365.35
	411.00				
		Rb	420.18	U	385.95
Na	330.23				
	588.99	Rh	437.48	V	311.07
	589.59				318.39
		S	180.73		411.17
Nb	316.34				437.90
	410.09	Sb	206.83		
	416.46			W	400.87
		Sc	424.68		429.40
Nd	430.35				
		Se	196.09	Y	371.10
Ni	225.38				377.43
	341.47	Si	251.61		
	349.29		288.15	Zn	213.85
					330.29
O	130.21	Sn	189.98		334.50
	777.19		242.16		377.21
			317.50		481.05
P	177.49		326.20		
	178.28			Zr	339.19
	185.94	Sr	407.77		360.11
	253.56				
		Ta	301.70		
Pb	220.35		362.66		
	405.77				

1. The signal from each photomultiplier is amplified (linear or logarithmic amplifier). A fast scanning frequency of the multiplexer is necessary to reach a good depth resolution (30,000 s^{-1} for the Siemens system in Fig. 8-3).
2. The use of high-resolution color display permits a real-time monitoring of the light intensity for each selected element so that the sputtering process can be interrupted at any time and the bottom of the crater examined by a complementary method (e.g., SEM, x-ray diffraction, ESCA).
3. The number of selected lines and the scanning frequency of the multiplexer are often very important, therefore the capacity of storage of the computer has to be large.

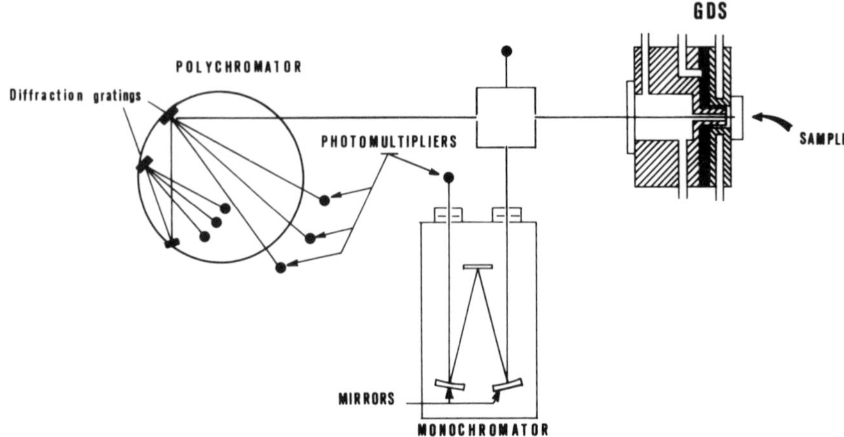

Figure 8-4. Optical system of the JY 50 GDS.

4. Complex and specific software is needed for checking, modifying, smoothing the element profiles, and if possible quantifying the results.

8.4. Analytical Results

In a glow discharge lamp the surface analysis of the sample is possible because of the fact that the material is removed layer by layer in a cathodic sputtering process. The light intensity of each line is simultaneously recorded. Thus, in the strict sense of the word, the result is not a surface analysis because the sample is continuously consumed. In fact, GD-OES is a depth profiling analysis. When the sputtering process is interrupted at the end of the analysis the first recorded result is the change with time of the intensities of analytical lines of selected elements. This qualitative analysis is often sufficient, especially when comparisons are made. To illustrate the possibilities of GD-OES for qualitative analysis, an example is given in Fig. 8-5 and the information that is extractable from such a record is summarized below. The sample was exposed for a corrosion test in a crevice with caustic species concentrated from river water pollution: liquid sampling from the crevice during the test indicated that the solution contained a large amount of sodium.

GD-OES showed the presence of a surface layer made of interpenetrated oxides and deposits:

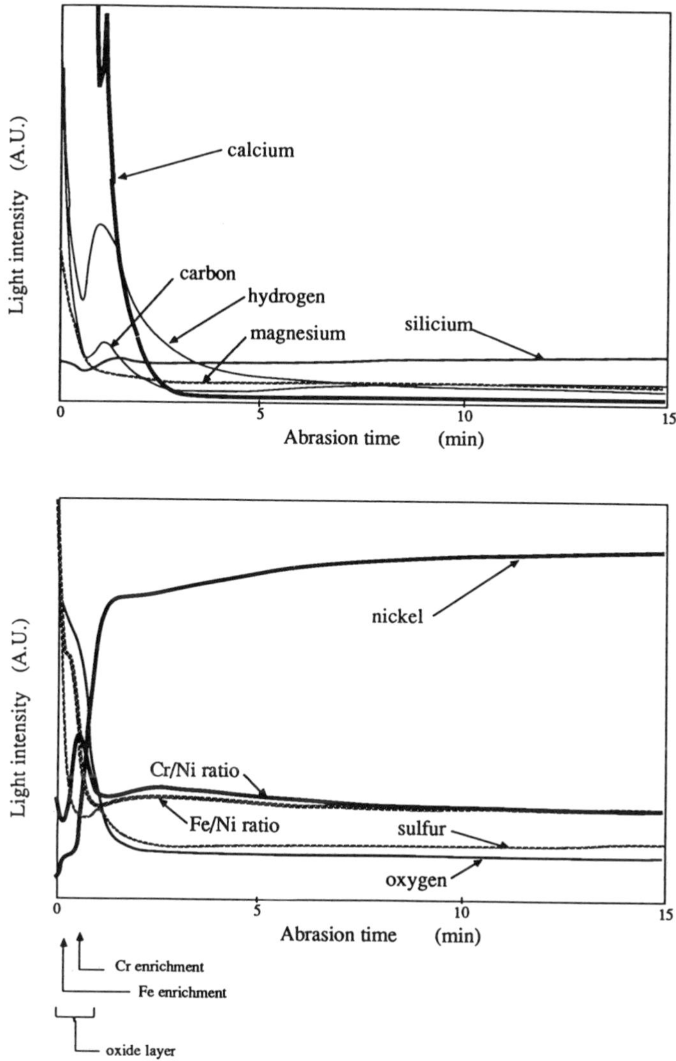

Figure 8-5. Qualitative surface analysis by GD-OES: study of corrosion products on Inconel.

1. The presence of metallic oxides is demonstrated by the presence of a layer with well-defined levels of the main metallic elements of the base material (profiles of Ni, Cr, and Fe) associated with the presence of oxygen. The abrasion of this layer takes about 1.5 min and corresponds to a thickness on the order of a few micrometers. Nickel is the main metallic component of the oxide but a chromium enrichment is

located at the metal/oxide interface and an iron enrichment occurs at the top of the layer (see Fe/Ni and Cr/Ni plots in Fig. 8-5).
2. The deposits are rich in calcium, and contain some magnesium.
3. A small amount of sulfur is present in the surface layers. It is mainly located in the near-surface deposit but there is also a small peak close to the metal/oxide interface; in this interface region oxygen is also present.
4. Hydrogen and carbon peaks are also present: their peaks are close to the metal/oxide interface but they are not exactly correlated. Oxygen could be associated with some of the above species.
5. Although the sensitivity of GD-OES for sodium is quite poor, this element appears to be present and perhaps correlated with carbon, hydrogen and/or sulfur (the profile of sodium is not shown in Fig. 8-5).

If we compare these results with those of other samples exposed to corrosion, it is possible to draw numerous metallurgical conclusions. If one needs more quantitative results, some modifications of the curves are necessary:

Time has to be correlated to depth.
Intensities of analytical lines have to be converted into concentrations.

8.5. Typical Operating Parameters for Surface Analysis

If the material, the anode, and the carrier gas are constant, the glow discharge conditions are described by three important parameters: current intensity, burning voltage, and gas pressure. These factors are not independent; if two of them are selected, the third one is then automatically determined. The operating parameters frequently used for bulk analysis are summarized in Table 8-3.

For surface analysis it is necessary to select softer conditions to improve depth resolution and to avoid local melting. In most cases the current intensity is lower than 100 mA and the burning voltage is chosen between 400 and 800 V. Constant-voltage operation is seldom employed because of

Table 8-3. Operating Conditions for Bulk Analysis

Current intensity: 80–200 mA
Burning voltage: 800–1600 V
Gas Pressure: a few millibars
Constant-voltage or constant-power modes

insufficient control of the discharge in the beginning. Such a problem is mainly encountered when the oxide layer at the surface of the material is an insulator (e.g., Ti, Zr, Al oxides). The discharge is then not stable because charge accumulation occurs near the surface. A few studies have been performed with a constant-voltage mode in the case of materials with a low melting point such as Sn, Sb, and Zn.

If the thickness of the layer is an important quantity, another system can be interesting. It consists of a mass flowmeter and a PID regulator (*p*roportional *i*ntegral *d*erivative pressure controller). With this apparatus it is possible to maintain constant both the intensity of current and the burning voltage by monitoring continuously the flow of gas. Thus, in a constant-current mode the voltage does not change even if the electrical characteristics of the sputtered layers are completely different. Currently it is not possible to regulate the voltage during the very few first seconds (5 s) of the discharge. An example of such a mode of operation is illustrated in Figs. 8-6 and 8-7. In this case the principal advantages of the system concern the profile of secondary elements like O, C, H. For example, without regulation of the voltage the profile of oxygen changes rapidly at the interface. In fact, this artifact is only due to the variation of the voltage. With the correction system the profile of oxygen does not vary in the interfacial zone.

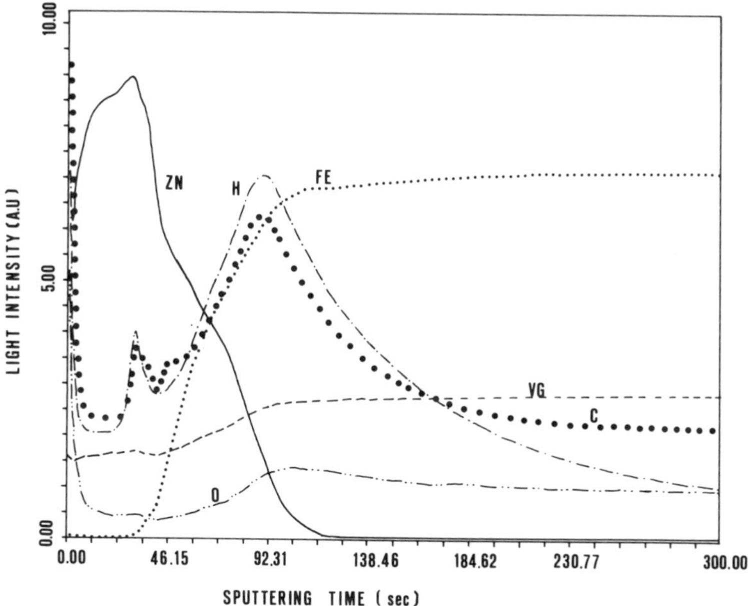

Figure 8-6. Analysis of a zinc coating on a steel sheet with the constant-current mode. $I =$ 75 mA, $V_g \approx 1200$ V in steel.

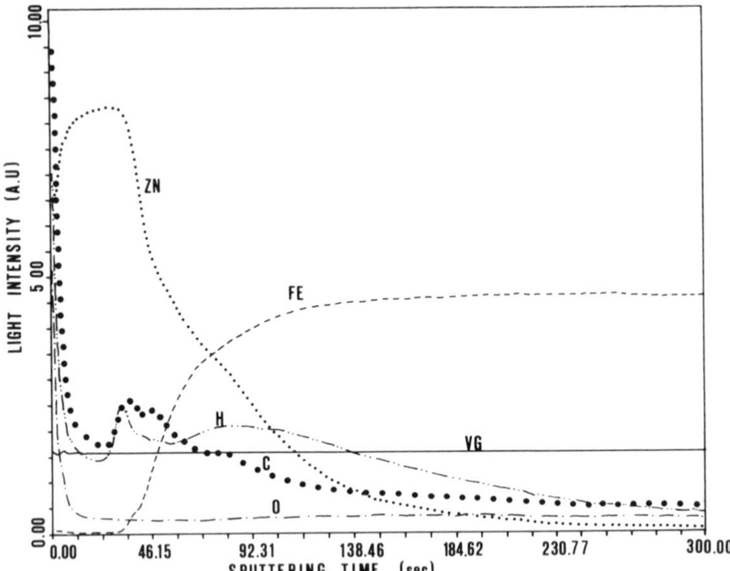

Figure 8-7. Analysis of a zinc coating on a steel sheet with the constant-current mode and regulation of the burning voltage by the PID controller. $I = 75$ mA, $V_g = 1200$ V in steel and in the zinc coating.

8.6. Shape of the Crater

The size of the crater is determined by the anode geometry. In Fig. 8-8, three types of craters are shown (two circular anodes and an oval one). In the first publications,[21,23,32] the authors mentioned that the bottom of the crater was not plane and that metal deposits were observed at the periphery. The influence of the type of lamp has been discussed.[32] In fact, these results are not complete because they were obtained with an anode the diameter of which was 8 mm and with a high lamp power. More recent studies have demonstrated that these conditions are not well suited for surface analysis.[33] These conclusions have been drawn by a group of French laboratories working together in a Groupement pour l'Avancement des Méthodes Spectroscopiques (GAMS) commission. In fact, the shape of the crater changes with the diameter of the anode, and with the operating parameters as can be seen in Fig. 8-9. Bouchacourt[34] tried to draw for low-carbon steel some areas on the diagram $I = f(V_g)$ where typical shapes are found (Fig. 8-10). When the conditions are mild, near the "normal" glow discharge region, the bottom of the crater is not completely sputtered. With high intensities of current the results are similar to those of earlier publications. A flat bottom

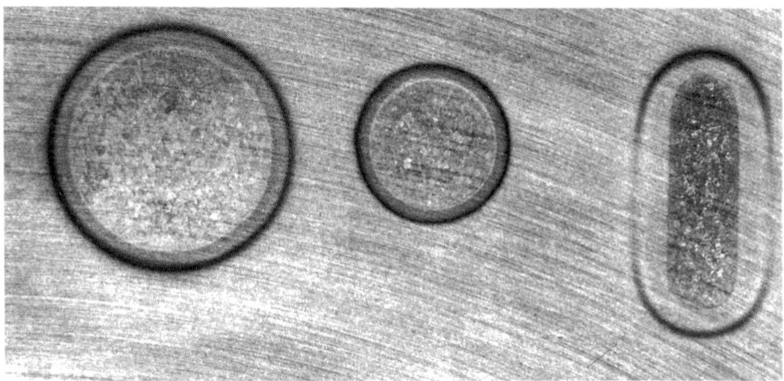

Figure 8-8. Different types of craters as a function of anode diameter (ϕ). (Left) $\phi = 0.8$ cm; (center) $\phi = 0.4$ cm; (right) $L = 0.8$ cm, $I = 0.2$ cm.

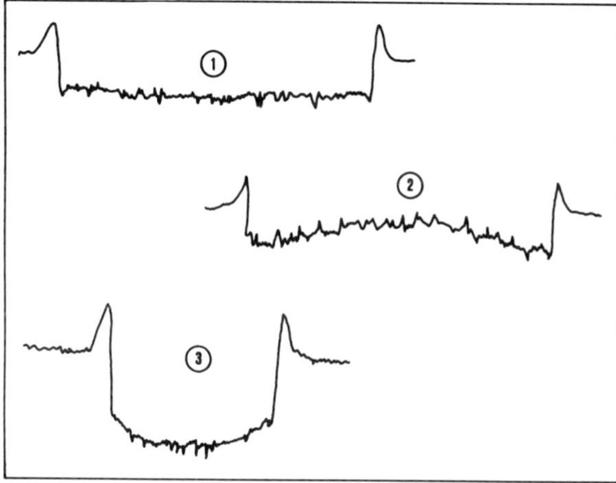

Figure 8-9. Influence of the anode diameter and of the operating parameters on the shape of the crater (iron matrix). (1) Diameter of the anode $\phi = 0.8$ cm, $V_g = 1000$ V, $I = 75$ mA. (2) $\phi = 0.8$ cm, $V_g = 1200$ V, $I = 100$ mA. (3) $\phi = 0.4$ cm, $V_g = 600$ V, $I = 25$ mA.

is obtained with a current below 40 mA and a voltage that is high enough to have an "abnormal" discharge. These examples demonstrate that the conditions are to be optimized for the material and the type of anode. As seen in Figs. 8-9 and 8-10, a roughening of the surface takes place during sputtering. The roughness of the bottom increases linearly with time. This is one of the factors limiting the depth resolution. It is clear that the choice of the operating parameters for depth analysis is a compromise between a good depth resolution and a high intensity of spectral lines.

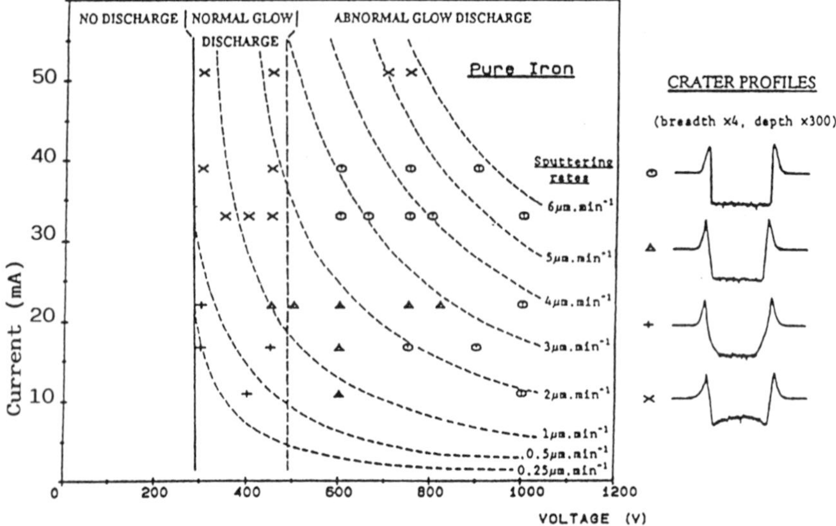

Figure 8-10. Current/Voltage dependence on crater shapes and sputtering rates. Pure iron; diameter of the anode $\phi = 0.4$ cm. Reproduced with permission from Ref. 34.

8.7. Sample Geometry

The shape of the lamp dictates the geometry of the samples. With the original Grimm lamp, and without special sample holders, the material must be flat and its dimensions must be larger than the seal, generally 20-mm diameter for an anode of 8-mm diameter. In more recent studies, several authors have used smaller anodes (4-mm and even 2-mm diameter) but with poorer light intensity. It is possible to use smaller specimens by mounting them in a conducting material, such as Ag or Cu, but the sputtered area must be larger than the diameter of the anode. The choice of the anode diameter is a compromise between lateral resolution and sensitivity. The analyses of circular samples have been performed in IRSID in France with some special tools.[35,37] In Figs. 8-11 and 8-12, two examples of sample holders for the analysis of tubes or wires are given. In Fig. 8-13, depth profiling results for circular samples are illustrated. The tests have been carried out on drawn wires of small diameter ($\phi = 0.8$ mm) in stainless steel from industrial production. The in-depth profiles of the elements Fe, B, Ca, Cr, S, and C were recorded. Particular attention was given to boron and calcium, which take part in the composition of wire-drawing antifriction grease and remain as a contamination at trace levels. In contrast, with chemical or electrochemical cleaning, ion erosion in a glow discharge does not cause a "preferential cleaning" of the pore sites. SEM observations of the

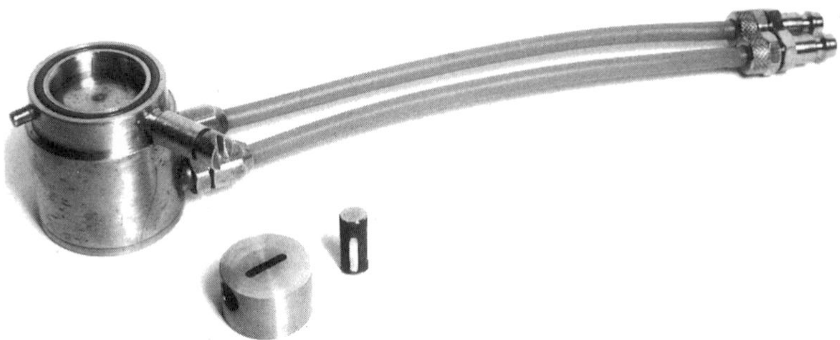

Figure 8-11. Special holder for the analysis of wires or tubes (diameter >2.2 mm).

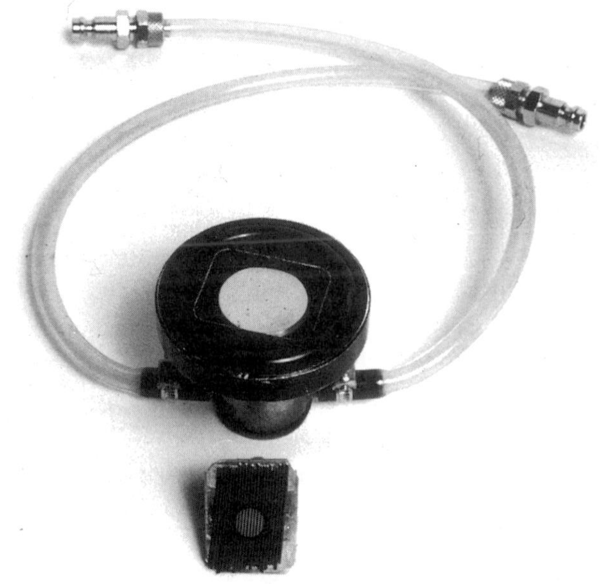

Figure 8-12. Special holder for the analysis of wires (diameter <2.2 mm).

crater at different depths show the presence of pollutants in the surface defects until their total elimination. The ion bombardment of these curved surfaces leads to the formation of a flat part on the upper generating lines of the wires; because of this phenomenon the in-depth resolution is not as

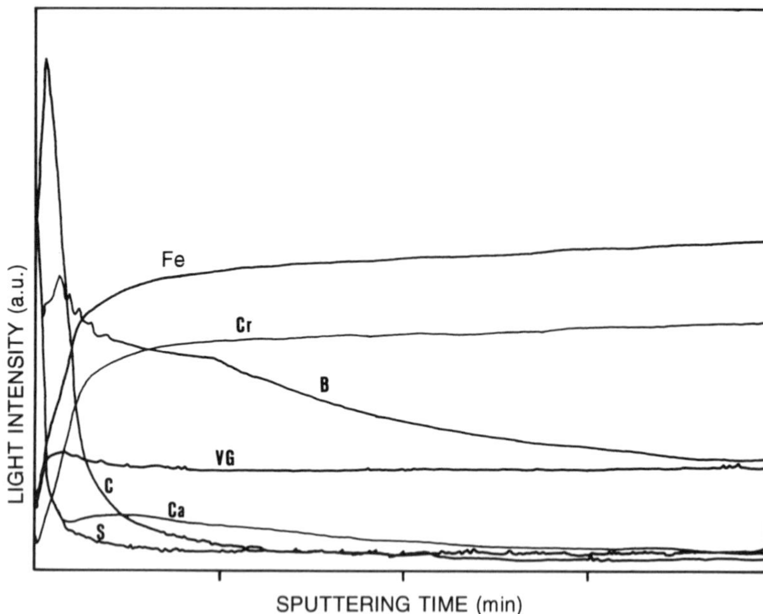

Figure 8-13. In-depth profile on the surface of a stainless-steel wire ($\phi = 0.8$ cm). Reproduced from Ref. 36.

good as for flat samples and the interface between two layers not so well defined. The results described in different publications[(35,37)] have shown the complementary nature of the SEM and GD-OES methods for the characterization of industrial surfaces. The results can be easily used for control purposes or for perfecting a manufacturing process.

8.8. Quantitative Surface Analysis with GD-OES

As has been demonstrated in the previous sections, applications of the GD-OES technique for surface analysis are numerous. But the results are most frequently presented as a relative intensity versus a sputtering time. To quantitatively determine the distribution of an element as a function of the distance from the surface, two problems have to be considered:

1. Sputtering rate of the target under argon ions accelerated in the cathode dark space
2. Relationship between concentration and light intensity

If the sample is homogeneous, quantitative analysis is very easy. The sputtered depth and the chemical content can be determined with reference to the bulk composition. Standardization of light intensities can be made with the help of reference samples. In a glow discharge there are few matrix effects because of the fact that the analysis is conducted under equilibrium conditions in a quasi-stationary state. This is true in depth profiling analysis when rather small changes of composition have to be investigated. Some applications of direct quantitative analysis have been described in the literature.(23,32,35) An example of calibration with the help of bulk analysis is given in Fig. 8-14. During cold-rolling a superficial enrichment of chromium appears by diffusion from the bulk to the oxide layer. Behind the peak of chromium a depleted zone can be observed. It is important to measure quantitatively this phenomenon to know the corrosion properties of the sheet.

If the sample matrix changes, the burning voltage and the subsequent sputtering rate are no longer constant and the problem of quantification is more complicated, as was pointed out very early by Jäger for bulk analysis.(38,39) In the field of surface analysis, many authors have now presented quantitative results with either a practical method or a more theoretical approach.

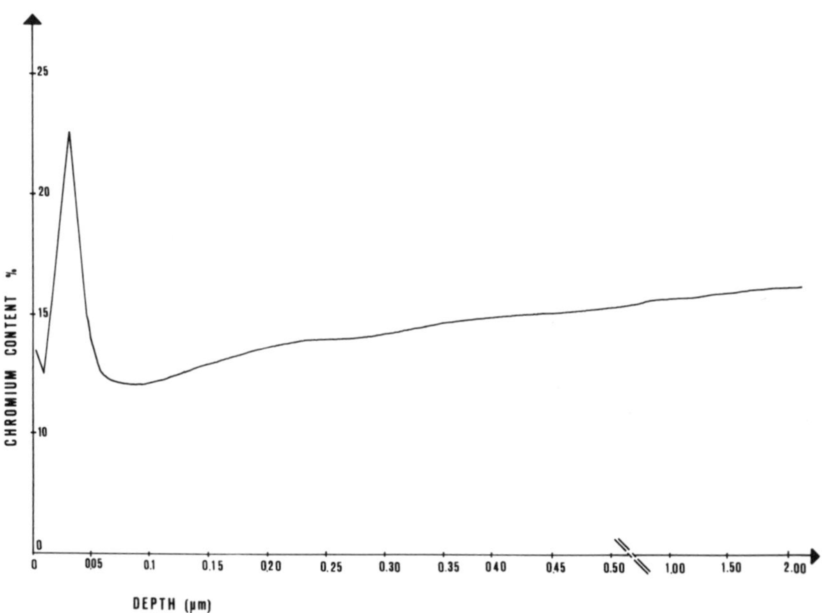

Figure 8-14. Study of chromium-depleted zones on a ferritic steel after cold-rolling.

8.8.1. Factors Affecting Sputtering Rate

8.8.1.1. Practical Considerations

a. Measurement of Eroded Depth. In the case of a homogeneous sample, calibration of the depth is most frequently obtained by weighing after sputtering for a definite time or by measuring the crater depth with a roughness meter or an optical method. To know the sputter behavior of a layer the same approaches can be used especially in the case of thick films. Some experimental results are given in the literature.(8,23,40,41) For thin layers the measurement is more complicated. In the case of oxides a different approach has been described.(42) Determination of the sputter rates is obtained by comparison with those of high-grade metals under the same excitation conditions. The sputter rate of an oxide can be calculated from the ratio of the measured to the calculated intensity of the oxide when the sputter rate of the corresponding metal is known. This method is used to estimate the sputtering rate of oxides and nitrides on steels. In another study, GD-OES was used to analyze the thin oxide film grown in dry air on the surface of a chromium–molybdenum low-alloy steel.(42) The thickness measured through calibration by x-ray interferometry was found to be about 2.5-6.5 nm. In Table 8-4, some experimental results obtained at IRSID are presented.

b. Influence of the Homogeneity of the Sample. It has also been demonstrated experimentally that the in-depth resolution is limited by the formation of structures at the bottom of the crater.(32,33,43) These phenomena are the result of many factors such as the discharge conditions and nature of the carrier gas. But the most important parameters are the chemical and metallurgical properties of the sample. An example is given in Fig. 8-15 to illustrate the influence of the size of grains in the crystal structure of the layer, in this case a zinc coating on a steel sheet. Sometimes the interpretation of results is further complicated by these phenomena and by the roughness

Table 8-4. Experimental Sputtering Rate for Some Materials

Material	Sputtering rate (μm/min)	Material	Sputtering rate
Carbon steel	1.7	Brass	6.2
Stainless steel	2.0	Zircaloy	1.7
Nickel alloy	2.3	Vanadium	1.2
Titanium sheet	0.8	Molybdenum	1.2
Aluminum sheet	1.4	Zinc	18

[a]Burning voltage, 800 V; current intensity, 50 mA.

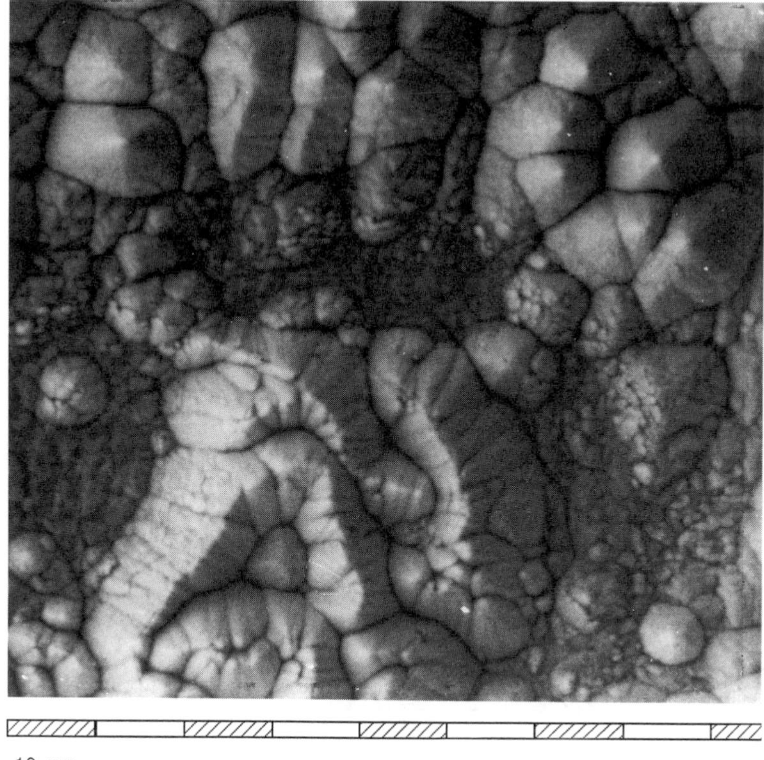

Figure 8-15. SEM observation of the bottom of a crater for a zinc coating.

created by the sputtering process especially near the interfacial region. The depth resolution is limited by the fact that signals originating from different layers may appear simultaneously. These problems do not occur only in GD-OES but in all methods using an ionic sputtering of the samples. Some authors have found that the gradually deteriorating depth resolution can be described by a dynamic instrument function.[40]

Preferential sputtering can occur when the eroded layer contains inclusions. The physical and chemical properties of these particles are generally very different from those of the matrix sample. The sputtering process is changed and cones are formed on the sample surface. Such observations have been made by Jäger and Blum[44] in the case of gold and brass samples. Fig. 8-16 gives another example of these problems. It can be seen that the formation of cones during the sputtering of this zinc coating is caused by the presence of inclusions of oxides containing silicon and aluminum. These particles are insulating and more resistant to sputtering than the matrix. The

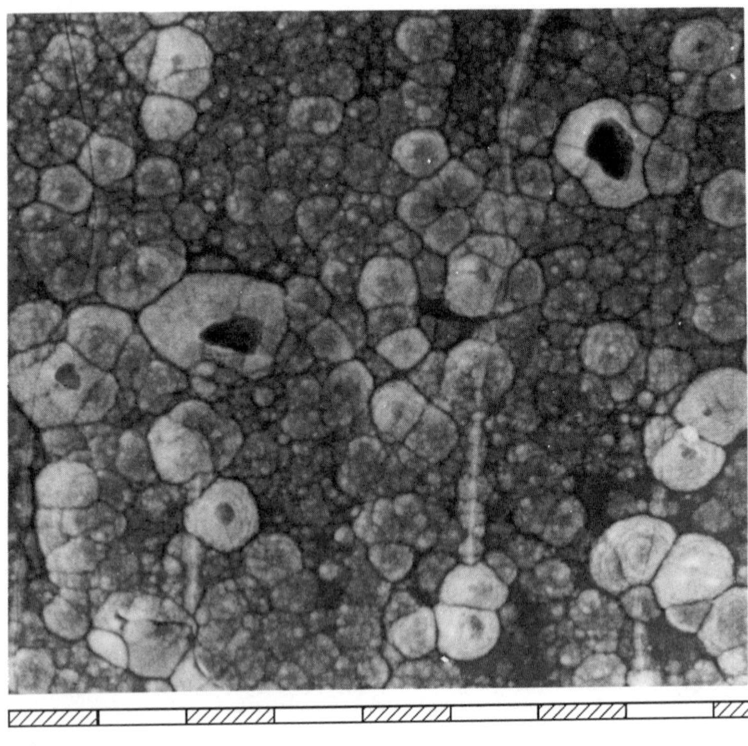

Figure 8-16. Formation of cones during the sputtering process (SEM analysis of a zinc coating).

removal of the top of the cones may be achieved by sputtering, but some data from electron micrographs indicate that the top is sometimes broken off as soon as the cone reaches a certain height. These phenomena, too, are not only observed in GD-OES but also with other methods using ionic sputtering. Greene has reported similar results in another type of discharge.[45]

c. Influence of the Temperature of the Sample. The influence of this parameter has been discussed for low-melting-point matrices such as zinc, tin, and lead.[46] In fact, it is well known that the sputtering rate increases rapidly when the temperature reaches approximately 100°C under the melting point. Different cooling systems have been studied to reduce this phenomenon. At the moment the best results are obtained with a direct circulation of water on the back of the sample. In Fig. 8-17, the effect of the application of a cooling system during the analysis of galvanized steel sheets

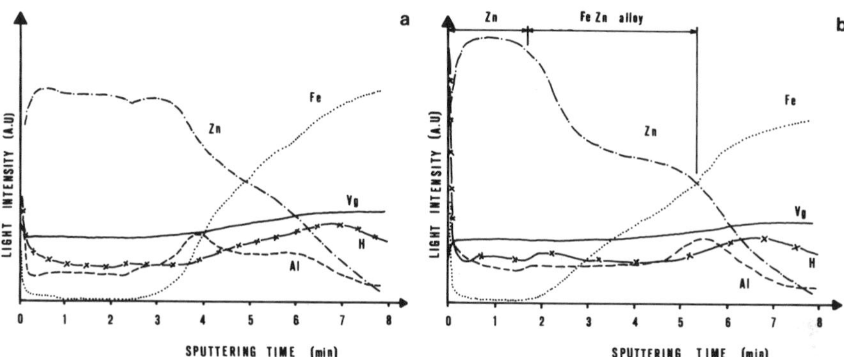

Figure 8-17. Analysis of the surface of a galvanized steel sheet. (a) Without a cooling system; (b) with a cooling system. V_g, burning voltage. Reproduced from Ref. 46.

is illustrated. With this system the results are considerably improved in the case of thin sheets.

8.8.1.2. Study of the Sputtering Rate Variation with Respect to Discharge Parameters

The variations of the burning voltage investigated at IRSID for several materials concern the range from 400 to 1600 V. An example is given in Fig. 8-18. In this case the current was fixed at 25 mA and the diameter of the anode was 4 mm. The erosion speed variation (which is proportional to the sputtering yield), at a constant current, is a linear function of V_g. Several authors have investigated this dependence and have obtained the same results.[8,34,47] The curves in Fig. 8-18 prove the existence of an energetic erosion threshold, an important parameter that surely governs the sputtering in the field of low energies.

The results presented in Fig. 8-19 were also obtained at IRSID. The erosion speed variation as a function of I is linear in the range considered (25 to 200 mA). The working voltage was kept constant (by pressure regulation) at 1000 V at each point.

8.8.1.3. Experimental Results

As discussed previously, the sputtering rate of a given material is often easy to measure. To quantify the results, it is necessary to know the sputtering yield, which is the average number of eroded atoms per incident ion. In the field of glow discharge in-depth analysis, there are few purely theoretical approaches to predict the sputtering yield of a target. In fact, authors

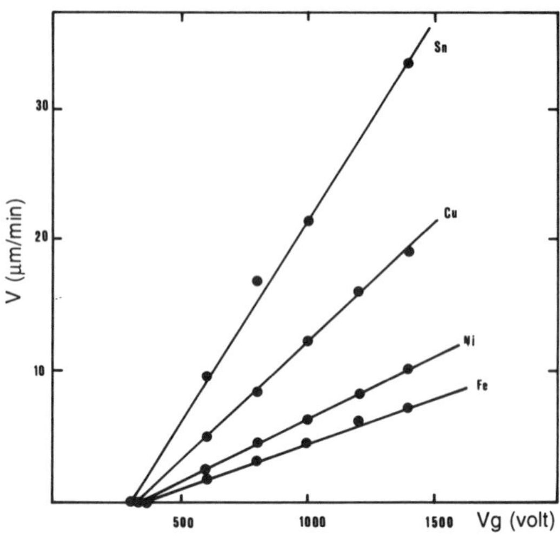

Figure 8-18. Variation of the sputtering rate as a function of the cathodic fall V_g. Diameter of the anode, 0.4 cm; intensity of current, 0.025 A. Reproduced from Ref. 51.

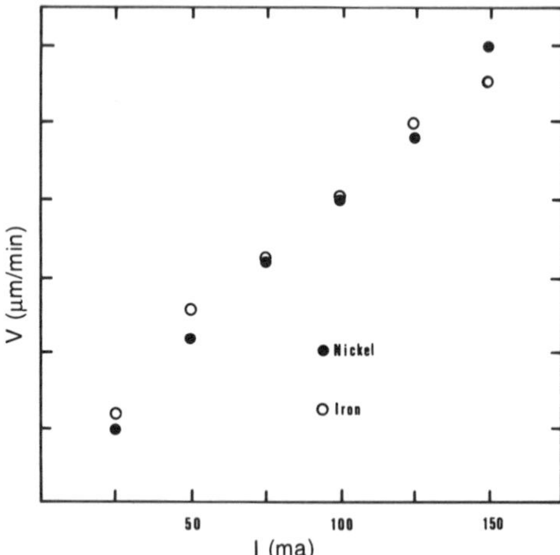

Figure 8-19. Variation of the sputtering rate as a function of the current intensity. Constant-voltage mode ($V_g = 1000$ V); diameter of the anode, 0.8 cm. Reproduced from Ref. 51.

commonly obtain from the measured eroded mass an equation leading to experimental sputtering yield,[8,40,47–49] as was discussed in Chapter 2.

The sputtering yield S is given by[8]

$$S = 10^{-6} qNe/Mi^{+} \quad (8\text{-}1)$$

where q is the sputtering rate ($\mu g/s^{-1}$), N the Avogadro number ($mole^{-1}$), e the electronic charge (C), M the atomic weight, and i^+ the ionic current (A). The ionic current is related to the total current by

$$i^+ = i/1 + \gamma \tag{8-2}$$

where γ is the average number of secondary electrons emitted by the target under the impact of a colliding ion. Many authors, especially Boumans,[(8)] have not found precise values for γ. Thus, in the literature, sputtering yields are reported in terms of $S/1 + \gamma$ or with a value of γ equal to 0.1 for all of the energies and all materials under argon bombardment. In more recent work on glow discharge lamps,[(50,51)] the following empirical relationship has been used to express the secondary electron emission yield γ:[(52)]

$$\gamma = 0.032\ (0.78E_i - 2\varphi) \tag{8-3}$$

where φ is the work function of the solid.

The results of these calculations are shown in Table 8-5 for 11 elements.

With these values the experimental sputtering yield can be calculated. The example of results given in Table 8-5 is obtained with the following operating parameters: diameter of the anode, 0.4 cm; current intensity, 0.025 A; burning voltage, 1000 V.

8.8.1.4. Theoretical Predictions—Modeling

The approach described below is a summary of the model developed by L. Ohannessian at IRSID.[(51)]

a. Principle. This target study can be divided into two steps: (1) energetic distribution of colliding ions, atoms, and neutrals and (2) physical mechanisms of the collision cascades leading to the sputtering. The authors have assumed a linear electric field in the dark space, computed the length of this space, and then the distribution of ions and neutrals at the target

Table 8-5. Calculated Value of the Secondary Electron Emission Yield for Different Materials and Experimental Sputtering Yield

	Target										
	Al	Ti	V	Fe	Ni	Cu	Mo	Ag	Sn	Au	Pb
γ	0.127	0.164	0.152	0.096	0.074	0.111	0.12	0.118	0.11	0.053	0.148
S_{exp}	0.45	0.27	0.36	0.54	0.74	1.54	0.32	2.49	1.09	1.53	2.44

[a]From Ref. 51

surface. The Sigmund theory offers a basic formalized system for the sputtering phenomena resulting from this bombardment.

b. Calculation of the Length of Cathode Dark Space and the Energetic Distribution of Ions and Neutral Atoms. The calculation of the length of the cathode dark space L remains within the hypothesis that the electric field undergoes a linear decrease from the cathode to the border which separates the dark space from the plasma.

The following equation has been used by Ohannessian:

$$L = 1.71 \times 10^{-4} R^{4/5}(1+\gamma)^{2/5} V_g^{3/5} T^{1/5} / I^{2/5} P^{1/5} \qquad (8\text{-}4)$$

with L in cm, R the anode radius in cm, I the current in A, V_g the burning voltage in V, P the pressure in Torr, and T the temperature of carrier gas in K. For a voltage of 1000 V, a pressure of 6 Torr, and a current density of 0.2 A/cm^2, L is equal to 0.03 cm. Allis *et al.*[(54)] mention a value of the same degree of magnitude in the same field of the selected operating parameters.

After having characterized the cathode dark space (area of argon acceleration) by its length L, the distribution of ions and neutral atoms can be calculated. In a pioneering work, Davis and Vanderslice[(55)] measured the energy distributions of ions bombarding the cathode in an abnormal glow discharge. Chouan and Collobert[(56)] offered an exact calculation of the distribution function in the case of a continuous diode discharge. Ohannessian has developed her calculation from the work of Abril *et al.*[(57)] These authors take the calculation back up by considering all of the particles: ions and neutral atoms. Abril *et al.* set the following hypothesis:

1. All of the ions come from the region situated between the cathode fall area and the area of emission of negative glow.
2. The predominant collisions in the dark space are charge transfer collisions where a fast ion leaves its charge to a rest neutral in order to provide a rapid neutral and rest ion.

$$Ar_1^+ + Ar_2 \rightarrow Ar_1 + Ar_2^+$$

The complete calculation is described by Ohannessian.[(51)] It is necessary to point out that the distribution of neutral atoms, taking into account the energy losses through successive impacts in the dark space, is different from that calculated by Abril *et al.* The amount of neutrals with low energy (some eV) is substantial; it is assumed that all the mobile neutral atoms will reach the cathode after a direct path, which is not true for those created too far from the cathode.

c. Sputtering Yield. 1. Theoretical expression of sputtering yield. From Sigmund's[53] theoretical expression established in 1969, Matsunami and Yamamura[58] derived an initial empirical formulation of the sputtering yield of monoatomic targets that takes the sputtering threshold energy into account. At the same time Yamamura *et al.*[59] established, by the compilation of experimental results, the following formulation:

$$S(W) = 0.42\alpha^* QKS_n(\varepsilon)/U_S[1 + 0.35U_S S_e(\varepsilon)][1 - (W_{th}/W)^{1/2}]^{1/4} \quad (8\text{-}5)$$

where $S(W)$ is the sputtering yield, α^*, Q, and W_{th} are empirical parameters, U_S is the sublimation energy in eV, $S_n(\varepsilon)$ and $S_e(\varepsilon)$ are reduced stopping cross sections, either elastic or inelastic, respectively (derived from LSS theory[60]), and ε is the reduced energy.

$$\varepsilon = 0.03255 M_2 W/Z_1 Z_2 (Z_1^{2/3} + Z_2^{2/3})^{1/2}(M_1 + M_2) \quad (8\text{-}6)$$

$$K = SN/S_n = 8.478 Z_1 Z_2 M_1/(Z_1^{2/3} + Z_2^{2/3})^{1/2}(M_1 + M_2) \quad (8\text{-}7)$$

$$\alpha^*(M_2/M_1) = 0.08 + 0.164\,(M_2/M_1)^{0.4} + 0.0145(M_2/M_1)^{1.29} \quad (8\text{-}8)$$

W_{th} is the threshold voltage:

$$W_{th} = U_S[1.9 + 3.8(M_2/M_1)^{-1} + 0.134(M_2/M_1)^{1.24}] \quad (8\text{-}9)$$

Therefore, this is the sputtering yield expression that has been used:

It is based on the principle of a colliding mechanism, resulting from works by Sigmund.
In addition, it takes the sputtering threshold energy into account.
Finally, it results from the recent compilation of experimental studies.

2. Comparison of experimental and predicted results. Energy distributions of colliding argon species at the target were calculated—both ions and neutral atoms. The formulation by Matsunami and Yamamura[58] is inserted into each program of distribution calculation. Thus, a mean value of the sputtering yield is obtained under the impact of ions on the one hand and fast neutrals on the other. The mean value of the number of atoms eroded because of colliding ions S_t is finally obtained by taking the contribution of neutrals into account (the latter being in the region of 50% of the sputtering yield).

Table 8-6. Total Theoretical Sputtering Yield for Different Materials

	Material									
	Al	Ti	V	Fe	Ni	Cu	Mo	Ag	Sn	Au
S_{ion}	0.23	0.11	0.15	0.28	0.28	0.53	0.12	0.78	0.28	0.46
$S_{neutral}$	0.27	0.12	0.16	0.39	0.38	0.85	0.14	1.52	0.53	0.82
S_{total}	0.5	0.23	0.31	0.67	0.66	1.38	0.26	2.3	0.81	1.28

For $T = 380\text{K}$, $V_g = 1000\text{V}$, $P = 6$ Torr, and $J_t = 0.2\ \text{A/cm}^2$, we come across the results shown in Table 8-6. The comparison between theoretical and experimental results is summed up in Table 8-7 and is represented in Fig. 8-20.

The experimental value of S is obtained by integrating the results of the secondary electron emission yield. It is possible to introduce different temperature values of the carrier gas in the calculation of the theoretical sputtering yield. It is recalled, in fact, that this parameter is not known precisely for given discharge conditions (as it is not measured). Thus, for a set of operating conditions where the voltage and current are known, and where the pressure of the gas is determined theoretically, we observe the influence of the temperature on the correlation of experimental and theoretical results (Table 8-7 and Fig. 8-20). The best correlation is obtained at 445 K (Fig. 8-21). Fig. 8-22 is a different illustration of the theoretical variations of the sputtering yield as a function of the carrier gas temperature. The importance of this experimental parameter on the sputtering rate is proven here.

3. Conclusion. From the original work of Ohannessian the results obtained are summed up by putting them into their analytical context:

1. The role played by the fast neutrals formed by charge exchange in the dark space and their contribution to the sputtering phenomenon is proven. The number of neutral atoms striking the target is much higher than that of ions (probably ten times more), but their energy is much lower (lower than the threshold energy of sputtering for a

Table 8-7. Comparison between Experimental and Theoretical Results of Sputtering Yield for Different Values of Temperature

		Al	Ti	V	Fe	Ni	Cu	Mo	Ag	Sn	Au
	S_{exp}:	0.45	0.27	0.36	0.54	0.74	1.54	0.32	2.49	1.09	1.53
S_{theory}	320 K	0.41	0.19	0.26	0.55	0.54	1.16	0.22	1.95	0.69	1.05
	380 K	0.5	0.23	0.31	0.67	0.66	1.38	0.26	2.3	0.81	1.28
	445 K	0.59	0.28	0.37	0.78	0.76	1.58	0.31	2.63	0.92	1.43
	512 K	0.69	0.32	0.43	0.91	0.87	1.82	0.37	2.95	1.05	1.6

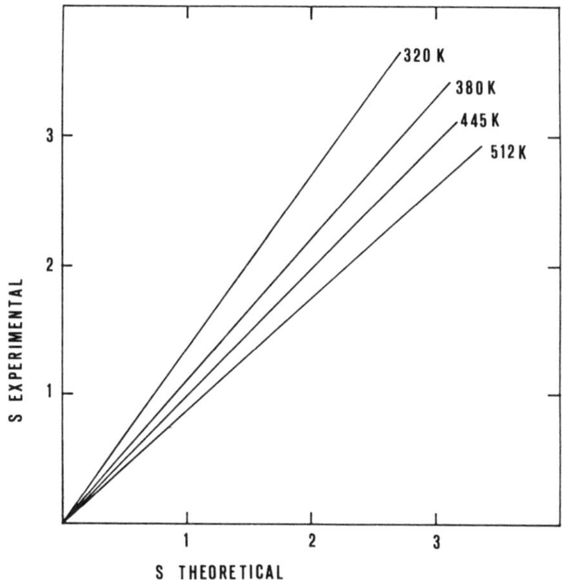

Figure 8-20. Investigation of the correlation between theoretical and experimental sputtering yields for different values of gas temperature. Reproduced from Ref. 51.

large fraction of atoms). Thus, their contribution to the total sputtering yield is an average of ≈50%.

2. A bibliographical investigation of the secondary electron emission phenomenon carried out elsewhere[51] enables calculation of its yield and the contribution of electrons to the total current. Using these data the experimental yield of the sputtering is obtained.
3. A good correlation exists between the experimental and theoretical sputtering yields for pure materials investigated under given discharge conditions. Therefore, it is possible to employ the realistic precalculations of the sputtering yield of monoatomic solids.

8.8.2. Concentration Calibration

For a full quantitative analysis, emission intensities have to be converted into concentrations. In the case of a homogeneous material the conversion is very easy but in general the problem is more complicated. Fig. 8-23 shows the spectral intensity–time profile of a multilayer coating prepared by chemical vapor deposition on a carbon steel. The outer layer consists of vanadium carbide and the inner layer, chromium carbide. The concentrations of carbon calculated and measured by other methods are 17 and 8% in the vanadium

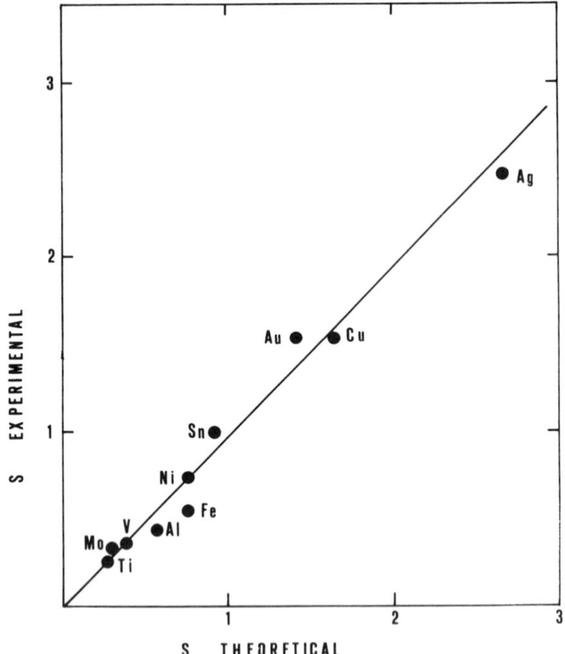

Figure 8-21. Correlation between $S_{\text{experimental}}$ and $S_{\text{theoretical}}$. $T = 445$ K; $V_g = 1000$ V; $J_t = 0.2$ A/cm^2. Reproduced from Ref. 51.

and chromium carbide, respectively. It is clear from this example that content is not proportional to profile intensity. If the apparent thickness of the two layers are compared with measurements by SEM, it appears that the horizontal axis of the spectral intensity–time profile is not proportional to depth. The same results have been obtained in Japan for two-layer Zn–Fe electroplating.[(61)]

8.8.2.1. Practical Approaches

It has been demonstrated that the main parameters influencing the intensities in the constant-current mode are the burning voltage and the sputtering rate. If the intensities in Fig. 8-23 are corrected point by point with these two parameters, the carbon content in the different carbides is quite good. This type of correction is very simple and gives a first estimation of the concentration of the main constituents. But it is too empirical for general application.

Different authors have proposed methods to quantify surface pollutants or enrichments by integrating the spectral intensity. Good correlations have

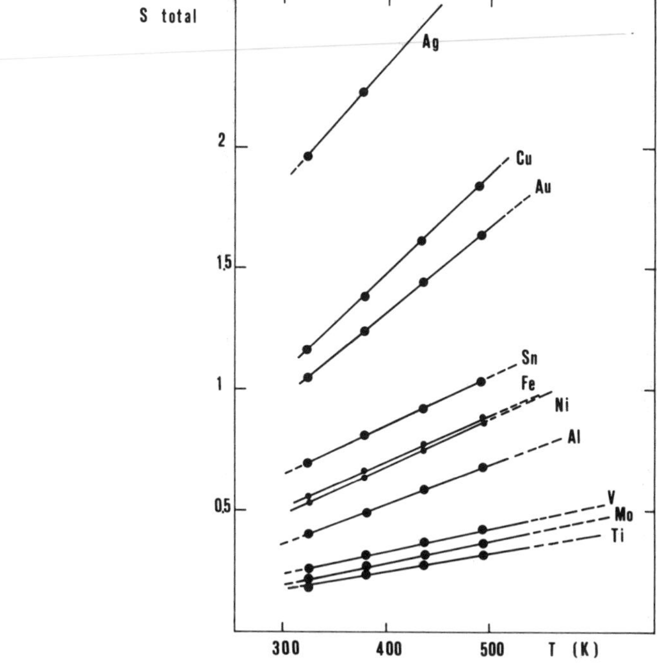

Figure 8-22. Theoretical variations of sputtering yield as a function of carrier gas temperature. $V_g = 1000$ V; $J_t = 0.2$ A/cm^2. Reproduced from Ref. 51.

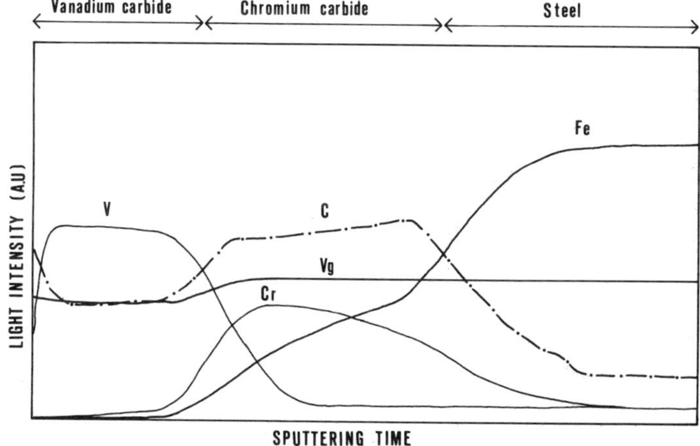

Figure 8-23. Qualitative analysis of a multilayer coating. V_g, voltage.

been obtained for example in the case of carbon content in surface or enrichment of manganese after annealing. The calibration is realized by chemical analysis: carbon in surface by the Ford method or combustion in high-purity oxygen[62] and Mn by ICP-OES.[63] If the sample matrix changes, the associated variations complicate the problem of quantification and less empirical methods must be used.

8.8.2.2. Theoretical Approach

In recent years, many authors, especially in Japan, have proposed more or less sophisticated methods for the quantification of in-depth profiles.[40,61,64–66] These methods are based on the assumption that the emission yield, R_{nm} (emission intensity/sputtering rate), of a spectral line m of the element n is independent of the matrix. Several investigations[61,67] have demonstrated that R_{nm} is practically independent of the matrix composition, but changes with voltage and current. The method described below has been developed in IRSID (Institut de la Recherche Sidérurgique, France).[43,49,67]

The intensity $JA^{(\lambda)}$ of the spectral line of an element A located at the surface of the sample eroded by cathode sputtering is given by:

$$JA^{(\lambda)} = K^{\mathrm{A}} R^{\mathrm{A}(\lambda)} \gamma_{\mathrm{M}}^{\mathrm{T}} I C_{\mathrm{A}} \tag{8-10}$$

where C_{A} is the atomic concentration of element A; K^{A} is the collection efficiency of the line $\lambda(\mathrm{A})$, which is an instrumental factor depending on the spectrometer operating parameters; $R^{\mathrm{A}(\lambda)}$is the emission yield of the element A for the line λ, which is independent of the matrix[49] and represents the number of excited atoms per eroded atoms; $\gamma_{\mathrm{M}}^{\mathrm{T}}$ is the total sputtering yield of the matrix (average number of eroded atoms per incident ion); and I is the current intensity of the lamp.

Equation (8-10) is correct if the sputtering steady state is reached. If the composition of the matrix changes during analysis (e.g., multilayer coatings) a modification of the electrical properties occurs and the burning voltage changes (in constant-current mode). As the emission yield depends on the voltage and the current, this means that the operating parameters have to be chosen in such a way that both voltage and current remain essentially constant during the sputtering time. This can be achieved in constant-current mode by selecting a burning voltage between 500 and 700 V. Thus, the discharge conditions will be approximately the same for all the matrices and $R^{\mathrm{A}(\lambda)}$ will be considered as a constant ($K^{\mathrm{A}}R^{\mathrm{A}} = K_{\mathrm{A}}$).

An extensive investigation of sputtering yield has been made by Ohannessian.[50,51] At constant current, variations with the voltage and the composition of the matrix have been indicated. Thus, γ is an important parameter that determines the shape of the profiles. For iron and zinc the following expression has been established:[67]

$$\gamma_1^T/V_1 - V_0 = \gamma_2^T/V_2 - V_0 \tag{8-11}$$

where γ_1^T (γ_2^T) is the total sputtering yield for the voltage V_1 (V_2) and V_0 is the threshold voltage ($V_0 \approx 300$ V). If the sputtering yield of the matrix is approximately constant throughout the analysis, Eq. (8-10) is used directly. An example of this case is the surface analysis of a sample treated by ion implantation. The spectral intensity is proportional to the concentration in depth. If the total quantity of the implanted element is known, it is very easy to find the concentration at each depth. An example of such an application is presented in Fig. 8-24 (implantation of phosphorus in an iron matrix).

If the sputtering yield changes with time, Eq. (8-10) can be written for each element A, B, C . . . of the matrix. Thus, it is possible to estimate the atomic fraction of the elements at any erosion depth and for the area of the specimen submitted to the discharge, i.e., a few square millimeters, through the expression:

$$XA_{at} = C^A/\Sigma_i C^i = (J_A/K_A)/(\Sigma_i J_i/K_i) \tag{8-12}$$

Practically, the ratio of the coefficient K for each element with respect to that of one major element is used:

$$XA_{at} = (J_A K_B/K_A)/(J_B + J_A K_B/K_A + JCK_B/K_C + \cdots) \tag{8-13}$$

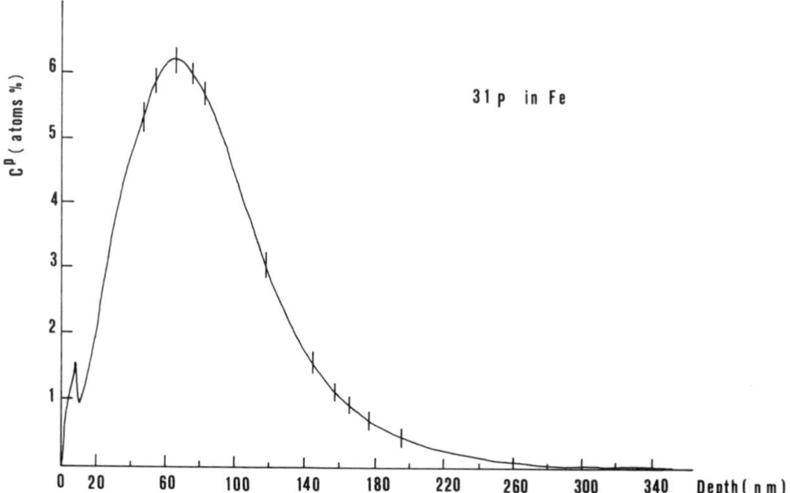

Figure 8-24. Concentration profile of implanted phosphorus in iron. ^{31}P in Fe: 125 keV, 5×10^{14} P/cm^2. Reproduced from Ref. 43.

It is sufficient to know the ratios K_B/K_A, K_B/K_C, ... They may be determined with the aid of an analysis with the same operating parameters on binary system B–A, B–C, ... of known composition or any kind of system containing B–A, B–C, and other elements.

a. Application to the Study of a Zinc Coating. This example has been chosen to illustrate the different shapes of the zinc profiles as a function of the discharge conditions. The sample is a zinc-electroplated steel sheet. Figures 8-25, 8-26, and 8-27 illustrate the different types of profiles obtained for zinc as a function of the operating conditions (burning voltage and current intensity).

In the first example (Fig. 8-25) the discharge parameters are approximately equivalent in the coating and in the steel ($V_g \approx 500$ V), so that the variations of the emission yield are minimal. The zinc profile decreases almost regularly from the coating to the substrate with a little "shoulder" in the interface region. The operating parameters are:

- $I = 50$ mA (constant-current mode)
- Diameter of the anode = 7 mm
- V_g varies from 455 V in the zinc coating to 535 V in the steel

The light intensity of the zinc line (J^{Zn}) decreases rapidly when the signal of iron (J^{Fe}) appears. The shoulder in the interface region is observed

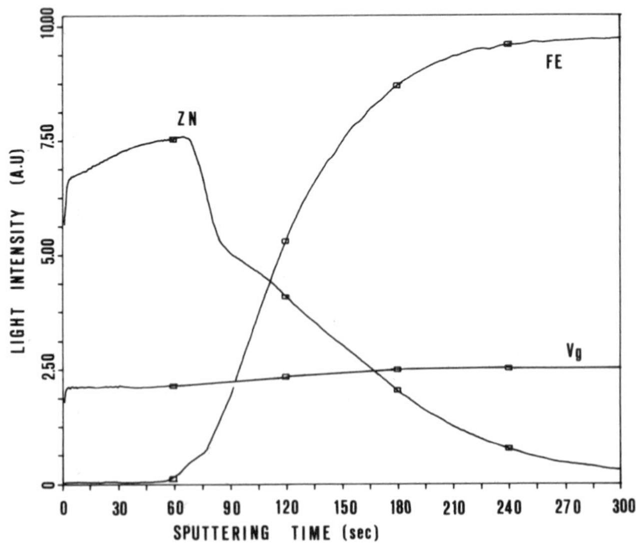

Figure 8-25. Analysis of a zinc coating in a constant-current mode. $I = 50$ mA; $V_g \approx 500$ V in the steel. Reproduced from Ref. 43.

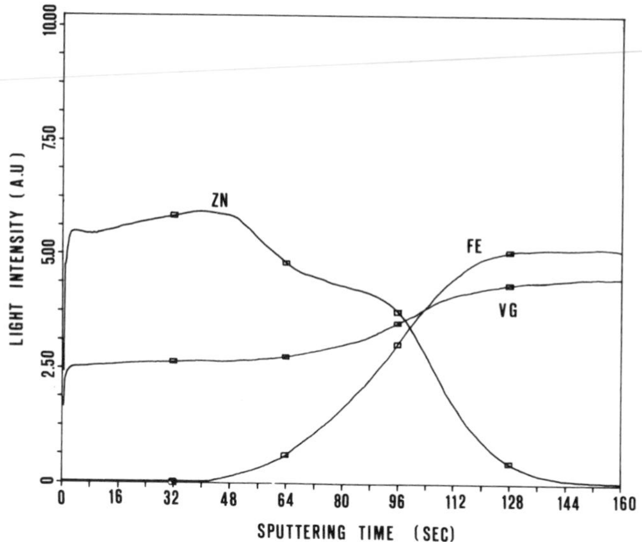

Figure 8-26. Analysis of a zinc coating in a constant-current mode. $I = 50$ mA; $V_g \approx 1000$ V in the steel. Reproduced from Ref. 43.

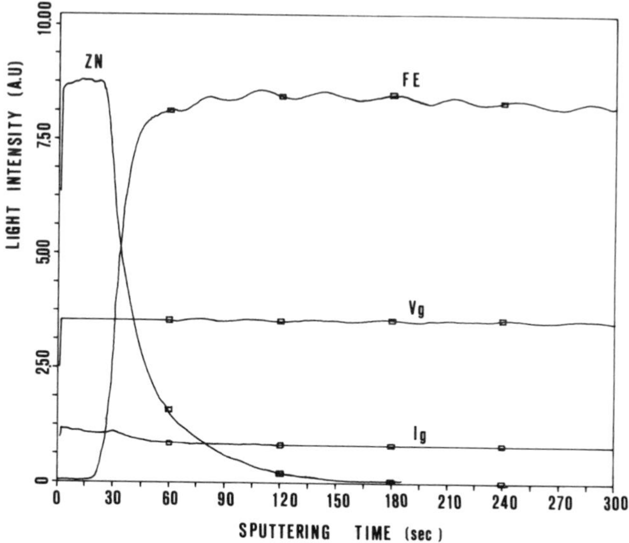

Figure 8-27. Analysis of a zinc coating in a constant-voltage mode. I_g, current intensity; $V_g \approx 750$ V. Reproduced from Ref. 43.

for a rather long time of erosion. The zinc content is calculated from Eq. (8-13) at each depth taking into account only zinc and iron in the composition of the coating. In Fig.8-28, it is observed that the concentration profile of zinc decreases regularly. Because of the roughness of the bottom of the crater, the concentration profile spreads largely.

If I and K^{Zn} are considered as constant, the variation of γ^T as a function of the depth may be described by:

$$J^{Zn}/C^{Zn} = a\gamma^T$$

(a is determined with a pure zinc sample) and corrected from the variation of the voltage with the aid of Eq. (8-11) ($V_0 \approx 300$ V). Thus, the variation of γ^T is calculated for a constant voltage (curve γ_v^T). It can be seen (Fig. 8-28) that the sputtering yield is then divided by 2 when the zinc concentration changes from 100% to 80%. If the zinc amount is then divided by 2, γ^T is divided by 3.8. Blaise has also found that the sputtering yield varies very quickly for two-phase systems.[68]

The important slope of the zinc profile is the result of the change in the sputtering yield which changes with the zinc content. In the interface region (at constant current) the voltage increases between the coating and the steel,

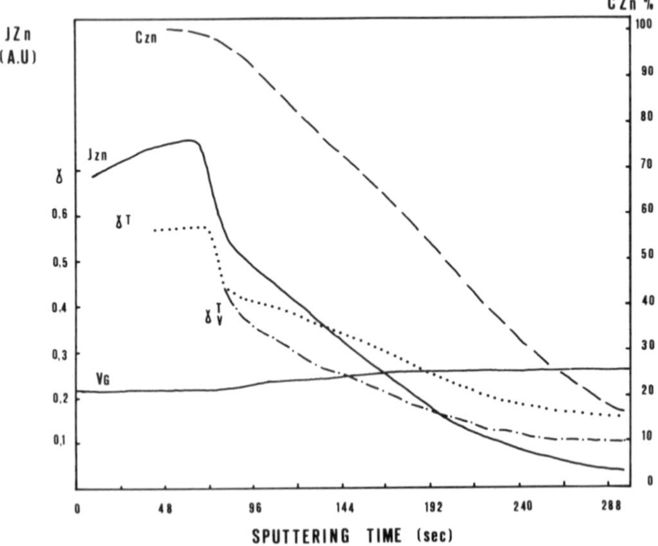

Figure 8-28. Variation of the sputtering yield as a function of zinc content. γ^T, sputtering yield; γ_v^T, sputtering yield calculated for a constant voltage; C_{Zn}, zinc content; J_{Zn}, intensity of the zinc line. Reproduced from Ref. 43.

thus the variation is less important and it can be observed that a small shoulder appears on the profile of zinc (J^{Zn}) as soon as the voltage increases.

If the constant-current mode is used (with a higher voltage), the variation of this last parameter increases between the coating and the substrate. Thus, γ^T decreases more slowly and the shoulder in the interface region is more important. The shape of the profile J^{Zn} looks like the curve recorded for a two-layer coating (Fig. 8-26). Finally, if a constant-voltage mode is selected ($V_g = 750V$ in Fig. 8-27), the current intensity decreases in the intermediate zone between zinc and iron. γ^T and J^{Zn} thus decrease more rapidly in this region and the shoulder observed in the first example disappears.

It is necessary to mention that, if the discharge parameters are very different for the two materials, the variations of the emission yield can also lead to another modification of the shape of the profile.

b. Study of the Enrichments in a Passive Layer. Quantitative analysis in the case of very thin films is much more difficult because of the changes in burning voltage at the beginning of the discharge. If the emission yield is considered as a constant during the analysis, the atomic fractions of the different elements in the matrix can be calculated from expression (8-13). Thus, it is possible to compare the evolution of the ratio of the element (A) to another one (B) between the substrate (S) and the superficial layer (L):

$$(X^A/X^B)_L/(X^A/X^B)_S \tag{8-14}$$

This determination can be calculated without calculating the ratio K_A/K_B. In fact, it is possible to write:

$$(C^A/C^B)_L/(C^A/C^B)_S = (J^A/J^B)_L/(J^A/J^B)_S \tag{8-15}$$

Such an approach has been applied to the determination of the enrichment of chromium and silicon in passive layers on stainless steels. The results obtained by this method and by the determination of the atomic fractions are very similar. The differences between the two modes of quantitation are less than 10%.[69]

As discussed earlier, the conversion of erosion times into depths is easy for a thick layer or a homogeneous material but more difficult in the interface region and for thin films where the burning voltage is not very well defined. To establish a general method of quantitative in-depth profiling (concentrations and depths), more fundamental work remains to be done. Specific approaches are now frequently proposed.[40,41,43,49,64–67,70]

8.9. Main Applications in Surface Analysis

Compared with the main techniques for surface analysis and in-depth profiling, GD-OES is fast and relatively easy to use. This is why these systems are used not only for research and development, but also for quality control. The primary applications are in the metallurgical, automotive, and nuclear industry fields. The list of applications is very long, so we give only a few of the technically most important works:

- Surface states after pickling, degreasing, and annealing(21,23,31,35,71)
- Oxide scales on hot-rolled steel(23,32,40)
- Passive layers, depleted chromium zones, and anodic oxides(42,71–74)
- Corrosion products(75)
- Transfer layers(76)
- Chromate films(77,78)
- Zinc and zinc alloy coatings on steel(31,40,61,65–67,79,80)
- PVD and CVD coatings(35,76)
- Ionic implantation profiles(35,40,76)
- Surface analysis of titanium(70)

It is not possible to describe all of them so we have chosen to summarize in the following section some examples in the field of coatings.

8.9.1. Analysis of Coatings and Surface Treatment by GD-OES

In this type of application, the thicknesses mostly concerned vary from 1 to 150 μm: glow discharge spectrometry is then practically the only method suitable for the study and control of the layers formed. GD-OES allows the continuous analysis of the first 150 μm within a reasonable time; there is no loss of information as may happen with the techniques that carry out analyses of a very superficial layer alternated with ion or chemical abrasions (discontinuous profiles). As for thicker layers (up to 300 μm), the sample may be recentered after elimination of the deposits on the edge of the crater which cause anode–cathode short circuits.

8.9.1.1. Hot Nitriding Treatment on Tool Steels

The quantitative evaluation of the concentrations of the modifying elements has been made by means of bulk test samples such as oxides, nitrides, and carbides. The sputtering rate and the voltage being practically identical in iron and in the iron nitrides analyzed, this estimation is fairly good in the absence of a compact oxide layer. In the case of nitriding in "tenifer," the values are certainly much less precise because of the high oxygen content in

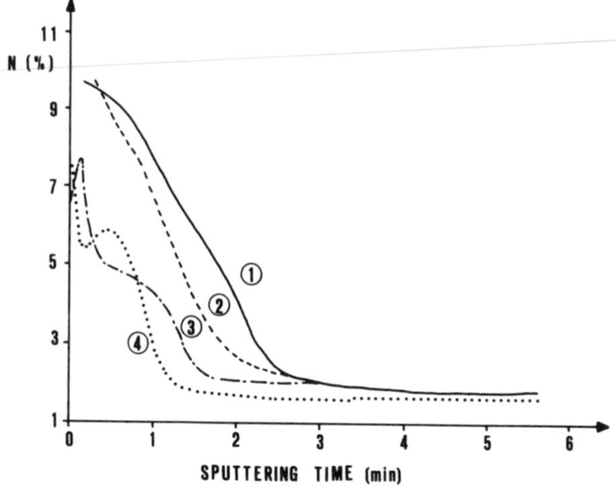

Figure 8-29. Nitrogen concentration profiles obtained with various tool steel surface treatments. (1) "Tenifer" (molten salts); (2) gaseous nitridation; (3) ionic nitridation (ε); (4) ionic nitridation (γ') Reproduced from Ref. 81.

the combination layer. In all cases the quantitative estimation is very good for the zone of diffusion (evolution of the profile of one element in a homogeneous material). The tracings of concentration profiles of nitrogen, oxygen, and carbon are shown in Figs.8-29, 8-30, and 8-31, respectively.

a. Nitrogen Concentration Profile. (Fig. 8-29). Taking into account the calibrations carried out, the maximum nitrogen content in the case of "tenifer" and of gaseous nitriding seems to exceed 10 %, which is quite reasonable

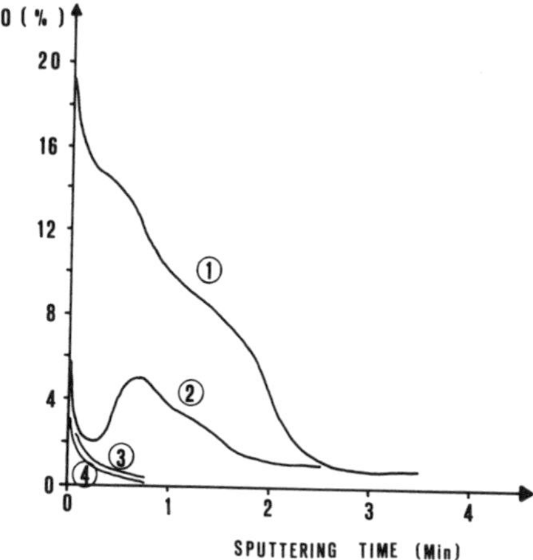

Figure 8-30. Oxygen concentration profiles obtained with various tool steel surface treatments. (1) "Tenifer" (molten salts); (2) gaseous nitridation; (3) ionic nitridation (ε); (4) ionic nitridation (γ') Reproduced from Ref. 81.

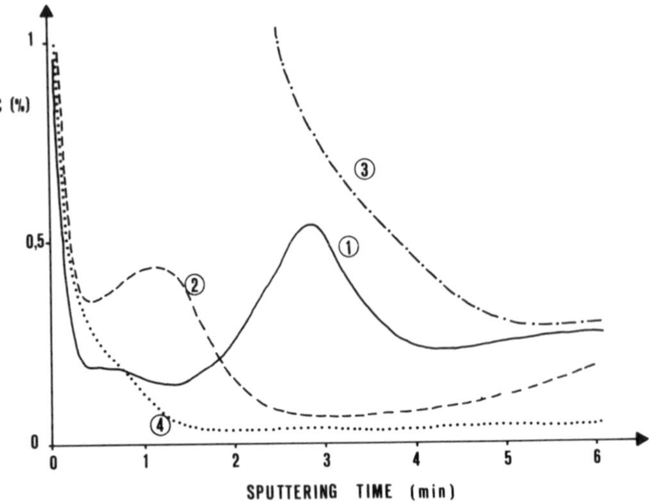

Figure 8-31. Carbon concentration profiles obtained with various tool steel surface treatments. (1) "Tenifer" (molten salts); (2) gaseous nitridation; (3) ionic nitridation (ε); (4) ionic nitridation (γ'). Reproduced from Ref. 81.

for obtaining ε nitrides. On the contrary, in the case of the sequenced treatment of ionic nitriding, the carbonitride ε contains about 8 % nitrogen, which is logical taking into account the presence of carbon. The level observed in that case in the underlayers, is about 5% nitrogen, corresponds in all probability to the γ' nitride formed during the first sequence of the treatment.

b. Oxygen Concentration Profile. (Fig. 8-30). If one makes exception of a very slight superficial contamination, the two ionic nitriding treatments are characterized by the absence of oxygen. In the case of gaseous nitriding and above all the "tenifer," we observe a very large superficial enrichment in oxygen which relates to an approximate thickness of 8 μm for the "tenifer" and of 5 to 6 μm for the gaseous nitriding. In parallel with this phenomenon, and over the same thicknesses starting from the surface, an increase is noted of the content in oxidizable alloy elements such as chromium and vanadium.

c. Carbon Concentration Profile. (Fig. 8-31). Ionic nitriding with a treatment sequenced to obtain an ε configuration presents a carbon profile that rises up to more than 2% on the surface, which corresponds in the iron–nitrogen–carbon diagram to the composition of the carbonitride ε with 8% nitrogen. Ionic nitriding with the production of γ' nitride, if superficial contamination is excepted, brings about a large decarburization over a depth estimated at more than 50 μm.

Nitriding in a "tenifer" salt bath could eventually bring about a superficial enrichment in carbon, but the profile of this element is very strongly disturbed on the surface by the presence of oxygen (see Fig.8-30). Gaseous nitriding under an ammonia atmosphere also brings about a decarburization over a depth estimated at 40 μm, and the carbon profile on the surface is disturbed by the presence of oxygen (see Fig.8-30).

These investigations have made it possible to explain differences of behavior due to friction and wear in a steel Z38 CDVO5 treated by the three methods.[81]

8.9.2. Ionic Implantation

The superficial layer concerned in this treatment is much thinner than in the previous example: most of the time it is less than 150 nm with the maximum concentration of the implanted element ranging from 30 to 60 nm. During the last few years, many studies have demonstrated the improvement of the resistance to wear by ionic implantation of several materials. Different ion species have been tried (e.g., C, B, N, P, V, Ti, Co, Sn, Ni, Cr) with greater or lesser success. Nevertheless, it is nitrogen that is most often used. Two remarkable facts have been observed:

1. A notable improvement of the resistance to wear
2. A persistence of this effect on depths of wear very much deeper than the implanted zones (up to 100 times more)

GD-OES is perfectly suitable for the study and the control of the process (Fig.8-32). It makes it possible to track down a large number of elements:

- Nitrogen (quantity implanted and depth involved)
- Oxygen and carbon: checking of the superficial pollution layer which may prevent the penetration of the nitrogen into the material.
- Pollutants (Ca, Si, Al, B) brought by the initial preparation of the surface (e.g., machining, polishing under water). Control of these elements, for which the sensitivity is high with GD-OES, makes it possible to complete the interpretation of the phenomena: information on the porosity of the superficial layer and showing up as "thermal" effects or as erosion by sputtering by comparing the composition of the surface before and after treatments.

On a more fundamental plane, one can see that the results show an excellent correlation with those obtained by the nuclear method using the

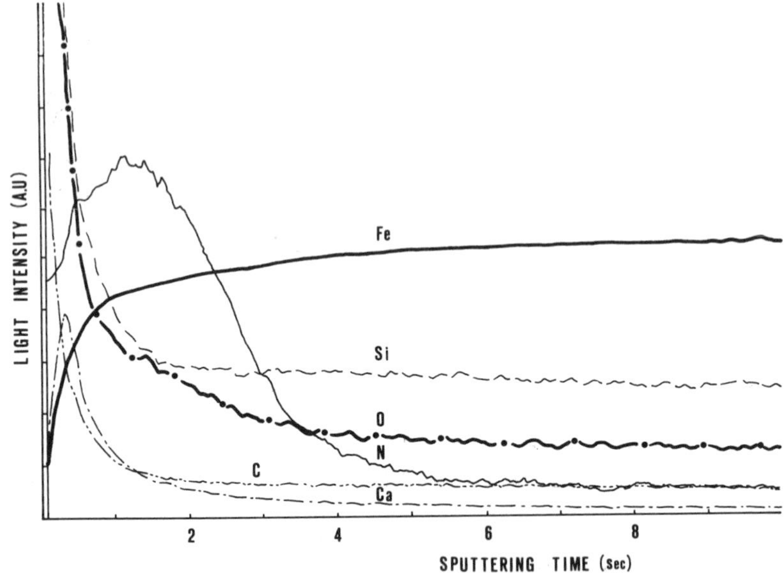

Figure 8-32. Ionic implantation of nitrogen in steel. ^{15}N:40 keV, 5×10^{17} N/cm^2.

resonant reaction $^{15}N(P, \alpha\gamma)^{12}C$ (for the analysis of N) which is nondestructive but much more difficult to use (Fig.8-33).

8.10 Limits and Advantages of GD-OES

8.10.1. Limits

The main negative properties of GD-OES are the following:

1. As for all the existing methods using ionic sputtering for in-depth profiling (e.g., AES, ESCA, SIMS), direct measurement of the eroded depth is not possible (see Section 8.8.1). The roughness of the surface increases during the course of sputtering and the shape of the bottom of the crater depends on the operating parameters and on the structure of the metals. This fact limits the depth resolution in the interface region especially if the studied layer is thick.
2. GD-OES has no lateral resolution. The eroded area commonly has a diameter of 0.4 or 0.8 cm.
3. The method does not give any direct information on chemical bonds. But the ratio between the intensity of different elements permits

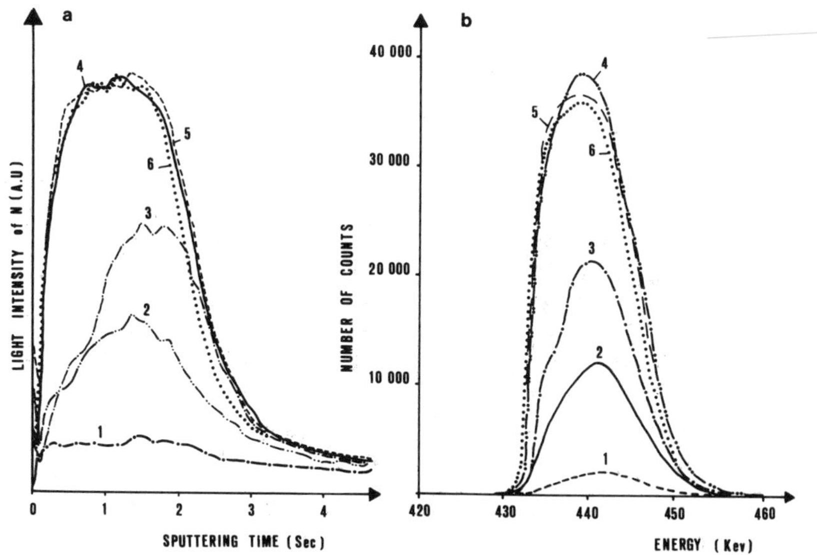

Figure 8-33. Comparison between GD-OES and nuclear method—ionic implantation of N. $^{15}N/cm^2$: $1 = 10^{16}$; $2 = 5 \times 10^{16}$; $3 = 10^{17}$; $4 = 2 \times 10^{17}$; $5 = 4 \times 10^{17}$; $6 = 6 \times 10^{17}$. (a) GD-OES; (b) nuclear method. Reproduced from Ref. 76.

drawing rapidly some conclusions on these bonds. Quantitative analysis may be difficult because the analysis is made at nonequilibrium conditions.

4. At the moment the sample must be electrically conductive or at least semiconductive. In a recent work, Duckworth and Marcus[(82)] introduced a new discharge tube using radio frequencies for the analysis of nonconductive materials.

8.10.2. Advantages

The main properties of GD-OES are as follows:

1. Almost all elements can be determined (O, N, and H in particular). But with argon as a carrier gas the sensitivity is poor for some of them, e.g., F, Cl, and I.
2. High vacuum is not required, so the technique is fast and rather easy to use.
3. Quantitative analysis in a homogeneous matrix is very easy because there are few matrix effects in the plasma and because calibration curves are commonly linear.

4. The initial and running costs are low relative to other surface methods.
5. The main advantage for industrial products is the ability to draw profiles from a few nanometers to several tens of micrometers in a short time.

8.11. Future Trends

Considerable progress has been made in recent years with the development of commercial spectrometers with a larger spectral range and suitable computers. GD-OES is now used either in research centers or directly in the plants for quality control (especially in Japan). In many aspects, improvements are possible. First it is necessary to point to the fact that the lamp in all commercial spectrometers is very similar to the Grimm tube. In the near future, new sources will appear, e.g., regulation of the flow of carrier gas by a mass-flowmeter (I and V_g constant) and the use of microwaves or rf lamps to increase sensitivity or to analyze nonconductive materials. The field of GD-OES will become larger in materials study. Because the specificity of the technique is the glow discharge tube, improvements in the lamp will be the most important changes. Other carrier gases may be used to improve the sensitivity. Progress is also needed in optical systems to increase the spectral range with the use of combined spectrometers (monochromator and polychromator). Two types of systems will be developed: a sophisticated spectrometer for research and a simplified apparatus for quality control of specific surfaces (e.g., coatings).

Compared with other techniques, the matrix effects in GD-OES are relatively simple, and quantitative in-depth profiling is thus likely to develop in the near future. Different mathematical models will emerge either for specific applications or for a more general approach. The number of GD-OES systems in the production plants will increase especially in the metallurgical and automotive industries.

References

1. W. Grimm, *Spectrochim. Acta* 23B (1968) 443.
2. M. Dogan, K. Laqua, and H. Massman, *Spectrochim. Acta* 26B (1971) 631; 27B (1972) 65.
3. H. de La Follie, 8th Spektrometertagung, Linz, 1970.
4. W. Grimm, 8th Spektrometertagung, Linz, 1970.
5. H. Jäger and R. P. Butler, 16th CSI, Heidelberg, 1971, Preprints, Vol. 2, p. 204, Adam Hilger, London, 1971.

6. M. E. Ropert, 16th CSI, Heidelberg, 1971, Preprints, Vol. 2, p. 214, Adam Hilger, London, 1971.
7. M. Dogan, Ph.D. thesis, University of Munster, 1970.
8. P. W. J. M. Boumans, *Anal. Chem.* 44 (1972) 1219.
9. R. Berneron, J. Cretin, and J. P. Moreau, Report IRSID RI 320, 1972.
10. M. E. Pillow, *Spectrochim. Acta* 36B (1981) 821.
11. A. I. Drobyshev and Y. I. Turkin, *Spectrochim. Acta* 36B (1981) 1153.
12. D. S. Gough and J. V. Sullivan, *Analyst* 103 (1978) 888.
13. N. P. Ferreira, J. A. Strauss, and H. G. Human, *Spectrochim. Acta* 38B (1983) 899.
14. R. A. Kruger, B. M. Bombelka, and K. Laqua, *Spectrochim. Acta* 35B (1980) 581.
15. M. Ancey, R. Berneron, and P. Parniere, *R.C. Met. Phys.* (1971) 582.
16. C. J. Belle and J. D. Johnson, *Appl. Spectrosc.* 27 (1973) 118.
17. J. E. Greene and J. M. Whelan, *J. Appl. Phys.* 44 (1973) 2509.
18. J. E. Greene, F. Sequeda Osorio, and B. G. Streetman, *Appl. Phys. Lett.* 25 (1974) 435.
19. J. E. Greene, F. Sequeda Osorio, and B. R. Natarajan, *J.Vac. Sci. Technol.* 12 (1975) 366.
20. H. Schneider and H. Shuman, Kernforschungszentrum Karlsruhe, Report KFK 2009, 1974.
21. R. Berneron and J. P. Moreau, 18th CSI, Grenoble, 1975, Preprints, Vol. 1, p. 263, GAMS, Paris, 1975.
22. R. Berneron and J. P. Moreau, CECA EUR 6279 FR, 1978.
23. R. Berneron, *Spectrochim. Acta* 33B (1978) 665.
24. M. Laporte, *Décharge électrique dans les gaz*, Armand Colin, Paris, 1948.
25. F. M. Penning, *Electrical Discharges in Gases*, p. 41, Philips Technical Library, Eindhoven, 1957.
26. J. A. C. Broekaert, *J. Anal. At. Spectrom.* 2 (1987) 537
27. R. Plesch, Siemens Analysentechnische Mitteilungen N2, p. 292.
28. A. Bengtson, L.Danielson, and A. Nordgren, 23rd CSI, Amsterdam, 1983.
29. D. A. Skoog and D. D. West, *Principles of Instrumental Analysis*, Saunders College, Philadelphia, 1980.
30. K. Laqua, W. D. Hagenah, and H. Waechter, *Z. Anal. Chem.* 225 (1967) 142.
31. A. Bengtson and L. Danielson, 6th International Conference on Thin Films, Stockholm, 1984.
32. K. Laqua, A. Quentmeier, and D. Demeny, Proceedings of 10th Jubilee National Conference on Atomic Spectroscopy, Veliko Turnovo, 1982.
33. J. Pontet and H. Hocquaux, Actes des journées de la Commission Décharge luminescente (now 1988), GAMS, Paris, 1988.
34. M. Bouchacourt, Actes des journées de la Commission Décharge Luminescente, GAMS, Paris, 1988.
35. H. Hocquaux, *Met. Corros. Ind.* 58 (1983) 693.
36. H. Hocquaux, J. P. Colin, *et al.*, Siemens Colloquium "Caractérisations de surface par SDL," Paris, 1984, unpublished work.
37. H. Hocquaux and P. Hunault, 1989 Pittsburgh Conference, Paper No. 686, unpublished work.
38. H. Jäger, *Anal. Chim. Acta* 58 (1972) 57.
39. H. Jäger, *Anal. Chim. Acta* 60 (1972) 303.
40. A. Bengtson and M. Lundholm, *J. Anal. At. Spectrom.* 3 (1988) 879.
41. K. H. Koch, M. Kretschmer, and D. Grunenberg, *Mikrochim. Acta* (*Wien*) II (1983) 225.
42. A. M. Brass, L. Nevot, M. Aucouturier, and R. Berneron, *Corros. Sci.* 24(1) (1984) 49.
43. J. C. Charbonnier, H. Hocquaux, J. P. Moreau, and D. Nizery, Mesucora 88, GAMS, Paris, 1988.
44. H. Jäger and F. Blum, *Spectrochim. Acta* 29B (1974) 73.
45. J. E. Greene, *J. Vac. Sci. Technol.* 15 (1978) 1718.

46. R. Berneron, B. Chetreff, J. P. Colin, P. de Gelis, H. Hocquaux, F. Hoffert, and S. Mathieu, 24th CSI, Garmisch Partenkirchen, 1985, unpublished work.
47. J. P. Moreau, CNAM thesis, Paris, 1975.
48. N. Laegreid and G. V. Wehner, *J. Appl. Phys.* 32 (1961) 365.
49. J. Pons-Corbeau, *Surf. Interface Anal.* 7 (1985) 169.
50. H. Hocquaux, L. Ohannessian, and J. Tousset, 24th CSI, Garmisch Partenkirchen, 1985, unpublished work.
51. L. Ohannessian, Ph.D. thesis, Lyon, 1986.
52. R. A. Bargolia, E. V. Alonso, J. Ferron, and A. Oliva Florio, *Surf. Sci.* 90 (1979) 240.
53. P. Sigmund, *Phys. Rev.* 184 (1969) 183.
54. W. P. Allis, G. Fournier, and D. Pigache, *S. Phys. (Paris)* 38 (1977) 915.
55. W. D. Davis and T. A. Vanderslice, *Phys. Rev.* 131 (1963) 219.
56. Y. Chouan and D. Collobert, *S. Phys. (Paris)* 43 (1977) 915.
57. I. Abril, A. Gras-Marti, and J. A. Valles-Abarca, *Phys. Rev.* 28 (1983) 3677.
58. N. Matsunami and Y. Yamamura, *At. Data Nucl. Data Tables* 31 (1984) 2.
59. Y. Yamamura, N. Matsunami, and N. Itoh, *Radiat. Eff.* 71 (1983) 65.
60. J. Lindhard, N. S. Scharff, and H. E. Schiott, *Mat. Fys. Medd. Dan. Vid. Sellsk.* 33 (1963) 14.
61. K. Takimoto, K. Nishizaka, K. Suzuki, and T. Ohtsubo, *Nippon Steel Tech. Rep.* 33 (1987) 28.
62. W. Kreschmer, K. H. Koch, and D. Grunenberg, *Stahl Eisen* 104 (1984) 193.
63. P. Netter, Report IRSID RI 1004, 1985.
64. K. H. Koch, D. Sommer, and D. Grunenberg, *Mikrochim. Acta Suppl.* II (1985) 137.
65. Y. Ohashi and Y. Konushi, *Tetsu To Hagane* 69(12) (1983) 1052.
66. K. Suzuki, K. Nishizaka, and T. Ohtsubo, *Tetsu To Hagane* 70(4) (1984) 295.
67. J. Pons-Corbeau, J. P. Cazet, J. P. Moreau, R. Berneron, and J. C. Charbonnier, *Surf. Interface Anal.* 9 (1986) 21.
68. G. Blaise, Colloquium on Sputtering, Les Arcs, France, 1985.
69. J. Pons-Corbeau, Report IRSID SCA 85/205, unpublished work.
70. K. Suzuki, T. Ohtsubo, and T. Watanabe, *Nippon Steel Tech. Rep.* 33 (1987) 36.
71. R. Berneron and H. Hocquaux, *Methodes Usuelles de Caractérisation des Surfaces*, p. 91, Eyrolles, Paris, 1988.
72. R. Berneron, J. C. Charbonnier, R. Namdar-Irani, and J. Manenc, *Corros. Sci.* 20 (1980) 899.
73. B. Alexandre, R. Berneron, J. C. Charbonnier, and J. Manenc, *Surf. Technol.* 9 (1979) 317.
74. F. Gaillard, M. Romand, H. Hocquaux, and J. S. Salomon, *Surf. Interface Anal.* 10 (1987) 163.
75. M. Dernat, H. Hocquaux, and L. Ohannessian, *L'eau, l'industrie, les nuisances* 86 (1984) 44.
76. H. Hocquaux and R. Leveque, *Galvano-Organo:Traitements de Surface* 750 (1984) 837.
77. K. Suzuki, S. Yamazaki, and T. Ohtsubo, *Tetsu To Hagane* 3 (1987) 565.
78. A. Quentmeier, H. Bubert, R. P. H. Garten, H. J. Heinen, H. Puderbach, and S. Storp, *Mikrochim. Acta Suppl.* II (1985) 89.
79. M. Iwai, M. Terada, H. Sakai, and S. Nomura, *Tetsu To Hagane* 11 (1986) 1759.
80. Y. Matsumoto, N. Fujino, and S. Tsuchiya, *Trans. ISIJ Jpn.* 1 (1987) 891.
81. R. Leveque, H. Michel, and M. Gantois, International Colloquium on Tool Steels, Saint-Etienne, France, 1977, unpublished work.
82. D. C. Duckworth and R. K. Marcus, *Anal. Chem.* 61 (1989) 1879.

9

Discharges within Graphite Furnace Atomizers

James M. Harnly, David L. Styris, and Philip G. Rigby

9.1. Introduction

In recent years, the concept of using electrical discharges within graphite furnace atomizers as atomization/excitation sources for atomic emission spectrometry has been established and evaluated.[1-31] The excitation in these sources is principally dependent on the discharge and is not a direct result of the electrothermal heating of the furnace. Consequently, this unique concept has been designated furnace atomization nonthermal excitation spectrometry (FANES). This can lead to some confusion since "nonthermal" also implies that the line radiance and profile associated with the discharge do not conform to Maxwell–Boltzmann statistics.[32] While it is certain that the discharge is necessary for excitation, there is considerable uncertainty whether all the discharges described in this chapter can be generally categorized as nonthermal. We shall nevertheless continue to use FANES to describe the concept, bearing in mind the above ambiguity.

To date, direct current (dc) and radio frequency (rf) discharges have been examined in this source. In both cases the analyte is thermally vaporized

James M. Harnly • United States Department of Agriculture, Agricultural Research Service, Beltsville Human Nutrition Research Center, Nutrient Composition Laboratory, Beltsville, Maryland 20705. **David L. Styris** • Battelle, Pacific Northwest Laboratory, Richland, Washington 99352. **Philip G. Rigby** • School of Biological and Chemical Sciences, University of Greenwich, Woolwich, United Kingdom.

Glow Discharge Spectroscopies, edited by R. Kenneth Marcus. Plenum Press, New York, 1993.

into the discharge region, so the volatilization/atomization processes associated with the graphite furnace are decoupled from the excitation process in the discharge.

The graphite furnace atomizer is well suited to this dual function concept. It is a very efficient atomization source for trace metal determinations. For most commercial furnaces, aqueous samples (1 to 100 μl) are deposited, dried, and then vaporized into a semirestricted furnace volume of less than 1 cm^3 to produce a highly concentrated atom cloud. The transport efficiency of analyte species into the analytical zone is close to 100% for the graphite furnace as opposed to 1–5% for other conventional atomizers (e.g., air–acetylene flame or inductively coupled plasma).[5] Residence times of analyte atoms in the analytical zone in the furnace are approximately two orders of magnitude greater than those found in other atomizers.[5] Consequently, gas-phase analyte densities are orders of magnitude more concentrated in the furnace. The furnace also possesses an inherent capability of providing matrix modifications *in situ*, since chemical and thermal pretreatment of the sample can be incorporated into the heating program.

Historically, the graphite furnace (GF) has been used primarily as an atom reservoir for atomic absorption spectrometry (GF-AAS). The dimensions of the furnace provide the longer optical path suitable for absorption measurements. The furnace can reach 3300 K, although routine operation is usually restricted to maxima of about 3000 K to enhance furnace life. At these temperatures, atomization is achieved for most molecular species, and thermal ionization is negligible.[33] GF-AAS exhibits excellent detection limits (from tens to tenths of picograms)[34] but the technique lacks the multielement capabilities inherent in other approaches such as inductively coupled plasma–atomic emission and mass spectrometries (ICP-AES and ICP-MS).

The graphite furnace has also been employed as an emission source (GF-AES)[35,36] in an attempt to use the inherent multielement advantage of the simplified optics of the emission mode and the available technology of multichannel spectrometers. Effective excitation, however, can only be achieved for transitions above 300 nm, i.e., those elements having excitation energies less than the energies associated with the high-energy tail ($\sim$2 eV) of the Maxwell–Boltzmann energy distribution of the thermionic electrons or the Planck distribution for thermal radiation at temperatures near 3000 K. Hence, for elements having ground state transitions at wavelengths less than 300 nm, the GF-AES detection limits are much poorer than those for GF-AAS.

The FANES concept extends the excitation energy range of the graphite furnace to elements with transitions below 300 nm and thus gives the furnace true multielement attributes. This concept was first proposed by Falk[1,2] who used a low-pressure, glow discharge with the furnace as a hollow

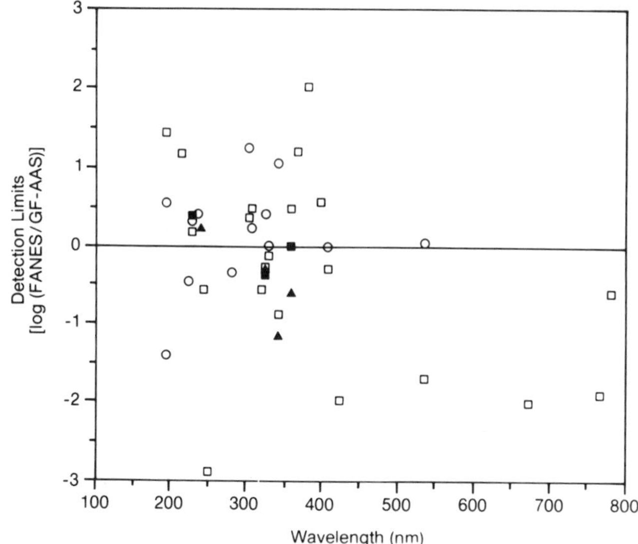

Figure 9-1. FANES detection limits taken from Table 9-1, normalized with respect to conventional GF-AAS detection limits, as a function of wavelength. (□) Falk *et al.*, HC-FANES[15]; (▲) Naumann *et al.*, HC-FANES[17]; (○) Sturgeon *et al.*, rf-FANES[27]; (■) Harnly *et al.*, HA-FANES.[25]

cathode (HC) and a point or a ring anode located external to the furnace; the discharge, in Ar or He, filled the furnace volume. This source employs operating parameters (1–25 Torr, 10–100 mA) and exhibits performance characteristics similar to classical HC lamps. As shown in Fig. 9-1, the reported detection limits are, on the average, a factor of three lower than those for GF-AAS.[5] We refer to this source as HC-FANES in future discussions in this chapter in order to differentiate from later source developments.

Ballou *et al.*,[18] Harnly *et al.*[25] and Riby *et al.*.[29] used a somewhat different design concept and reversed the relative polarities associated with HC-FANES. The furnace comprised the anode and the cathode was a central graphite rod running the length of the furnace. The discharge in this design is a bright corona constricted around the cathode. Operating pressures (5–200 Torr in Ar and 10–600 Torr in He) are considerably higher than those normally expected for dc glow discharges. The ability to maintain a stable plasma at these pressures may result from the smaller cathode areas associated with this design and, therefore, higher current densities. This type of hollow anode (HA) system is designated as HA-FANES in the discussions that follow. The detection limits compare well with HC-FANES and conventional GF-AAS (Fig. 9-1).

The most recent developments in this field are the rf glow discharge sources described by Blades and co-workers[20,24] and by Sturgeon and co-workers.[21,27,28,30,31] These sources have the same geometry as those used in HA-FANES, but are operated in He at atmospheric pressure using rf generators as power sources. The detection limits compare well with HA- and HC-FANES and conventional GF-AAS. The spectrometric technique that uses these sources has been designated furnace atomization plasma excitation spectrometry (FAPES). For consistency in nomenclature in this chapter, these systems will be referred to as rf-FANES.

The excitation sources that have just been described, HC-, HA-, and rf-FANES, employ electrical discharges operated under conditions markedly different from each other. In addition, all three sources are markedly different from classical glow discharge conditions. First, all three FANES discharges have been optimized for the excitation process only, not for sputtering, etching, or plating. Sample volatilization is accomplished thermally and the rate of volatilization can be as much as three orders of magnitude greater than sputtering rates. The sputtering process remains a significant process, but as an interfering reaction leading to premature appearance and loss of analyte. Second, all three FANES discharges involve "hot" electrodes having temperatures ranging from 800 to 2800 K, depending on the FANES source and the element being determined. Above 1800 K the generation of thermionic electrons becomes important and can significantly reduce the discharge potential. Discharges under these conditions have been referred to as low-voltage arcs.[44–47] Finally, HA- and rf-FANES sources are operated at considerably higher pressures than those used for most classical glow discharge systems. These higher pressures result in longer residence times for the analyte.

It is the purpose of this chapter to instill an awareness of and to provide some physical perspective on this new and growing development in glow discharge spectrometry and to share with the reader the present understanding associated with the various FANES sources. To achieve this the experimental results obtained for the three systems described above will be reviewed. Comparisons will be made of their analytical attributes, their geometries, and the nature of their discharges, and finally physical interpretations will be presented and discussed.

9.2. System Designs

An overview of the physical designs (Table 9-1) and operating parameters (Table 9-2) of HA-, HC-, and rf-FANES is presented in this section. In Table 9-1 rf-FANES has been further partitioned according to the designs of Blades *et al.*[20,24] and of Sturgeon *et al.*[21,27,28,30,31]. In

Table 9-1. Furnace Designs Used for FANES

	HC-FANES	HA-FANES	rf-FANES	
			Blades	Sturgeon
Furnace				
Manufacturer	Carl Zeiss	Laboratory built	Thermo Jarrell–Ash	Perkin–Elmer
Type of furnace	Conventional	Integrated contact cuvette (Ringsdorff-werke)	Conventional	Conventional
Dimensions[a]	28 × 6.5 mm	19 × 6.5 mm	28 × 6.5 mm	28 × 6.5 mm
Electrical contacts	Longitudinal	Lateral	Longitudinal	Longitudinal
Support gas control	Cont. pump	Static	Cont. flow	Cont. flow
Furnace enclosure	Vacuum seal	Vacuum seal	Vacuum seal	Positive pressure
Plasma				
Power source	dc	dc	rf	rf
Cathode	Furnace	Axial rod	Axial rod	Axial rod
Anode	External ring	Furnace	Furnace	Furnace
Pressure	1–140 Torr (Ar, He)	70–200 Torr (Ar) 100–600 Torr (He)	760 Torr (He)	760 Torr (He)

[a]Length × diameter.

Table 9-2. Operating Conditions

	HC				HA	rf	
	F[a]	N	D	L	H	B	S
Temperatures (°C)	—	1850–2350	—	—	1845	2000	1100–2700
Heating rate (°C s^{-1})	2000	2000	2000	2000	1100	1800	2000
Back. corr.	no	no	no	yes	yes	no	no
Platform	no	no	yes	yes	no	no	yes
Matrix modifiers	no	no	yes	no	no	no	yes
Measurements	ht.	ht.	ht.	ht. and area	ht. and area	ht. and area	ht. and area
Spectrometer	MR & HR[b]	HR	MR	MR & HR	HR	MR	MR
Support gas	Ar & He	Ar	Ar	Ar & He	Ar & He	He	He
Pressure (Torr)	1–4	40–140	10–20	1–30	190(Ar) 600(He)	760	760
Current (mA)	20–30	55–90	30–80	20–90	70–200	N.A.	N.A.
Potential (V)	400–600	N.A.	N.A.	300–600	200–300	N.A.	N.A.
Max. power (W)	18	N.A.	N.A.	54	60	20	75

[a]B, Blades *et al.*[(20,24)]; D, Dittrich *et al.*[(7,11,13,16,22,23,26)]; F, Falk *et al.*[(1–6,9,10,12,15)]; H, Harnly *et al.*[(18,25,29)]; L, Littlejohn *et al.*[(8,14,19)]; N, Naumann *et al.*[(17)]; S, Sturgeon *et al.*[(21,27,28,30,31)].

[b]MR, medium-resolution monochromator; HR, high-resolution monochromator.

Table 9-2, the operating parameters are presented for the seven major investigators in the field.

Design similarities can be seen for all four FANES devices (Table 9-1) (Fig. 9-2 to 9-5). The furnaces are composed of pyrolytically coated graphite, are resistively heated, and, except for the integrated contact cuvette (ICC) used with HA-FANES, are of conventional design, i.e., they are identical to those used routinely for GF-AAS, cylinders 6 mm in internal diameter and 28 mm long. The ICC is also a cylinder 6 mm in diameter, but it is only 19 mm long and has wings attached externally to each side (Fig. 9-5). The axial electrodes of HA- and rf-FANES are also made of pyrolytically coated graphite.

Design differences can be seen in the furnace contacts (conventional versus ICC) and the gas containment (four different approaches). The shapes of the furnaces were described above. Conventional furnaces are clamped at both ends of the cylinder and the electrical current flows longitudinally

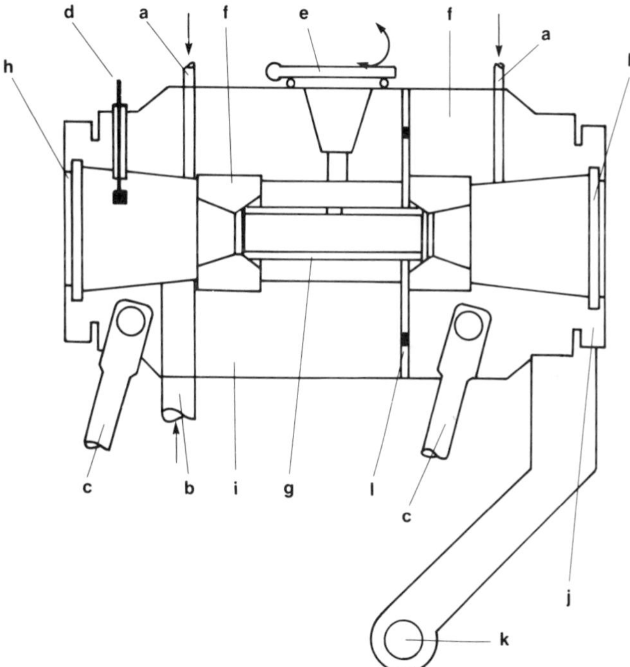

Figure 9-2. Schematic diagram of the HC-FANES source of Falk *et al.*[5] (a) Carrier gas port; (b) pump port; (c) electrical connector to heating transformer; (d) anode; (e) removable lid for sample injection; (f) graphite electrode; (g) graphite furnace and hollow cathode; (h) window; (i) water-cooled vacuum vessel; (j) water-cooled part of the vacuum vessel and rotation arm for changing the graphite tube; (k) pivot; (l) gasket.

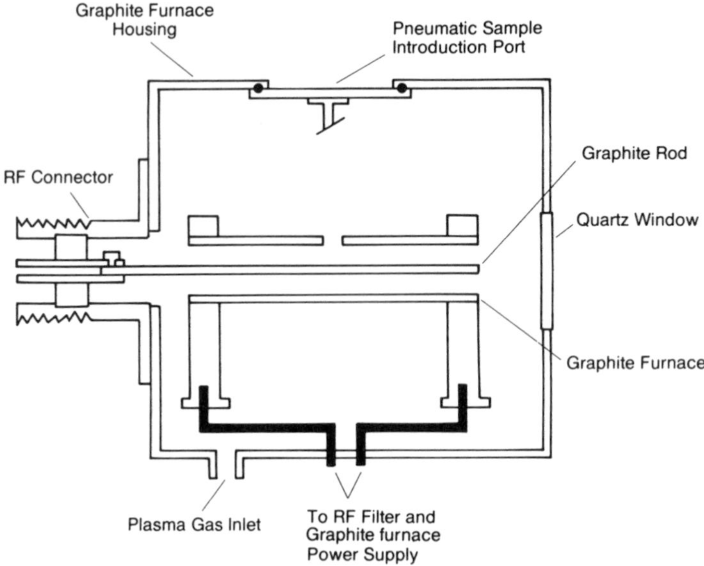

Figure 9-3. Schematic diagram of the rf-FANES source used by Liang and Blades.[20]

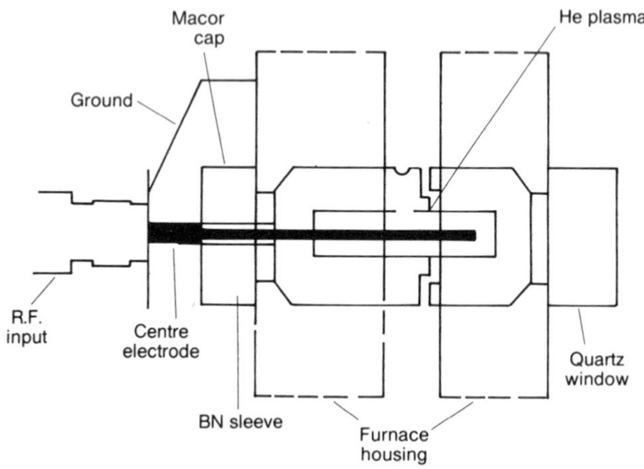

Figure 9-4. Schematic diagram of rf-FANES source used by Sturgeon *et al.*[21]

(Figs. 9-2 to 9-4). The ICC is clamped on the wings (Fig. 9-5) and the electrical current flows transversely. This transverse flow produces a uniform temperature over the length of the furnace.[37] For conventional furnaces there is a significant temperature decrease between the center and each end.[38,39] This temperature gradient can produce nonuniform atomization

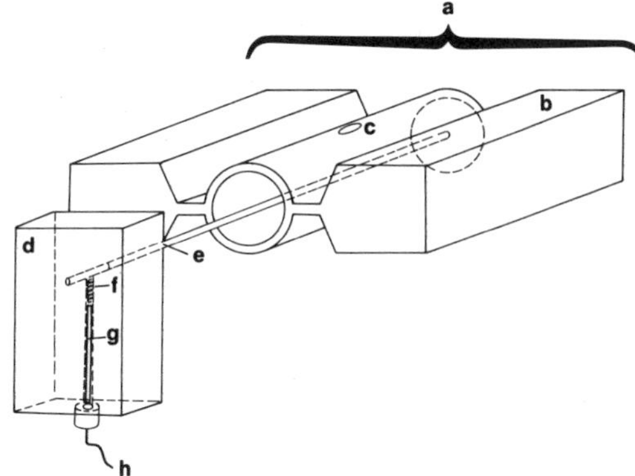

Figure 9-5. Schematic diagram of HA-FANES. (a) Integrated contact cuvette; (b) contacts for furnace power supply; (c) sample loading port; (d) Macor support block for cathode; (e) cathode; (f) spring; (g) contact pin; (h) insulated wire from the dc discharge high-voltage power supply.[18]

and excitation effects and can be a source of interferences, as will be discussed in a later section.

The gas containment designs of the four FANES devices are considerably different. The HC-FANES devices employed by Falk,[1–6,9,10,12,15] Littlejohn,[8,14,19] Dittrich,[7,11,13,16,22,23,26] and Naumann[17] are of the same design. These devices resemble the conventional Perkin–Elmer (Ridgefield, Conn.) design with vacuum seals added in the appropriate locations. HC-FANES has been operated between 1 and 140 Torr, with the majority of the work being done between 1 and 30 Torr. The vacuum seals are not of a high quality, and continuous pumping is generally required to maintain a constant reduced pressure. For some systems, a continuous gas flow is necessary to stabilize the discharge. The leak rates and the gas flowrates necessary to stabilize the plasmas appear to vary for each system.

The HA-FANES source has been constructed in a six-way vacuum cross.[18,25,29] A constant pressure of a few Torr can be held without pumping. The HA-FANES is run in the static mode, i.e., flow of the support gas is stopped for the atomization step. The optimum conditions for most elements have been found to be around 200 Torr of Ar and 400–600 Torr of He.[29]

Both rf-FANES devices are run at atmospheric pressure with He as the support gas. Sturgeon *et al.*[21,27,28,30,31] use a conventional Perkin–Elmer furnace (Model HGA-500). External air is prevented from reaching the interior of the furnace by the positive pressure of the support gas through an internal flow through the furnace, which can be halted during the atomization cycle, and a continuous external flow around the furnace. The low mass of He makes it necessary to maintain high support gas flows both internally and externally throughout the atomization cycle. Infusion of atmospheric air has been a problem as will be discussed later.

The device employed by Blades *et al.*[20,24] is a conventional Thermo Jarrell–Ash (Franklin, Mass.) furnace head (Model IL655) that has the capability of operating at greater than atmospheric pressure. The furnace is designed to prevent leakage outward, not inward. Considerably lower flows are used than reported by Sturgeon *et al.* It is not clear whether the He flow is needed to prevent infusion of the atmosphere or to enhance the discharge.

The operating parameters of each of the FANES systems will be discussed in detail in appropriate sections of the chapter.

9.3. Fundamental Considerations

While the basic properties of glow discharges operated with "cold" electrodes have been investigated extensively, there has been considerably less effort directed toward understanding such discharges in the presence of "hot" electrodes. We shall not endeavor to review such work, but will instead address some of the basic characteristics of FANES sources themselves. The reader should be aware of Falk's excellent review of HC-FANES[15] and is directed to recent reviews by Slevin and Harrison,[40] Caroli,[41] Pillow,[42] and Broekaert[43] on "cold" HC discharges and the existing literature on "hot" cathode discharges.[44–47]

9.3.1. Sputtering

Although sputtering is no longer the primary means of generating the analyte gas phase, it is still one of the most important processes within the discharge. Sputtering phenomena are divided into two classes: "chemical" and "physical." Chemical sputtering involves incident ions that react chemically with the cathode to produce a volatile species. This can certainly be important when sufficient reactive impurities are contained in the support gas, but it is more controllable than the inherent physical sputtering associated with FANES. We will therefore discuss in this section only physical sputtering processes as related to FANES systems. More general details of sputtering in glow discharges have been discussed most adequately in Chapter 2.

Physical sputtering is a multiple sequence collision process involving a cascade of translating subsurface target atoms ("knock-ons"). This cascade can itself interact with the surface, in which case sufficient momenta can be transferred from knock-ons to surface species for sputtering to occur. Wehner[48] has shown that for C the sputter yields (atoms sputtered per incident bombarding particle) are about 0.06 and 0.1 for 400 eV He and Kr, respectively. If sputtering of analyte-containing species is the result of a transfer of the collisions in the graphite to the analyte, we should generally expect even smaller yields because of the mass mismatch between C atoms and the analyte atom or molecule species. This mismatch prevents maximum energy transfer between the cascading C atoms (mass m_1) and the analyte species (mass m_2) on the surface. This can be seen for elastic collisions where the maximum fraction of the energy transferred is given by $4m_1m_2/(m_1 + m_2)^2$. However, a cascade collision may be initiated directly in the analyte crystallites on the surface of the furnace. Sputtering will then occur directly from incident particle bombardment of the crystallite, i.e., collision cascades are induced directly into the crystallite to induce sputtering. The smaller the crystallite, the greater the surface-to-volume ratio is, so the probability of the cascading knock-on atoms interacting with the surface to produce sputtering should increase. A greater sputtering yield is, therefore, expected for the smaller crystallites. However, some of the crystallite species must couple with the graphite substrates; the smaller the crystallite, the greater the fraction of crystallite atoms involved in the coupling. The general result is that cascades created directly within the crystallite will produce cascades in the substrate. The reverse is also true—cascades in the substrate will produce cascades in the crystallite. The extent of this cascade interchange depends on how well the mass impedance between crystallite substrate atoms match one another.

Falk[(12)] estimated the sputtering rate in the HC-FANES system, assuming a sputter yield of 0.2, to be approximately three orders of magnitude smaller than the rate of thermal vaporization. But as discussed above, the sputter yields in small crystallites may be enhanced because of crystallite size. Other effects may offset any increased yields, however. Kaminsky[(49)] makes the point that the mean free paths associated with the glow discharge give rise to multiple collisions in the gas that can result in formation of molecular and atomic ions of sputtered target species. If these ions diffuse to the cathode fall region, they will return to the cathode and be redeposited or possibly induce further (self) sputtering. Von Hippel[(50)] had indicated that 90% of the sputtered species diffuse back to the cathode in a glow discharge. The sputter yields in this case would indeed be smaller than the thermal vaporization yields in the HC-FANES system.

Analyte sputtering does not occur in the dc HA-FANES system unless, of course, samples are deposited on the central probe. Sputter rates from the probe, by the above arguments, will still be insignificant compared with thermal vaporization rates unless self-sputtering becomes significant. Certainly, the larger radial electric fields associated with the HA geometry will provide the greater ion energy and would thus enhance the self-sputtering yields. This will be of little consequence unless these yields are greater than unity, in which case the sputtering rate could conceivably increase monotonically.

Sputtering of analyte in the rf-FANES geometry should be somewhat more significant. The secondary electrons can gain significant energy by resonating in the rf field. Thus, ionization should increase and more ions should become available to induce sputtering. But just as for the dc case, redeposition and self-sputtering must be considered. These processes could be more severe because of the increased field strength and increased probability for ionization of the sputtered species that enter the glow region. Consider, however, the mobilities of the ions. These are so small at 1 atm that the ion can be considered immune to the presence of the rf field. Hence, neither self-sputtering nor redeposition is involved, and analyte sputtering in an rf discharge will be insignificant compared with the thermal vaporization.

9.3.2. Sheath Dimensions

The principal regions of interest associated with the FANES-type plasmas are the cathode fall region (sheath) adjacent to the cathode and the negative glow that neighbors this region. The positive space charge that accumulates in the sheath contributes to a steep voltage drop in this region. Most of the discharge voltage drop occurs here with the drop being related to the space charge density. As observed from the more conventional dc discharges, the radial extent of both regions (glow and sheath) diminishes with increasing pressure. For the negative glow this is a result of decreasing electron mean free path; the range of excitation decreases with increasing pressure. To understand the relationship between the sheath and pressure requires a more detailed consideration of ion transfer into the sheath. Bohm[(51)] has shown that for ions to enter the sheath region they must have a velocity (v_0) greater than $(kT_e/m_i)^{1/2}$, T_e being the electron temperature, m_i the mass of the ion, and k the gas constant. This so-called Bohm sheath criterion simply means that the electron density gradient in the sheath is less than the ion density gradient produced by the acceleration of the electrons and ions in the sheath,[(52)] i.e., $\nabla^2 V < 0$, where V is the voltage at any point within the discharge. Prior to entering the cathode fall region, the ions must accelerate through a potential drop ΔV in order to acquire a velocity greater than v_0. Equating kinetic energy gained to the potential energy gives

$$1/2 m_i v_0^2 = e\Delta V = eE\Delta r \tag{9-1}$$

where, for discussion purposes, the electric field E is assumed uniform over the radial distance Δr. Note that ion temperature and collisions in the sheath are neglected. Using the Bohm sheath criterion, this distance becomes

$$\Delta r = kT_e/eE \tag{9-2}$$

This is the minimum distance an ion must travel in a radial field, E, in order to contribute to sheath formation. This distance, which is a quasi-neutral transition region between the plasma and the positive space charge region, accounts for the largest part of the sheath extension.[53] The much smaller Debye shielding length associated with the sheath will not be considered here.

It has been shown by von Engel[54] that T_e is inversely proportional to the pressure p, so

$$\Delta r \propto 1/Ep \tag{9-3}$$

Therefore, as a result of differences in the electron and ion density gradients that establish the Bohm sheath criterion and because of the electron temperature pressure relationship to p, the cathode sheath thickness diminishes with increasing pressure.

The successful operation of the HA-FANES as an emission source at pressures greater than 10^4 Pa is probably due to the relatively small influence of thermionic emission from the relatively small central cathode. This thermally isolated cathode lags the tube temperature and will, from the Richardson–Dushman equation, exhibit a thermionic electron emission current density (J_{ha}) that is well below that of the tube wall. The temperature lag is dependent on the ohmic heating produced by the discharge current. The thermionic current density (J_{hc}) for the HC-FANES is, in contrast, that of the wall itself. The relative differences in thermionic current (i) for the two cases will be even greater because of the smaller cathode surface area associated with the cathode used in the HA configuration. The cathode in the HC-FANES system is the tube itself and, therefore, attains a higher temperature earlier in the temperature ramp than does the cathode in the HA-FANES system. The resulting larger thermionic emission induces an electronic conductivity for the hot HC geometry that exceeds that of the HA case. Therefore, in the constant-current mode of operation, the discharge voltage for the "hot" HC-FANES plasma will fall below that of the "hot" HA-FANES plasma, assuming the same furnace temperature programs for both cases. The plasma cannot be sustained if this discharge voltage falls below that which can accelerate secondary electrons to ionizing energies faster than elastic collisions associated with higher operating pressures can soak up this energy. That is to say, the electron mean free path must remain greater than the Debye length. The HA-FANES plasma can be quenched more readily by the thermionic emission if the pressure is sufficiently high. The constricted glow region of the higher-pressure discharges requires that the monitoring of photons from this source be done very near the cathode surface. The HA geometry is more suitable for achieving this because the temperature of the central cathode lags behind that of the furnace wall. The strong dependence of radiative emissive power on temperature suggests that, for identical temperatures and ramp rates for the two cases, the spectral interferences from cathode optical emission should be smaller for the HA case. High-pressure operation of the HC-FANES is therefore tube temperature limited because of radiative emission as well as thermionic emission. The HA-FANES is also limited by tube temperature, of course, but the temperature limit is greater in this case because the cathode temperature lags behind that of the tube.

9.3.3. Discharge Characteristics

The principal research efforts to elucidate discharge characteristics associated with various diode configurations have been devoted to cold cathodes where cathode emission is a secondary electron emission phenomenon. Discharges involving thermionic cathodes have been investigated to a much lesser degree. The reader is referred to Hernquist and Johnson's excellent summary of the early work on these "hot" cathodes.[44] Wittig[45] observed an HC effect (rapidly increasing discharge current accompanied by a decreasing voltage) in low-pressure (2 torr) "hot" HC discharges. The minimum voltage necessary to sustain a given current was found to be dependent on the cathode temperature. Salinger and Rowe[46] showed that thermionic electrons that enter the glow in these discharges approach a near-Maxwellian energy distribution through mutual interactions. Excitation and ionization is induced by the electrons in the high-energy tail of this distribution. It has been reported that cumulative ionization processes by collisions between excited atoms are important.[47,48] Consequently, it is possible to sustain a discharge even when the discharge potential drop is well below the lowest excitation energy of the gas atoms. Wittig[45] explained that this is because cumulative ionization is proportional to the square of the plasma density. If this density is increased, keeping current constant, it should be expected that the ionization efficiency will increase rapidly and therefore allow a decrease in the electron temperature.

Characteristics of FANES-type discharges are controlled by the type of gas, gas pressure, voltage, and furnace temperature. The applied voltage controls the current, which controls the charge distribution according to the mobility of the charged species. The electric field is produced by this charge distribution and satisfies Poisson's equation. The solution of this equation for an HA geometry indicates the field is highly nonuniform.[54] Maximum field strength occurs at the cathode and decreases rapidly to become a constant value of $(2i/\mu)^{1/2}$, where i is the current and μ is the mobility, at larger distances from the cathode. Unfortunately, because of the nonsymmetry of the HC-FANES source, a closed form solution to Poisson's equation is not possible for this case.

Consider the principal differences between the HA- and the HC-FANES discharges. The HC system, because of its larger cathode area, will be more strongly influenced by photoelectron emission from the cathode into the plasma. But during the tube heating both systems can reach temperatures where electrons from the cathode are supplied primarily from thermionic emission. This supply satisfies Richardson–Dushman's equation for the current density

$$j_e = AT^2 \exp[-(e\phi - e\sqrt{eE})/kT] \tag{9-4}$$

where ϕ is the work function, e the electronic charge, and E the applied field needed to account for the influence of external fields on this emission (Schottky effect). The theoretical value of the constant A is about 120 A/cm^2-deg^2.[55] The temperature of the cathode in the HC-FANES source increases at a greater rate than that of the HA-FANES because, for the HA source (1) the cathode is thermally isolated and (2) thermal radiation from the anode to the cathode will be negligible until the anode temperature reaches approximately 1300 K. The cathode temperature can therefore lag behind the tube temperature by more than 1000 K during a rapid heating ramp of the tube. It is not unreasonable to assume that the thermionic emission into the HC plasma will, because of the higher cathode temperature and larger cathode surface area, be considerably greater than that achieved for the HA plasma. This is true only if the field term in the exponential of the Richardson–Dushman equation can be neglected, otherwise the applied field can control the thermionic emission. But the applied field is a complicated function of the cathode dimensions and depends strongly on the geometry. It is therefore not possible to predict *a priori* which of the two geometries will exhibit the greater thermionic emission for all tube temperatures. Note that Langmuir[56] has shown that if an anode voltage V is applied in the vicinity of an electron emitter in vacuum, the emission current (i) follows the relation

$$i = sV^{3/2} \tag{9-5}$$

where s is a constant that depends on geometry. This derivation is for vacuum and is, therefore, a space charge-limited situation. The thermionic emission becomes mobility limited at the higher pressures; emission current is then proportional to V^2.[53] The discharge current derived by von Engel for the HA geometry shows this square dependence on the applied voltage, so the current can be considered mobility limited in this case. Now, Schottky effects become significant for fields on the order of 10^6 V/m. Thermionic emission current densities for the two geometries are, therefore, equal for equal cathode temperatures if the sheath fields are less than 10^6 V/m. The assumption that the HC configuration exhibits the greater thermionic emission current is then valid for these lower field strengths or for applied potentials less than 500 V. Note that 500 V applied to a 1-mm-diameter cathode will provide a 10^6 V/m field when space charge is absent; the presence of positive space charge such as that responsible for the cathode fall only increases the field.

The "hollow cathode effect," or enhanced electron density in an HC glow, is responsible for the enhanced glow observed in HC systems.[57] Even greater enhancement should be expected from the HC-FANES plasma because of the stronger thermionic emission associated with this system. These thermal electrons leave the cathode sheath, enter the glow, and become trapped in the cylindrical region imposed by the retarding radial fields of the sheath. The probability for elastic and for inelastic gas-phase collisions is, therefore, enhanced, which further enhances the electron density. In contrast, the electrons emitted by the cathode in the HA geometry travel in radially symmetric fields. Most of the fast electrons lose their energy by impacting the outer cylindrical anode, not in elastic collisions that will ionize or excite the gas.

It is noted that the dc FANES discharges differ from the more conventional planar glow discharge. In fact, von Engel[54] classified the HA geometry as a glow discharge without zones near the anode, i.e., without the positive zones observed for the planar case. The HC plasma differs from the planar glow discharge in that it exhibits an enhanced glow, the "hollow cathode effect" that is discussed above.

The rf discharges described by Sturgeon *et al.*[21,27,28,30,31] and by Blades *et al.*[20,24] are of the HA configuration. These rf corona discharges are not yet understood. The rf modulation of the (sheath) voltage and sheath thickness must control the phase and intensity of the current, but a mathematical description has not been developed.

The high power input associated with rf discharges at elevated pressure implies enhanced ionization. Generally speaking, the impedance of the rf plasma decreases as the frequency of the applied voltage increases. It has been proposed by McDonald and Tatenbaum[58] that the rf field can drive electrons to energies sufficient for ionization if elastic collisions in the gas are in phase with this field, i.e., if collision-induced reversal of the electron motion is in phase with the applied field. Energy lost by electrons through ionizing collisions can then be reestablished by the primary and secondary electrons through their interaction with the field, and the process continues. This explains why the minimum operating pressure (maximum mean free path) increases (decreases) with increasing frequency. Chapman reports the suggestion of Holland *et al.*[59] that electrons can impact the electrodes to produce secondaries that are rapidly accelerated across the positive sheath into the discharge. A resonance effect (multipacting) can then occur that increases the electron energy, provides an efficient electron supply, and induces efficient ionization.[58] It has also been suggested by Keller and Pennebacker[60] that electrons may also gain energy from the modulated edge of the positive sheath, where sheath momentum is shared with the electrons as they are reflected in resonance.

9.3.4. Voltage–Current Characteristics

Figures 9-6 and 9-7 show the current versus discharge voltage curves for the HA- and the HC-FANES. In both cases the general slopes of these curves decrease with pressure. The curves imply operation in the abnormal discharge region since the magnitude of the slopes are finite. Such a discharge extends over the entire inner surface of the cathode and any increase in current results in an increase in current density (j) and discharge voltage (V). The slope for the HA-FANES is about 18 times greater than that for

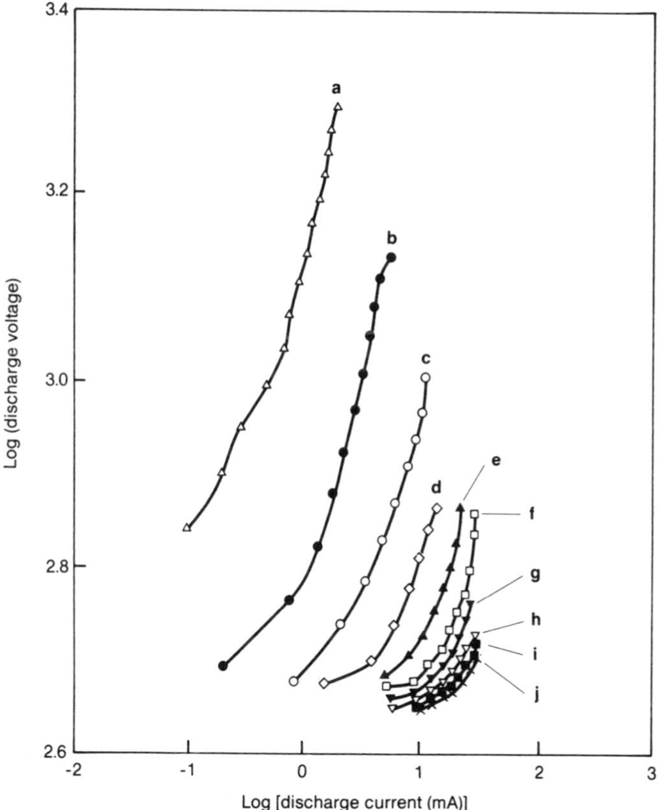

Figure 9-6. Current–voltage characteristics for HA-FANES operated with a 1-mm-diameter cathode and no furnace heating at pressures of: (a) 10, (b) 30, (c) 50, (d) 70, (e) 90, (f) 110, (g) 130, (h) 150, (i) 170, and (j) 190 torr.[25]

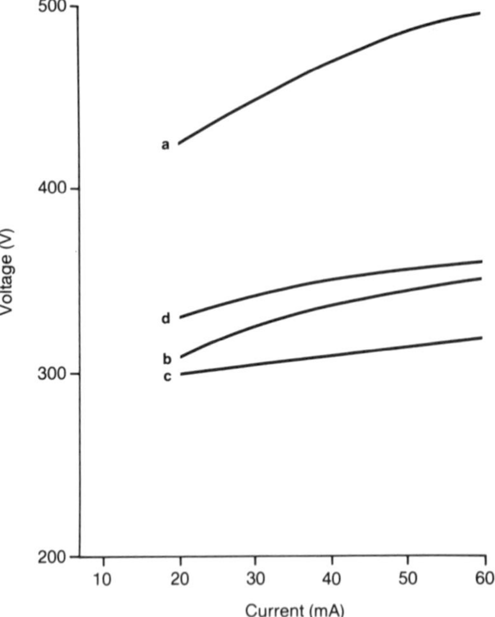

Figure 9-7. Current–voltage characteristics for HC-FANES with no furnace heating at pressures of: (a) 0.75, (b) 3.8, and (c) 7.5 Torr Ar and (d) 6.8 Torr He.[15]

the HC-FANES for currents between 20 and 30 mA at nearly identical pressures (7.5 and 10 Torr, respectively). If this ratio is multiplied by the ratio of the respective cathode areas (0.1), it is seen that the $(\partial V/\partial j)p$ for the HA-FANES is about 1.8 times that of the HC-FANES when operation is in the above pressure range. This derivative represents the reciprocal of the electrical "conductivity," σ, of the gas. Hence, for the above pressure

$$\sigma_{\mathrm{hc}}/\sigma_{\mathrm{ha}} = \rho_{\mathrm{hc}}\mu/\rho_{\mathrm{ha}}\mu = \rho_{\mathrm{hc}}/\rho_{\mathrm{ha}} \sim 1.8 \tag{9-6}$$

where ρ is the charged density and μ is the mobility. This result confirms that it is the charge density difference associated with maintaining a given current in the two different geometries that is responsible for the differences in current responses between the HA and HC cases. This might be expected since the radial current per unit length is given by

$$i = 2\pi r\rho(r)v(r) = 2\pi r\rho(r)\mu E(r) \tag{9-7}$$

where $v(r)$ and $E(r)$ are the radial velocity and electric field, respectively and are highly dependent on system geometries. Equation (9-7) can be used with the Poisson equation in cylindrical coordinates to obtain i in terms of the voltage, V, applied to the HA system.[54] The current for the HA-FANES is then described by

$$i_{ha} = (V - V_0) V_0 \mu / [r \ln(R/r)] \tag{9-8}$$

where R and r are the anode and cathode radii for this geometry, and V_0 is the breakdown (or running) potential for the HA discharge. Figure 9-6 shows departure from this predicted linearity with voltage. End effects are probably responsible since the calculation is made for an infinitely long tube and thus neglects such effects.

From Eq. (9-8):

$$V/i = [r \ln(R/r)]/\mu V_0 = (1/\mu E_0) \ln(R/r) \tag{9-9}$$

The denominator on the right-hand side of Eq. (9-9) is, by definition, the drift velocity in the field E_0, and the drift velocity is inversely proportional to pressure.[54] Thus, taking the pressure derivative of the logarithm of Eq. (9-9) gives:

$$\partial/\partial p \ln(V/i) = k\partial/\partial p \ln p = k1/p \tag{9-10}$$

The observed decreases in (log $V -$ log i) with increasing pressure (Fig. 9-6) are therefore a result of the effects of pressure on mobility and on the mobility-dependent breakdown field E_0.

The HC geometry is, unfortunately, not an amicable geometry for closed form derivations of $V(i)$. But the data of Fig. 9-7 imply a linear $V(i)$ relationship exists and that the slope decreases with increasing pressure, similar to the HA case. It is expected then that the slope has a V_0 dependence that is similar to that described for the HA geometry in Eq. (9-9). The radial (geometric) dependence cannot, of course, exhibit this similarity. It should be noted that V_0 is dependent on the uniformity of the electric field and decreases with increasing deviation from this uniformity.[54] But the HA geometry produces a hyperboliclike field distribution, so it should be expected that

$$V_{0,(\text{hollow cathode})} > V_{0,(\text{hollow anode})} \tag{9-11}$$

This implies, from Eq. (9-9), that

$$(\partial V/\partial i)_{p,\ \text{hollow anode}} > (\partial V/\partial i)_{p,\ \text{hollow cathode}} \tag{9-12}$$

which is the observed geometric effect. It is concluded then that differences observed for the $V(i)$ curves in the two cases are caused by the dependence of the current on breakdown voltage, which is dependent on field uniformity and hence on the geometries involved.

9.4. Operational Characteristics

9.4.1. Analytical Signals

The analytical emission signals associated with FANES systems are dependent on furnace-induced volatilization and atomization, on the discharge excitation, and, to some extent, on discharge-induced atomization. None of these processes is well understood. Furnace atomization is a well-documented phenomenon, but low-pressure discharge-induced excitation processes are, in general, not nearly as well characterized. Emissions from these discharges are dependent on the support gas, the gas pressure, the gas temperature, the discharge current, and the cathode/anode geometries. Each of the first four factors will be discussed and the cathode/anode geometries will be compared throughout. Literature on glow discharges is extensive, and pertains mostly to use of these discharges in sputtering processes at low pressures ($<$10 Torr). Little information is available for "hot" cathodes or anodes and operation at higher pressures.

9.4.1.1. Support Gas

Ar and He are the only support gases that have been used to date with FANES devices. For HA- and HC-FANES, the two gases can be readily interchanged. A major analytical consideration is the higher excitation energy of He. For HA-FANES it is also possible to operate at much higher pressures with He (600 Torr) than with Ar (200 Torr). For rf-FANES, He is used almost exclusively. Both Blades and Sturgeon have reported difficulty in initiating an Ar rf plasma at atmospheric pressure.

9.4.1.2. Discharge Pressure

The operating pressure of any of the FANES devices would appear to represent a compromise between optimum conditions for furnace atomization and glow discharge excitation. Higher pressures provide longer residence times in the furnace and, thus, larger integrated signals. Lower pressures provide longer mean free paths and, consequently, contribute to higher average kinetic energies for free electrons in the potential gradient of the discharge. This kinetic energy increase can manifest itself as an increased capability to ionize and to excite higher energy levels by electron impact.

First, consider the effect of pressure on furnace atomization. The graphite furnace, as characterized by absorption measurements, induces transient atomization of the analyte that is dependent on the convolution of the supply and loss functions. The supply function is dependent primarily on the chemical nature of the analyte, the interaction of the analyte with the chemical matrix and the furnace surface, and the heating rate and final temperature of the furnace. The loss function, for operation in the static mode (gas flow is terminated during the atomization step), is usually determined by the rate of gaseous diffusion of the analyte from the furnace. Under ideal, isothermal conditions, the loss function remains constant during the presence of the analyte and the appearance function has no effect on the integrated signal. The conventional furnace differs from ideality, however, because of the lack of spatial and temporal isothermality.[38,39] Consequently, for GF-AAS, considerable emphasis is placed on maintaining a consistent supply function (i.e., the same for samples and standards) in order to ensure accurate determinations.

As stated above, the analyte loss is usually controlled by diffusion, provided the sample matrix does not have a large mass. The diffusion rate is inversely proportional to the pressure. The low operating pressure of HC-FANES (routinely 1–30 Torr) increases the diffusion (loss) rate and decreases the analyte residence time and the integrated analytical signal, as compared with operation at atmospheric pressure. HC-FANES systems generally use continuous pumping and a constant support gas flow to maintain the desired pressure and to reduce the effect of entrained air. It is not clear whether the analyte residence is limited by diffusion or convection.

HA-FANES operates in the static mode (no pumping or internal gas flow) with pressures of 150–200 Torr of Ar or 300–600 Torr of He. The analyte residence times are much closer to those found for conventional graphite furnace atomization than HC-FANES.

rf-FANES operates at atmospheric pressure in He and would be expected to have residence times similar to conventional GF-AAS if Ar is used as the support gas. Instead, rf-FANES uses exclusively He, which has a greatly reduced mass compared with Ar. Sturgeon *et al.*[21,27,28,30,33] use a constant flow of He through the furnace in order to exclude air entrainment. This flow has not altered the residence times.

Now consider the effect of pressure on glow discharge excitation and, more specifically, HC excitation. The excitation process and, consequently, the emitted intensities of low-pressure, HC discharges are well documented as being highly pressure dependent.[40–42,49,61,62] This is apparent from an examination of the emission intensities of the support gas (Fig. 9-8). At low pressures the maximum intensity from the discharge is observed on the cylindrical axis. The intensity increases (not shown in Fig. 9-8) and the maximum region broadens radially as the pressure increases. Eventually, as

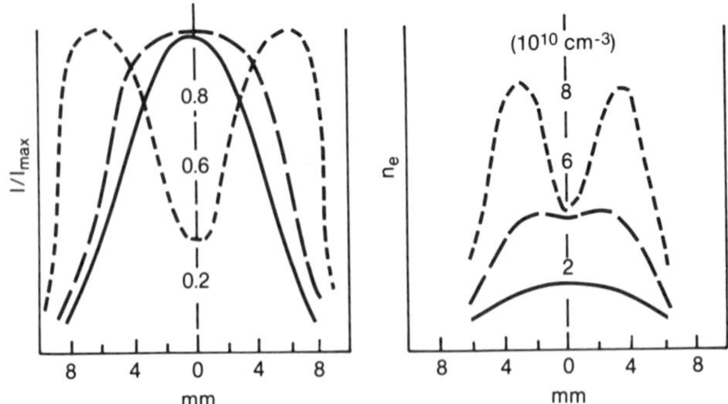

Figure 9-8. Radial distribution of intensity ratio, I/I_{max}, of the spectral line of argon at 772.4 nm and of electron density, n_e, in the negative glow of a cylindrical cathode 2 cm in diameter and 10 cm long at a discharge current of 50 mA in Ar at: (——) 0.1 Torr, (– – –) 1.0 Torr, and (- - - -) 2.0 Torr.[61]

the pressure continues to increase, the maximum moves away from the center and toward the cathode wall. A similar pattern is observed for the electron density (Fig. 9-8). As the pressure increases, the electron mean free path decreases and the energy from the electrons, accelerated across the dark space, is dissipated in inelastic collisions closer to the cathode wall. This finding supports the assumption that electron collisions are primarily responsible for excitation and that the electron density distribution will determine the maximum region of emission.

The exact position of the maximum emission region will depend on the pressure and the mass of the support gas. The pressure dependence of the intensity integrated over the entire HC cross-sectional area has not been reported.[42] From Fig. 9-8, it is clear that for a given pressure, there is a maximum observation point within the cathode. For a fixed observation point, at the cylindrical axis of the HC, the emitted intensity can be expected to increase, pass through a maximum, and then decrease as a function of pressure. The higher pressures, however, will give rise to lower diffusion rates and longer residence times. Thus, it is not clear how fast the intensity at the axial position will decline.

Falk[12] has reported results for the support gas emission that are consistent with the data in Fig. 9-8. For observations made at the central axis of the furnace, the intensities for Ne(I) and Ar(I) increase monotonically from 0.5 to 6.0 Torr, and then decrease between 7 and 30 Torr. Falk[12] also showed that between 0.5 and 8.0 Torr the emitted intensities from He(II), 333.49 nm, and Ar(II), 454.51 nm, decreased with increasing pressure. Transitions requiring high-energy electrons appear to be more sensitive to increased pressure.[12,42,61] The number of fast electrons (19–26 eV) shows a sharp maximum between 1 and 3 Torr.

It should be noted that the HC-FANES is most conveniently viewed along the central axis. All of the data reported by Falk were obtained by viewing the central position. This serves to minimize the background from blackbody emission from the furnace wall at higher atomization temperatures (above 2300 K). No attempt was made, for any of the HC-FANES studies, to characterize the spatial dependence of the emission signal or to correlate the total emitted intensity with the various parameter changes.

The plots of emission intensities for a series of Fe lines from an Fe HC lamp as a function of pressure in Fig. 9-9 show the expected behavior described above. A broad emission maximum is observed between 0.1 and 10 Torr.[40,62] It must be remembered, however, that these data reflect the maximization of the combined sputtering and excitation processes.

It is expected that a nonsputtered analyte species will show a similar spatial dependence as the fill gas. Falk *et al.*[15] have reported experimental detection limits for 22 elements using pressures ranging from 1 to 5 Torr for HC-FANES. Detailed intensity versus pressure data have not been reported. It must be assumed that there was a maximum for most elements within this pressure range, otherwise a higher pressure range would have been used. Conversely, Falk *et al.*[15] showed that for C(I), 247.8 nm, above 1800 K, the emission intensities increased steadily over a range of 4–22 Torr in Ar and 7–30 Torr in He. Above 1800 K, C is thermally volatilized from the furnace surface and can therefore be considered a nonsputtered species.

Naumann *et al.*[17] have reported HC-FANES results that differ considerably from those reported by Falk *et al.* The optimum pressures for Co, Cr, Cu, and Ni, in the single-element mode, were 102, 121, 40, and 116 Torr,

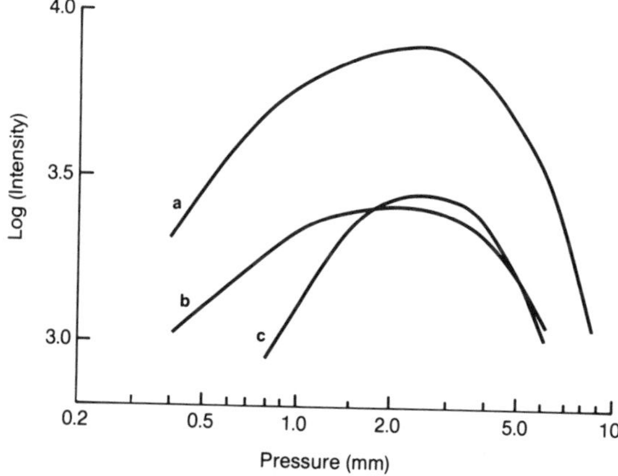

Figure 9-9. Dependence of spectral intensity, I, on the fill gas pressure of Ne in a hollow cathode discharge at a constant current of 90 mA for: (a) Fe, 268.7 nm; (b) Fe, 368.2 nm; and (c) Fe, 379.4 nm.[62]

respectively. Unfortunately, a description of the optics used and of the observation region of the discharge was not provided. Considering the tendency of the maximum negative glow region to move toward the cathode wall with increased pressure (Fig. 9-8), it would appear that these authors viewed a large fraction of the off-axis region of the plasma. Overall, the signal response as a function of pressure is still unclear for HC-FANES.

Optimum operating conditions for HA-FANES[18,25] differ considerably from those for HC-FANES. The HA geometry eliminates any true HC effect. High-energy electrons accelerated across the potential drop of the cathode dark space pass through the negative glow region, strike the furnace wall or anode, and are lost. The spatial orientation of the glow discharge to the cathode surface as a function of pressure, however, is similar to that of the HC configuration or conventional planar glow discharges. As pressure increases and the electron mean free path decreases, the region of maximum emission intensity moves closer to the cathode surface (Fig. 9-8).

Ballou *et al.*[18] have shown that for Ar at low pressures (<5 Torr) the discharge uniformly fills the furnace. As pressure is increased to approximately 15 Torr, the discharge shrinks dramatically until it constitutes a corona surrounding the cathode. As shown in Fig. 9-8 the region of maximum intensity is close to the cathode wall. Increasing the pressure up to 200 Torr serves to brighten and more clearly define the limits of the negative region close to the cathode. The maximum signal-to-noise ratio (SNR) is found within 1 or 2 mm of the cathode surface at all but the lowest pressures. Similar results are observed for He, except that a dramatic shift in pressures is observed. A diffuse discharge is observed in He at pressures as high as 100 Torr and the discharge shrinks and brightens as pressures are increased up to 600 Torr.

Table 9-3 shows the spatial dependence for the Cd and Cu analytical signals in HA-FANES reported by Harnly *et al.*[25] at 30 Torr of Ar. The

Table 9-3. Peak Area versus Viewed Region of the Discharge[a]

Region viewed of corona[b]	Peak area (I-s)	
	Cd	Cu
1. Above	199. ±0.4	686. ±0.8
2. Top	191. ±0.5	1233. ±1.7
3. Center	152. ±0.4	78. ±0.6
4. Bottom	103. ±0.3	706. ±0.7
5. Left	155. ±0.3	1283. ±0.9
6. Right	151. ±0.4	611. ±1.8

[a]Results obtained at 70 Torr, 50 mA, atomization temperature of 1845°C, with a 1.0-mm-diameter cathode. Precisions determined as 3 sigma of the baseline noise.
[b]See Fig. 9-10 for diagram of viewed region.

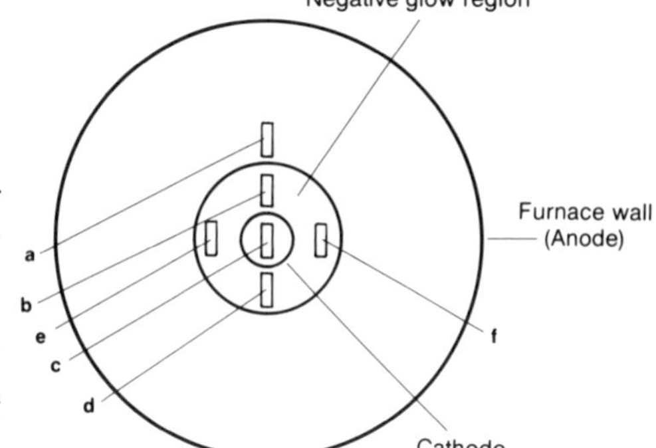

Figure 9-10. Schematic diagram of the end-on view of the HA-FANES source. The optics were adjusted so that the entrance slit (100 μm wide and 500 μm high) viewed the six positions shown: (a) above, (b) top, (c) center, (d) bottom, (e) left, and (f) right of the cathode.[25]

regions examined are shown in Fig. 9-10. The analytical signals decrease dramatically outside the corona. In general, the region of the maximum analyte signal also gives the best SNR.

The effect of pressure on the integrated analytical signals for Cd and Cu in Ar using HA-FANES is shown in Fig. 9-11.[29] The optics remained focused at position "e" (Fig. 9-10) for the entire experiment. In general, the signals increase linearly with increasing pressure. This linear increase in the integrated signal can be explained solely by the decrease in the diffusion

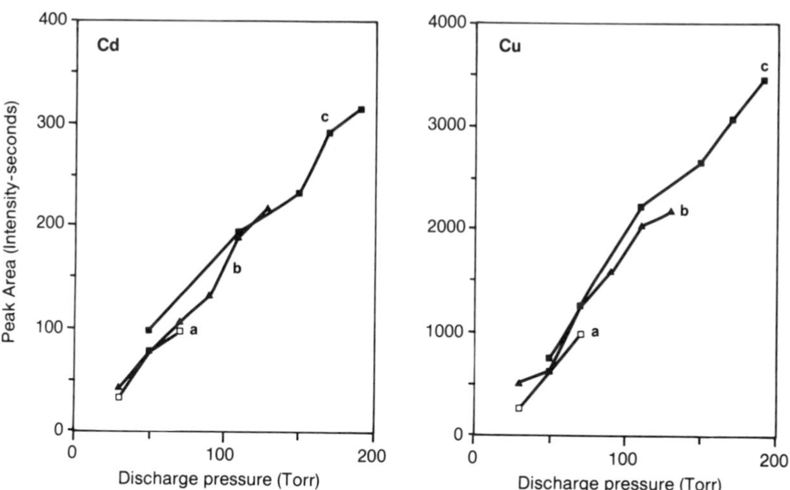

Figure 9-11. Peak area for 2 ng of Cd and of Cu in HA-FANES with atomization at 2100 K and viewing position "e" (see Fig. 9-10) as a function of the Ar pressure at discharge currents of: (a) 20, (b) 50, and (c) 70 mA.[25]

coefficient. A series of currents was used since there is a finite range of effective currents for each pressure (see Section 9.4.1.4.).

The SNR for Cr in Ar as a function of pressure and discharge current is shown in Fig. 9-12. The shapes of the SNR plots in Fig. 9-12 are determined primarily by the stability of the discharge, i.e., the background emission. A comparison of the maximum of each plot reveals that the best SNR is found at 160 Torr.

rf-FANES, operated at 1 atm of He, looks very similar to HA-FANES. A bright region surrounds the central electrode and a less intense plasma fills the remainder of the furnace volume. Although the rf potential is bipolar, the self-biasing nature of the rf discharge, which arises from the difference in the electron and ion velocities, appears to provide a negative bias to the central electrode. Emission signals for the rf-FANES are obtained from the region just outside the corona surrounding the central electrode. There have been no measurements of the spatial dependence of the analyte signal.

Shorter residence times have been observed for rf-FANES than for conventional GF-AAS. This is most likely a result of the larger diffusion coefficients for the analyte in He relative to Ar. Sturgeon *et al.*[21,27,28,30,31] reported a full width at half-height of 230 ms for Mn as compared with approximately 400 ms for atomization from the wall of a conventional furnace with 1 atm of Ar. Sturgeon also employs a constant internal flow of

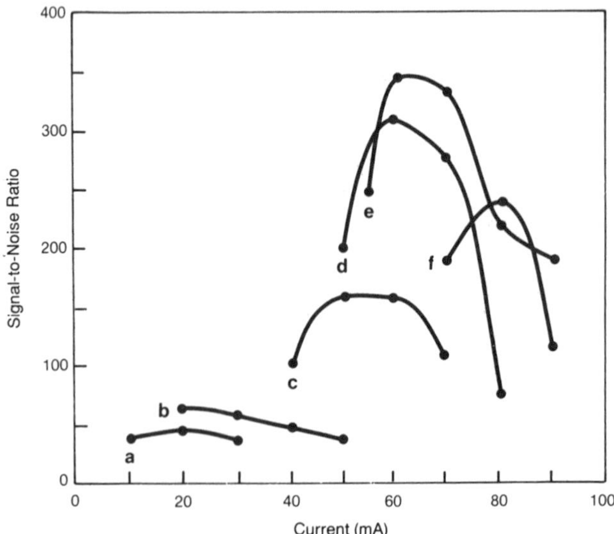

Figure 9-12. SNR for 100 pg of Cr as a function of the discharge current at an atomization temperature of 2500 K and pressures of: (a) 20, (b) 40, (c) 80, (d) 120, (e) 160, and (f) 160 Torr.[29]

He to prevent entrainment of atmospheric air. The analyte signal shapes, however, do not vary for He flows from 75 to 325 ml min^{-1}. This suggests that analyte loss is not influenced by the He flow and is still diffusion limited, probably because of the low atomic mass of the He.

9.4.1.3. Atomization Temperature

In general, the atomization temperature for all the FANES sources is determined by the volatility of the element being determined. The lowest temperature is used that provides complete volatilization of the analyte. With respect to the excitation process, the critical temperature is 1800 K. Below 1800 K, a "cold" cathode requires a potential of several hundred volts to sustain currents of 20–100 mA. Above 1800 K, the potential necessary to sustain the same current with a "hot" cathode drops precipitously (Fig. 9-13) because of the evolution of thermionic electrons (as discussed in Section 9-3).

As established for GF-AAS, the atomization temperature refers specifically to the furnace wall temperature. It is generally synonymous, after a few seconds into the atomization cycle, with the gas temperature, which initially lags behind the rapidly rising wall temperature by several hundred

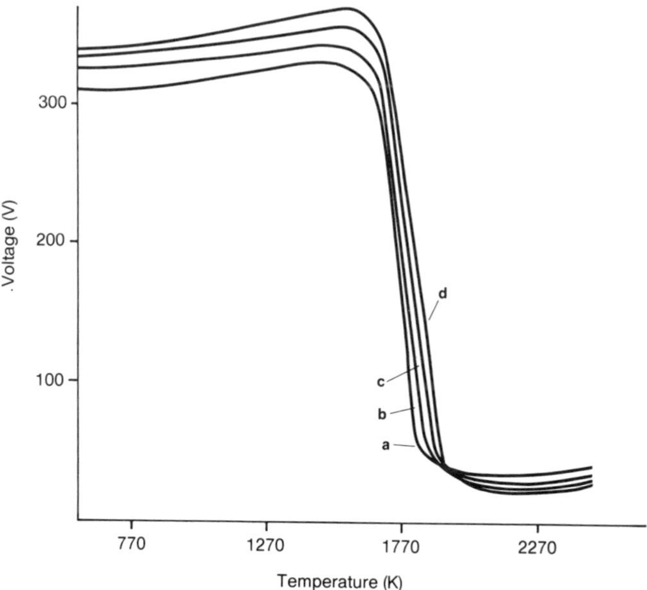

Figure 9-13. Discharge voltage of HC-FANES at 3.8 Torr Ar as a function of the cathode temperature for constant currents of: (a) 20, (b) 30, (c) 40, and (d) 60 mA.[15]

degrees.[39] The rate of diffusion of the analyte from the furnace is proportional to $T^{3/2}$. Consequently, the atomization temperature is generally kept as low as possible to enhance residence time and the integrated signal.

The temperature necessary for complete volatilization varies with the element and also varies significantly with the operating pressure of the specific FANES system. In general, HC-FANES requires the lowest temperatures, HA-FANES moderately higher temperatures, and rf-FANES, at atmospheric pressure, uses temperatures comparable to those of conventional furnace systems.

As mentioned above, cathode temperatures in excess of 1800 K can have a significant influence on the discharge because of the considerable quantity of thermionic electrons that are emitted. This has been well documented for HC-FANES by Falk *et al.*[15] At 1800 K the electrical conductivity of the gas experiences a rapid increase and the potential across the discharge rapidly decreases (Fig. 9-13).[15] A similar effect is observed on the emission intensity of the support gases (Fig. 9-14). Below 1800 K, the effect of increasing the temperature is a slight enhancement of emission intensity. This is likely caused by the reduction of gas density and an associated increase in the electron mean free path.[15] The magnitude of the decrease of the emission intensity of He at 1800 K in Fig. 9-14 appears to be less

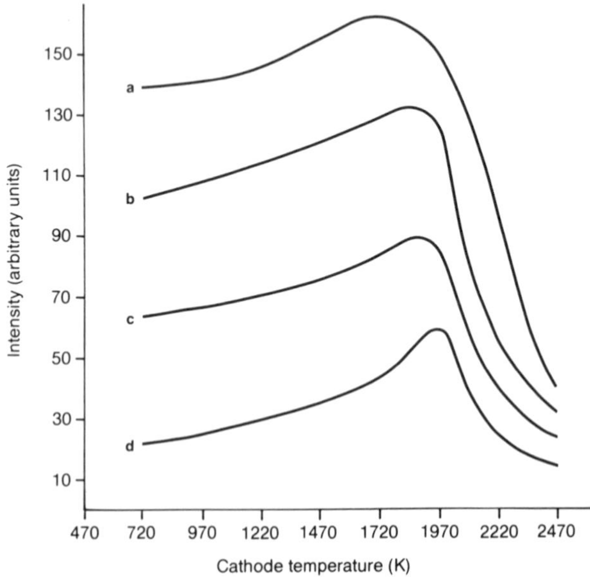

Figure 9-14. Intensity of the He 318.774-nm line in HC-FANES as a function of the cathode temperature at a discharge current of 40 mA and pressures of: (a) 6.8, (b) 9.8, (c) 20.2, and (d) 30.0 Torr.[15]

significant at higher pressure. Similar potential–temperature profiles have been observed for HA-FANES, however, at much higher pressures.

The analytical importance of the thermionic effect for FANES is not clear. Falk *et al.*[(15)] observed no effect of the thermionic electrons on the detection of a wide range of elements by HC-FANES. It can be assumed, however, that at the reduced pressures (1–30 Torr) of HC-FANES, all elements were volatilized at temperatures less than 1800°C. Naumann *et al.*[(17)] reported atomization temperatures for Co, Cr, Cu, and Ni ranging from 2100 to 2600 K, at pressures from 40 to 140 Torr. For each element, the detection limits were very competitive with Falk *et al.*[(15)] and GF-AAS, i.e., the detection limits of elements atomized at temperatures greater than 1800 K did not deteriorate.

No discernible effect of thermionic electrons on analytical determinations by HA-FANES has been observed. Like HC-FANES, however, the cathodic atomization temperatures of HA-FANES are less than 1800 K. Although the dramatically higher pressures of HA-FANES (up to 200 and 600 Torr of Ar and He, respectively) require furnace temperatures up to 2500 K, complete volatilization of the analytes (tested to date) has been achieved before the temperature of the central cathode reached 1800 K. This difference in temperatures arises from the temperature lag of the cathode. Welz *et al.*[(39)] have reported that a platform in a graphite furnace takes about 2 s to reach 1800 K when the furnace power supply is set for a 2800 K atomization step. The central electrode can be expected to exhibit a greater lag than the platform since it is completely removed from the wall, is attached to a large cool mass outside one end of the furnace, and is heated only radiatively and convectively. Consequently, the analytical signal for Cr, atomized at 2500 K, has returned to baseline before a decrease in the discharge potential is observed.[(29)]

It is difficult to monitor the voltage potential across an rf plasma with existing rf power supplies. Sturgeon *et al.*,[(21)] however, have reported an increase in reflected power for rf-FANES with increasing furnace temperature. Sturgeon suggested that this may be related to thermionic electron emission. The increase in reflected power is delayed, occurring about 3 s after initiating a 2800 K atomization step. This time lag may be associated with the central electrode temperature lag. The increase in reflected power is, therefore, a possible result of thermionic electrons emitted by the central electrode at 1800 K.

Blades and co-workers have not mentioned a similar increase in reflected power. This is most likely due to the automatic impedance matching circuit associated with their power supply.

The central cathode of HA- and rf-FANES introduces some analytical signal features that are not seen for transient GF-AAS signals. The temperature lag of the central electrode produces double peaks of the emission spectra in the atomization step.[24,27,29] Analyte atomized from the wall condenses on the cooler central electrode. Continued heating of the furnace leads to eventual reatomization of the analyte from the cathode; hence, the double peaks. Double peaks are observed for the less volatile elements (Cu and Cr) at low discharge currents. At higher currents, only a single peak is observed. The process of atomization, condensation, and reatomization may still prevail, but the temperature difference between the cathode and the furnace wall is sufficiently small that the peaks are not distinguishable.

For HA-FANES operation, the temperature of the cathode at the start of the atomization step is a function of the discharge current, resulting from ohmic heating of the cathode. Because of the low mass of the rod-shaped cathode, the temperature can exceed 1300 K (dull-red color temperature) at moderate currents (60 mA). Thus, at the start of the atomization step the cathode temperature is higher than that of the furnace wall. Upon atomization, however, the wall heats rapidly and quickly exceeds the cathode temperature. For a 3-s atomization step at 2300 K, the wall reached the set temperature after only approximately 1.25 s. The cathode temperature at the end of the 3-s cycle has just reached 1800 K, i.e., the voltage drop across the discharge started just prior to the end of the cycle.

The cathode of the HA-FANES source has been used as a platform for sample deposition instead of deposition on the furnace wall. Preliminary data for Cu showed the peak maximum and width to be approximately the same for deposition of the sample on the cathode or on the wall. Use of the cathode as a platform produced double peaks for Cd. In this case one peak occurred prior to the atomization step after the discharge had been initiated. Ohmic heating of the cathode by the discharge current resulted in thermal atomization and nearly quantitative transfer of Cd to the cold furnace wall. A second peak for Cd was then observed when the furnace was heated during the atomization step.

9.4.1.4. Discharge Current

It is difficult to draw any generalities with respect to the discharge current for the FANES devices. The two best-characterized devices, HA- and HC-FANES, display different integrated signal-discharge current profiles. HC-FANES signals increase monotonically with increasing current while HA-FANES signals are independent of the current above a threshold. There is insufficient information for rf-FANES to support any conclusions.

Falk *et al.*[15] have observed consistent increases in the peak heights of the analytical signals with increasing discharge currents for HC-FANES. Littlejohn[19] has reported similar results. The analytical signal is proportional to i^n, where i is the discharge current and n is an integer representing the number of collisions involved in the excitation process. Examining the support gas emission, it was observed that ^{3}He (318.8 nm) had n values of approximately 2 and 1 for hot and cold cathodes, respectively.[15] For Ar (451.1 nm), n had values of approximately 0.5, 1, and 2, depending on pressure and temperature. It was suggested that the value of 0.5 pointed to a loss mechanism that was current dependent. Similar results were observed for six atomized elements (Fig. 9-15). For Al, n was approximately 2.0; for Cr and Ni, approximately 1; and for Co, Cu, and Fe, about 0.5.

The integrated analytical signals for HA-FANES, at significantly higher pressures than HC-FANES, show a completely different dependence on the discharge current (Fig. 9-16).[25] Integrated Cd and Cu signals, obtained at 70 Torr of Ar, initially increase with increasing current and then reach a plateau. Initially the signals were proportional to i^n with n approximately equal to 1 and 0.5 for Cd and Cu, respectively. The SNRs (Fig. 9-16) are relatively comparable at all but the lowest and highest currents for both elements. Cr showed similar results at pressures up to 200 Torr. At each pressure, the integrated Cr signals did not vary as a function of current.

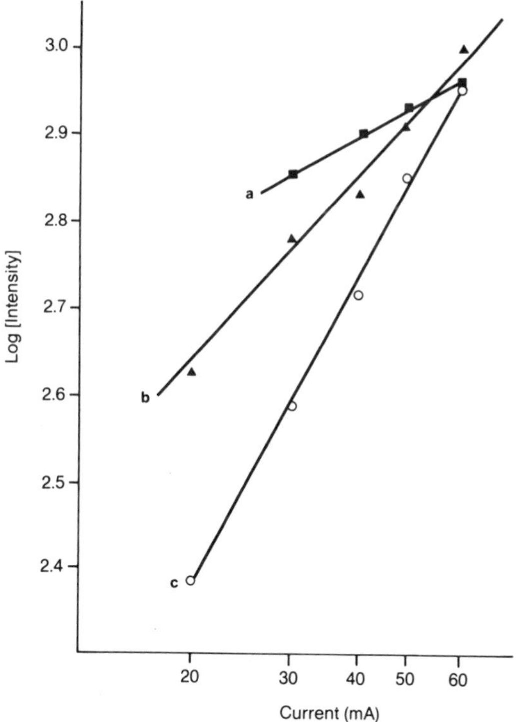

Figure 9-15. Dependence of the emitted analytical intensities in HC-FANES on the discharge current for: (a) Co, Cu, and Fe; (b) Cr and Ni; and (c) Al.[15]

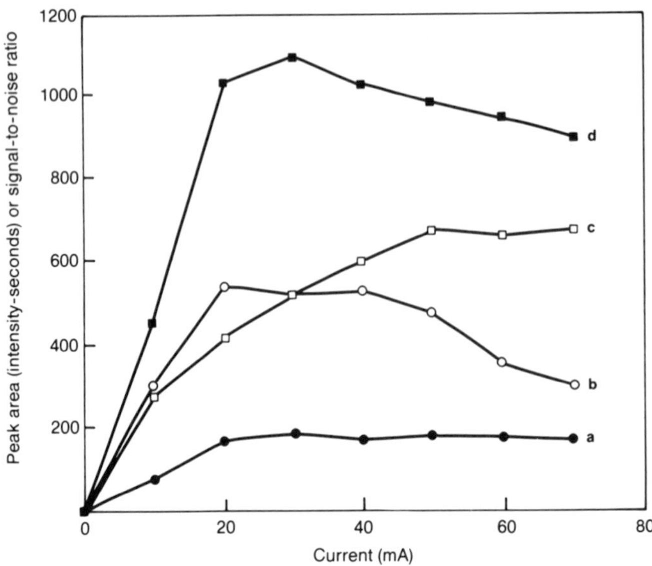

Figure 9-16. Dependence of peak areas (a and b) and signal-to-noise ratios (c and d) for 2 ng of Cd and Cu, respectively, on the discharge current in HA-FANES with atomization at 2100 K, 70 Torr Ar, and viewing position "e" (see Fig. 9-10).[25]

The response shapes shown in Fig. 9-12 are primarily a reflection of the background noise as a function of current.

For HA-FANES, the usable range of discharge currents is determined by the pressure. Higher currents are achievable at higher pressures but the usable range of currents is reduced. At any given pressure, the lower current limit is the current necessary to cover the entire cathode surface with the discharge.[15,25] At this minimum current, the discharge is just on the threshold of the abnormal region.[40–42] From this minimum current, higher currents can only be achieved with an increase in the current density and increased voltage. The current necessary to reach the abnormal discharge mode increases with pressure. The upper current limit is the highest current that can be used without excessive arcing within the system. This upper current limit is determined by the electrical insulation of the system and the contamination of the support gas. For HA-FANES, the upper current limit was characterized by a series of "minidischarges" between the anode (furnace wall) and the point at which the cathode enters the Macor support block (Fig. 9-5).[25] These "minidischarges" appear as random and rapidly flickering arcs. The main discharge appears uninterrupted.

The limits of the current ranges described above further clarify the data in Figs. 9-11 and 9-12. In Fig. 9-11, one current level was not possible over

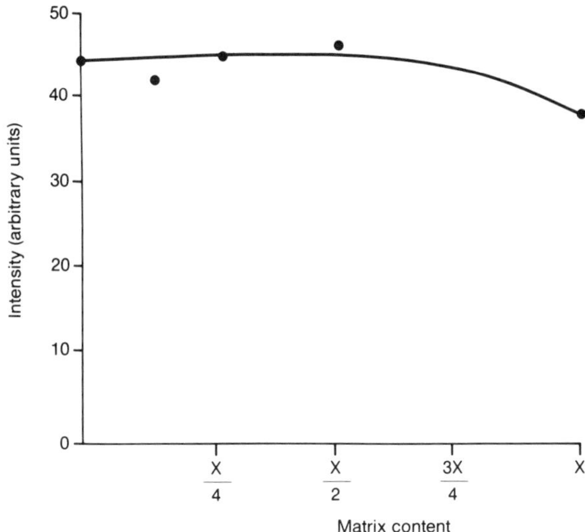

Figure 9-17. Dependence of the emission intensity of 100 pg of Cd (228.8 nm) on the combined chloride matrix concentration, X, in HC-FANES atomized at 1800 K in 19 Torr Ar with a discharge current of 30 mA. X = 1% m/V NaCl, 1% m/V KCl, 0.5% m/V $MgCl_2$, and 0.5% m/V $CaCl_2$.[8]

the range of pressures examined. Discharge currents of 50 and 70 mA caused instability at the lower pressures while currents of 30 and 50 mA were insufficient to reach the abnormal discharge mode at the higher pressures. In Fig. 9-12, the integrated signals, at a given pressure, showed no variation with current, but the background emission levels grew noisier toward the upper and lower current limits. Consequently, the maximum SNRs are found at intermediate current levels.

Two different cathode diameters have been employed for HA-FANES.[25] The analytical signals appear to be dependent on the current density. Comparable analytical signal levels for 1.0-mm- and 0.5-mm-diameter cathodes were found for Cd and for Cu at currents of 50 and 25 mA, respectively.

The rf-FANES sources have been operated at 20–100 W.[20,21,24,27,28,30,31] This compares with 60 W dissipated power for HA-FANES and 20 W for HC-FANES. The lower power limit is that which is necessary to maintain the plasma, and the upper limit, like that for the low-pressure discharges, is determined by the appearance of the arcing between the electrodes. In general, the analytical signal increases with increasing power. Sturgeon reported the doubling of the Cd signal magnitude when power was increased from 50 to 75 W.[27]

9.4.2. Noise Sources

Falk[1] predicted theoretically that the SNRs of nonlocal thermodynamic equilibrium systems, such as HC-FANES, are greater than those where LTE exists. In practice, Falk *et al.*[5] reported that the limiting noise and, consequently, the detection limits of HC-FANES are determined not by the recombinant continua but by extraneous molecular bands. These bands provide intense and structured background and arise from molecules from ambient gases that have entered the furnace through flaws in the vacuum seal. Continuous pumping is required to maintain a constant pressure during the atomization cycle. Although no specific molecules were identified by Falk, the most likely species are those that Sturgeon *et al.*[21] identified for rf-FANES.

The rf-FANES source described by Sturgeon *et al.*[21,27] uses a commercially available furnace that is not sealed against ambient atmosphere, but depends on a positive internal fill-gas pressure. A positive internal (125 ml min^{-1}) and external (1.5 liters min^{-1}) He flow is used during the atomization cycle. Sturgeon identified CO, OH, NH, and CN in the background spectra.

The CO and OH bands are not observed in HA-FANES emission spectra, but CN and NH bands have been identified. The high vacuum enclosure used for HA-FANES is superior to the vacuum enclosures of the other sources discussed in this chapter. Slight changes in pressure have been observed over extended periods, but these periods are orders of magnitude greater than atomization cycle times. Consequently, the atomization step is initiated under static conditions (constant pressure with no gas flow). The presence of CN bands appears to reflect an inherent problem arising from frequent exposure of the source to ambient air when the sample is introduced into the furnace. There is no apparent change when ultrapure Ar is substituted for the less pure and less expensive grade routinely used.

At higher analyte concentrations the reproducibility of the signals is controlled by analyte flicker. The furnace atomization process is a major source of this flicker. It appears that plasma excitation processes do not introduce a measurable increase in signal variance when moderate current levels are chosen. At higher current levels, where "minidischarges" are observed, a significant decrease in reproducibility is noticed. The associated increase in noise level is responsible for the deterioration of the SNR.

9.4.3. Interferences

Data on specific interferences for any of the FANES sources are limited at the present stage of development. It is conceivable that chemical, or matrix, interferences will be more severe for FANES than for conventional

GF-AAS since the technique combines the sophistication of the graphite furnace atomizer with the complexity of sustaining a discharge process at high temperatures and at reduced or at atmospheric pressure. For this reason it is probable that chemical–matrix interferences will be more severe for FANES than for conventional GF-AAS. The possibility exists, however, that collisions with high-energy electrons from the glow discharge may provide the means of diminishing some of the common furnace atomization interferences. Unquestionably, spectral interferences for FANES will be more severe than for GF-AAS, but less severe than for ICP-AES because of the differences in the electron energies associated with the three sources.

The greatest potential for matrix interferences for any of the FANES devices is the lack of stability of the discharge process throughout the atomization cycle. Ideally, every atom must be subjected to the same excitation potential or isoelectric conditions during its residence in the furnace. It has been shown that the discharge potential is dependent on the discharge pressure and furnace temperature (see Sections 9.4.1.2 and 9.4.1.3). Consequently, if isothermal and isobaric conditions prevail during the atomization process, the integrated signal will depend only on the analyte concentration. In reality, furnace temperatures vary temporally and spatially (for the non-ICC furnaces of HC- and rf-FANES) and large localized pressure differentials are produced within the furnace because of the rapid heating process. In addition, the sample matrix may introduce large numbers of ions into the furnace atmosphere during the atomization step. Thus, the probability of isoelectric FANES operation seems unlikely.

At this time there is no evidence to demonstrate failure or dramatic changes in the discharge during the atomization cycle for any of the devices. Indirect evidence suggests that the various FANES discharges may not be as robust as hoped. This indirect evidence is the apparent need for the use of the method of standard additions to achieve accurate determinations in known materials. Falk *et al.*[15] have reported the successful determination of Na in Al alloys, Ag in Au metal, and Cd in whole blood. In each case, the method of standard additions was necessary to obtain accurate results. Sturgeon *et al.*[27] found it necessary to use the method of standard additions to determine Cd and Pb in marine sediment, dogfish muscle, and lobster hepatopancreas reference materials. Acceptable accuracy has yet to be reported from analyses based on calibrations with aqueous or dilute acid standards.

With respect to gas-phase interferences, the reduced pressure discharges exhibit two distinct differences from GF-AAS. First, lower atomization temperatures are associated with the reduced pressure. These result in lower support gas temperatures and significantly fewer gas collisions. Reduced gas collisions suggest the possibility of increased chemical interferences from undissociated analyte molecules. Second, longer electron mean free paths are associated with reduced pressure. This allows for increased collisions between nonfragmented analyte molecules and high-energy electrons. Depending on the dissociation cross section relationship to this energy, analyte molecule dissociation can be enhanced.

Falk *et al.*[6] and Littlejohn *et al.*[8] reported that the flux of high-temperature electrons in HC-FANES is useful in reducing halide interferences. They demonstrated that, for the determination of Cd, HC-FANES tolerated concentrations of Na, K, Ca, and Mg chlorides (1.0% of NaCl and KCl, and 0.5% of $MgCl_2$ and $CaCl_2$) that are two orders of magnitude greater than those acceptable for GF-AAS atomization from the wall (Fig. 9-17). In another study, Falk *et al.*[15] reported that a concentration of 0.025% NaCl was necessary to produce a 20% suppression of Cu and that NaCl concentrations in excess of 0.065% were required to produce 20% suppression of Co, Cr, Fe, and Ni signals. No suppression of the Al signal was observed at NaCl concentrations extending to 0.25%. This study also correlated the severity of the interference with the correspondence of the atomization temperatures (overlap of the gas phases) of the analyte and the interferent. The NaCl, having a low atomization temperature, most severely affected the most volatile element, Cu. Effects of the NaCl on the other elements diminished with decreased volatility. The same study determined that recoveries of the same six elements were 100 ± 5% in up to 10% HNO_3. At 70% HNO_3, recoveries were within 100 ± 20%. These recoveries are reasonable considering the acid concentration.

Blades and co-workers[24] reported that a concentration of 0.029% NaCl produced a 20% suppression of the Ag signal in rf-FANES. The effect of $NaNO_3$ was even greater with a concentration of 0.021% $NaNO_3$ yielding a 20% suppression of the Ag signal.

Another source of interference is the loss of analyte prior to initiation of the atomization step. This is of greater concern for HC-FANES because of the lower pressures. Falk *et al.*[15] reported early losses for standards and samples. Dittrich *et al.*[13,16] reported similar losses and improvements in sensitivities and detection limits through the use of matrix modifiers. They credited the improved signals to reduced preatomization analyte losses.

Analyte loss caused by sputtering prior to thermal atomization is unique to the FANES process. This is of greatest concern for HC-FANES because the sample is placed directly on the cathode. Littlejohn *et al.*[8] reported preatomization loss of Cd. Early loss of sulfur as carbon sulfide for standard acid sulfide samples has also been observed. A significant loss of P is observed for rf-FANES if the discharge is ignited prior to initiating the atomization temperature ramp. The severity of this type interference is dependent on gas pressure, and the time interval between start of the discharge and start of atomization.

Sputtering cannot contribute to analyte losses in HA-FANES since the analyte is deposited on the anode. An arc discharge from the cathode to the sample deposition site has been observed at the initiation of the discharge in older furnaces subjected to several hundred firings. It is not clear whether the dried sample perturbed the graphite furnace sufficiently to act as an arc admission point (electric field concentrator) or whether the pyrolytic coating had been roughened in this area to expose a more conductive surface. Reproducibility of the analytical signals appears to be worse when this phenomenon occurs. New furnaces remedy the problem.

Preatomization sputtering losses for rf-FANES appear unlikely. Mean free paths for the cations are small at atmospheric pressure and the low mass of He makes it an inefficient primary ion for sputtering. Arcing to the sample deposition site at the initiation of the discharge has been reported. This problem was alleviated by using a graphite platform within the furnace. The edges of the platform probably served as field concentration sites for discharge initiation.

Emission spectra of the FANES devices are less complex than for ICP-AES, but they are more complex than the simple absorption spectra observed for GF-AAS. Even the broad support gas continuum can present a problem. The magnitude of this continuum increases as a function of increasing temperature. Thus, the background varies throughout the temperature atomization ramp. There is, therefore, a definite need for real-time background correction.

Wavelength modulation has been successfully used with HA-FANES.[18,15,29] The narrow duration of the analytical signals for FANES suggests that the frequency of modulation (56 Hz) used for GF-AES and GF-AAS is not sufficiently rapid.[18,25] Falk *et al.*[4] employed three-step square-wave modulation at 130 Hz, and sine-wave modulation at 200 Hz has been used with HA-FANES. The square wave has a theoretical SNR advantage of a factor of 1.8 relative to sine-wave modulation.[63]

Falk *et al.*[4] reported that wavelength modulation as compared with intensity modulation yielded worse detection limits for a series of spectral resolutions. The intensity modulation, however, did not provide a background correction. If a background value is subtracted, the limiting noise will increase by a factor of 1.4, assuming a quadratic addition of independent noise sources. Correcting the intensity-modulated detection limits by this factor yields comparable values for both methods.

This study just cited concluded that the best SNRs occurred at a resolution of $1\text{–}2 \times 10^4$. This resolution corresponds to a slit width of 200 μm for the echelle grating used. A consistent improvement of the SNRs with larger slit widths was found for four elements. This is unexpected for either the statistical or fluctuation noise-limited cases. The improvement may be associated with the larger viewed region of the furnace.

Structured spectral interferences as opposed to broadband interferences arise from the support gas and strong molecular spectra from entrained air. The structured overlap can be reduced through use of high-resolution spectrometers. Wavelength modulation will also be more susceptible to structured interferences because of the wider spectral range that is viewed. Interferences for Cr at 357.9 nm (CN band) and at 425 nm [Ar (I) line] were observed for a medium-resolution monochromator with wavelength modulation. Sturgeon *et al.*[27] reported an interference from Fe at the Pb line of 283.3 nm for rf-FANES. Sturgeon indicated that ICP wavelength tables were useful for predicting the presence of potential spectral interferences but were not appropriate for rf-FANES with respect to the relative intensities.

9.4.4. Figures of Merit

9.4.4.1. Detection Limits

A tabulation of reported detection limits for the FANES systems are shown in Table 9-4. Detection limits for conventional GF-AAS have been included for comparison. All results were obtained under conditions optimized individually for each element. The operating conditions under which the data in Table 9-4 were acquired are shown in Table 9-2. All values have been corrected to be consistent with a definition of three standard deviations (3σ) of the background noise for the detection limit.

For the limited number of elements for which a comparison is possible, the similarities among the detection limits for the various FANES methods are remarkable considering the disparate operating parameters. It can be seen that the detection limits for Ag, As, Bi, Cd, Co, Pb, Se, and Zn are very similar. B, Cr, Cu, Fe, Ni, Tl, and V (for HC- and rf-FANES) show large discrepancies. It must be remembered that the results for HA- and rf-FANES are very preliminary.

All the FANES detection limits can best be characterized as "comparable" to those for GF-AAS, with one notable exception. Falk *et al.*[5] have shown, using a statistical approach, that the HC-FANES detection limits are better than those for GF-AAS by a factor of three. A direct comparison of the data in Table 9-4 shows that, relative to GF-AAS, the detection limits of HC-FANES are better by a factor of three for eight elements (predominantly the alkalis and alkaline earths), worse by a factor of three for seven elements (predominantly the nonvolatiles), and comparable (less than a factor of three different) for eight elements. The rf-FANES detection limits reported by Sturgeon[21,27,28,30,31] and compared to GF-AAS are better for two elements, worse for three elements, and comparable for eight elements.

The notable exception mentioned above is B. HA- and HC-FANES detection limits for B are 750 times lower than that for conventional GF-AAS. The reason for the large difference is not completely understood,

Table 9-4. Metal Detection Limits (pg)

	FANES							
	HC				HA	rf		
	F[a]	N	D	L	H	B	S	GF-AAS
Ag	0.6			4.		0.3	0.5*	0.8*
Al	22.							6.*
As						103.*	30.*	
Au	4.							15.*
B	2.			80.	2.			1500.*
Be							5.*	2.*
Bi	45.						25.	15.
Ca	0.08							8.
Cd	0.6			1.	2.		0.9*	0.4*
Co		5.		4.				3.*
Cr	6.	0.5			0.8			2.*
Cu	1.	1.[b]			0.5		5.	2.
Dy			200.					57.
Er			220.					100.
Eu			100.[c]					24.
Fe	7.			0.7			50.	3.*
Ga				0.6				
Ho			440.					
K	0.04							3.
Li	0.03							3.
Lu			10,800.					
Mg				3.				
Mn							0.9*	2.*
Mo	600.			32.				6.
Na	0.06							<75.
Ni	2.	1.[d]					164.	8.*
Pb	4.			12.			7.2*	4.*
Rb	2.							8.
Sc			30.					
Se	800.						1700.	45.
Sm			70.[c]					240.
Sn							10.*	30.*
Tb			3,100.					
Tc			90.*				13.	
Ti	1200.							75.
Tm			300.					
Tl	0.3						17.*	15.*
V	8.			184.				30.
Y			2,600.					13,000.
Yb			20.					2.
Zn	3.			4.				0.8*

[a]F, Falk *et al.*[15]; N, Naumann *et al.*[17]; D, Dittrich *et al.*[13,16,26]; L, Littlejohn *et al.*[14]; H, Harnly *et al.*[25,29]; B, Blades *et al.*[24]; S, Sturgeon *et al.*[28,30]

[b]327.4 nm.

[c]Ion line.

[d]352.4 nm.

*Matrix modifier used.

although initial data suggest that collisions with the high-energy electrons break down B_2O_3, which is lost from the furnace using straight furnace atomization with GF-AAS.

9.4.4.2. Reproducibility

The reproducibility is defined as the relative standard deviation for the analyte signals well above (greater than ten times) the detection limit. Reproducibility runs from 2% to 4% and is element dependent for GF-AAS using autosamplers. At these levels (greater than ten times the detection limit), the reproducibility is dependent on the concentration, i.e., signal uncertainty is dependent on signal strength. The source of signal error is generally attributed to the uncertainty of the sample position after the drying process. Platforms in the furnace restrict the sample position but produce smaller, broadened signals.

Standard deviations reported by Falk *et al.*(15) for HC-FANES are 2–3%. Harnly *et al.*(25) reported values of 2–6% for the HA-FANES. Sturgeon *et al.*(21,27) reported rf-FANES relative precisions ranging from 3 to 10% at atmospheric pressure. Although the upper values in these last two cases (6 and 10%, respectively) may seem high relative to GF-AAS, it must be remembered that these values were obtained using manual sample delivery without the use of platforms. The existing data show that the discharge only has an adverse effect on analytical precision at the extremes of the current range (see Section 9.4.1.4).

9.4.4.3. Dynamic Range

The available data suggest that the dynamic ranges for the FANES devices are not comparable to those observed for ICP-AES (five to six orders of magnitude). The major disadvantage for FANES is that the source is not optically thin. The furnace is 19–28 mm long and is occupied predominantly by ground state atoms that are capable of absorbing the emitted photons. This situation is made worse for HC- and rf-FANES by the nonuniformity of the temperature along the furnace length. The temperature at the end of the furnace can be several hundred degrees cooler than in the middle.(38,39) This gives rise to condensation near the ends of the furnace and further enhances self-absorption processes. This effect was so severe for GF-AES that for some elements it is impossible to observe ground state transitions that are not self-reversed, even at low concentrations.(64) This is not a problem for HA-FANES, which uses isothermal furnaces (ICC). This furnace is shorter and has a uniform temperature along its length. The nonoptimum geometry may be partially remedied for HC-FANES by operation at reduced pressures (1–5 Torr).

Working ranges for HC-FANES for Ag, Na, and Pb cover 5–6 orders of magnitude of concentration with less than a 10% deviation from linearity.[3,5] Data for Cd show curvature occurring, however, after 2 orders of magnitude.[6] Dittrich *et al.*[16] reported calibration ranges of 3 orders of magnitude for the rare earth elements. Dynamic ranges for rf-FANES also show reduced linearity. Sturgeon has observed linear ranges of 2–4 orders of magnitude, and Blades *et al.*[24] reported a usable concentration range for Ag of 4 orders of magnitude with linearity over 2–3 orders of magnitude. Initial data for HA-FANES suggest dynamic ranges of 3–4 orders of magnitude at discharge pressures of 70–200 Torr in Ar.

9.5. Applications

Considering the body of literature to be found for the various FANES devices, there are relatively few applications of the technique. To a large extent, this has been due to the general lack of availability of the instrumentation. In the last two years, however, the number of operational FANES devices has tripled. Hopefully, this trend will continue and the number of applications-oriented research papers will increase dramatically.

9.5.1. Metal Determinations

One of the earliest systematic applications of HC-FANES was the determination of Cd in deproteinized whole blood.[6] The results, obtained using the method of standard additions for two different samples, were in excellent agreement with those obtained by GF-AAS (97 and 99% recovery). These determinations were made using 19 Torr Ar support gas pressure, a discharge current of 30 mA, and an atomization temperature of 900 K, fairly standard conditions for HC-FANES (Table 9-2).

A study of the effect of the metal chloride concentrations on the analytical determination was discussed previously (Section 9.4.3). The data suggested that the inorganic component (as chlorides) was unlikely to provide an interference. Spike recoveries in the deproteinized whole blood, however, gave only approximately 20% recovery. It was concluded that other inorganic or organic components were responsible for the low recoveries and necessitated the use of the method of additions.

Falk *et al.*[9] determined Cu, Fe, and Ni in grass and corn (Soviet Union reference materials) and orchard leaves (SRM 1571, USA). These determinations were made simultaneously using solid samples and a specially designed furnace. The results were very acceptable with elemental recoveries of 92% to 103% in the orchard leaves and 92% to 140% in the grass and corn standards. Argon was used as the support gas at 25.4 Torr with a discharge current of 30 mA and an atomization temperature of 2700 K.

In another simultaneous multielement study, Falk *et al.*[15] determined Al, Co, Cr, Cu, Fe, and Ni doped into deionized water and HNO_3. The recoveries were generally low for the deionized water and were attributed to the preatomization loss of the elements (see Section 9.4.3). These results led to a detailed study on the optimum drying and ashing conditions. The subsequent recoveries in HNO_3 were within $\pm10\%$ of 100% for HNO_3 concentrations up to 15%. The analytical conditions used were 15.4 Torr of Ar as the support gas, a discharge current of 30 mA, and an atomization temperature of 2300 K.

Dittrich *et al.*[16] recently employed HC-FANES for the determination of 11 rare earth elements (Dy, Er, Eu, Ho, Lu, Sc, Sm, Tb, Tm, Y, and Yb) in the single-element mode using both atomic and ionic transitions. The detection limits ranged from 20 to 530 pg for all the elements except Y, Tb, and Lu, which were 2600, 3100, and 10,800 pg, respectively (Table 9-4). In the presence of other rare earth elements the emitted intensities for each of the elements were suppressed. This placed the detection limits in a rare earth matrix at 10 to 1000 ng. Ar was used as the support gas at 12 Torr with a discharge current of 30 mA and atomization temperatures of 2100 and 3000 K.

Dittrich *et al.*[13] also examined the determination of ^{99}Tc using FANES. They obtained a detection limit of 90 pg using $Ni(NO_3)_3$ as a matrix modifier and a tungsten platform. The matrix modifier helped prevent preatomization losses and losses caused by molecular formation. The platform reduced the formation of carbides. Optimum atomization conditions were obtained for 11 Torr Ar, a discharge current of 30 mA, and a temperature of 2900 K.

In another interesting study, Dittrich *et al.*[26] determined Sb using a hydride HC-FANES method. The use of cold (external trap with liquid nitrogen) versus hot (in the graphite furnace) trapping was investigated along with atomization with and without cooling the furnace after the trapping. A detection limit of 14 pg was obtained using a hot trap and immediate atomization (no cooling) with an Ar pressure of 10 Torr, a discharge current of 60 mA, and an atomization temperature of 1400 K. The detection limit for GF-AAS is cited as 22 pg with straight atomization and matrix modification.

Using rf-FANES, Sturgeon *et al.*[27] determined Cd and Pb in three reference materials from the National Research Council of Canada: marine sediment (BCSS-1), dogfish mussel (DORN-1), and lobster hepatopancreas (TORT-1). Aqueous calibration proved satisfactory for Cd, using peak area measurements, but for Pb it was necessary to employ the method of standard additions to obtain accurate results. Recoveries, with the method of standard additions, ranged from 83% to 105% for peak area and 104% to 121% for peak height. These results were obtained using a He support gas at 760 Torr, an atomization temperature of 1400 K, and 50 and 75 W plasma powers for Cd and Pb, respectively.

9.5.2. Nonmetal and Molecular Determinations

One of the attractive features of the FANES devices is the ability to determine the nonmetals under conditions similar to those used for metals. Dittrich *et al.*(7,11) investigated the atomic emission of Br, Cl, and F and their molecular emission as In and Mg compounds. The molecular emission detection limits (Table 9-5) are factors of 2 to 30 times better than those for atomic emission. The detection limit for Cl was slightly better as a Mg compound while the In complex was preferable for F. The atomic emission detection limits agree with those reported by Littlejohn(14) but the detection limit for Cl was considerably worse than that reported by Falk *et al.*(15) The molecular emission detection limits for HC-FANES were also found to be slightly better than those obtained using molecular absorption with furnace volatilization.(7) The analytical sensitivity for each of the halides was found to be significantly suppressed by the presence of the other halides. These data were obtained using either Ar or He at 15 Torr, a 30 mA discharge current, and atomization temperatures of 2100 K for In and 2400–2700 K for Mg.

Dittrich and Fuchs(22,23) investigated the possibility of determining P in the atomic and molecular emission modes. For atomic emission, the HC-FANES detection limit (600 pg) was improved by a factor of 7 with the use of La as a matrix modifier (90 pg). The improved detection limit with the matrix modifier was attributed to greater atomization efficiency, i.e., reduced losses from preatomization evaporation and from molecule formation. The optimal atomization parameters for P were Ar at 15 Torr, 40 mA, and

Table 9-5. Nonmetal Detection Limits (pg)

	HC^a					
	F	D		L		
Element	Atom	Atom.	Mole.	Atom.	Mole.	GF-AAS
Br	—	12,000	6000*	23,000	—	—
Cl	120	8,000	240*	6,800	5100	—
F	—	—	250*	38,000	—	—
I	—	9,000	—	4,800	—	—
P	—	90*	700*	210	—	4500*
S	—	—	—	—	4000	—
Si	—	—	—	7,400	—	—

^{a}HC, HC-FANES; F, Falk *et al.*(15); D, Dittrich and Fuchs(11,22,23); L, Littlejohn.(14)
*Matrix modification.

2100 K. The optimal parameters for P in the presence of La were Ar at 15 Torr, 80 mA, and 2300 K. Dittrich and Fuchs[23] also examined the molecular emission of P as PO and HPO. While the detection limits for both PO and HPO were found to be better than those for GF-AAS, they were not as good as those found for the atomic emission of P in the presence of La. For these molecules the discharge current was found to represent a compromise; increased current led to better excitation but also led to greater dissociation of the molecules. For PO, the most intense emission was observed at 15 Torr of Ar, 80 mA, and 1700 K.

Littlejohn has examined the feasibility of determining S as CS. No signals were detectable for the atomic S. The CS band at 257.9 nm proved the most sensitive. A major problem was the change in the background continuum as a function of temperature. The emission bands were too broad to permit use of wavelength modulation. A detection limit of 4 ng was obtained.

9.6. Simultaneous Multielement Determinations

A predicted advantage for any of the FANES devices, even if the detection limits are only comparable to GF-AAS, is the potential for multielement determinations. The simplified optical requirements of the emission mode and the available technology for multichannel spectrometers would seem to place multielement furnace determinations within reach. At this time the future of multielement FANES is uncertain. This is the result of the lack of data characterizing the dependency of the analytical signals and the SNR on the atomization temperature and the discharge pressure and current. Perhaps even more critical, data that characterize the compromise in the accuracy of the determinations (the presence of interferences) with respect to the operating parameters are not available.

Simultaneous multielement determinations have been investigated by Falk *et al.*[15] in two different experiments. The first study determined six elements (Al, Cr, Cu, Fe, Mn, and Ni) doped in deionized water; this permitted the effect of matrix interferences to be ignored. The second study determined six elements (Al, Co, Cr, Cu, Fe, and Ni) doped in HNO_3 and in NaCl.

In the first study the compromise atomization parameters were a 4-s atomization at 2100 K, Ar support gas at 19 Torr, and a discharge current of 30 mA. A conventional multielement spectrometer was used with the adjunct computer-controlled data acquisition system. The recovery of 5 and 50 μg liter^{-1} additions ranged from 60% to 124%, with all but two of the values falling between 81% and 97%. Fe provided the highest recovery

(124%) and Al the lowest (60%). The generally low recoveries were attributed to excessive pretreatment temperatures and system evacuation before complete drying was accomplished.

The second study, in HNO_3 and NaCl, employed almost the same atomization conditions (a support gas pressure of 15 Torr was used instead of 19 Torr) and the same spectrometer and data acquisition system. The drying and ashing temperature programs were carefully optimized for the element to be analyzed. The multielement detection limits ranged from 5 to 30 times worse than the single-element values previously reported. These poor detection limits were the result of the long integration time (1.5 s) of the data acquisition system, which was an order of magnitude too large for most elements. Optimum time gating for each element was not possible with the adjunct computerized data acquisition system (the detection system was designed for another source). The relative precisions of the elements ranged from 2% to 3% for the short term (within experiment) and from 4% to 10% for the long term (day-to-day). These values are consistent with results observed for single-element operation. The linear range was 2.5 to 3.5 orders of magnitude for all six elements using the poorer detection limits, and 3.5 to 4.5 orders of magnitude if extrapolated to the best detection limits. The recoveries of the six elements in HNO_3 and NaCl have been discussed in detail in Section 9.4.3. The severity of the interferences were related to the elemental volatility and seemed little affected by the compromise multielement atomization conditions.

Naumann *et al.*[17] performed optimization studies for HC-FANES using Co, Cr, Cu, and Ni. Optimum single-element conditions were found to be temperatures of 2123, 2161, 2310, and 2623 K, pressures of 102, 141, 40, and 116 Torr, and currents of 59, 90, 90, and 90 mA for Co, Cr, Cu, and Ni, respectively. The compromise multielement conditions, optimized from the SNR, were an atomization temperature of 2315 K, an Ar pressure of 120 Torr, and a discharge current of 68 mA. The compromise detection limits were degraded by less than a factor of 2. No analyses were undertaken using these parameters.

Initial single-element results for HC- and rf-FANES suggest that the optimum parameters found for the nonmetals are not significantly different from those for the metals.

It is intriguing to consider the possibility of simultaneous, state-of-the-art detection limits for both metals and nonmetals. An accurate assessment, however, of the simultaneous multielement capabilities of the FANES must await further studies. These studies must include a wider selection of elements including the nonmetals, a spectrometer system optimized for the rapid FANES signals and associated wavelengths, and determinations in a large variety of reference materials.

Acknowledgment

Pacific Northwest Laboratory operated for the U.S. Department of Energy at the Battelle Memorial Institute under Contract DE-ACO6-76RLO 1830.

References

1. H. Falk, Einige theoretiche uberlegungen zum vergleich der physikalischen grenzen thermischer und nicht-thermischer spektroskopischer strahlungsquellen, *Spectrochim. Acta* 32B (1977) 437.
2. H. Falk, E. Hoffmann, I. Jaeckel, and C. Ludke, Atomic emission trace analysis by non-thermal excitation, *Spectrochim. Acta* 34B (1979) 333.
3. H. Falk, E. Hoffmann, and C. Ludke, FANES (furnace atomic non-thermal excitation spectrometry)—a new emission technique with high detection power, *Spectrochim. Acta* 36B (1981) 767.
4. H. Falk, E. Hoffmann, C. Ludke, J. M. Ottaway, and S. K. Giri, Furnace atomization with non-thermal excitation—experimental evaluation of detection based on a high-resolution echelle monochromator incorporating automatic background correction, *Analyst* 108 (1983) 1459.
5. H. Falk, E. Hoffmann, and C. Ludke, A comparison of furnace atomic non-thermal excitation spectrometry (FANES) with other atomic spectroscopic techniques, *Spectrochim. Acta* 39B (1984) 283.
6. H. Falk, E. Hoffmann, C. Ludke, J. M. Ottaway, and D. Littlejohn, Studies on the determination of cadmium in blood by furnace atomic non-thermal excitation spectrometry, *Analyst* 111 (1986) 285.
7. K. Dittrich, B. Hanisch, and H.-J. Stark, Molecule formation in electrothermal atomizers: Interferences and analytical possibilities by absorption, emission and fluorescence processes, *Fresenius Z. Anal. Chem.* 324 (1986) 497.
8. D. Littlejohn, J. Carroll, A. M. Quinn, J. M. Ottaway, and H. Falk, Comments on the characteristics of an atomizer for furnace atomic non-thermal excitation spectrometry (FANES), *Fresenius Z. Anal. Chem.* 323 (1986) 762.
9. H. Falk, E. Hoffmann, C. Ludke, and K. P. Schmidt, Untersuchungen zur direktanalyse fester pflanzlicher stoffe mittels FANES (furnace atomic non-thermal excitation spectrometry), *Spectrochim. Acta* 41B (1986) 853.
10. H. Falk and J. Tilch, Atomization efficiency and over-all performance of electrothermal atomizers in atomic absorption, furnace atomization non-thermal excitation and laser-excited atomic fluorescence spectrometry, *J. Anal. At. Spectrom.* 2 (1987) 527.
11. K. Dittrich and H. Fuchs, Molecular non-thermal excitation spectrometry (MONES): A procedure for the determination of non-metals using diatomic molecules in the non-thermal (FANES) atomizer; Part 1. Determination of fluoride and chloride ions by magnesium fluoride and magnesium chloride MONES, *J. Anal. At. Spectrom.* 2 (1987) 533.
12. H. Falk, Hollow-cathode discharge within a graphite furnace: Furnace atomic non-thermal excitation spectrometry (FANES), in: *Improved Hollow Cathode Lamps for Atomic Spectroscopy* (S. Caroli, ed.), pp. 74–118, Ellis Horwood Ltd., Halstead Press, New York, 1987.
13. K. Dittrich, T. Glaubauf, H. Fuchs, and K. Mauersberger, Analytical applications of furnace atomization non-thermal excitation spectrometry (FANES) and molecular non-thermal excitation spectrometry (MONES); Part 2. Determination of technetium-99 by FANES and electrothermal atomization atomic absorption spectrometry, *J. Anal. At. Spectrom.* 3 (1988) 89.

14. D. Littlejohn, Graphite furnace atomic emission spectrometry—the rediscovery of a technique, *Anal. Proc.* 25 (1988) 217.
15. H. Falk, E. Hoffmann, and C. Ludke, Experimental and theoretical investigations relating to FANES, *Prog. Anal. Spectrosc.* 11 (1988) 417.
16. K. Dittrich, G. Eismann, and H. Fuchs, Analytical applications of furnace atomization non-thermal excitation spectrometry (FANES) and molecular non-thermal excitation spectrometry (MONES); Part 3. Determination of rare earth elements by electrothermal atomization atomic emission spectrometry (ETA-AAS), FANES and furnace ionization non-thermal excitation spectrometry (FINES), *J. Anal. At. Spectrom.* 3 (1988) 459.
17. B. Naumann, B. Knull, F. Kerstan, and J. Opfermann, Multivariate optimization of simultaneous multi-element analysis by furnace atomic non-thermal excitation spectrometry (FANES), *J. Anal. At. Spectrom.* 3 (1988) 1121.
18. N. E. Ballou, D. L. Styris, and J. M. Harnly, Hollow-anode plasma excitation source for atomic emission spectrometry, *J. Anal. At. Spectrom.* 3 (1988) 1141.
19. D. Littlejohn, Becoming absorbed and excited in atomic spectrometry, *Anal. Proc.* 26 (1989) 92.
20. D. C. Liang and M. W. Blades, An atmospheric pressure capacitively coupled plasma formed inside a graphite furnace for atomic emission spectroscopy, *Spectrochim. Acta* 44B (1989) 1059.
21. R. E. Sturgeon, S. N. Willie, V. Luong, S. Berman, and J. G. Dunn, Furnace atomization plasma emission spectrometry (FAPES), *J. Anal. At. Spectrom.* 4 (1989) 669.
22. K. Dittrich and H. Fuchs, Analytical applications of furnace atomic non-thermal excitation spectrometry (FANES) and molecular non-thermal excitation spectrometry (MONES); Part 3. Determination of trace amounts of phosphorous by FANES, *J. Anal. At. Spectrom.* 4 (1989) 705.
23. K. Dittrich and H. Fuchs, Analytical applications of furnace atomic non-thermal excitation spectrometry (FANES) and molecular non-thermal excitation spectrometry (MONES); Part 5. Study of the MONES of PO and HPO for the determination of trace amounts of phosphorous, *J. Anal. At. Spectrom.* 5 (1990) 39.
24. D. L. Smith, D. C. Liang, D. Steel, and M. W. Blades, Analytical characteristics of furnace atomization plasma excitation spectrometry (FAPES), *Spectrochim. Acta* 45B (1990) 493.
25. J. M. Harnly, D. L. Styris, and N. E. Ballou, Furnace atomic non-thermal excitation spectrometry with the furnace as a hollow anode, *J. Anal. At. Spectrom.* 5 (1990) 139.
26. K. Dittrich, B. Radziuk, and B. Welz, Investigations of the determination of chloride and bromide by furnace atomic non-thermal excitation spectrometry and furnace ionic non-thermal excitation spectrometry, *Spectrometry* 6 (1991) 465.
27. R. E. Sturgeon, S. N. Willie, V. T. Luong, and S. S. Berman, Determination of cadmium and lead in sediment and biota by FAPES, *J. Anal. At. Spectrom.* 5 (1990) 635.
28. R. E. Sturgeon, S. N. Willie, V. Luong, and S. S. Berman, Figures of merit for furnace atomization plasma emission spectrometry, *Anal. Chem.* 62 (1990) 2370.
29. P. G. Riby, J. M. Harnly, D. L. Styris, and N. E. Ballou, Emission characteristics of chromium in hollow anode-furnace atomization non-thermal excitation spectrometry, *Spectrochim. Acta* 46B (1991) 203.
30. R. E. Sturgeon, S. N. Willie, V. T. Luong, and S. S. Berman, Application of platform and palladium modification techniques with furnace atomization plasma emission spectrometry, *J. Anal. At. Spectrom.* 6 (1991) 19.
31. R. E. Sturgeon, S. N. Willie, V. T. Luong, and S. S. Berman, Characteristic temperatures in a FAPES source, *Spectrochim. Acta* 46B (1991) 1021.
32. C. T. J. Alkemade, T. J. Hollander, W. Snelleman, and P. J. T. Zeegers, *Metal Vapors in Flames*, Pergamon Press, Elmsford, N.Y., 1982.
33. R. E. Sturgeon and S. S. Berman, Analyte ionization in graphite furnace–atomic absorption spectrometry, *Anal. Chem.* 53 (1981) 632.

34. W. Slavin, *Graphite Furnace AAS—A Source Book*, Perkin–Elmer Corp., Norwalk, Conn., 1984.
35. J. M. Ottaway and F. Shaw, Carbon furnace atomic-emission spectrometry: A preliminary appraisal, *Analyst* 100 (1975) 438.
36. M. S. Epstein, T. C. Rains, and T. C. O'Haver, Wavelength modulation for background correction in graphite furnace atomic emission spectrometry, *Appl. Spectrosc.* 30 (1976) 324.
37. W. Frech, D. C. Baxter, and B. Hutsch, Spatially isothermal graphite furnace for atomic absorption spectrometry using side-heated cuvettes with integrated contacts, *Anal. Chem.* 58 (1986) 1973.
38. R. E. Sturgeon and S. S. Berman, Determination of the efficiency of the graphite furnace for atomic absorption spectrometry, *Anal. Chem.* 55 (1983) 190.
39. B. Welz, M. Sperling, G. Schlemmer, N. Wenzel, and G. Marowsky, Spatially and temporally resolved gas phase temperature measurements in a Massman-type graphite tube furnace using coherent anti-Stokes Raman scattering, *Spectrochim. Acta* 43B (1988) 1187.
40. P. J. Slevin and W. W. Harrison, The hollow cathode discharge as a spectrochemical emission source, *Appl. Spectrosc. Rev.* 10 (1975) 201.
41. S. Caroli, Low-pressure discharges: Fundamental and applicative aspects, *J. Anal. At. Spectrom.* 2 (1987) 661.
42. M. E. Pillow, A critical review of spectral and related physical properties of the hollow cathode discharge, *Spectrochim. Acta* 36B (1981) 821.
43. J. A. C. Broekaert, State of the art of glow discharge lamp spectrometry, *J. Anal. At. Spectrom.* 2 (1987) 537.
44. K. G. Hernquist and E. O. Johnson, Retrograde motion in gas discharge plasmas, *Phys. Rev.* 98 (1955) 1576.
45. H. L. Wittig, Hollow cathode discharge with thermionic cathodes, *J. Appl. Phys.* 42 (1971) 5478.
46. S. N. Salinger and J. E. Rowe, Monte Carlo simulation of the low-voltage arc mode in plasma diodes, *J. Appl. Phys.* 39 (1968) 3933.
47. R. M. Martin and R. E. Rowe, Experimental investigations of the low-voltage arc in noble gases, *J. Appl. Phys.* 39 (1968) 4289.
48. G. K. Wehner, *J. Appl. Phys.* 31 (1960) 1392.
49. M. Kaminsky, *Atoms and Ionic Impact Phenomena on Metal Surfaces*, p. 143, Academic Press, New York, 1965.
50. A. von Hippel, Kathodenzerstaubungsprobleme III zur theorie der kathodenzerstaubung, *Ann. Phys.* 81 (1976) 1043.
51. D. Bohm, Minimum ionic kinetic energy for a stable sheath, in: *The Characteristics of Electrical Discharges in Magnetic Fields* (A. Guthrie and R. K. Walkerling, eds.), pp. 77–86, McGraw–Hill, New York, 1949.
52. F. F. Chen, *Introduction to Plasma Physics*, Plenum Press, New York, 1949.
53. B. Chapman, *Glow Discharge Processes*, Wiley, New York, 1980.
54. A. von Engel, *Ionized Gases*, pp. 214–228, Oxford University Press, London, 1955.
55. C. P. Herring and M. H. Nichols, Thermionic emission, *Rev. Mod. Phys.* 21 (1949) 185.
56. I. Langmuir, The effect of space charge and residual gases on the thermionic current in high vacuum, *Phys. Rev.* 2 (1913) 450.
57. P. F. Little and A. von Engel, The hollow cathode effect and the theory of glow discharges, *Proc. R. Soc. London Ser. A* 224 (1954) 209.
58. S. D. McDonald and S. J. Tatenbaum, High frequency and microwave discharges, in: *Gaseous Electronics* (M. M. Hirsh and H. G. Oskam, eds.), Vol. 1, pp. 173–217, Academic Press, New York, 1978.
59. L. Holland, W. Steckelmacher, and Y. Yarwood, *Vacuum Manual*, pp. 384–385, Spon, London, 1974.

60. J. M. Keller and W. B. Pennebacker, Electrical properties of rf sputtering systems, *IBM J. Res. Dev.* 23 (1979) 3.
61. V. P. Gofmeister and Y. M. Kagaan, Mechanism of excitation in a hollow cathode in argon, *Opt. Spectrosc.* (*USSR*) (*Engl. Transl.*) 26 (1969) 379.
62. H. M. Crosswhite, G. H. Dieke, and C. S. Legagneur, Hollow iron cathode discharge as a source for wavelength and intensity standards, *J. Opt. Soc. Am.* 45 (1955) 270.
63. T. C. O'Haver, Waveform effects in wavelength modulation spectrometry, *Anal. Chem.* 49 (1977) 458.
64. J. E. Marshall, D. Littlejohn, J. M. Ottaway, J. M. Harnly, N. J. Miller-Ihli, and T. C. O'Haver, Simultaneous multi-element analysis by carbon furnace atomic-emission spectrometry, *Analyst* 108 (1983) 178.

10

Laser-Based Methods

Kenneth R. Hess

10.1. Introduction

The development of cost-effective laser systems has generated a host of laser hyphenated techniques that have been introduced into the analytical laboratory. These hyphenated techniques take advantage of the laser's ability to deliver a high photon flux, high photon energies, and a narrow, tunable photon wavelength range to optimize specific atomization/excitation/ionization processes in an analytical procedure. A laser system coupled to a glow discharge is one such hybrid technique that offers unique opportunities for both diagnostic and analytical investigations. This chapter will serve as an introduction and overview of several reported methodologies that have advantageously combined laser systems and glow discharges.

The analytical utility of the glow discharge results in part from a two-step method of forming analytical species of interest. In the first step, the solid cathode material is sputtered from the surface by the impacting ions of the rare gas employed in the discharge. The sputtering process is largely nonselective, with relative elemental atomization rates for this step varying by less than an order of magnitude. In addition, it has been postulated that under steady-state conditions the elemental sputter rates become equivalent, as discussed in Chapter 2. Overall, the sputtering process creates an atomic population that is highly representative of the bulk cathode composition, providing an excellent atom source for atomic spectroscopy. The second step

Kenneth R. Hess • Department of Chemistry, Franklin and Marshall College, Lancaster, Pennsylvania 17604

Glow Discharge Spectroscopies, edited by R. Kenneth Marcus. Plenum Press, New York, 1993.

of the process involves the collisional excitation/ionization of the atomic population, generating excited-state sputtered atoms for atomic emission and ions for mass spectrometry. The ionization step is also relatively nonselective, providing an ionic population that is representative of the atomic population.

A laser system may be coupled to the discharge in order to enhance either of these two steps. First, the high power and spatial resolution of the laser may be advantageously employed to ablate material from the sample cathode, enhancing sample atomization for subsequent excitation and ionization in the discharge. Alternatively, the laser may ablate material from a secondary sample (not the cathode) into the discharge for excitation and ionization. This method would not require the sample to be conducting, unlike the cathode of a normal dc glow discharge, which is required to be electrically conducting. This expands the possible analytical applications of glow discharge excitation and ionization to nonconducting materials without employing rf discharges.[1,2] Second, the high photon energy and tunability of the laser may be employed as a method for enhancing the excitation processes in the discharge. Coupling of the laser to the discharge for subsequent excitation of the discharge sputtered atoms has been effectively demonstrated in a variety of application areas including atomic fluorescence, optogalvanic effect spectroscopy, metastable atom depopulation studies, resonance ionization, and degenerate four-wave mixing experiments. Either method of interfacing laser systems to the glow discharge source, enhanced atomization or enhanced excitation, provides opportunities to investigate fundamental discharge processes, with several of the laser/discharge techniques found also to be of analytical utility. Basic discussions of laser enhancement of atomization and laser enhancement of excitation, with representative applications, are provided in this chapter as an introduction to the topic.

10.2. The Laser

Currently, there are many different types of laser systems available for interaction with the glow discharge. The key characteristics that provide the exceptional utility of the laser include: (1) the high spatial coherence and directional properties of the beam, which allow easy manipulation and focusing of the beam, providing high photon densities and spatial resolution, (2) a high photon flux, which allows processes with low cross sections, such as multiphoton events, to be investigated, (3) high photon energies combined with the high photon flux, which will allow high energy densities for processes requiring photon-induced damage, such as laser ablation, (4) with the use of a dye laser and frequency doubling crystal, monochromatic, tunable photon wavelengths from the UV to IR are available to optimize specific excitation or atomization processes in an analytical procedure, and (5) the laser pulse duration can be very fast, down to the femtosecond range, allowing very fast chemical and physical processes to be investigated.

Detailed descriptions of the fundamentals of laser operation are beyond the scope of this chapter and are covered in many excellent sources.[3–9] Essentially, lasers result from the excitation of a gain medium to a condition where a high energy level is populated to a greater extent than at a lower level, termed a population inversion. Electrons in the higher energy level are then stimulated to relax simultaneously, releasing a large number of photons with the same wavelength at the same time, generating the coherent laser pulse. The method for creating the population inversion and the medium employed are varied, which gives rise to the large number of potential laser systems, including gas lasers (e.g., excimer systems, N_2, CO_2, copper vapor), solid-state lasers (Nd:YAG, ruby), semiconductor lasers, and liquid dye laser systems. Each of these systems has its own characteristic power levels, pulse rates and durations, wavelength ranges, and expense. With this array of choices of laser instrumentation, many different lasers have been employed in conjunction with the glow discharge. The specific choice of a laser system to use is dependent on the specific application or investigation desired, and each type of laser will have its own advantages and disadvantages.

The details of how to make such a choice are far too extensive to elaborate on in this chapter, but a few general comments can be made. For supplemental atomization in a glow discharge, the laser should be optimized to generate a high power density. The repetition rate can be important if the signal is to be accumulated or if a sequential scanning system, such as a quadrupole mass spectrometer, is employed. Wavelength is of less importance. For supplemental excitation, a broad range of wavelength tunability is a great asset for selectively exciting particular atomic transitions. A high power system is advantageous so as to saturate transitions and increase the sensitivity of the techniques. A high pulse rate will allow for a higher duty cycle and enhance the analytical utility of the system. Further information on the various laser systems, their particular attributes, and their applications can be found in the previously referenced sources.

10.3. Supplemental Atomization

Although glow discharge sputtering effectively creates an atomic population representative of the conductive cathode material, there are unique advantages to using a high-powered laser as a source for supplemental or secondary atomization. The ability to create an atomic population independent of direct discharge processes is beneficial for the analysis of nonconducting materials and for investigations of fundamental glow discharge processes such as the mechanisms of ionization present in the discharge and the fundamental basis of relative sensitivity coefficients. These fundamental studies of glow discharge processes have been the main application of the laser ablation/glow discharge technique and will be the primary focus in this section.

10.3.1. Introduction to Laser Ablation

When a high-power laser beam is focused onto a sample surface, a laser-generated plume of material can result. This plume consists of ground-state atoms and multiatom clusters of the sample material, and excited-state atoms or ions of the sample. The atoms formed in such an ablation process have found application as a reservoir for atomic absorption measurements[10,11] and atomic fluorescence,[12–14] while the excited-state atoms have been monitored by atomic emission[15–17] and the ions extracted into mass spectrometers for mass spectral analysis,[18–21] including the LAMMA instruments.[22–24] In these latter methods (AE and MS), the laser is serving as the method of both atomization and excitation. In addition, laser ablation has been advantageously employed as a solid sample introduction system for a variety of subsequent ionization and excitation sources, including inductively coupled plasma (ICP) emission,[25–29] ICP mass spectrometry,[30–34] microwave discharges,[35] dc plasma devices,[36] arc/spark emission sources,[37] resonance ionization methods,[38,39] and ion cyclotron resonance chambers for the investigation of atom/ion reactions.[40] These methods may be extended to the glow discharge source, with the laser serving as an ablation source for injecting solid material into the glow discharge for subsequent excitation and ionization, or for direct laser vaporization and ionization of the cathode material which can assist in fundamental studies of discharge processes.

10.3.2. Mechanisms of Laser Ablation

The mechanism of laser ablation is complex and not clearly defined, with research into laser/solid interactions continuing.[41–49] The dominant mechanism of the laser/solid interaction is believed to be a thermal heating of the sample originating from the exchange of energy from the photons absorbed by the solid and the solid's electrons, which then exchange energy with the bulk lattice.[50] The basic idea behind laser ablation is to dump a large amount of energy into the sample surface in a very short period of time, rapidly raising the temperature of the solid and vaporizing material from the surface. Any parameters that impact the rate at which energy is absorbed by the material will influence the effect of a laser pulse on a solid. These parameters are most often divided into two categories: properties of the solid and characteristics of the laser.

There are several properties of the material that can influence the amount of energy dumped into the solid during the laser pulse. These include the material's reflectivity and absorption coefficient for the wavelength of incident photons, and the thermal properties of the material such as volume specific heat, latent heat, and thermal conductivity. The reflectivity and absorption coefficient of material can impact the number of photons absorbed by the sample at lower power densities. As the power level is increased to levels generally encountered in laser ablation studies, the sample ultimately acts as a "blackbody" and absorbs all of the incident photons, limiting the influence of the reflectivity and absorption coefficient on the degree of ablation. The additional material parameters that can influence the extent of laser interaction are those that control the rate of laser heating of the material. The volume specific heat of the material will determine the amount of energy required to raise the surface temperature of the sample to the threshold of vaporization, with the latent heat determining the excess energy required to undergo the phase change and vaporize the material. The lower these values are, the lower the energy required for vaporization and the greater impact of the laser pulse. Thermal conductivity is an important parameter in determining how much heat can dissipate through the solid. The greater the thermal conductivity, the harder a material is to ablate. At sufficiently high power densities, enough energy will be available at the sample surface to rapidly vaporize the material regardless of the volume specific heat, latent heat, or thermal conductivity and these material parameters will lose their importance. Ultimately, at sufficient laser power densities, material properties have little effect on the ablation process.

Laser characteristics that can impact the ablation process include wavelength, pulse duration, and laser power density. At low laser power densities ($<10^5$ W/cm^2), there is a laser wavelength dependence on the ablation process. This dependence originates from a relationship between reflectance and λ, with the reflectance increasing with increasing λ. After the laser power density is sufficiently large to cause the sample to behave as a blackbody ($>10^8$ W/cm^2), there is almost no wavelength dependence on the ablation process. The laser pulse length can influence the amount of energy dissipated into the material, with a longer pulse allowing the thermal conductivity of the sample to transfer absorbed energy into the bulk sample and limit the heat built up on the surface. Short pulses (nanosecond–microsecond) are normally used in an effort to limit the energy losses caused by thermal conductivity and to optimize the power density of the laser.

Laser power density appears to be the most important laser characteristic to impact the ablation process, with three distinct regions of laser interaction occurring. At power levels $<10^4$ W/cm^2, simple laser heating of the surface occurs. When the laser power density is increased to the range of 10^4–10^6 W/cm^2 the sample surface begins to melt with some differential boiling of elements with low vapor pressures. When the power density reaches 10^6–10^8 W/cm^2 the sample undergoes laser vaporization with a strong dependence on the thermal conductivity of the sample observed.[51] As the laser power density is increased above 10^8 W/cm^2, the surface temperature of the sample rises to the vaporization temperature rapidly enough to limit melting and the fractionalization of low-melting elements. Essentially, a vapor front is set up that moves through the material, generating a thermal shock wave that causes rapid heating of the sample and further vaporization.[42,52] Most of the material removed is in atomic form, with some clusters, large aggregates, and a large number of electrons.[42,43] At power densities above $\approx 10^8$ W/cm^2, the rate of sample removal becomes limited by the rate at which atoms leave the surface of the vapor. Continuing to increase the laser power above this level does not appear to enhance surface removal since the surface becomes shielded by the plasma, limiting the interaction with laser photons.[53,54]

Laser ionization is also dependent on laser power densities with little direct ionization observed below $\approx 10^8$ W/cm^2, and that which is observed is thermal in origin and shows a strong dependence on the ionization potential of the element.[54] If the power density is sufficiently high ($>10^9$ W/cm^2), the atoms cannot leave the vapor before further interactions with the incident laser photons occur. The laser light is then absorbed by these atoms, generating a vapor plasma that results in excitation and ionization.[55] At this point, ionization is no longer thermal and may occur by several mechanisms including multiphoton ionization and interaction with electric fields generated by the laser.[42] Electrons present in the plasma can also undergo a reverse bremsstrahlung process (avalanche ionization) that can significantly increase the energy of free electrons, causing a high degree of electron impact ionization.[42,51,54] At power densities of 10^9 W/cm^2 and above, nearly uniform atomization with 100% ionization of the sample occurs with little dependence on either sample or laser parameters. This results in relative sensitivity coefficients for ionization that are close to unity.[18,42,51,53,54] At higher power densities, the plasma will absorb more of the laser photons, resulting in the creation of multiply charged ions. Upper limits on the power densities used for laser ablation/ionization are in the 10^{10}–10^{11} W/cm^2 range. Some control of the ablation process is available by varying the laser power density. Lower laser power densities ($\approx 10^8$ W/cm^2) will result in the production of a largely atomic population of the sample material, while higher power densities ($>5 \times 10^8$ W/cm^2) will result in ionization of the ablated material.[56] Depending on the type of investigation, either of these power regimes may be employed. Further details on the mechanisms of laser ablation may be found in the referenced sources, and especially a recent book devoted to applications of laser ablation.[57]

10.3.3. Laser Ablation Coupled to Glow Discharge Devices

Coupling the laser ablation atomization process to the discharge for subsequent ionization may provide several advantages over direct laser ablation/ionization of the sample. Since the laser is used only to atomize the sample, significantly lower laser power levels are required, possibly reducing laser complexity. In addition, the shot-to-shot irreproducibilities inherent in laser operation will not have as great an effect on the low-power atomization process as they will on the high-power ionization processes, and therefore the shot-to-shot reproducibility of the technique may be enhanced relative to laser atomization/ionization alone. A combination of laser ionization and glow discharge ionization may also result in an enhanced ionization efficiency and enhanced sensitivity. Two fundamental configurations for coupling laser ablation to the glow discharge can be envisioned. Either a secondary sample can be positioned adjacent to a discharge plasma that is sustained by another cathode and material ablated into the discharge, or the laser may be used to ablate directly material from the discharge cathode. Illustrations of these two methods are provided from doctoral research at the University of Virginia.[58,59] To the best of the author's knowledge, these are the only such studies of laser ablation into glow discharge devices currently available.

10.3.3.1. Laser Ablation of an Auxiliary Sample

In this configuration, the sample is positioned adjacent to the glow discharge plasma, which is sustained by a separate cathode, as illustrated in Fig. 10-1. The laser beam is focused onto the sample surface and material is ablated into the discharge for subsequent excitation and ionization. Several potential problems must be taken into consideration with this configuration. First, the ablated material must be transported from the laser impact region to the negative glow of the glow discharge. Parameters that can impact the transport processes include discharge gas pressure and the distance between the sample and the negative glow region of the discharge. A lower pressure will allow greater diffusion of sample material into the negative glow; however, because of the operating pressure regime of the discharge, lower limits on discharge pressure exist, below which the discharge voltage becomes too large for stable discharge operation. An attempt to minimize the sample distance from the cathode should also be made, but placing the ablated sample too close to the cathode can result in arcs and other discharge instabilities as well as sputter deposition of the cathode material onto the ablated sample. Judicious use of a discharge cathode with a low sputter yield, i.e., tantalum, can reduce the number of cathode sputtered atoms in the discharge

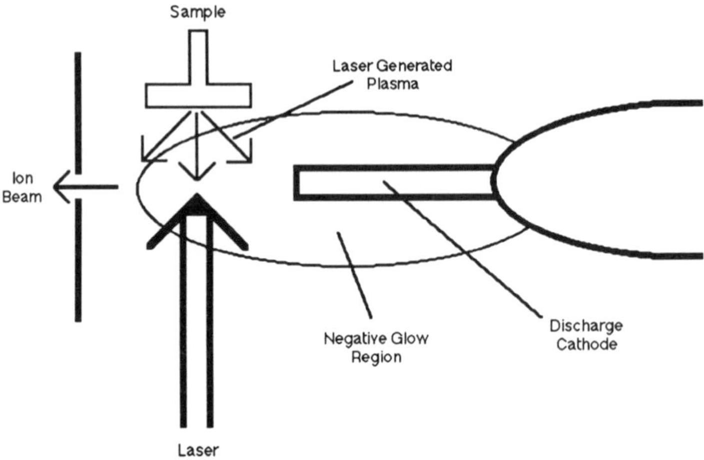

Figure 10-1. Diagram of the experimental configuration employed during laser ablation of a sample positioned adjacent to the discharge plasma.

and lower redeposition and other potential interferences caused by the cathode material. Another configuration envisioned to alleviate these problems would involve a separate ablation cell with the ablated material being swept through an insertion tube into the discharge by the discharge gas, similar to the configurations employed for the introduction of ablated material into inductively coupled plasmas.[25–34] In this case, the ablation step and discharge excitation/ionization steps are performed in separate chambers and each can be optimized independently. With the exception of a recent paper discussed in Section 10.3.5, no experimental results from this type of configuration involving a glow discharge are available. The results of supplemental ionization discussed in the following two sections were obtained on a glow discharge mass spectrometer using a sample placed in the discharge source chamber and have been discussed in somewhat more detail in other references.[58,59]

Initial investigations were performed to ascertain whether any discharge-generated ion signal of the ablated material could be observed. In these studies, the output from a XeCl excimer pumped dye laser operating at the peak wavelength for the rhodamine 590 dye was focused onto a 4-mm copper disk positioned adjacent to the negative glow region of the discharge. The laser beam was defocused to the point at which the laser-produced ion signal (no discharge present) just disappeared. At this point, the power density of the beam is sufficient for atomization without ionization and ions formed from glow discharge processes should be readily discernible. Figure 10-2 depicts an analog signal profile of the ^{63}Cu ion signal with and without

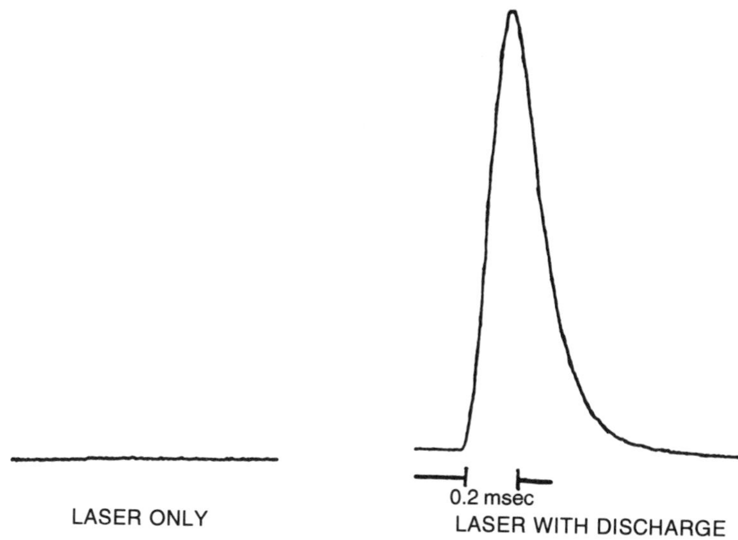

Figure 10-2. Temporal profile of the ^{63}Cu ion signal produced by laser ablation of a copper disk placed adjacent to the discharge plasma, with and without a discharge. Argon discharge, 0.4 Torr, 4 mA, Ta cathode. [Reprinted with permission from: K. R. Hess and W. W. Harrison, Laser ablation and ionization studies in a glow discharge, in: *Lasers and Mass Spectrometry* (D. M. Lubman, ed.), p. 213, Oxford University Press, London, 1990.]

a supplemental glow discharge operated at 4 mA and 0.4 torr argon. As can be seen, the discharge is ionizing copper material ablated from the cathode by the laser.

In order to analyze the laser-generated ions, a data acquisition scheme is employed to open a data gate during the time period of laser ablation, with a second background gate opened during the time period when the laser is off. Subtracting the background gate from the laser data gate will provide an ion signal consisting only of the ions resulting from laser ablation and discharge ionization. As previously mentioned, discharge pressure is one parameter that can directly influence the amount of ablated material that reaches the negative glow region for supplemental excitation and ionization. The ablated ion signal was observed to maximize at low pressure, 0.3 Torr. Unfortunately, discharge ionization processes in this type of source have been observed to maximize at ≈0.8 Torr[58] so that the pressure most suitable for ablation of material into the discharge is not the optimum pressure for further discharge ionization of that material. This limits the extent of ionization observed for the ablated material, limiting the sensitivity of the technique and any potential analytical applications. To improve this situation, a dual chamber source would be required in which each step in the process could be optimized.

As an illustration of the limited analytical utility of this method, a National Institute of Standards and Technology #410 steel was used as the ablated sample and a mass spectrum of the resulting laser-ablated/discharge-ionized material obtained. The mass spectrum over the iron mass region is presented in Fig. 10-3. There are two major problems with this spectrum. First, the relative insensitivity of the method can be judged by considering the ^{52}Cr peak which is present in the sample at a level of $\approx 2\%$. Second, the isotopic abundances and relative peak heights for the elements are not correct. The ^{52}Cr peak is substantially higher than expected relative to the ^{56}Fe peak for a steel sample. This is most likely caused by a deterioration of the sample as the laser ablates material from the sample surface, generating a laser-produced crater. By the time the mass scan reaches mass 56, a significant alteration of the sample surface has occurred and the ablated material is no longer reaching the discharge and the ion signal is significantly reduced. As the mass of the monitored ions increases, the signal decreases, resulting in skewed isotopic abundance measurements. Methods for alleviating this problem include the use of a mass spectrometer that does not scan sequentially but rather analyzes the entire mass range from a single event, such as a time-of-flight instrument, ion trap, or Fourier transform mass spectrometer, and rotating the sample to provide a fresh surface for each laser/sample interaction.

One of the major advantages of employing the laser as a supplemental method of atomization is the potential ability to analyze nonconducting

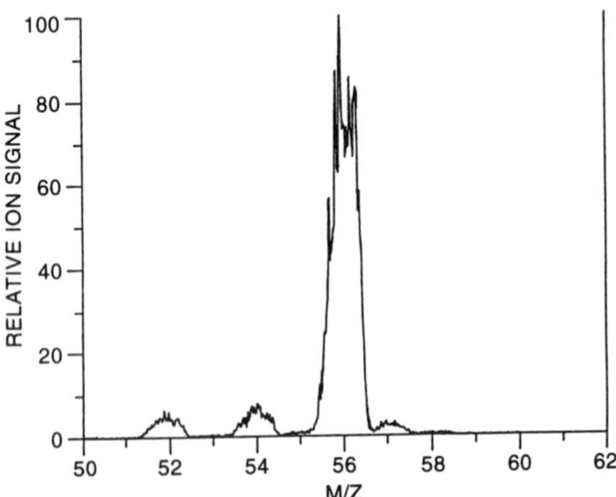

Figure 10-3. Mass spectrum produced by laser ablation of a NIST #410 steel sample into an adjacent glow discharge. Argon discharge, 0.4 Torr, 4 mA, Ta cathode. [Reprinted with permission from: K. R. Hess and W. W. Harrison, Laser ablation and ionization studies in a glow discharge, in: *Lasers and Mass Spectrometry* (D. M. Lubman, ed.), p. 214, Oxford University Press, London, 1990.]

materials. As an example, sample disks 2.0 cm in diameter of $CuCO_3$ and $CuSO_4$ were prepared from the powdered material using a press and die. These disks were then mounted adjacent to the discharge which was operating with a tantalum cathode at 4 mA and 0.4 Torr argon. The laser was defocused to a power density sufficient to cause only atomization, and the laser-ablated/discharge-ionized ion signal was collected over the copper mass range. The results are presented in Fig. 10-4. As with the NIST #410

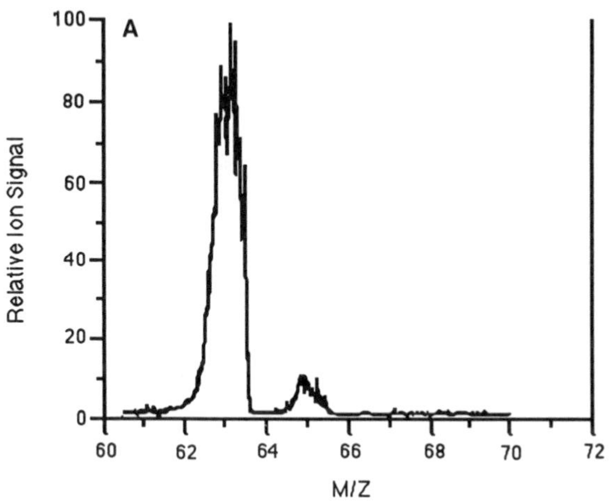

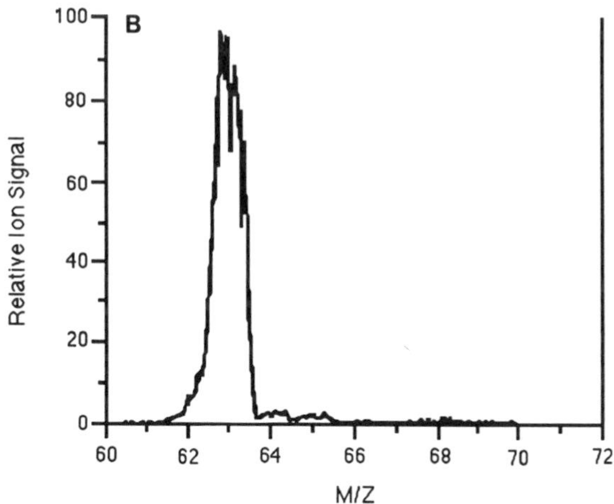

Figure 10-4. Mass spectra produced by laser ablation of (A) $CuSO_4$ and (B) $CuCO_3$ pressed powdered samples into an adjacent glow discharge. Argon discharge, 0.4 Torr, 4 mA, Ta cathode.

metal sample, ions from higher masses, which are recorded later in time with the quadrupole, are significantly reduced relative to their expected contributions. Because of cratering effects, there is a time dependence on the ablation process, with the laser becoming less efficient at removing material as the number of incident laser pulses increases. This generates a lower ion signal from the ^{65}Cu isotope which is measured a number of laser shots after the ^{63}Cu. Instrumental methods for addressing this problem, time-of-flight mass spectrometer and rotating sample, have been discussed previously. Ultimately, although ablation of a sample into a discharge for subsequent ionization has been demonstrated, the method is far from analytically useful and requires modification to reduce the effects of the time-dependent behavior of the ion signals. One possible method for reducing the cratering effects would be to ablate the discharge cathode directly, allowing discharge sputtering and redeposition processes to "resurface" the sample between laser pulses. However, this method has the disadvantage of the inability to analyze nonconducting material.

10.3.3.2. Laser Ablation of the Cathode Material

In this configuration the laser beam is focused directly onto a conducting sample that also serves as the discharge cathode, as illustrated in Fig. 10-5. The material ablated from the sample surface will directly diffuse into

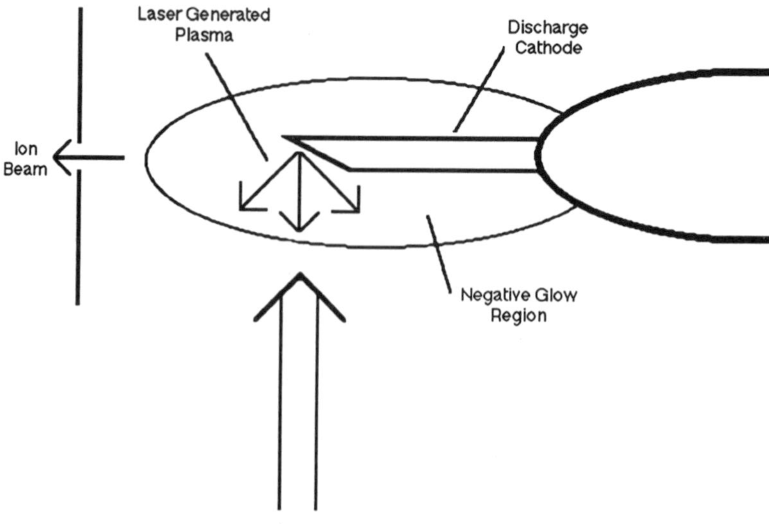

Figure 10-5. Diagram of the experimental configuration employed during direct laser ablation of a discharge cathode.

the negative glow region, more closely simulating the action of sputtered material and reducing the impact of transport processes on the introduction of material into the plasma. In addition, the discharge will continue to sputter and erode the cathode surface, mitigating to some extent the effects of the laser-produced crater, possibly enhancing the stability of the ion signal from the laser-ablated material.

Experimentally, the laser was focused onto the surface of a copper cathode (angled at 45°) with a laser power density sufficient to cause atomization without laser ionization. The glow discharge was then struck for ionization of the ablated material. The laser impact results in the rapid release of a substantial density of electrons and cathode material, causing an instantaneous reduction in the resistance of the discharge. This rapid reduction in resistance leads to arcs in the discharge, generating noise spikes that interfere with the computerized data acquisition system, prohibiting the acquisition of mass spectra due only to laser ablation/discharge ionization. These noise problems limited the experimental investigations of this configuration, and the only data available under these circumstances were analog time profiles of the ^{63}Cu ion signals taken from an oscilloscope with an XY recorder.

Figure 10-6 provides the analog time profiles for a ^{63}Cu ion signal at various discharge pressures. The sharp initial peak is the noise signal generated by the ablation of material into the plasma, corresponding to the firing of the laser. As can be observed, there is a laser-dependent ion signal on top

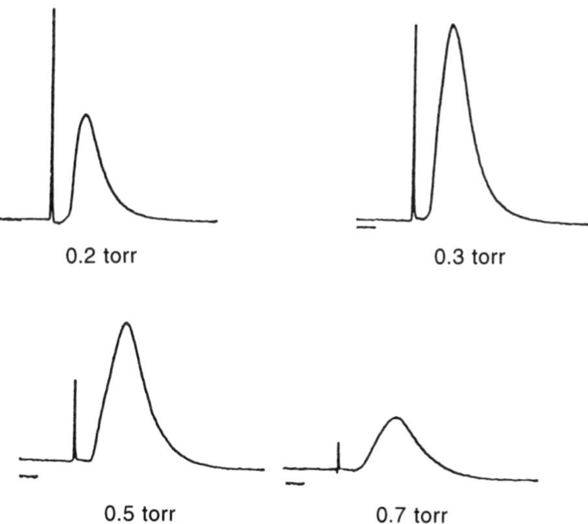

Figure 10-6. Analog time profiles of a ^{63}Cu ion signal generated from the ablation of a copper cathode during discharge operation versus discharge pressure. Argon discharge, 2 mA.

of a background ion level due to the glow discharge. The available results do show that the ablation of an atomic population from a discharge cathode can be subsequently ionized in the discharge. With further instrumental developments to limit the effect of noise spikes, mass spectra of the ablated material could be obtained. Future investigations in this direction may lead to some analytical utility of the method, but currently the main application of laser ablation/glow discharge excitation is as a method of investigating fundamental discharge processes.

10.3.4. Fundamental Discharge Investigations

The ability of the laser to ablate material independent of the discharge provides an opportunity to directly inject an atomic population into the discharge to study various discharge processes. Illustrative examples of fundamental investigations of the glow discharge employing laser ablation are presented in the following sub-section.

10.3.4.1. Mechanism of Formation of ArM^+ Species

The formation of an atomic population independent of the discharge can also provide insight into fundamental discharge mechanisms. As an example, the laser has been used to investigate the mechanism of formation of species of the type ArM^+, where Ar is the discharge gas and M is the sputtered metal atom. These argides are common in low-pressure glow discharge sources and can generate a signal as high as 10% of the base ion signal from the bulk metal.[60] These ions are common isobaric interferents in glow discharge mass spectrometry and several methods of overcoming these interferences have been proposed, including monitoring of a different isotope of the element of interest (if possible), variation of the discharge gas to shift the mass of the rare gas–metal ion,[61] use of a collision cell to dissociate these molecules,[62,63] and selective laser resonance ionization.[64] Information on the mechanism of formation of these ions would be of benefit in adjusting discharge conditions or developing methods for reducing these interferences.

Two mechanisms of formation of the ArM^+ species have been proposed: associative ionization

$$Ar^* + M^0 \rightarrow ArM^+ + e^- \tag{10-1}$$

where Ar^* is a metastable state of the argon, and a three-body collision mechanism[60]

$$M^+ + 2Ar^0 \rightarrow ArM^+ + Ar^0 \tag{10-2}$$

These mechanisms are argon pressure dependent, with the associative ionization process dominating at low pressures and the three-body collision process dominating at pressures of approximately 1.0 Torr and above.[(60)] The pressure regime in which a standard diode geometry discharge operates, 0.3–1.5 Torr, is such that a contribution from both mechanisms may be expected for the formation of ArM^+. Laser ablation of a sample in the discharge chamber under an argon pressure identical to that used for a discharge will allow the formation of M^+ with Ar^0. This should permit the formation of ArM^+ through the three-body collision mechanism in the absence of any associative ionization. This assumes little Ar* formation with laser ablation in a low-pressure argon environment, an assumption that is supported by the lack of argon emission from laser-generated plasmas under a 1 Torr argon environment.[(65,66)] If a glow discharge is then operated under the same conditions, both associative and three-body collision mechanisms can play a role in the formation of the ArM^+ and this signal would be expected to increase relative to the base signal from the metal ion. Overall, through laser ablation, mechanism (10-2) can be created in the relative absence of mechanism (10-1), allowing the extent of each mechanism to be determined, provided the signals are normalized in some manner. Mechanism (10-1) is indirectly dependent on the M^+ signal. The amount of ArM^+ formed by this mechanism is dependent on the Ar^* and M^0 populations, which, in turn, directly impact the M^+ signal observed for this discharge configuration.[(58)] Mechanism (10-2) is directly dependent on the M^+ population. Since both mechanisms are related to the metal ion signal, this signal serves as a good standard against which the laser ablation and glow discharge ArM^+ signals can be normalized. A ratio of M^+/ArM^+ should remove the metal atom and ion concentration dependence on the amount of ArM^+ formed with both the ablation process and the discharge. Since the discharge allows a contribution from both mechanisms and laser ablation only a contribution from the three-body collision process, a comparison of the M^+/ArM^+ ratios for each process will provide a relative value for the extent of ArM^+ formed by each mechanism.

These processes were studied by ablating a copper pin that also served as a sample cathode under different pressures of argon, while monitoring the $Cu^+/ArCu^+$ ratio.[(58,59)] The results are presented in Table 10-1. The ratio

Table 10-1. $Cu^+/ArCu^+$ Ratios during Laser Ablation and Discharge Operation at Various Pressures of the Argon Fill Gas.

Pressure	Ablation	Discharge	Ablation/discharge ratio
0.25 torr	529.1	371.7	1.42
0.45 torr	483.4	366.8	1.32
0.60 torr	857.6	689.3	1.24
0.80 torr	646.9	582.6	1.11

of the results for the discharge (both mechanisms present) to the results for the ablation [mechanism (10-2) only] provides an estimation of the relative importance of each ArM^+ formation process. For example, at 0.25 torr the ablation-only measurements show a ratio of Cu^+ to $ArCu^+$ of 529 while for the discharge this ratio is 372. This shows there is a larger $ArCu^+$ signal with the discharge, as would be expected since both mechanisms of formation are contributing to the observed $ArCu^+$ signal. The $ArCu^+$ signal for the discharge is approximately 30% larger than for ablation, indicating that the associative ionization mechanism contributes approximately 30% to the total observed $ArCu^+$ signal in the discharge, with the three-body collision process contributing the other 70%. The relative importance of the three-body process is plotted as a function of pressure in Fig. 10-7 and shows that as the pressure increases, so does the extent of the three-body collision process. The formation of $ArCu^+$ becomes completely dependent on the three-body mechanism at approximately 1 Torr, as was previously postulated.[60]

10.3.4.2. Sputter Redeposition Studies

A second type of experiment that illustrates the potential application of laser ablation for fundamental studies of discharge processes involves the investigation of relative sensitivity factors (RSFs). RSFs are a measure of the relative difference in elemental sensitivity caused by differences in the net

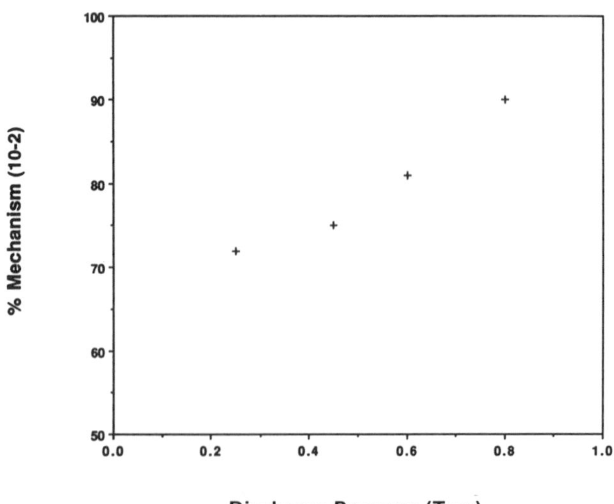

Figure 10-7. Relative extent of the three-body collision mechanism for the formation of ArM^+ species in the discharge versus discharge pressure. (See text for more detail.)

ionization rate of the various species arising from various physical, chemical, or instrumental factors. Variations in the sputter atomization rates for different elements, different ionization rates due to the different elemental ionization potentials, and differential mass throughput through a quadrupole can cause the observed differences in ionization rate. For rapid semiquantitative analysis of the entire periodic chart, RSFs of approximately 1 would work best and allow direct interelement comparisons of relative ion signals. RSFs in the glow discharge have been observed to be on the order of 3–5,[67,68] indicating little selective enhancement of elemental ion signals. Elemental sputter yields with argon as the discharge gas are known to vary within this range,[69,70] indicating the observed RSF values arise from differences in sputtering, with uniform elemental ionization efficiencies. Experiments employing a pulsed discharge with laser ablation have provided evidence that observed differences in ionization of sputtered elements are in fact due mostly to differences in elemental sputter yields and not to a selective ionization process.[58,59]

Direct ablation of the discharge cathode during operation of the discharge is hampered by the sudden pulse of material into the discharge which can generate electrical arcs and interfere with the electronic components of the system. If a pulsed discharge is employed, and the laser pulses timed to impact the cathode surface between discharge pulses, laser ablation of the cathode can be performed with no deleterious effects. Figure 10-8 illustrates the temporal response of ion signals observed for the pulsed discharge/laser ablation combination. Each discharge pulse sputters the cathode surface and

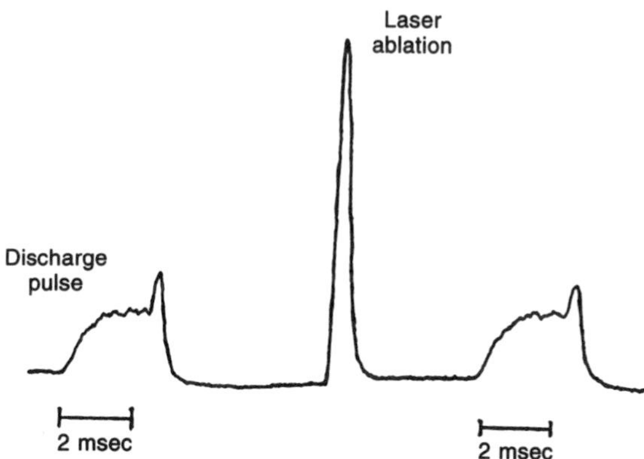

Figure 10-8. Temporal response of discharge/ablation ^{63}Cu ion signals during laser ablation between the pulses of a pulsed glow discharge. [Reprinted with permission from: K. R. Hess and W. W. Harrison, Laser ablation and ionization studies in a glow discharge, in: *Lasers and Mass Spectrometry* (D.M. Lubman, ed.), p. 215, Oxford University Press, London, 1990.]

creates an atomic population whose elemental composition will be dependent on the relative sputter yields of the elements present in the cathode. A significant portion of the sputtered material, estimated to be 99% at 1.0 Torr,[71,72] is then redeposited back on the cathode surface through collisions with gas atoms and ions present above the sample surface. At the termination of the discharge pulse, the cathode will be covered by a thin layer of the redeposited material. This redeposited material is then ablated and ionized by the subsequent laser pulse.

If the material ablated results from the redeposition of the sputtered species, then the ablated ion signal can be used to determine the composition of the redeposited material. Elements with a higher sputter yield in the discharge should show an enhanced concentration in the redeposited material relative to the bulk concentrations, while those species that do not sputter readily will show no enhancement in the redeposited material. The elemental ratios (RSFs) from the ablated ion signal should be similar to the discharge sputter yields and different from the bulk concentration ratios. If the RSFs in glow discharge mass spectrometry arise from differences in sputter rates and not differential ionization, then the glow discharge RSFs should match those from the ablated redeposited material.

A NIST reference steel sample, #410A, was employed to investigate the respective RSFs as discussed above. Table 10-2 presents the certified ratio of ^{62}Ni to ^{63}Cu along with the ratio observed for laser ablation only, laser ablation with a pulsed discharge, and the glow discharge only. As can be seen, the ratio for laser ablation closely matches the certified ratio, indicating nonselective ionization of the material by laser ablation. At these power densities, complete ionization of ablated material is expected.[18,42,51,53,54,56] When the laser is timed to impact the cathode between the pulses of the discharge, the ^{62}Ni/^{63}Cu ratio changes. Since the laser power is sufficient for complete ionization of the ablated material, this indicates the cathode surface composition has been altered by redeposition of the sputtered discharge species. The ratio observed in the glow discharge is approximately the same as that observed with laser ablation of the redeposited material, evidence to support the conclusion that the discharge ionizes sputtered material nonselectively and that the observed glow discharge RSF values arise from

Table 10-2. ^{62}Ni/^{63}Cu Ion Ratios Observed in Various Ion Sources for NIST #410 Steel

Ion Source	^{62}Ni/^{63}Cu ratio
Ablation only	0.25
Ablation between discharge pulses	0.43
Discharge only	0.39
Certified NIST value	0.24

differences in elemental sputter yields. Further experimental evidence that the redeposited material is being ablated can be found in other sources.[58,59]

10.3.5. Analytical Investigations

The combination of laser ablation/discharge excitation of atomic emission analysis has been reported recently.[73] This paper, reportedly the first to investigate the analytical utility of the combination of laser vaporization followed by excitation in a glow discharge, was directed to overcome disadvantages of each method when employed independently. Emission spectroscopy based solely on laser vaporization and excitation allows microsampling and the ability to analyze both conducting and nonconducting materials, but suffers from a lack of precision, with poor shot-to-shot reproducibility largely due to fluctuations in the power density of the beam and sample cratering effects, and poor linearity of analytical curves. Glow discharges, on the other hand, are stable with good precision and employ a low-pressure inert gas to reduce linewidths and background interferences. However, discharges suffer from the inability to perform microanalysis and the need for a conducting sample. The combination of laser vaporization as a method for the creation of an atomic population followed by excitation in the discharge for atomic emission would then take advantage of each technique's strengths while reducing their shortcomings.

In this work, a Q-switched Nd: YAG laser (250 mJ/pulse, 10-ns pulses) was focused to 1.5 mm on an aluminum alloy. The vaporized material was then transported by a stream of argon 5 mm into a supplemental hollow-cathode glow discharge with an aluminum cathode. The influences of a variety of experimental factors, such as current, argon flow rate, cathode size, and discharge pressure, were studied to maximize the analytical potential of the system. The reported results showed relatively good precision and linearity, with relative standard deviations of 2–7% in the determination of a variety of transition metals at concentrations of 0.01–0.20%. Future studies on specific applications and possible source modifications may further improve the technique, making it an analytically useful method for the analysis of nonconducting materials or for applications requiring high spatial resolution of the analysis.

10.4. Supplemental Excitation

The plasma region of a low-pressure glow discharge device contains a diverse population of species which may interact with laser photons, providing a variety of methods to monitor supplemental laser excitation in the discharge plasma. There are sputtered atoms of the cathode material, molecules such as dimers and molecular clusters, metastable atoms of the rare gas, and ions from the sputtered material and discharge gas, all of which can be excited by laser photons. This excitation can lead to emission of a photon, direct ionization of the species, or changes in the fundamental properties of the discharge due to enhanced or reduced ionization. These effects may be monitored optically, electronically, or mass spectrometrically, generating a variety of analytical laser/discharge excitation methodologies. A brief introduction to several of these methodologies is provided in this section.

Glow discharges have several advantages for their use as atom reservoirs for atomic spectroscopy. First, the cathodic sputtering process creates an atomic population directly from the solid cathode material, eliminating the need for dissolution and possible solvent effects, contamination, and dilution of the sample. The sputtering process is stable, reproducible, and the atomic population can be easily controlled through the manipulation of discharge current. Second, as opposed to flames or ICP devices, most discharges operate in a reduced pressure (<10 Torr) environment of an inert gas. The low-pressure inert environment reduces the number and variety of background molecular species formed by collisional processes, although some still exist. The relatively low-pressure discharge also reduces collisional pressure broadening of atomic transitions, enhancing the resolution of spectroscopic techniques. A reduction in source pressure, and the associated reduction in collision rates, will also significantly reduce quenching reactions, allowing spectroscopy of transient excited-state species. Finally, discharge devices operate effectively in a pulsed mode which allows synchronization with a pulsed laser, increasing the effective duty cycle of the system. Proper timing sequences will allow time-resolved laser interactions with the discharge, permitting laser interaction at any point in a discharge pulse. A particular advantage of the time-resolved interactions is the ability to make spectroscopic measurements after the discharge excitation processes cease and the corresponding discharge emission has decayed, but before the atomic population has diffused from the laser interaction volume. This provides an effective method for reducing background emission interferences from the glow discharge, enhancing the measurements generated by laser interaction.

10.4.1. Atomic Fluorescence

Atomic fluorescence measurements employing glow discharges were the first reported combination of lasers and discharge devices and laser-excited atomic fluorescence spectroscopy (LEAFS) techniques continue to be the most widely employed analytical laser/discharge interaction. The fundamentals of atomic fluorescence methods are covered in several excellent sources,[74–84] including Chapter 3 of this volume, and will only be treated briefly here. The method is based on the absorption of radiation by an atomic vapor, producing an excited-state atomic population which subsequently relaxes by emitting characteristic radiation which is monitored spectroscopically. Depending on the wavelengths of excitation and relaxation, several types of atomic fluorescence are possible. A few of the most common pathways are illustrated graphically in Fig. 10-9. In Fig. 10-09A, the $\lambda_{excitation}$ and $\lambda_{emitted}$ are the same and this process is referred to as a resonant pathway. In Fig. 10-9B, $\lambda_{excitation}$ is shorter than $\lambda_{emitted}$, giving rise to a Stokes direct line fluorescence process. In Fig. 10-9C, $\lambda_{excitation}$ is longer than $\lambda_{emitted}$ with the supplemental excitation to level 1 provided by the discharge plasma, a process called anti-Stokes direct line fluorescence. Finally, in Fig. 10-9D, the emitted photon is observed after another relaxation process (level 2 to level 1), termed Stokes stepwise line fluorescence. These are only a few of a number of potential fluorescence pathways available with a combination of discharge and photon excitation, and other sources provide a more complete

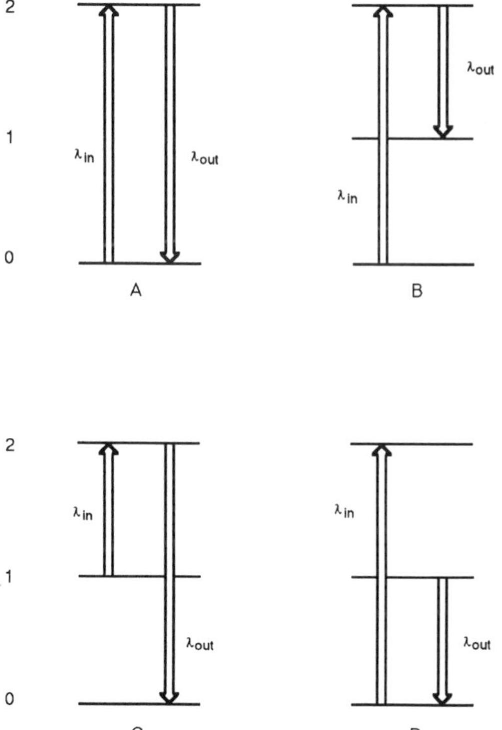

Figure 10-9. Diagram of several possible atomic fluorescence pathways. (See text for more detail.)

discussion of the potential mechanisms.[74–84] Since fluorescence measurements involve two wavelength-dependent resonant steps, excitation and emission, the method is very selective and not as prone to background interferences as is atomic emission spectroscopy. In addition, the two-step process significantly decreases the spectral complexity of the resulting emission spectra since fluorescence photons will only be observed for certain excitation transitions. The technique is also inherently very sensitive, reaching toward single atom detection, as discussed in a recent paper.[85]

The basic instrumentation for fluorescence measurements is diagrammed in Fig. 10-10. The system essentially consists of a source for excitation photons, an atom reservoir, a monochromator for selection and isolation of the emitted photons, and a gated detection system such as a boxcar averager. The detection system is normally placed at 90° with respect to the excitation source to reduce the effects of scattered radiation from the incident beam. In addition, scattered radiation can be reduced by employing a nonresonant method of fluorescence in which the excitation wavelength differs from the emitted wavelength.

Advantageous properties of the laser excitation sources include high power for efficient optical saturation of the excitation transition, high wavelength resolution to enhance method selectivity, tunability to increase excitation versatility, and the ability to operate at a high pulse repetition rate or continuously to enhance the technique's duty cycle. Various types of laser systems clearly meet most, if not all, of these characteristics. Specific discussions of the choice of laser systems for LEAFS are provided in other

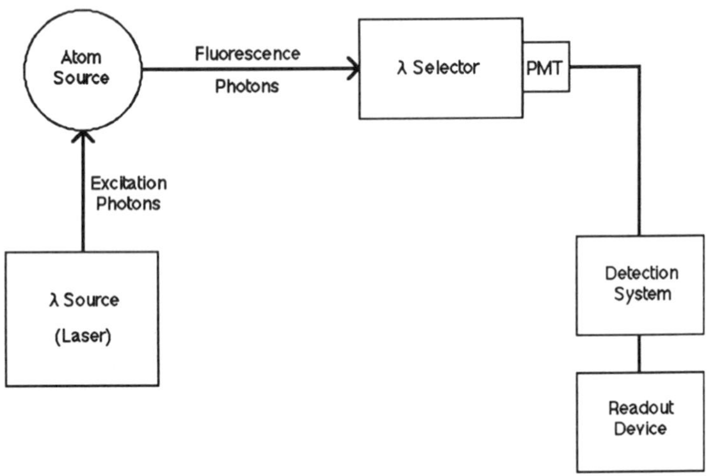

Figure 10-10. Basic instrumentation employed for atomic fluorescence measurements illustrating the fundamental components of the system.

sources.[82,83] Advantageous properties of the atom source include the ability to produce a large and stable atomic population with little background emission, the ability to be pulsed synchronous to the laser to enhance the duty cycle of the method, an inert environment to prevent chemical reactions and lower molecular background, and preferably a low pressure to limit collisional quenching of the excited state, pressure broadening of the atomic transitions, and laser light scatter. Again, details on source selection are covered in other references[82,83] and will not be presented here but it should be noted that pulsed glow discharges appear to be very attractive sources for LEAFS.[85]

10.4.1.1. Analytical Applications

Analytical applications of LEAFS have developed significantly since their origins in the 1970s with most applications employing flames or electrothermal atomizers as atomic sources. Many of these applications are covered in a recent review[83] and Omenetto[86] has provided an excellent summary of the developments of LEAFS with various atomic sources. The use of glow discharges for atomic fluorescence measurements evolved naturally from their use as sources in atomic absorption and atomic emission spectroscopy.[87] Early applications of glow discharge atomic fluorescence used a dual discharge design in which separate discharges served as the excitation radiation source and as the reservoir of the atomic vapor. Limits of detection for these initial investigations were reported to be on the order of 20–400 ppm for a variety of elements in an iron matrix, with a reproducibility of ±1%.[88] Detection limits from this early work were improved, to levels of 1–100 ppm for various elements in alloys of iron, aluminum, and copper and an initial discussion of the possibility of using fluorescence for depth profiling of metal samples was also provided in a follow-up report.[89] One reason for the enhancement in detection limits originated by pulsing the discharge and carefully timing the excitation source and detector response in such a manner as to discriminate against background emission from the glow discharge atomizer, improving the signal-to-noise ratios in the experiments. Utilization of the discharge's ability to be pulsed by synchronizing excitation in the period after the decay of atomizer emission has become the standard method of glow discharge operation for LEAFS. The availability of tunable dye laser systems in the 1970s resulted in a further improvement in sensitivity and versatility of LEAFS, with more recent applications of the technique employing lasers as excitation sources in fluorescence techniques.

Two articles from J. D. Winefordner's group at the University of Florida appeared in the mid-1980s detailing the analysis of lead[90] and indium.[91] The first of these reports employed a pulsed discharge with laser excitation applied 300 ms after the termination of the discharge. The discharge cathodes were either NIST reference copper rods containing certified lead concentrations of 0.4–128 μg/g, or aqueous solutions of lead dried onto graphite electrodes. Output from a nitrogen-pumped dye laser was frequency doubled to provide 283.31-nm excitation photons, with the lead's fluorescence radiation at 405.78 nm isolated by a 0.5 m monochromator. Results from the analysis of the NIST material suffered from inexplicably poor precision, with relative standard deviations of 8–20% observed, especially for two of the samples. Detection limits based on 3σ of the background were calculated to be approximately 0.1 μg/g. When lead solution residues were analyzed on a graphite electrode, the behavior was similar to that observed for the copper cathodes, only with substantially improved precision. The detection limits for these analyses were calculated to be 20 pg of lead, illustrating the analytical potential of glow discharge sources for LEAFS. In the second paper, indium samples were prepared by dissolving pure indium metal in nitric acid and depositing 10 μl of the resulting solution on a cathode prepared from either graphite or copper.[(91)] Excitation of the indium employed a nitrogen pumped dye laser, frequency doubled to the 303.94-nm In(I) transition. Typical pulse energies of the 15-mm unfocused beam traversing the discharge were on the order of 5 μJ. Nonresonant indium fluorescence was observed at 325.61 nm with limits of detection of indium determined to be 8 ng at 3σ. When the copper electrodes were employed as the substrate for the deposited solution, the results were not quite as good, with a greater deviation in the data and a reported detection limit of 11 ng. One reason postulated for the decreased analytical performance with the copper electrodes is the potentially higher background fluorescence from molecular species formed from sputtered cathode material. Copper has a higher sputter atomization rate than graphite and is known to form a variety of molecular species in discharges. Fluctuations in the background emission from these molecular species could increase the deviations in the data and reduce the analyte number densities due to molecule formation, possibly reducing the sensitivity of the technique. The composition of the discharge cathodes seems to have an effect on the LEAFS signals detected and should be considered when using solutions deposited on cathodes in the glow discharge.

More recent investigations have shown the LEAFS technique to provide remarkable sensitivity, possibly approaching single atom detection. In a 1989 paper, Winefordner *et al.* used a commercial hollow cathode lamp as an atomization source for LEAFS analysis of lead, calculating reasonable theoretical limits of detection to be 1.8 *a*g within the laser beam volume.[(92)] This study employed a copper vapor laser to pump a dye laser, which was subsequently frequency doubled to 283.3 nm, as an excitation source for the lead atoms. Fluorescence emission from the lead was monitored at 405.7 nm.

Using the fundamental collection and detection efficiencies of the instrumentation, a system responsivity to lead was calculated to be 1.64×10^{-8} V/pulse-atom ± 50%. Based on this system responsivity and the laser probe volume, 1 V of signal was calculated to correspond to an atomic density of 7.77×10^9 atoms/cm^3. This value was checked by substituting the hollow cathode lamp with a quartz cell containing 0.2 mg lead heated to 646 K, providing accurate vapor pressure data from which atomic density could be calculated. A lead density of 7.7×10^8/atoms/cm^3 provided a fluorescence signal of 0.0194 V which will correspond to 4×10^{10} atoms/cm^3 for a 1-V signal. A final estimate of system responsivity was given as 5.3×10^{-9}V/atom. Based on background noise considerations and this system responsivity, a detection limit of 1100–6000 lead atoms per probe volume was calculated. These studies clearly illustrate the potential for the detection of extremely low numbers of lead atoms with glow discharge LEAFS and provide the foundation for a subsequent paper in which experimental detection limits for lead and iridium were compared with a calculated intrinsic detection limit.[(93)]

In this subsequent investigation, 50-μl samples of lead and iridium solutions were deposited in a disposable graphite cup which served as the discharge cathode for the sputter atomization of the dried solution. The discharge was pulsed and the excitation pulse from a frequency-doubled, excimer-pumped dye laser was timed to interact with the atomic population 100 μs after the discharge was extinguished. At this point in the discharge pulse sequence, no emission from the discharge was observed, lowering the background and decreasing the detection limit by about a factor of three. The complete instrumental system is diagrammed in Fig. 10-11. The boxcar integrator gate was set to collect data during the time period of the lead fluorescence signal. The theoretical detection limit for lead was calculated to be 25 atoms per probe volume. When extrinsic factors such as noise and probing efficiencies are taken into account, the detection limit for lead is calculated to be 300 atoms in the probe volume. Although impressive, these calculated detection limits in the probe volume are of limited value and more conventional sample detection limits were measured using dried solution residues. These values are provided in Table 10-3. These experimental values are higher than those calculated from the theoretical considerations, indicating incomplete atomization of the sample or inefficient probing of the sample. When atomization and probing efficiencies were considered in the theoretical calculations, an intrinsic detection limit in the sample of 1.6×10^6 atoms, or 0.5 fg, was determined. This is only about three orders of magnitude lower than the experimental value of 500 fg, with the difference most likely originating from difficulties in complete atomization and efficient experimental probing. Further modifications of the system to reduce background fluorescence and increase collection efficiency (both spatially and

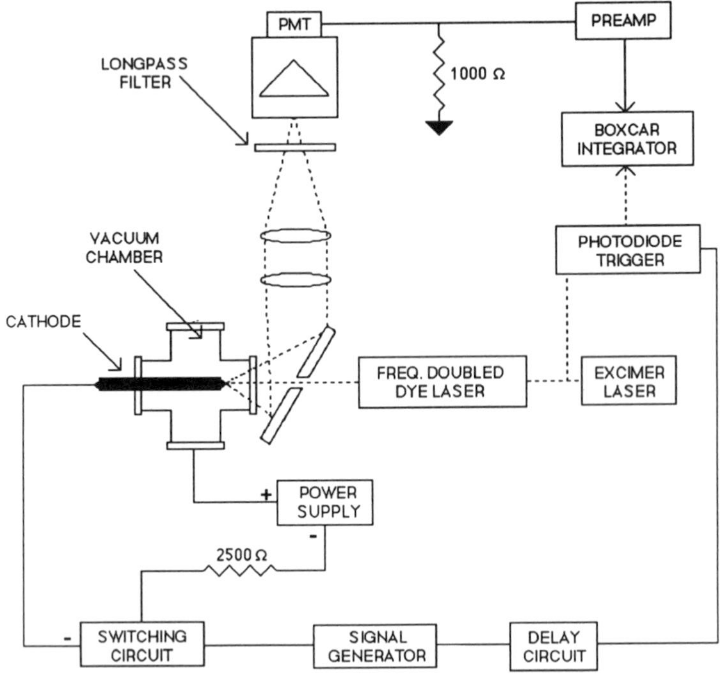

Figure 10-11. Instrumentation employed for the determination of indium employing a pulsed discharge. [Reprinted with permission from: M. Glick, B. W. Smith, and J. D. Winefordner, *Anal. Chem.* 62 (1990) 158. Copyright 1990 American Chemical Society.]

Table 10-3. Limits of Detection for Pb and Ir[a]

Element	λ_{ex} (nm)	λ_{em} (nm)	LOD[b] (ppb)	LOD[c] (pg)
Pb	283.3	405.8	0.1	0.5
Ir	285.0	357.4	6	20

[a]Taken from M. Glick, B. W. Smith, and J. D. Winefordner, *Anal. Chem.* 62 (1990) 160. Copyright 1990 American Chemical Society.
[b]Calculated by using 50-μl samples.
[c]Calculated by using 2-μl samples.

temporally) are expected to generate experimental detection limits that approach the intrinsic detection limits even closer, providing a method for elemental analysis that truly does approach single atom detection.

Other recent analytical applications of LEAFS with glow discharges include a report on the possibility of using the technique for depth profiling of microelectronic materials, reporting sodium analysis in molybdenum at a limit of detection of 24 ng/g,[(94)] and the analysis of iron in brass at a level of 150 ng/g.[(95)] In the second of these articles, the authors indicate a substantial source of noise is laser scatter from clusters of the sputtered material, limiting the sensitivity of the technique. Evaluation of cluster formation processes and the plasma conditions that minimize cluster formation will also have to be considered in order to approach the intrinsic detection limits of the technique.

An alternative application of glow discharges to laser-induced fluorescence spectroscopy does not rely on the discharge's ability to sputter atomize the cathode material, but rather advantageously employs the discharge to convert gaseous species to forms that can be monitored by fluorescence methods. These methods are particularly attractive for the analysis of species whose major transitions from the ground state lie in the vacuum ultraviolet range, inaccessible to laser wavelengths. The discharge plasma can overcome this problem by one of two methods. The first employs the discharge to preexcite the sample species to a metastable level whose transitions are accessible to laser fluorescence excitation. Examples provided for this method included the analysis of neon in helium and argon, with detection limits of 0.3 ppb and 100 ppm reported, and ionization of N_2 to excited state N_2^+ which is then analyzed to about 10 ppb in helium.[96] The second method uses chemical reactions in the discharge plasma to convert the analytical species to a form that is amendable to laser-induced fluorescence methods. Examples of this method included the analysis of N_2 and NO after oxidation to NO_2 in the plasma, with detection limits of approximately 1 ppb for analysis in helium, argon, and air.[96] The potential of the discharge for conversion of species to more analytically useful forms can be expected to be more fully exploited in the future.

10.4.1.2. Diagnostic Studies by Laser-Induced Fluorescence

In addition to analytical applications, laser-induced fluorescence can also be used as a diagnostic tool for fundamental investigations of glow discharge plasmas. The technique is particularly attractive as a diagnostic tool because of the method's ability to monitor both ground and excited-state populations of atoms, ions, or molecules and its high degree of spatial and temporal resolution. Information available from fluorescence measurements can also be used to estimate the absolute number densities of atomic species.[97] More comprehensive discussions of the diagnostic applications of laser fluorescence are contained in other references,[82–84,86] and only a few representative applications are presented here. Although several of these illustrative applications do not employ glow discharge devices as the atom reservoir, they are included in this section since the diagnostic methodologies discussed are applicable to glow discharge systems as well.

One of the primary applications of laser-induced fluorescence for diagnostic studies has been to study combustion processes.[80] Fluorescence methods can detect a variety of flame or plasma species, including radicals such as OH, O, and H, and various molecular compounds such as CO_2 and NO_2, with high sensitivity and spatial resolution. The fast photon pulses provided by the laser will also allow detection of rapidly reacting intermediates in flames or plasmas. Fluorescence measurements can also be employed to determine flame or plasma excitation temperatures by measuring the relative signal intensities of fluorescence transitions.[(82,98)] There are a number of possible experimental variations of the methodology, each with a set of specific equations and conditions to be used for the determination of temperature. An excellent tabulation of the details of many of these methods is available, and should be consulted for further information on the specific requirements for this type of diagnostic application of laser fluorescence.[(99)]

A particular advantage of laser fluorescence measurements is the spatial resolution of the method. Since only the volume of the plasma intersected by the laser beam and the solid angle of detection will provide fluorescence signals, laser interaction provides an excellent method for spatially profiling plasma species, both atomic and molecular. An illustrative example of the technique is provided by Gillson and Horlick[(100,101)] who used atomic fluorescence measurements to spatially map the population of ground-state neutral atom and ion species in an ICP. Information available from such extensive mapping includes locating the regions of the plasma that will contain the most useful analytical information, investigations of the effects of instrumental parameters such as argon flow rate and torch power on observed signals, and insight into how and where plasma processes such as recombination occur. Applications of spatial profiling of species by laser-induced fluorescence in a glow discharge have been reported by Winefordner *et al.*[(102)] These studies provided information on the population densities of various species within the discharge plasma, determining the best region of the discharge to view for analytical information. In addition, this paper reported on the use of temporal fluorescence monitoring of sodium atoms as a method of determining the diffusion rate of sodium atoms through the argon discharge.

Closely related to the determination of diffusion rates are experiments designed to determine the gas velocities in plasmas.[82] An illustrative method for the determination of the velocity of individual atoms using laser saturated fluorescence is contained in a publication by She *et al.*[(103)] In this example, a laser beam is split into two parallel beams which interact with the sample cell. If the concentration of atoms in the cell is such that less than one resonant atom is contained within the viewing region of the system, then this atom will interact with each laser beam as it travels across the sample cell. This should give rise to two bursts of fluorescence, separated by the time it takes the atom to traverse the distance between the beams. From these measurements, the atom's velocity can be determined.

The spatial resolution can be further enhanced using a two-photon excitation fluorescence method where the intersection of two lasers in a volume is required for the generation of a fluorescence signal, termed fluorescence dip spectroscopy.[104,105] In this technique, two laser beams with a spatial and temporal overlap are directed into the sample region. If one laser wavelength is tuned to an atomic transition, the second laser wavelength can be scanned to further excite the atom from the energy level reached in the first excitation step. When the second photon is resonant with a transition originating from the first excited level, the fluorescence from the first excitation step will decrease due to the depopulation of the level by the absorption of the second photon. In addition to enhanced spatial resolution afforded by the overlapping laser beams, theoretical studies have shown that fundamental information on the second transition, such as the absorption oscillator strength, can also be derived from the experimental measurements.

The capability of the fluorescence method to analyze molecular species in glow discharges has been illustrated by Winefordner *et al.* who reported on the use of fluorescence techniques to investigate the spectroscopy of lead dimers sputtered in a low-pressure glow discharge device.[106,107] In addition to the fundamental information on lead dimer spectroscopy obtained by these studies, information on the sputtering and dimer formation processes in glow discharges was also obtained. These studies indicate that the mechanism of molecule formation in the plasma is cluster formation with some molecules sputtered from the cathode surface.[107] The intensity of the molecule fluorescence was observed to be dependent on the operating parameters of the glow discharge, indicating the formation processes are sensitive to discharge conditions. Glow discharge mass spectrometry is known to suffer from isobaric interferences generated by clusters and molecular species formed in the discharge. Fundamental information on how, and under what discharge conditions, these molecules are formed would be useful in attempts to reduce isobaric interferences. Molecular fluorescence studies such as these could be a useful method for obtaining this fundamental information. In addition to molecule formation in the discharge, fluorescence measurements have also been employed for fundamental investigations of sputtering processes and the effect of target oxidation on these processes employing a pulsed argon ion beam and stainless steel samples.[108] Other illustrative applications of laser fluorescence for fundamental studies of various plasmas include the use of atomic fluorescence to investigate possible chemical effects of a reactive atmosphere of oxygen as compared with argon in laser ablation of copper and lithium targets[109] and the use of LEAFS to study the effect of easily ionized elements on the atomic densities in dc plasma devises.[110]

Laser-induced fluorescence has also been valuable for fundamental investigations and characterizations of chemical vapor deposition (CVD) glow discharge devices. These devices have found significant application for the production of thin-film coatings of silicon for semiconductor devices. Ho and Breiland used laser-excited fluorescence to observe an intermediate species of Si_2 in a CVD device as a thin film of silicon was prepared from a silane discharge system.[111] Observation of this intermediate provided insight into the mechanisms of deposition of the silicon. Further laser-induced fluorescence investigations of the silane discharge system have been reported by Yamamoto *et al.* who used the technique to spatially profile various species in a dc glow discharge device.[112] The densities of both ground-state and excited-state species were monitored as a function of discharge current and spatial location in the discharge, and absolute densities of the species were calculated from the experimental measurements. These experiments also generated data allowing for the calculation of the radiative lifetime of silicon atoms. These lifetimes were monitored versus pressure to show that the lifetime of the excited atomic state was less than the collisional rate in the discharge, indicating the absence of collisional quenching by the buffer gas atoms.

Laser excitation can provide a method to simultaneously create an excited-state population of atoms or molecules at a specific point in time in glow discharge plasmas. If instrumental effects such as the time constant of the detection system are known, they can be deconvoluted from the fluorescence decay curve of the excited species, providing information on collision rates and state lifetimes. Complete details of these methods can be found in other sources.[82,86,113,114]

If a laser is sufficiently intense that saturation of the excitation step in the fluorescence process occurs, information on transition probabilities and quantum efficiencies can be developed. An illustrative example of such a method has recently been described for the determination of the atomic transition probability of the argon 430.0-nm transition.[115] In this experiment, the population of a $1s_5$ level of argon is determined by a self-absorption method. A laser tuned to 419.1 nm then saturates the transition to the $3p_8$ level. Based on the population of the $1s_5$ level, the population of the $3p_8$ level can be determined, provided the transition is saturated. The fluorescence at 430.0 nm is then observed, and based on the number density of the upper $3p_8$ level and the intensity of the fluorescence emission, a transition probability for the 430.0-nm transition can be calculated.

Although an attractive method for the characterization of plasmas, relatively few direct references of diagnostic applications of fluorescence spectroscopy with glow discharge devices have been reported. Most of the examples provided above employed flames, ICPs, or CVD devices. With the expanding application of glow discharge devices in various analytical methodologies, more diagnostic studies of the discharge will be performed. These studies can be expected to advantageously employ laser fluorescence methods, and an increase in the reported diagnostic applications of laser fluorescence in glow discharges can be anticipated.

10.4.2. Optogalvanic Effect Spectroscopy

As an alternative to optical monitoring of laser/discharge interactions, changes in the electrical properties of the discharge may also be used to detect optical transitions, a technique commonly referred to as optogalvanic spectroscopy. General descriptions of the method are provided elsewhere[116–120] and a thorough review of optogalvanic effect (OGE) spectroscopy has recently appeared.[121] Briefly, the current flow through the discharge is essentially an electrical circuit that obeys Ohm's law. The current through the system is dependent on both the applied voltage and the discharge resistance, with the discharge resistance determined by parameters such as discharge pressure and the net ionization rate in the plasma. Alterations in the ionization rate of the discharge will lead to variations in discharge resistance and changes in the electrical properties of the discharge circuit. Enhancements in the ionization rate of the plasma will result in a decrease in discharge resistance and an increase in discharge current if the system is operating in a constant-voltage mode. If the system is operating under constant-current conditions, then the discharge voltage will decrease with the enhancement of the plasma ionization rate. Alternatively, the plasma ionization rate can be decreased, resulting in an increase in discharge resistance and a corresponding decrease in discharge current (with constant voltage) or increase in discharge voltage (with constant current). The current/voltage characteristics of the discharge may then be monitored by a simple resistor/capacitor circuit as illustrated in Fig. 10-12. Laser photons provide one method of altering the ionization rate of the discharge, and in combination with the electronic detection system, provide the foundation for OGE spectroscopy. Advantages of electrical detection of optical transitions include possible reductions in noise from stray light, the ability to monitor nonluminescent species or transitions that are normally difficult to observe optically, and the ability to generate an electrical signal directly without the need of an optoelectrical transducer.

Optogalvanic signals resulting from increases in discharge resistance were historically the first observed.[122] These signals correspond to net decreases in the ionization rate of the discharge and arise principally through a reduction in the population of the rare gas metastable atoms. The metastable gas atoms are relatively long-lived excited-state atoms whose radiative decay to the ground state is forbidden. They play a significant role in the net ionization rate of the discharge through two processes. First, they serve as an excited-state precursor to ionization, reducing the energy required for ionization. Second, metastables are an integral part of the Penning ionization process in which there is a potential energy transfer from the metastable atom to a discharge species, resulting in ionization provided the metastable energy is greater than the ionization potential of the colliding species (see Chapter 2). The metastable argon levels are the $^3P_2$ at 11.55 eV and the $^3P_0$

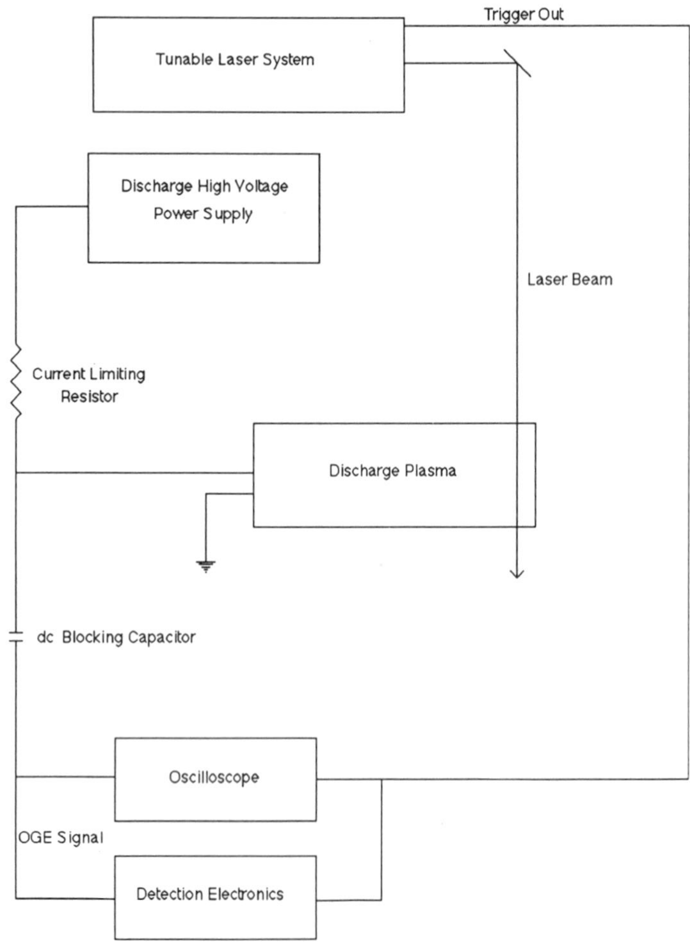

Figure 10-12. Generic instrumental arrangement employed for optogalvanic effect spectroscopy. $R = 2.2\ \text{k}\Omega$, $C = 1\ \mu\text{F}$.

at 11.72 eV, while the neon metastable levels are the $^3P_2$ at 16.62 eV and the $^3P_0$ at 16.71 eV. Upon laser interaction, absorption of a photon by a metastable rare gas atom can occur resulting in the excitation of the rare gas atom into a nonmetastable state from which it can radiatively decay, decreasing the net population of metastable atoms. The net depopulation of metastable atoms will then reduce the overall rate of ionization, leading to a decrease in discharge current under constant-voltage conditions.

Enhancements in the ionization rate of the plasma are generally obtained by use of the laser to excite a species to a higher energy level. One proposed mechanism for enhanced ionization is a simple model where the excited species undergoes ionization more readily than the unexcited species. For example, if the dominant mechanism of ionization of iron species in a glow discharge was electron impact, only those iron atoms that undergo collisions with electrons possessing >7.9 eV (first ionization potential for ground-state iron) would be ionized. The average electron energies in glow discharge devices have been reported to be on the order of 1–4 eV, with a fraction possessing higher energies as approximately defined by a Maxwell–Boltzmann distribution.(123,124) If an average electron energy of 2 eV is assumed, 4.8% of the electrons would have energies above the ionization threshold of iron.(124) If the iron atom now absorbs radiation corresponding to its transition at 344.06 nm, it will be excited to an energy level ≈3.6 eV above the ground state. The iron atom is now only 4.3 eV below its ionization potential, and approximately 27% of the electrons have the requisite energy for ionization,(124) an increase of sixfold as compared with the number of electrons able to ionize iron from the ground state. Since there is a substantially greater population of electrons with sufficient energy to cause ionization, the probability of ionization from excited state has increased relative to the ground state, and an increase in the net ionization rate of the discharge would be observed. Under this mechanism, the enhancement should be dependent on both the energy of the absorbed photon and the ionization potential of the species.

Keller and Zalewski, working with a hollow cathode discharge, did not find an ionization potential dependence on the monitored OGE signal, leading them to propose that a second mechanism for the generation of the enhanced OGE signal was operative.(125,126) This second mechanism involves use of the atomic system merely as a method of transferring laser energy to the electron population in the discharge by superelastic collisions, thereby increasing the electron energy distribution of the plasma and lowering discharge impedance. In electrical plasmas such as hollow cathode glow discharges, the electron temperature and the atomic excitation temperatures are intimately tied together due to the equilibrium established by the many electron/atom collisions. When an atom is excited by the laser, the electron collision frequency is sufficient to prevent a significant change in energy level populations. Instead, through electron/atom collisions the laser energy is redistributed among all of the electrons, leading to a slightly higher average electron energy and a decrease in discharge resistance. Dreze *et al.* simultaneously monitored the OGE effect and variations in atomic emission from a variety of uranium and xenon transitions when a U(I) transition was illuminated by laser photons.(127) They found a small enhancement in the measured excitation temperature of the discharge along with an increase in emission from U(I), U(II), and Xe(I) transitions, indicating a more universal mechanism of enhanced ionization than the simple excitation of uranium.

These results seem to confirm the postulated mechanism of energy transfer to the electron population. A recent evaluation of rate constants for various possible collisional processes potentially responsible for OGE signals in a high-pressure Hg discharge have also indicated that lowered impedance effects observed when atomic transitions are matched by the laser are due to superelastic collisions involving electrons, resulting in increases in the net electron energy of plasma and enhanced ionization.[128] This process is highly dependent on the electron density and collision rate in the plasma, only being dominant when the superelastic collision rate constant is greater than the rate constant for ionization from the excited state, and may not be operative under all discharge conditions, so that the ionization enhancement mechanism may be operative in some discharge sources. Another report has indicated that a pressure pulse caused by the laser interaction with the discharge may be responsible for the OGE effect,[129] but this effect was originally postulated for observed signals in an I_2 discharge in which the I_2 can dissociate into 2I with laser interaction, increasing the pressure and lowering discharge resistance,[130] and does not appear to be applicable to atomic systems. The exact mechanism of the OGE effect remains unclear, and further detailed theoretical discussions and applications of laser-induced optogalvanic spectroscopy can be found elsewhere.[121] As an introduction to the application of the method, a few specific examples of OGE spectroscopy are presented here.

10.4.2.1. Optogalvanic Studies of Discharge Ionization Processes

Monitoring the optogalvanic signal as a function of discharge parameters and discharge gases can provide fundamental information on the ionization processes occurring in a glow discharge device. Several such examples of mechanistic studies using OGE spectroscopy have been reported, a few of which are discussed here as a representative introduction to the method. Zalewski and co-workers[131] carried out a mechanistic study of the laser-induced impedance changes in a neon hollow cathode. They found a variety of transitions originating from the $2p$ neon states, all of which resulted in resistance decreases (enhanced ionization) upon laser irradiation. The magnitude of the resistance decrease was always proportional to the product of the wavelength, degeneracy, and oscillator strength of the transition, indicating the impedance change due to each of these atomic transitions is proportional to the probability for absorption of a laser photon. In addition to transitions originating in the $2p$ levels, transitions originating in the $1s$ neon metastable level were also observed. These transitions gave rise to the strongest signals observed and displayed both increases and decreases in resistance, depending on the transition. The magnitude of these changes did not correlate to the λgf values as did the $2p$ transitions. These impedance changes originate from the depletion of the metastable population by excitation from the 1*s* level to the 2*p* levels, and would be expected to show a dependence on the ability of the excited state to decay back to the ground state. This dependence would be related to the branching ratio of the radiative decay of the excited $2p_x$ state to the various $1s_x$ states and the quantum yield for the decay of the $1s_x$ state back to the ground state. Quantitative values of these effects are difficult to determine, but qualitative agreement was found by Zalewski *et al.*[(131)] and these investigations provide evidence for the role of metastable atoms in the ionization process of the glow discharge. Similar studies of the OGE effect in a neon glow discharge were carried out by Doughty and Lawler who also found a variety of transitions that exhibited both positive and negative resistance changes upon laser irradiation.[(132)]

Smyth and Schenck[(133)] also used the OGE to investigate the ionization mechanisms present in a neon hollow cathode lamp. They observed 75 transitions in the 572–654 nm region, a majority (57) of which showed decreased discharge resistances corresponding to transitions originating in the 2*p* states. The other transitions (18) are 1*s*–2*p* adsorptions and these showed a combination of increased and decreased resistances, dependent on the transition and discharge current. All 18 transitions exhibited decreased resistances (ionization enhancements) at high discharge currents. They explained these results in terms of competing mechanisms of ionization present in the hollow cathode discharge. At low discharge currents, the metastable neon atoms play a disproportionate role in the ionization of the discharge. Depopulation of these states by 1*s*–2*p* transitions will then decrease the net ionization in the plasma, increasing the discharge resistance. As the current is increased, electron impact ionization becomes dominant relative to the metastable dependent processes. Under these conditions, absorption of a laser photon by the neon atom will enhance ionization by electron impact with a resultant increase in the ionization rate of the discharge and a decrease in discharge resistance. In a follow-up paper, Smyth *et al.*[(134)] used a mass spectrometer to directly monitor the neon ion signal upon laser irradiation in an effort to correlate directly the observed OGE signals with changes in Ne ionization in the plasma. As expected, they found the Ne^+ signals to increase for the 2*p* transitions (which resulted in a decrease in discharge resistance) and found the Ne^+ to decrease for the metastable depopulation transitions originating from the 1*s* levels. These investigations were extended to the mass spectrometric monitoring of discharge species other than the discharge gas, such as sputtered material and background gases. Smyth *et al.*[(135)] and Bentz[(136)] monitored OGE effects and the ion signals mass spectrometrically from a neon hollow cathode discharge operating with a carbon cathode. Upon irradiation of the neon $1s_5$–$2p_4$ transition (594.48 nm), ion signals from $^{20}Ne^+$, $^{20}Ne^+$,$^{20}Ne_2^+$, $^{12}C^+$, and $^{14}N^+$ all decreased by approximately

20%. Further studies were performed with a copper cathode, monitoring both the OGE signals and decreases in ion signal as the discharge current was varied. The $^{63}Cu^+$ exhibited a decrease of 6.6% at 4 mA, which fell to no change at 40 mA. The OGE signal corresponded to an increase in discharge resistance, and also decreased as discharge current was increased. These studies provided evidence that at low discharge currents in hollow cathode lamps, the primary mechanism of ion formation was through metastable impact (Penning ionization), shifting to ionization by electron impact at the higher discharge currents.

Similar investigations employing a coaxial cathode glow discharge (pin-type geometry) were performed to allow a comparison of depopulation effects for argon versus neon in order to provide further insight into the role of Penning ionization in the discharge.[137] If the Penning process is important, species with ionization potentials less than the metastable energy of the discharge gas should show decreases in ion signal with laser depopulation, while those species with ionization potentials greater than the metastable energy will be unaffected by metastable depopulation. In these investigations, discharge ion signals were monitored with a quadruple mass spectrometer as the laser was tuned to rare gas transitions which result in the depopulation of the metastable states (594.48 nm for neon, 696.54 nm for argon). Upon laser irradiation in a neon discharge, those species with ionization potentials less than the metastable energy of 16.62 eV (OH^+, H_2O^+, N_2^+, Cu^+, Zn^+) showed ion signal decreases. In addition, the Ne^+ signal also decreased with laser irradiation due to removal of metastable atoms which serve as a precursor to neon ionization. Other species which are believed to be ionized through an associative ionization mechanism, such as $CuNe^+$, N_2H^+, Cu_2^+, and Zn_2^+, also show decreases, possibly due to the reduced ion populations resulting from metastable depopulation. A difference mass spectrum for a brass cathode in a neon discharge, resulting from subtraction of signals with the laser off from those with the laser on, is presented in Fig. 10-13. This "negative mass spectrum" clearly illustrates the reduced sputtered ion intensity with metastable depopulation. When argon was employed as the discharge gas, the results were substantially different for those species whose ionization potentials are greater than the argon metastable energy (11.55 eV). Species such as OH^+ (I.P. = 13.2 eV), H_2O^+ (I.P. = 12.8 eV), N_2^+ (I.P. = 15.6 eV), N^+ (I.P. = 14.55 eV), and O^+ (I.P. = 13.62 eV) did not exhibit decreases in ion signal with metastable depopulation. These experiments provided evidence for the assertion that Penning ionization is an important mechanism of ionization in the coaxial glow discharge.

10.4.2.2. Other Applications of OGE Spectroscopy

In addition to the study of discharge ionization processes, the OGE in glow discharges has found other applications in a variety of areas. One of

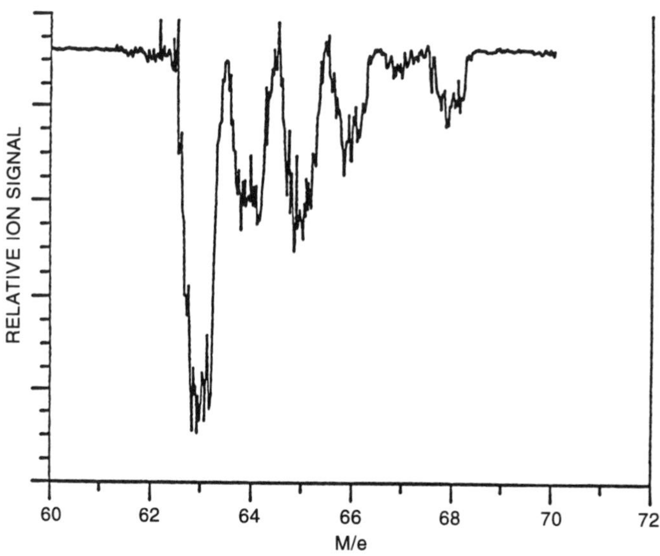

Figure 10-13. Example of a mass spectrum obtained by subtracting a glow discharge ion signal from an ion signal obtained during the time period of laser depopulation of neon metastable atoms at 594.48 nm. Neon discharge, 0.4 Torr, 3 mA, brass cathode. [Reprinted with permission from: K. R. Hess and W. W. Harrison, *Anal. Chem.* 60 (1988) 694. Copyright 1988 American Chemical Society.]

the first applications of OGE spectroscopy involved use of the technique as a method for wavelength calibration of laser systems.[138,139] Laser photons are directed into a commercial hollow cathode lamp which provides a variety of possible OGE transitions. As the laser wavelength is scanned across an atomic transition, of either the sputtered cathode material or the discharge gas, a resulting OGE signal can be detected. Accurate knowledge of the exact transition wavelengths provides the information necessary to accurately calibrate the laser scan control. In addition, the electrical detection of the OGE effect provides a direct electrical signal which can be used in conjunction with some type of a feedback mechanism to directly lock the laser on a specific transition for an extended period of time. Another clear advantage of this method is its relative simplicity, with no need for complex optical isolation and detection instrumentation.

Other fundamental studies of discharges in which optogalvanic spectroscopy has also been successfully employed include the generation of detailed, high-resolution, atomic spectra.[140] The OGE has also found application in a variety of hyperfine-structure studies,[141–143] as a method for the determination of term values of neon,[144] for measuring He and Ne line profiles,[145] for observing collisional mixing of atoms in the discharge,[146] as a method of detection in state selective laser photodissociation of TI_2,[147] and as a method for determining the threshold energy of sputtering.[148]

Another potentially useful application of the OGE for discharge diagnostic investigations employs Stark shifts and splittings for electric field strength measurements in hollow cathode discharges. Initial observations of a correlation between Stark shifts of a krypton OGE signal and proximity to the cathode, and hence electric field strength, were reported in 1983.[149] This technique was exploited in a detailed study of a krypton hollow cathode discharge in which a linear decrease in electric field versus distance from the cathode was observed. In addition, this study employed the Stark-shifted OGE signals to calculate the number density of charge carriers, and to investigate the interrelationship of various discharge parameters such as current, voltage, cathode dark space distance, and pressure.[150] The theory behind these measurements was subsequently outlined in another report[151] and the measurements extended to a neon discharge.[152]

Reports of direct analytical applications of optogalvanic spectroscopy are limited, although the method has shown some promise as a possible analytical method by observation of trace amounts of sodium in a lithium hollow cathode.[153] Keller *et al.* have also illustrated the potential analytical utility of the method by observing an OGE signal from ^{235}U in a hollow cathode lamp, with the ^{235}U level calculated to be $\approx 3 \times 10^9$ atoms/cm^3.[154] In another report, Keller and Zalewski studied noise and other considerations in OGE spectroscopy using a hollow cathode lamp and calculated a theoretical detection limit of $\approx 3 \times 10^6$ atoms/cm^3 for minority sputtered species and $\approx 10^2$ atoms/cm^3 for metastable species of the discharge gas.[125]

In a somewhat different analytical application, monitoring of the OGE in a secondary discharge cell served as a detector for laser atomic absorption measurements.[155,156] The instrumental setup is diagrammed in Fig. 10-14. In these applications, the laser is directed through a sample cell (flame) into the discharge containing the same species as the potential analyte which serves as the detector. The laser power must be in a range in which changes in power produce a linear response in the detected OGE signal. If this condition is met, then if no laser photons are absorbed by the sample, a baseline OGE signal is measured with the detector discharge. As analyte atoms or molecules are added to the sample cell, they will absorb a fraction of the incident laser light and the power density of light arriving at the detector discharge will be reduced. This reduction in power should result in a reduction in the monitored OGE signal that is proportional to the number of photons absorbed, which in turn is proportional to the concentration of sample species in the cell, providing the analytical utility of the technique. The method has been demonstrated for both atomic[155] and molecular species.[156]

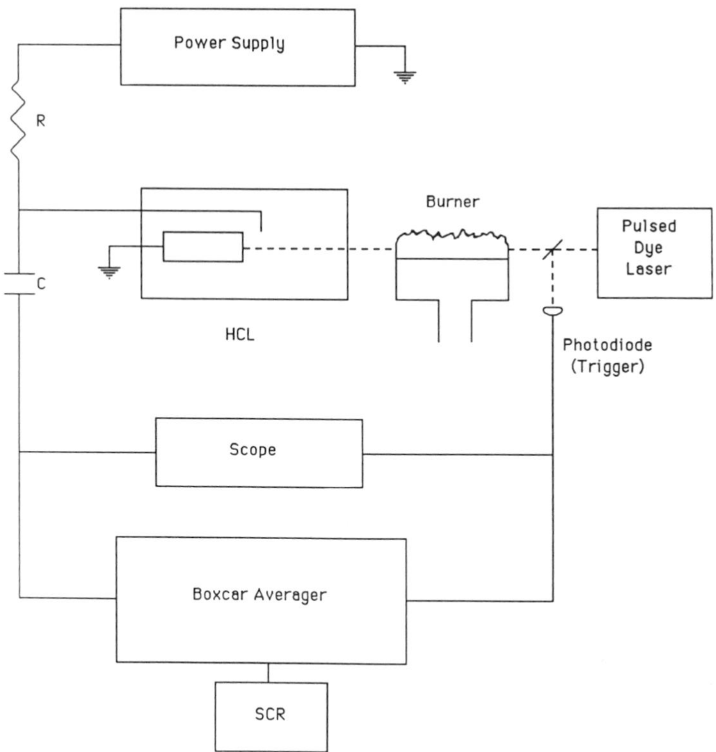

Figure 10-14. Diagrammatic representation of the experimental system employed for optogalvanic effect detection of laser atomic absorption: $R = 25\ \text{k}\Omega$, $C = 1\ \mu\text{F}$; SCR, strip chart recorder. [Reprinted with permission from: M. A. Nippoldt and R. B. Green, *Appl. Opt.* 20 (1981) 3207. Copyright 1981 Optical Society of America.]

Most of the previous discussions of OGE applications have involved atomic transitions, although the OGE can be observed for molecular transitions as well. Examples include a study of OGE spectra generated from the illumination of iodine[130] and the monitoring of the transient intermediate species, SiH, by the OGE in a (CVD) device.[157] Further applications of OGE spectroscopy are sure to arise as research in this area continues.

10.4.3. Laser-Enhanced Ionization

A method similar to OGE spectroscopy is laser-enhanced ionization (LEI). Rather than monitoring the change in discharge voltage or current with the interaction of laser photons, LEI methods directly measure electrons released when the ionization of the absorbing species is enhanced due to photon excitation. Since the electrons are measured directly, a discharge circuit is not required, permitting LEI to be performed in flames. A diagram of the general instrumentation employed for LEI measurements is presented in Fig. 10-15. There have been many reports of the LEI process, with a few illustrative examples provided in a variety of sources.[158–168]

The LEI process has been largely limited to flames although the technique could be performed in a glow discharge. The main difficulties with LEI in discharges deal with the difficult detection of the relatively few electrons released as a result of the ionization enhancement, relative to the large background electron population generated by current flow through the system. In addition, LEI assumes an ionization enhancement is observed due to excitation of a discharge species. This may not occur if ionization proceeds by the Penning ionization process, or if the observed reduction in discharge resistance with laser interaction is due to a change in the electron

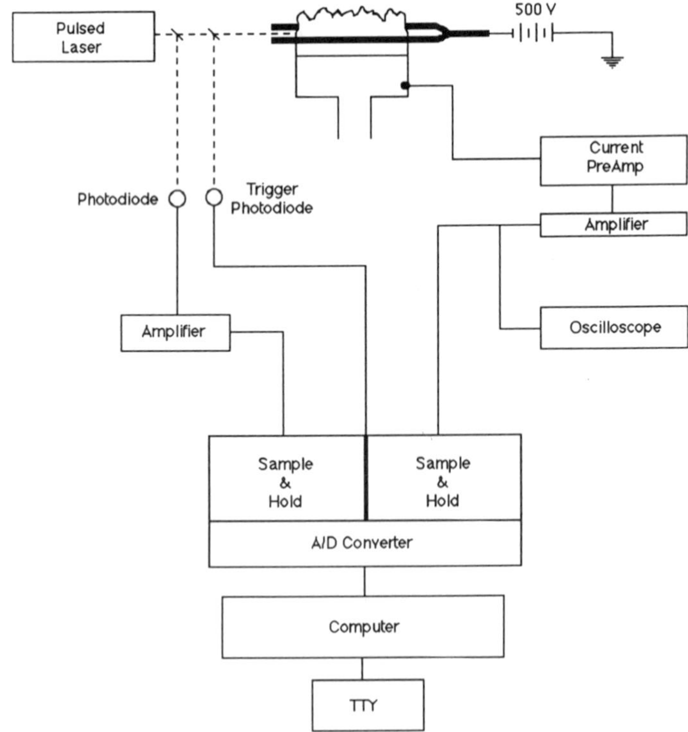

Figure 10-15. Representative instrumental arrangement employed for laser-enhanced ionization spectroscopy. [Reprinted with permission from: G. C. Turk, J. C. Travis, J. R. DeVoe, and T. C. O'Haver, *Anal. Chem.* 50 (1978) 818. Copyright 1978 American Chemical Society.]

energy distribution and not necessarily due to enhanced ionization, as discussed in Section 9.4.2. In fact, under certain circumstances, the OGE signal in discharge plasmas has not been found to depend on either the ionization potential of the atom or how close the terminal state lies to the ionization continuum,[125] as would be expected under a mechanism of enhanced ionization. This is in contrast to LEI signals in flames which have shown an ionization potential dependence.[169] Apparently, the atmospheric pressure flame provides a higher collisional environment for electron excitation and ionization than the low-pressure discharges, generating a distinct difference in OGE and LEI mechanisms and possibly serving as an explanation for the lack of LEI signals observed in plasma devices.

10.4.4. Resonance Ionization Mass Spectrometry

A laser may also be used for selective ionization enhancement of sputtered species in the discharge, possibly increasing both the sensitivity and the selectivity of glow discharge mass spectrometry. In these applications, a high-power tunable dye laser is focused into the negative glow region of the discharge, where the photons may interact with the substantial population of sputtered atoms. An atom may resonantly absorb one or more photons, generating either an ion directly or an excited-state atom which requires less energy for ionization, enhancing the ionization rate. Direct ionization due to the resonant absorption of laser photons is generally termed resonance ionization spectroscopy (RIS) if the ejected electron is detected, or resonance ionization mass spectrometry (RIMS) if the positive ion is monitored.

The resonance ionization process involves the sequential absorption of photons having energies which match bound–bound transitions, followed by a bound–continuum transition. The bound–bound transition renders the process highly wavelength selective, with only elements whose bound–bound transitions correspond to the wavelength of the laser preferentially ionized during the laser pulse. When combined with mass spectrometry, the method has two degrees of selectivity, wavelength and mass. In addition to possible enhanced sensitivity with laser ionization, the selective photoionization step in the RIMS process will allow the reduction of potential isobaric interferences present in normal mass spectra and provides unique analytical utility.

Many detailed descriptions and reviews of various resonance ionization techniques, including theoretical and practical discussions of the parameters necessary for application of the methodology, are available.[170–181] Briefly, in a collisionally active environment such as the glow discharge, many pathways exist for selective, photon-induced ionization of a species. Any combination of laser and discharge excitation that results in a net energy sufficient to cause ionization can be a viable pathway for resonance ionization. Those pathways involving only the absorption of photons are generally referred to as multiphoton ionization (MPI), while pathways that advantageously employ discharge energy in the ionization process are referred to as LEI methods. A few representative MPI pathways are diagrammed in Fig. 10-16, where ω represents a laser photon. In scheme 1 of Fig. 10-16, an atom, A, absorbs a photon, ω_1, causing excitation to a true energy level greater than halfway to the ionization potential of the atom. While in the excited state, the atom absorbs a second (coincident) photon of the same wavelength as the first, resulting in ionization. In scheme 2, the laser is frequency doubled to provide a photon of higher energy, $2\omega_1$, for the initial photon absorption to a true excited state, followed by the absorption of a nondoubled photon to cause ionization. If two lasers are available and their pulses interact the discharge at the same time, multiple color experiments can be designed where one laser is tuned to one atomic transition, and the second laser is tuned to a different transition originating from the excited state of the first transition, with multiple photon adsorptions resulting in ionization, as represented by schemes 3 and 4 in Fig. 10-16. Scheme 5 involves a two-photon absorption through a virtual level, followed by the absorption of a third photon for ionization. More detail on these pathways can be found elsewhere.[171] The key feature common to all of these pathways

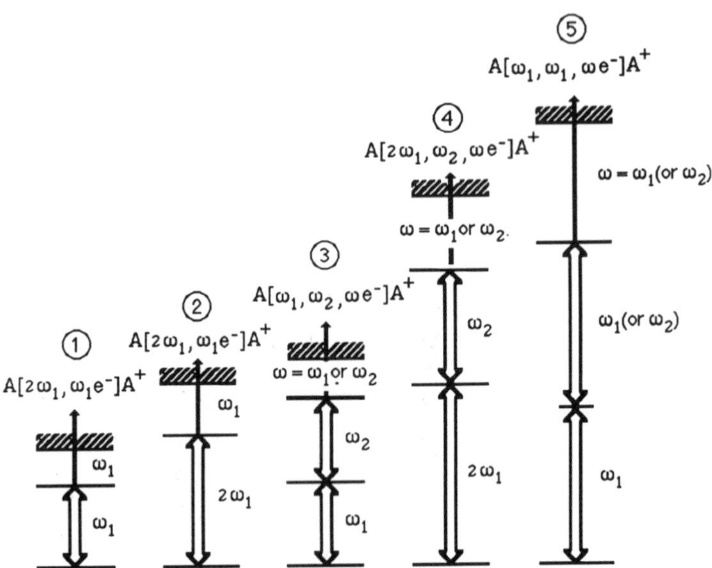

Figure 10-16. Example of various pathways available for resonance ionization. See text for details on each pathway. [Reprinted with permission from: G. S. Hurst, M. G. Payne, S. D. Kramer, and J. P. Young, *Rev. Mod. Phys.* 51 (1979) 774. Copyright 1979 American Physical Society.]

is the presence of at least one resonant step to provide selectivity. The exact pathway chosen is highly dependent on a multitude of instrumental factors such as laser power densities, collisional excitation rates in the atom source, absorption cross sections, the species under investigation, and so forth. A variety of illustrative examples of the resonance ionization method have appeared, employing a wide range of atomic sources including furnaces,[182] laser ablation,[183–185] filaments,[186–188] and ion guns.[189–191] Each of these applications is in a sense unique, with specific instrumental details that influence the methodology and pathway chosen. For the purposes here, the detailed discussion of resonance ionization is confined to the application of the glow discharge as the source of atoms for RIMS.

Resonance Ionization in the Glow Discharge

Resonance ionization in the glow discharge appears to have the potential to enhance the effectiveness of glow discharge mass spectrometry. Although the glow discharge provides an ample sputtered ion signal for mass spectrometric analysis, it has been estimated that <1% of the atomic population is ionized.[192] There remains a large nonionized atomic population so that efficient supplemental ionization by the laser may increase the sensitivity of glow discharge mass spectrometry. In addition, glow discharge mass spectrometry suffers from the potential for isobaric interferences from other elements and a host of molecular species. Many of these interferences can be discriminated against by using a high-resolution magnetic sector mass spectrometer, as in a commercial glow discharge mass spectrometer,[193] but many potential interferences require resolutions above those currently available.[194] The resonant nature of the laser ionization process provides a means for selectively ionizing an element and subtracting all nonresonant (steady state) ions, including contributions from molecular species and other sputtered atoms, discriminating against isobaric interferents. The enhanced selectivity of the RIMS process is an attractive feature for analytical investigations.

The glow discharge offers several possible advantages as an atom reservoir for resonance ionization. These include the simplicity of the discharge, its ability to create an atomic population representative of the solid cathode material, the low differential elemental sputter rates observed in the discharge, its stability, and the ability to pulse the discharge to couple more effectively with a pulsed laser system. Possible disadvantages include the relatively high pressure in region where ions are formed, resulting in a lowered extraction efficiency of laser-formed ions, the electric fields associated with the glow discharge, and the relatively high-background ion signal from the discharge which must be subtracted from the laser-produced signal.

Previous investigations of resonance ionization in a glow discharge have been reported in more detail elsewhere.[58,64,195,196] These investigations employed an excimer pumped dye laser system with a frequency doubler to generate photons that subsequently interacted with a coaxial discharge which was operated with argon fill at pressures of 0.3–2.0 torr, currents of 1–5 mA, and voltages of 300–3500 V.

When the dye laser is tuned to an appropriate atomic transition which results in enhanced ionization, a resonant, laser-generated ion signal is observed on top of the "background" ion signal from glow discharge ionization. These signals are illustrated in Fig. 10-17. The laser-generated signal may be isolated by setting a data sampling window over the time period of laser interaction, collecting the "laser + discharge ions," then setting a second "background" data gate during a time period when the laser is off, and finally subtracting the "background" signal from the "laser + discharge ions" signal. The ions observed after "background" subtraction should be only those formed during the laser pulse, directly attributable to laser-induced ionization, either through a multiphoton process or through an LEI process.

The ions formed by laser interaction will exhibit a wavelength dependence corresponding to atomic transitions of the element of interest. Fig. 10-18 is a wavelength ionization scan for iron in the glow discharge. The signals observed were obtained by setting the mass spectrometer on the ^{56}Fe isotope and scanning the dye laser wavelength. Each observed peak corresponds to an atomic transition of iron, as tabulated in Table 10-4. A partial energy diagram for iron is provided in Fig. 10-19 to illustrate the collisional and photon-induced ionization processes occurring for several of the observed signals. These transitions may originate from the ground state or from any one of several low-lying energy levels which have been populated

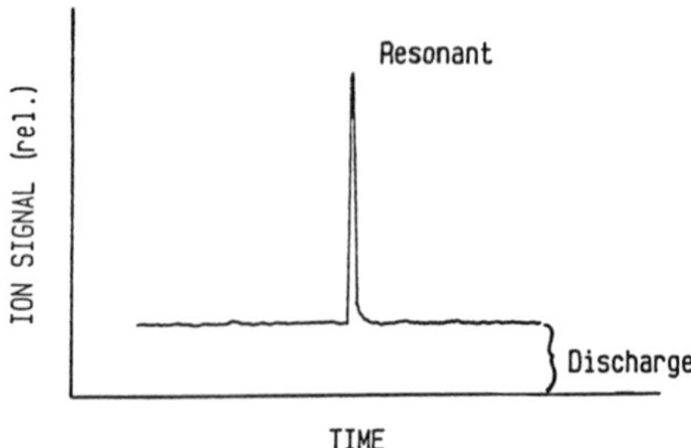

Figure 10-17. Temporal profile of resonant and nonresonant (discharge) ionization in a glow discharge.

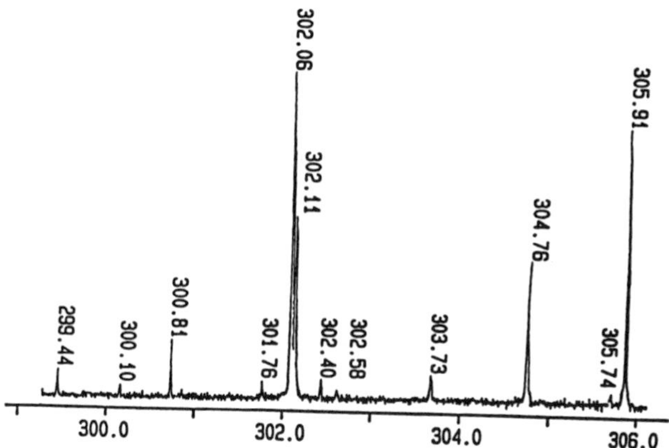

Figure 10-18. Wavelength scan of ^{56}Fe ions produced by resonant ionization in a glow discharge. Argon discharge, 0.3 torr, 2 mA, pure iron cathode.

Table 10.4. Iron Transitions Which Generated a Resonance Ionization Signal in the Glow Discharge for the Wavelength Region 299–306 nm

Wavelength (nm)	Transition energies (eV)
299.44	0.05–4.19
300.10	0.09–4.22
300.81	0.11–4.23
301.76	0.11–4.22
302.05	0.09–4.19
302.06	0.00–4.11
302.11	0.05–4.16
302.40	0.11–4.21
302.58	0.12–4.22
303.74	0.11–4.19
304.76	0.09–4.16
305.74	0.86–4.91
305.91	0.05–4.11

by collisional excitation in the glow discharge. The availability of supplemental excitation from the discharge expands the number of transitions amendable to resonance ionization and provides a multitude of pathways for ionization.

For the transitions observed in Fig. 10-18, simultaneous absorption of two photons will result in direct photoionization of the iron atom and a multiphoton ionization process. Alternatively, the absorption of a single photon may result in an excited-state iron atom which may subsequently be

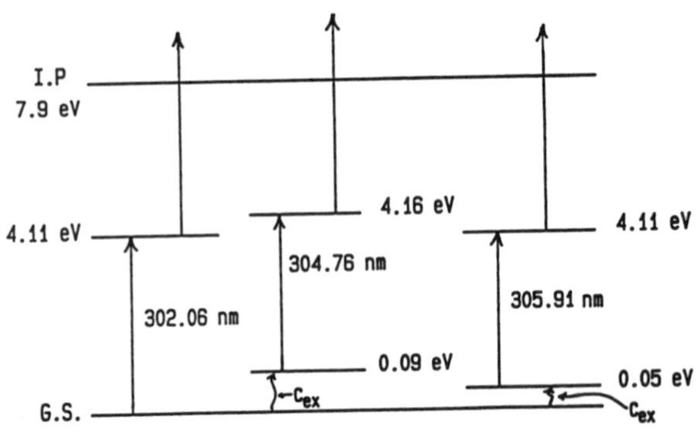

Figure 10-19. Partial iron energy level diagram for several of the transitions detected in Fig. 10-18 showing transitions which rely on collisional excitation (C_{ex}) in addition to laser photon absorption for ionization.

ionized by a collisional process in the glow discharge. If electron impact is a dominant mechanism of ionization in the discharge, the ionization of an excited-state species will be enhanced relative to a ground-state atom as discussed previously for OGE spectroscopy (Section 10.4.2). The net result in a single photon enhancement of ionization is indicative of the LEI process. The photon-induced ion signals observed in the discharge may arise from either the MPI or the LEI process and it is not certain which is responsible for the enhanced ionization observed. Energy dependence studies on the process have been performed in an effort to distinguish the two processes,[64] but these studies were inconclusive. In any event the signal is resonant providing an enhanced degree of selectivity and sensitivity.

A potential problem with resonance ionization in the glow discharge is the relatively high-pressure region in which the laser-generated ions are formed. The rare gas pressure necessary for discharge operation provides a collisionally rich environment which contributes to a decreased efficiency of resonant ion collection because of collisional scattering, recombination, and decreased extraction of the ions. In order to most efficiently collect ions formed by laser interaction, the laser beam must be positioned directly adjacent to the exit orifice of the discharge cell. As the beam is moved from the ion exit orifice, the laser-generated ion signal falls dramatically. A possible solution would be to construct a discharge system in which the laser beam interacts external to the discharge source, as illustrated in Fig.10-20. Since no discharge is operating on this side of the ion exit orifice, the pressure in this region may be reduced to a few millitorr or less, reducing the number of collisions and increasing the collision-free paths. A substantial population of sputtered atoms may diffuse through the ion exit orifice and subsequently interact with the laser to form ions, which may then be more efficiently

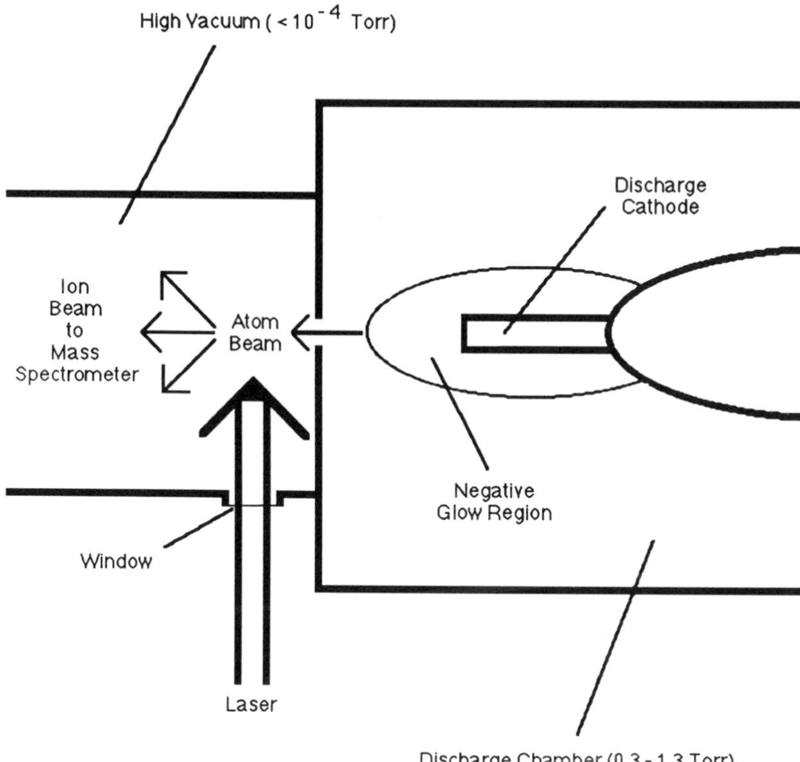

Figure 10-20. Schematic diagram of an experimental arrangement for resonance ionization of an atomic population external to the discharge source.

collected and analyzed. Preliminary work with such a system has been attempted, but no significant RIMS signal was observed.[196] The instrumental configuration for these studies was not optimized, with the long distance between the laser interaction zone and the entrance to the ion optics significantly reducing the collected ions. Also, atomic absorption signals showed a negligible atom population external to the source, necessitating system modifications to provide a higher atom throughput from the source. No further investigations of resonance ionization external to a discharge source have been reported, but this remains a possible area of future investigation.

Although the glow discharge/resonance ionization technique may not be optimized for analytical applications, previous results have shown the inherent selectivity of resonance ionization as a possible method for discriminating against isobaric interferences in glow discharge mass spectrometry. Figure 10-21A shows a portion of a glow discharge mass spectrum for a

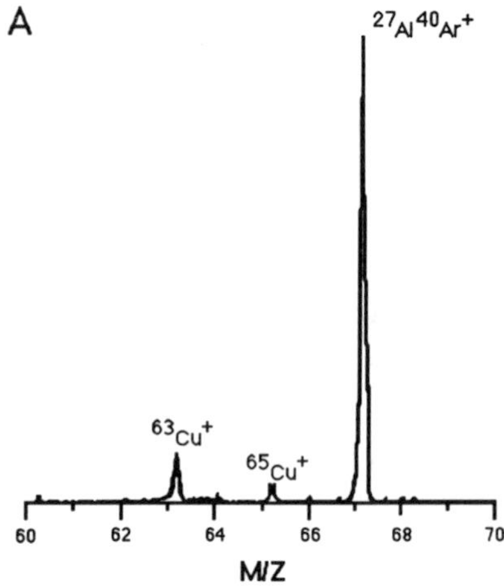

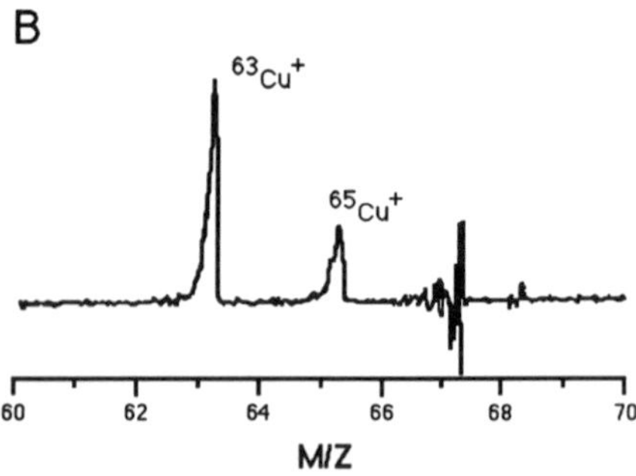

Figure 10-21. Representative example of the potential of resonance ionization to reduce isobaric interferences in glow discharge mass spectrometry. (A) Glow discharge mass spectrum showing ions from copper isotopes (Cu at 0.29% in an aluminum matrix) and a $^{27}Al^{40}Ar$ species. (B) Resonance ionization mass spectrum generated using a Cu(I) transition at 324.75 nm illustrating the reduction of the $^{27}Al^{40}Ar$ species. Argon discharge, 0.3 Torr, 2 mA. [Reprinted with permission from: K. R. Hess and W. W. Harrison, *Anal. Chem.* 58 (1986) 1700. Copyright 1986 American Chemical Society.]

NIST #603 aluminum sample. The total copper concentration in this sample is 0.29%. As can be seen, a large peak at m/z 67 arises from an $AlAr^+$ "argide" species. Species formed from a combination of argon and both the sputtered material and background gases such as nitrogen are commonly found in the glow discharge. They can often create problems in elemental analysis if the only isotope of the element of interest occurs at an m/z value corresponding to an argide species. Resonance ionization methods can selectively ionize the atomic species in the presence of the molecular compounds, generating a method to eliminate the isobaric interference.

Figure 10-21B shows the same portion of the aluminum mass spectrum employing resonance ionization of the copper atoms at 324.75 nm and subtracting the background glow discharge ions. In this spectrum, the copper isotopes are enhanced while the $AlAr^+$ has been essentially removed.

Another problem potentially eliminated by resonance ionization is the effect of a large spectral peak directly adjacent to a trace constituent, swamping any ion signal from the trace material. Although this is not a problem for high-resolution mass spectrometers, it can be a significant problem when lower-resolution quadrupole instruments are employed. For example, Fig. 10-22A shows a portion of a NIST #626 zinc glow discharge mass spectrum.

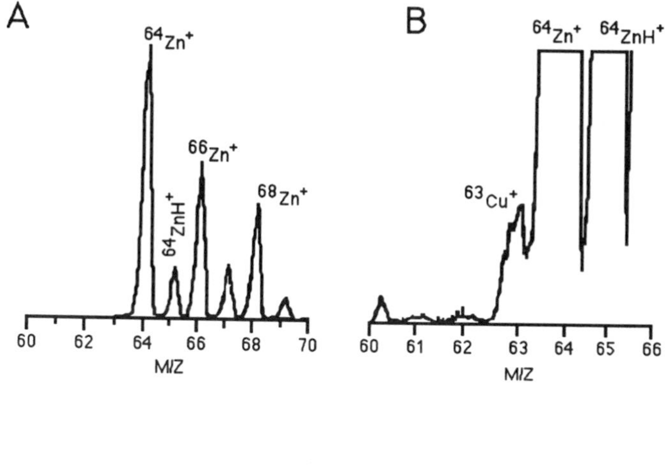

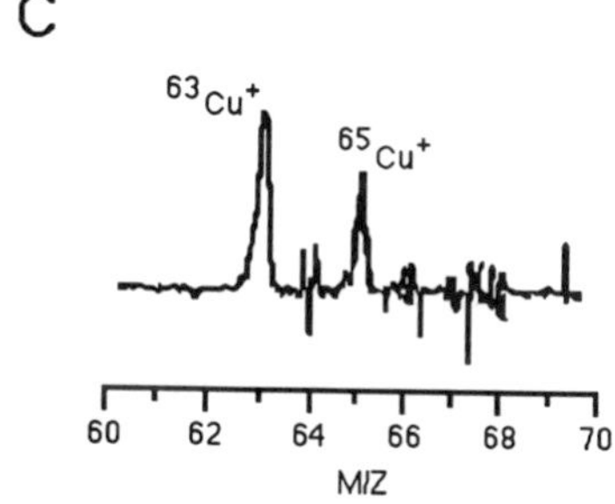

Figure 10-22. Illustrative example of the use of resonance ionization to measure small concentrations of copper (0.058%) in a zinc matrix (NIST #626 reference sample). (A) Low-gain glow discharge mass spectrum showing the zinc isotopes. (B) Expansion of the glow discharge mass spectrum in A showing the ^{63}Cu ion signal relative to the much larger zinc ion signals. (C) Resonance ionization mass spectrum of the copper at 324.75 nm. Argon discharge, 0.3 Torr, 2 mA. [Reprinted with permission from: K. R. Hess and W. W. Harrison, *Anal. Chem.* 58 (1986) 1701. Copyright 1986 American Chemical Society.]

Copper is present at 0.058% total in this sample, but the ^{65}Cu isotope is masked by a $^{64}ZnH^+$ interferent while the ^{63}Cu peak is barely resolved from the much larger ^{64}Zn signal, as illustrated in Fig. 10-22B. Resonance ionization of the copper atoms at 324.75 nm allows subtraction of the zinc peaks while continuing to ionize and collect the copper. In Fig. 10-22C, resonance ionization of copper in the zinc sample allows both copper isotopes to be visualized in their proper isotopic ratios. These are currently the only reported applications of resonance ionization in a glow discharge, though the combination of discharge atomization and laser ionization offers opportunities for further study.

10.4.5. Additional Laser-Based Techniques

In addition to the applications of glow discharges discussed previously, the plasmas have also found use as atom reservoirs for several specialized laser spectroscopic techniques. These methods have been employed largely for physical measurements of atomic properties such as transition linewidths or atomic lifetimes, rather than for analytical applications. Brief discussions of the techniques with representative literature citations are provided in the following sections.

10.4.5.1. Doppler-Free Spectroscopy

The combination of the discharge's ability to create a substantial and stable atomic population of most elements in a low-pressure environment, which limits collisional broadening and quenching of the atomic population, has resulted in their application as atomic sources for a variety of Doppler-free spectroscopic experiments. The elimination of the velocity broadening component of the atomic linewidth available with these investigations provides high-resolution atomic spectra which generate a wealth of spectroscopic information, including isotopic shifts and hyperfine coupling constants. An illustrative example of a Doppler-free spectrum as compared with the Doppler-broadened spectrum is provided in Fig. 10-23. This spectrum was obtained by the technique of saturation absorption spectroscopy.[84,197] A variety of other specific Doppler-free spectroscopic techniques are available and these are too numerous and experimentally complex to discuss in detail. Included in this list are intermodulated fluoroscence,[198] optical–optical double resonance,[199] two-photon absorption methods,[199] and laser polarization spectroscopy.[200] The previously discussed OGE method of recording atomic spectra has also been employed as a method for generating Doppler-free atomic spectra.[84,121]

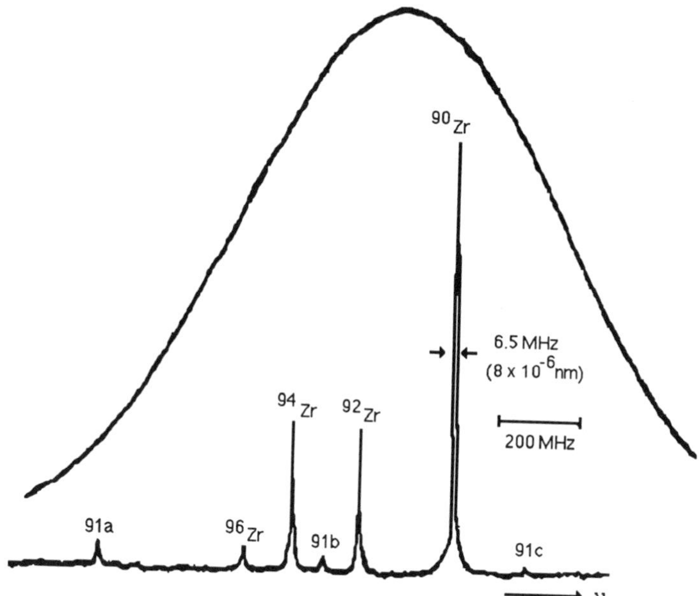

Figure 10-23. Comparison of a Zr 614.3-nm Doppler-free spectrum (lower trace) with a Doppler-broadened spectrum (top trace) employing a discharge as an atom source. The three strongest ^{91}Zr hyperfine components are labelled a, b, and c in the lower trace. [Reprinted with permission from: P. Hannaford and A. Walsh, *Spectrochim. Acta* 43B (1988) 1062. Copyright 1988 Pergamon Press.]

10.4.5.2. Laser Atomic Absorption

As the cost of tunable dye laser systems decreases and they become more widely available, lasers may find increasing application as a cost-effective alternative to hollow cathode lamps as line sources for atomic absorption experiments. The main advantage of a tunable laser system over hollow cathode lamps is the availability of narrow linewidths over a wide range of wavelengths, which would be suitable for multielement analysis as compared with the limited elemental application of specific hollow cathode lamps. The lasers also provide a substantial number of incident photons, are coherent, and easily directed, enhancing the spatial resolution of atomic absorption experiments. Other than the cost and complexity, the main disadvantages of the laser atomic absorption source are an incident power density that is too high to measure the small differences in power due to photon absorption and the generally poor shot-to-shot reproducibility of the laser. A representative example of laser atomic absorption with a discharge atom reservoir is the determination of thorium atom densities in a hollow cathode discharge as reported by Gagne *et al.*[201] In this work, an Ar^+ laser pumped a dye laser, whose output was reduced and adjusted by the use of optical filters. More recent developments in the field of semiconductor diode lasers could provide further applications of lasers to atomic absorption. The diode lasers are significantly less expensive, of lower power, and are generally more stable than other laser types, all attractive features for atomic absorption measurements. The main drawback is the very limited wavelength range currently available with diode lasers and the narrow tuning range of most systems. Recent publications have indicated that semiconductor diode lasers are better spectral line sources than hollow cathode lamps,[202,203] of course provided an atomic transition is accessible to the laser wavelength range. The current range of available wavelengths is approximately 700–900 nm, and atomic number densities can be calculated for atomic transitions which lie in this range.[204] Continued advances in the laser technology can be expected to extend the accessible range of diode lasers making them more attractive as line sources for atomic absorption investigations, both in flames and in glow discharge plasmas.

10.4.5.3. Degenerate Four-Wave Mixing

Low-pressure discharges have also been employed as an atom source for nonlinear spectroscopic techniques. Several degenerate four-wave mixing (DFWM) experiments have been reported in flames,[205–207] and recently the methodology has been employed in conjunction with a glow discharge source of atoms.[208,209] Detailed theoretical discussions of the DFWM technique are provided in these references and in a recent book,[210] and only a short introduction to the technique will be discussed here. The analytical signal in DFWM arises from the interaction of three coherent and degenerate (identical frequencies) laser beams, generating a fourth beam which is detected. A schematic of the overall instrumentation is provided in Fig.10-24, while Fig. 10-25 provides a simplified illustration of the laser interaction zone. The forward pump beam, l_f, and the probe beam, l_p, will overlap in a region of the source and since the two beams have identical wavelengths, an interference pattern can be generated, creating a spatial variation in the photon intensity of the overlap region. The interference pattern can act as a grating, so that when the third beam, the backward pump beam l_b, interacts with the region, a portion of the radiation can be Bragg-scattered to form the output phase conjugate beam, l_{pc}, which is subsequently detected. Conservation of momentum will require that the output beam is counterpropagating to the input beam, providing a beam with a field strength that is the complex conjugate of the pump beam. The analytical utility of the method originates from the dependence of l_{pc} on the intensities of the other three laser beams and the physical properties of the interacting medium. The medium provides

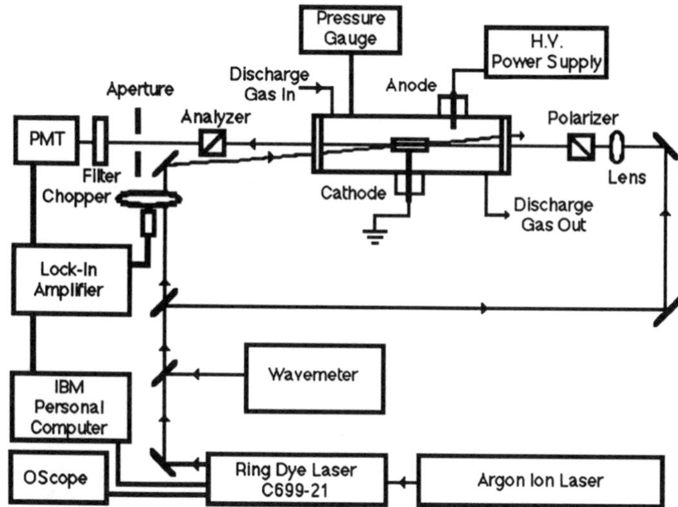

Figure 10-24. Instrumentation employed in degenerate four-wave mixing experiments. [Reprinted with permission from: G. A. Luena and W. G. Tong, *Appl. Spectrosc.* 44 (1990) 1669. Copyright 1990 The Society for Applied Spectroscopy.]

Figure 10-25. Laser beam alignment in degenerate four-wave mixing experiments. [Reprinted with permission from: J. M. Ramsey and W. B. Whitten, *Anal. Chem.* 59 (1987) 168. Copyright 1987 American Chemical Society.]

a significant contribution to the nonlinear refractive index, and hence the ability to scatter I_b, only when the light frequency is close to an allowed transition, providing the ability to generate spectra of atomic species in the medium. In addition, the signal is also proportional to the square of the concentration of absorbing species, providing the potential for quantitative analysis.[(205)]

There are several attractive features of the DFWM technique reported to be advantageous relative to other spectroscopic methods such as atomic fluorescence.[(205)] First, in theory no signal should exist without an absorbing species, providing a very low background against which to measure the signal. In addition, the signal photons are all coherent, allowing for very efficient detection, even over long distances or through fiber-optic cables, and for discrimination against background photons arising from incoherent processes, enhancing the signal-to-background characteristics of the method. Further discrimination from background photons can also be obtained through polarization of the laser beams, generating a polarized output from a "real" signal. The interaction of the forward and backward pump beams provides Doppler-free spectroscopic signals, providing a method for high-resolution spectroscopy. The high spectral resolution available with DFWM can be effectively employed to enhance the selectivity of the method and, with a low-pressure source which limits transition linewidths, to observe hyperfine splitting patterns of atomic transitions.[205] The DFWM interaction is essentially instantaneous, reducing problems associated with quenching in a collisional environment. The technique is relatively simple compared with other nonlinear methods, requiring only one laser and wavelength-independent phase-matching conditions. Due to the fact that analytical signals arise only from the region of overlapping beams, the method also has a very high degree of spatial resolution which is beneficial for spatial mapping of species in various sources.

The application of DFWM with a glow discharge atom reservoir has been reported by Tong *et al.* in two recent publications.[208,209] In the first of these reports, 1 mg of sodium was deposited onto the surface of a demountable hollow cathode by drying a sodium carbonate solution and the discharge was then struck. The DFWM signal from the sodium was monitored with sufficient resolution to observe the hyperfine splitting of the Na D_2 transition. Although no direct analytical studies of the method were performed in this work, the authors illustrate the potential analytical sensitivity of the method by commenting that the signal beam was visible to the naked eye when only 1 mg of sodium was placed in the cathode. The second of these papers provides further details and an expansion of the studies reported in the first paper. Possible future applications of the DFWM technique may include further analytical development of the method, an increased application in the area of high-resolution spectroscopy, and advantageously employing the high degree of spatial resolution afforded by DFWM in spatial mapping studies of discharge species.

10.5. Conclusions

The combination of a glow discharge's ability to generate a copious atomic population from a solid sample, a low-pressure operating regime, and the ability to be pulsed synchronous with a laser provides unique opportunities to exploit laser/discharge combinations for both fundamental and analytical investigations. Previous applications of glow discharges and laser interactions have included laser ablation, laser fluorescence, resonance ionization, OGE spectroscopy, and nonlinear absorption techniques. As the instrumentation for both lasers and discharges improves, and as novel spectroscopic techniques are developed, glow discharge sources will find increasing use as atom sources for laser spectroscopic measurements.

References

1. D. Duckworth and R. K. Marcus, *Anal. Chem.* 61 (1989) 1879.
2. M. R. Winchester and R. K. Marcus, *J. Anal. At. Spectrom.* 5 (1990) 575.
3. O. Svelto and D. C. Hanna, *Principles of Lasers*, Plenum Press, New York, 1976.
4. D. C. O'Shea, W. R. Callen, and W. T. Rhodes, *Introduction to Lasers and Their Applications*, Addison–Wesley, Reading, Mass., 1977.
5. J. C. Wright and M. J. Wirth, *Anal. Chem.* 52 (1980) 1087A.
6. W. Demtroder, *Laser Spectroscopy: Basic Concepts and Instrumentation*, Springer, Berlin, 1981.
7. P. W. Milonni and J. H. Eberly, *Lasers*, Wiley, New York, 1988.
8. J. Wilson and J. F. B. Hawkes (eds.), *Lasers: Principles and Applications*, Prentice–Hall, Englewood Cliffs, N.J. 1987.
9. L. J. Radziemski, R. W. Solarz, and J. A. Paisner (eds.), *Laser Spectroscopy and Its Applications*, Dekker, New York, 1987.
10. T. Ishizuka, Y. Uwamino, and H. Sunahara, *Anal. Chem.* 49 (1977) 1339.
11. R. Wennrich and K. Dittrich, *Spectrochim. Acta* 39B (1984) 657.
12. A. Quentmeier, W. Sdorra, and K. Niemax, *Spectrochim. Acta* 45B (1990) 537.
13. W. Sdorra and K. Niemax, *Spectrochim. Acta* 45B (1990) 917.
14. S. Grazhulene, V. Khvostikov, and M. Sorokin, *Spectrochim. Acta* 46B (1991) 459.
15. G. Dimitrov and T. Zheleva, *Spectrochim. Acta* 39B (1984) 1209.
16. K. J. Mason and J. M. Goldberg, *Anal. Chem.* 59 (1987) 1250.
17. Y. Ilda, Appl. Spec. 43 (1989) 229.
18. R. J. Conzemius and J. M. Capellen, *Int. J. Mass Spectrom. Ion Phys.* 34 (1980) 197.
19. H. Van Doveren, *Spectrochim. Acta* 39B (1984) 1513.
20. R. J. Cotter, *Anal. Chem.* 39B (1984) 1513.
21. F. P. Novak, K. Balasanmugam, K. Viswanadham, C. D. Parker, Z. A. Wilk, D. Matern, and D. M. Hercules, *Int. J. Mass Spectrom. Ion Phys.* 53 (1983) 135.
22. E. Denoyer, R. Van Grieken, F. Adams, and D. F. S. Natusch, *Anal. Chem.* 54 (1982) 26A.
23. R. W. Odom and B. Schueler, in: *Lasers and Mass Spectrometry* (D. M. Lubman, ed.), p. 103, Oxford University Press, London, 1990.
24. A. H. Verbueken, F. J. Bruynseels, R. Van Grieken, and F. Adams, in: *Inorganic Mass Spectrometry* (F. Adams, R. Gijbels, and R. Van Grieken, eds.), p. 173, Wiley, New York, 1988.
25. T. Ishizuka and Y. Uwamino, *Spectrochim. Acta* 38B (1983) 519.
26. A. Aziz, J. A. C. Broekaert, K. Laqua, and L. Leis, *Spectrochim. Acta* 39B (1984) 1091.
27. M. Thompson, S. Chenery, and L. Brett, *J. Anal. At. Spectrom.* 4 (1989) 11.
28. S. Darke, S. E. Long, C. J. Pickford, and J. F. Tyson, *J. Anal. At. Spectrom.* 4 (1989) 715.
29. S. Lin and C. Peng, *J. Anal. At. Spectrom.* 5 (1990) 509.
30. P. Arrowsmith, *Anal. Chem.* 59 (1987) 1437.

31. J. W. Hagar, *Anal. Chem.* 61 (1989) 1243.
32. J. Marshall, J. Franks, I. Abell, and C. Tye, *J. Anal. At. Spectrom.* 6 (1991) 145.
33. H. Pang, D. R. Wiederin, R. S. Houk, and E. S. Yeung, *Anal. Chem.* 63 (1991) 390.
34. E. R. Denoyer, K. J. Fredeen, and J. W. Hager, *Anal. Chem.* 63 (1991) 445A.
35. F. Leis and K. Laqua, *Spectrochim. Acta* 34B (1979) 307.
36. P.G. Mitchell, J. Sneddon, and L. J. Radziemski, *Appl. Spectrosc.* 41 (1987) 141.
37. W. van Deijcke, J. Balhe, and F. J. M. Maesser, *Spectrochim. Acta* 34B (1979) 359.
38. D. W. Beekman, T. A. Callcott, S. D. Kramer, E. T. Arakawa, G. S. Hurst, and E. Nussbaum, *Int. J. Mass Spectrom. Ion Phys.* 34 (1980) 89.
39. N. S. Nogar and R. C. Estler, in: *Lasers and Mass Spectrometry* (D. M. Lubman, ed.), p. 65, Oxford University Press, London, 1990.
40. B. S. Freiser, *Talanta* 32 (1985) 697.
41. J. W. Carr and G. Horlick, *Spectrochim. Acta* 37B (1982) 1.
42. B. E. Knox, in: *Dynamic Mass Spectrometry*, Vol. 2 (D. Price, ed.) p. 61, Heyden & Son, Philadelphia, 1970.
43. Y. P. Raizer, *Sov. Phys. Usp.* 8 (1966) 650.
44. D. Cremers and L. J. Radziemski, in: *Laser Spectroscopy and Its Applications* (L. J. Radziemski, R. W. Solarz, and J. A. Paisner, eds.) p. 351, Dekker, New York, 1987.
45. M. Thopson, S. Chenery, and L. Brett, *J. Anal. At. Spectrom.* 5 (1990) 49.
46. P. Arrowsmith, in: *Lasers and Mass Spectrometry* (D. M. Lubman, ed.), p. 179, Oxford University Press, London, 1990.
47. Z. Hwang, Y. Teng, K. Li, and J. Sneddon, *Appl. Spectrosc.* 45 (1991) 435.
48. A. Vertes, M. De Wolf, P. Juhasz, and R. Gijbels, *Anal. Chem.* 61 (1989) 1029.
49. L. Balazs, R. Gijbels, and A. Vertes, *Anal. Chem.* 63 (1991) 314.
50. N. Bloembergen, in: *Laser–Solid Interactions and Laser Processing*—1978 (S. D. Ferris, H. J. Leamy, and J. M. Poate, eds.), p. 1, American Institute of Physics, New York, 1979.
51. R. A. Bingham and P. L. Salter, *Int. J. Mass Spectrom. Ion Phys.* 21 (1976) 133.
52. J. F. Ready, *J. Appl. Phys.* 36 (1965) 462.
53. K. D. Kovalev, G. A. Maksimov, A. I. Suchkov, and N. V. Larin, *Int. J. Mass Spectrom. Ion Phys.* 27 (1978) 101.
54. I. Opauszky, *Pure Appl. Chem.* 54 (1982) 879.
55. J. T. Luxan and D. E. Parker, *Industrial Lasers and Their Applications*, Prentice–Hall, Englewood Cliffs, N.J., 1985.
56. K. P. Seltzer and J. J. Kunze, *Phys. Scripta 25* (1982) 929.
57. L. J. Radziemski and D. A. Cremers (eds.), *Laser-Induced Plasmas and Applications*, Dekker, New York, 1989.
58. K. R. Hess, Ph.D. dissertation, University of Virginia, Charlottesville, 1986.
59. K. R. Hess and W. W. Harrison, in: *Lasers and Mass Spectrometry* (D. M. Lubman, ed.), p. 205, Oxford University Press, London, 1990.
60. J. W. Coburn and W. W. Harrison, *Appl. Spectrosc. Rev.* 17 (1981) 95.
61. N. Jakubowski and D. Stuewer, 1990 Winter Conference on Plasma Spectrochemistry, St. Petersburg, Fla., 1990.
62. F. L. King and W. W. Harrison, *Int. J. Mass Spectrom. Ion Proc.* 89 (1989) 171.
63. D. C. Duckworth and R. K. Marcus, *Appl. Spectrosc.* 44 (1990) 649.
64. K. R. Hess and W. W. Harrison, *Anal. Chem.* 58 (1986) 1696.
65. K. Kagawa and S. Yokoi, *Spectrochim. Acta* 37B (1982) 789.
66. K. Kagawa, M. Ohtani, S. Yokoi, and S. Nakajima, *Spectrochim. Acta* 39B (1984) 525.
67. F. L. King and W. W. Harrison, *Mass Spectrom. Rev.* 9 (1990) 285.
68. W. Vieth and J. C. Huneke, *Spectrochim. Acta* 46B (1991) 137.
69. N. Laegrreid and G. K. Wehner, *J. Appl. Phys.* 32 (1961) 365.
70. G. Carter and J. S. Colligan, *Ion Bombardment of Solids*, American Elsevier, New York, 1968.

71. E. Nasser, *Fundamentals of Gaseous Ionization and Plasma Electronics*, Wiley–Interscience, New York, 1971.
72. H. Mase, S. Nakaya, and Y. Hatta, *J. Appl. Phys.* 38 (1967) 2960.
73. Y. Iida, *Spectrochim. Acta* 45B (1990) 427.
74. L. M. Fraser and J. D. Winefordner, *Anal. Chem.* 44 (1972) 1444.
75. N. Omenetto, *Anal. Chem.* 48 (1976) 75A.
76. N. Omenetto and H. G. C. Human, *Spectrochim. Acta* 39B (1984) 1333.
77. H. G. C. Human, N. Omenetto, P. Cavalli, and G. Rossi, *Spectrochim. Acta* 39B (1984) 1345.
78. S. J. Weeks and J. D. Winefordner, in: *Lasers in Chemical Analysis* (G. M. Hieftje, J. C. Travis, and F. E. Lytle, eds.), p. 159, Humana Press, Clifton, N.J., 1981.
79. J. C. Wright, in: *Lasers in Chemical Analysis* (G. M. Hieftje, J. C. Travis, and F. E. Lytle, eds.), p. 185, Humana Press, Clifton, N.J., 1981.
80. R. P. Lucht, in: *Laser Spectroscopy and Its Applications* (L. J. Radziemski, R. W. Solarz, and J. A. Paisner, eds.) p. 623, Dekker, New York 1987.
81. J. D. Ingle, Jr., and S. R. Crouch, *Spectrochemical Analysis*, Chap. 11, Prentice–Hall, Englewood Cliffs, N.J., 1988.
82. N. Omenetto and J. D. Winefordner, *Prog. Anal. At. Spectrosc.* 2 (1979) 1.
83. D. J. Butcher, J. P. Dougherty, F. R. Preli, A. P. Walton, G. Wei, R. L. Irwin, and R. G. Michel, *J. Anal. At. Spectrom.* 3 (1988) 1059.
84. P. Hannaford and A. Walsh, *Spectrochim. Acta* 43B (1988) 1053.
85. J. D. Winefordner, B. W. Smith, and N. Omenetto, *Spectrochim. Acta* 44B (1989) 1397.
86. N. Omenetto, *Spectrochim. Acta* 44B (1989) 131.
87. G. F. Kirkbright, *Analyst* 96 (1971) 609.
88. D. S. Gough, P. Hannaford, and A. Walsh. *Spectrochim. Acta* 28B (1973) 197.
89. D. S. Gough and J. R. Meldrum, *Anal. Chem.* 52 (1980) 642.
90. B. W. Smith, N. Omenetto, and J. D. Winefordner, *Spectrochim. Acta* 39B (1984) 1389.
91. B. M. Patel and J. D. Winefordner, *Spectrochim. Acta* 41B (1986) 469.
92. B. W. Smith, J. B. Womack, N. Omenetto, and J. D. Winefordner, *Appl. Spectrosc.* 43 (1989) 873.
93. M. Glick, B. W. Smith, and J. D. Winefordner, *Anal. Chem.* 62 (1990) 157.
94. S. Grazhulene, V. Khvostikov, and M. Sorokin, *Spectrochim. Acta* 46B (1991) 459.
95. J. C. Travis, G. C. Turk, R. L. Watters, Jr., L. Yu, and J. L. Blue, *J. Anal. At. Spectrom.* 6 (1991) 261.
96. A. Bol'shakov, N. V. Golovenkov, S. V. Oshemkov, and A. A. Petrov *J. Anal. At. Spectrom.* 5 (1990) 549.
97. M. J. Rutledge, B. W. Smith, and J. D. Winefordner, *Anal. Chem.* 59 (1987) 1794.
98. D. Sethi and H. Sethi, *Spectroscopy* 2 (1987) 26.
99. G. Zizak, N. Omenetto, and J. D. Winefordner, *Opt. Eng.* 23 (1984) 749.
100. G. Gillson and G. Horlick, *Spectrochim. Acta* 41B (1986) 1299.
101. G. Gillson and G. Horlick, *Spectrochim. Acta* 41B (1986) 1323.
102. C. van Dijk, B. W. Smith, and J. D. Winefordner, *Spectrochim. Acta* 37B (1982) 759.
103. C. Y. She, W. M. Fairbank, Jr., and K. W. Billman, *Opt. Lett.* 2 (1978) 30.
104. N. Omenetto, G. C. Turk, M. Rutledge, and J. D. Winefordner, *Spectrochim. Acta* 42B (1987) 807.
105. J. B. Simeonsson, B. W. Smith, J. D. Winefordner, and N. Omenetto, *Appl. Spectrosc.* 45 (1991) 521.
106. B. M. Patel, B. Smith, and J. D. Winefordner, *Spectrochim. Acta* 40B (1985) 1195.
107. B. M. Patel and J. D. Winefordner, *Appl. Spectrosc.* 40 (1986) 667.
108. A. Elbern and P. Mioduszewski, *J. Vac. Sci. Technol.* 16 (1979) 2090.
109. A. L. Lewis, II, and E. H. Piepmeier, *Appl. Spectrosc.* 37 (1983) 523.

110. M. S. Hendrick, M. D. Seltzer, and R. G. Michel, *Spectrochim. Acta* 41B (1986) 335.
111. P. Ho and W. G. Breiland, *Appl. Phys. Lett* 44 (1984) 51.
112. Y. Takubo, Y. Takasugi, and M. Yamamoto, *J. Appl. Phys.* 64 (1988) 1050.
113. P. Hannaford and R. M. Lowe, *Opt. Eng.* 22 (1983) 532.
114. P. Hannaford and R. M. Lowe, *Aust. J. Phys.* 39 (1986) 829.
115. A. Hirabayashi, S. Okuda, Y. Nambu, and T. Fujimoto, *Phys. Rev. A* 35 (1987) 639.
116. G. J. Beenen, B. P. Lessard, and E. H. Piepmeier, *Anal. Chem.* 51 (1979) 172.
117. D. S. King and P. K. Schenck, *Laser Focus* 18(3) (1982) 50.
118. J. E. M. Goldsmith and J. E. Lawler, *Contemp. Phys.* 22 (1981) 235.
119. J. C. Travis and J. R. DeVoe, in: *Lasers in Chemical Analysis* (G. M. Hieftje, J. C. Travis, and F. E. Lytle, eds.), p. 185, Humana Press, Clifton, N. J., 1981.
120. J. C. Travis, G. C. Turk, J. R. DeVoe, P. K. Schenck, and C. A. van Dijik, *Prog. Anal. At. Spectrosc.* 7 (1984) 199.
121. B. Barbieri, N. Beverini, and A. Sasso, *Rev. Mod. Phys.* 62 (1990) 603.
122. F. M. Penning, *Physica* (*The Hague*) 8 (1928) 137.
123. D. Fang and R. K. Marcus, *Spectrochim. Acta* 46B (1991) 983.
124. B. Chapman, *Glow Discharge Processes*, Wiley, New York, 1980.
125. R. A. Keller and E. F. Zalewski, *Appl. Opt.* 19 (1980) 3301.
126. R. A. Keller and E. F. Zalewski, *Appl. Opt.* 21 (1982) 3392.
127. C. Dreze, Y. Demers, and J. M. Gagne, *J. Opt. Soc. Am.* 72 (1982) 912.
128. J. Kramer, *J. Appl. Phys.* 64 (1988) 1758.
129. P. A. Fleitz and C. J. Seliskar, *Appl. Spectrosc.* 43 (1989) 293.
130. C. T. Rettner, C. R. Webster, and R. N. Zare, *J. Phys. Chem.* 85 (1981) 1105.
131. E. F. Zalewski, R. A. Keller, and R. Engleman, Jr., *J. Chem. Phys.* 70 (1979) 1015.
132. D. K. Doughty and J. E. Lawler, *Appl. Phys. Lett.* 42 (1983) 234.
133. K. C. Smyth and P. K. Schenck, *Chem. Phys. Lett.* 55 (1978) 466.
134. K. C. Smyth, R. A. Keller, and F. F. Crim, *Chem. Phys. Lett.* 55 (1978) 473.
135. K. C. Smyth, B. L. Bentz, C. G. Bruhn, and W. W. Harrison, *J. Am. Chem. Soc.* 101 (1979) 797.
136. B. L. Bentz, Ph.D. dissertation, University of Virginia, Charlottesville, 1982.
137. K. R. Hess and W. W. Harrison, *Anal. Chem.* 60 (1988) 691.
138. R. B. Green, J. C. Travis, and R. A. Keller, *Anal. Chem.* 48 (1976) 1954.
139. R. B. Green, R. A. Keller, G. G. Luther, P. K. Schenck, and J. C. Travis, *IEEE J. Quant. Elec.* QE-13 (1976) 63.
140. W. B. Bridges, *J. Opt. Soc. Am.* 68 (1978) 352.
141. G. Chevalier, J. M. Gagne, and P. Pianarosa, *J. Opt. Soc. Am. B Opt. Phys.* 5 (1988) 1492.
142. M. N. Reddy and G. N. Rao, *J. Opt. Soc. Am. B Opt. Phys.* 6 (1989) 1481.
143. R. Singh, G. N. Rao, and R. K. Thareje, *J. Opt. Soc. Am. B Opt. Phys.* 8 (1991) 12.
144. N. Ami, A. Wada, Y. Adachi, and C. Hirose, *Chem. Phys. Lett.* 153 (1989) 118.
145. T. F. Johnston, Jr., *Laser Focus* 14(3) (1978) 58.
146. N. Uchitomi, T. Nakajima, S. Maeda, and C. Hirose, *Opt. Commun.* 44 (1983) 154.
147. A. S. Naqvi, K. N. Ullah, and M. I. Rehmatullah, *Opt. Commun.* 80 (1991) 331.
148. S. Fujimaki, Y. Adachi, and C. Hirose, *Appl. Spectrosc.* 41 (1987) 1243.
149. K. Kawakita, T. Nakajima, Y. Adachi, S. Maeda, and C. Hirose, *Opt. Commun.* 48 (1983) 128.
150. S. Fujimaki, Y. Adachi and C. Hirose, *Appl. Spectrosc.* 41 (1987) 567.
151. C. Hirose and T. Masaki, *Appl. Spectrosc.* 42 (1988) 811.
152. N. Ami, A. Wada, Y. Adachi, and C. Hirose, *Appl. Spectrosc.* 43 (1989) 245.
153. R. B. Green, R. A. Keller, G. G. Luther, P. K. Schenck, and J. C. Travis, *Appl. Phys. Lett.* 29 (1976) 727.

154. R. A. Keller, R. Engleman, and E. F. Zalewski, *J. Opt. Soc. Am.* 69 (1979) 738.
155. M. A. Nippoldt and R. B. Green, *Appl. Opt.* 20 (1981) 3206.
156. M. A. Nippoldt and R. B. Green, *Anal. Chem.* 55 (1983) 1171.
157. W. G. Tong and R. W. Shaw, *Appl. Spectrosc.* 40 (1986) 494.
158. J. C. Travis, G. C. Turk, and R. B. Green, *Anal. Chem.* 54 (1982) 1006A.
159. G. C. Turk, J. R. DeVoe, and J. C. Travis, *Anal. Chem.* 54 (1982) 643.
160. N. Omenetto, T. Berthoud, P. Cavalli, and G. Rossi, *Anal. Chem.* 57 (1985) 1256.
161. M. J. Rutledge, M. E. Tremblay, and J. D. Winefordner, *Appl. Spectrosc.* 41 (1987) 5.
162. I. Magnusson, *Spectrochim. Acta* 43B (1988) 727.
163. K. C. Lin, S. H. Lin, P. M. Hunt, G. E. Leroi, and S. R. Crouch, *Appl. Spectrosc.* 43 (1989) 66.
164. K. C. Lin and Y. S. Duh, *Appl. Spectrosc.* 43 (1989) 20.
165. O. Axner and H. Rubinsztein-Dunlop, *Spectrochim. Acta* 44B (1989) 835.
166. O. Axner and S. Sjostrom, *Appl. Spectrosc.* 44 (1990) 144.
167. O. Axner, M. Norberg, M. Persson, and H. Rubinsztein-Dunlop, *Appl. Spectrosc.* 44 (1990) 1117.
168. K. C. Ng, M. J. Angebranndt, and J. D. Winefordner, *Anal. Chem.* 62 (1990) 2506.
169. G. C. Turk, J. C. Travis, and J. R. de Voe, *Anal. Chem.* 50 (1978) 817.
170. J. P. Young, G. S. Hurst, S. D. Kramer, and M. G. Payne, *Anal. Chem.* 51 (1979) 1050A.
171. G. S. Hurst, M. G. Payne, S. D. Kramer, and J. P. Young, *Rev. Mod. Phys.* 51 (1979) 767.
172. G. S. Hurst, *Anal. Chem.* 53 (1981) 1448A.
173. G. S. Hurst, *J. Chem. Educ.* 59 (1982) 895.
174. G. I. Bekov, V. N. Radayev, and V. S. Letokhov, *Spectrochim. Acta* 43B (1988) 491.
175. E. Sekreta, K. Owens, A. Engel, and J. P. Reilly, *Spectrochim. Acta* 43B (1988) 679.
176. J. D. Fasset and J. C. Travis, *Spectrochim. Acta* 43B (1988) 1409.
177. G. S. Hurst and M. G. Payne, *Spectrochim. Acta* 43B (1988) 715.
178. J. P. Young, R. W. Shaw, and D. H. Smith, *Anal. Chem.* 61 (1989) 1271A.
179. G. S. Hurst and M. G. Payne, *Principles and Applications of Resonance Ionization Spectroscopy*, Hilger, Philadelphia, 1988.
180. D. H. Smith, J. P. Young, and R. W. Shaw, *Mass Spectrom. Rev.* 8 (1989) 345.
181. C. H. Chen, M. G. Payne, G. S. Hurst, S. D. Kramer, S. L. Allman, and R. C. Phillips, in: *Lasers and Mass Spectrometry* (D. M. Lubman, ed.), p. 3, Oxford University Press, London, 1990.
182. A. S. Gonchakov, N. B. Zorov, Y. Y. Kuy zyakov, and I. O. Matveev, *J. Anal. Chem. USSR* (*Engl. Transl.*) 34 (1980) 192.
183. S. Mayo, T. B. Lucatorto, and G. G. Luther, *Anal. Chem.* 54 (1982) 553.
184. M. W. Williams, D. W. Beekman, J. B. Swan, and E. T. Arakawa, *Anal. Chem.* 56 (1984) 1348.
185. N. S. Nogar and R. C. Estler, in: *Lasers and Mass Spectrometry* (D.M. Lubman, ed.), p. 65, Oxford University Press, London, 1990.
186. D. L. Donahue, J. P. Young, and D. H. Smith, *Int. J. Mass Spectrom. Ion Phys.* (1982) 293.
187. J. P. Young and D. L. Donohue, *Anal. Chem.* 55 (1983) 88.
188. D. L. Donohue, J. P. Young, and D. H. Smith, *Appl. Spectrosc.* 39 (1985) 93.
189. D. L. Donohue, W. H. Christie, D. E. Goeringer, and H. S. McKown, *Anal. Chem.* 57 (1985) 1193.
190. J. D. Blum, M. J. Pellin, W. F. Calaway, C. E. Young, D. M. Green, I. D. Hutcheon, and G. J. Wasserburg, *Anal. Chem.* 62 (1990) 209.
191. J. E. Parks, in: *Lasers and Mass Spectrometry* (D. M. Lubman, ed.), p. 37, Oxford University Press, London, 1990.

192. W. Gerhard and H. Oechsner, *Z. Phys. B.* 22 (1975) 41.
193. D. J. Hall and P. K. Robinson, *Am. Lab.* 19 (1987) 74.
194. F. L. King, A. L. McCormack, and W. W. Harrison, *J. Anal. At. Spectrom.* 3 (1988) 883.
195. P. J. Savickas, K. R. Hess, R. K. Marcus, and W. W. Harrison, *Anal. Chem.* 56 (1984) 817.
196. P. J. Savickas, Ph.D. dissertation, University of Virginia, Charlottesville, 1984.
197. T. W. Hansch, I. S. Shahin, and A. L. Schawlow, *Phys. Rev. Lett.* 27 (1971) 707.
198. D. S. Gough and P. Hannaford, *Opt. Commun.* 55 (1985) 91.
199. R. J. McLean, P. Hannaford, H. A. Bachor, P. T. H. Fisk, and R. J. Sandeman, *Z. Phys. D.* 1 (1986) 253.
200. G. A. Luena and W. G. Tong, *Appl. Spectrosc.* 44 (1990) 1668.
201. P. Pianarosa, Y. Demers, and J. M. Gagne, *Spectrochim. Acta* 39B (1984) 761.
202. J. Lawrenz and K. Niemax, *Spectrochim. Acta* 44B (1989) 155.
203. R. Hergenroder and K. Niemax, *Spectrochim. Acta* 43B (1988) 1443.
204. T. E. Barber, P. E. Walters, and J. D. Winefordner, *Appl. Spectrosc.* 45 (1991) 524.
205. J. M. Ramsey and W.B. Whitten, *Anal. Chem.* 59 (1987) 167.
206. W. G. Tong, J. M. Andrews, and Z. Wu, *Anal. Chem.* 59 (1987) 896.
207. J. M. Andrews and W. G. Tong, *Spectrochim. Acta* 44B (1989) 101.
208. W. G. Tong and D. A. Chen, *Appl. Spectrosc.* 41 (1987) 586.
209. D. A. Chen and W. G. Tong, *J. Anal. At. Spectrom.* 3 (1988) 531.
210. R. A. Fisher (ed.), Optical Phase Conjugation, Academic Press, New York, 1983.

11

Laser-Based Diagnostics of Reactive Plasmas

Bryan L. Preppernau and Terry A. Miller

11.1. Introduction

There should be little doubt to readers of this volume that reactive plasma processing applied to the development of electronic devices and novel materials has considerable significance for current and future technology. Furthermore, it is fairly apparent that the design and control of plasma processes has been empirical in nature over the years, as is the case in the analytical glow discharge plasmas discussed in the previous chapters. However, we have now perhaps reached a point where it has become necessary to have a more comprehensive diagnostic and theoretical interpretation of discharge reactor systems in order to advance the promise of this technology.[1]

With the advent and promise of extraordinary computing ability and sophisticated plasma models, in principle one could describe in detail the complex interactions present in low-temperature, nonequilibrium processing plasmas. Thus, it would seem that in the very near future the empirical approach could soon be discarded, save for one important element to this problem. There is presently a lack of *in situ* nonintrusive high-resolution experimental diagnostic data for the major interactive components of processing plasmas as well as for the predominate driving mechanism, the

---

Bryan L. Preppernau and Terry A. Miller • Laser Spectroscopy Facility, Department of Chemistry, The Ohio State University, Columbus, Ohio 43210.

Glow Discharge Spectroscopies, edited by R. Kenneth Marcus. Plenum Press, New York, 1993.

electric field. Only through confirmation of the calculated models by correlation with diagnostic data can a comprehensive description and hence the possibility of logical design of plasma processes become a reality.

In this chapter we survey the current *in situ* plasma diagnostics with emphasis on those techniques that are laser-based. Given that the relative populations of component species in processing plasmas are for the most part dictated by the continuous velocity distributions of electrons, one cannot hope to directly measure all possible fractional energy distributions internal to each component species. Couple this difficulty with the added complexity presented by the addition of surface boundary conditions, and competing processes such as recombination, charge transfer, etc., one can quickly see that a full diagnosis of even the simplest of discharge systems is basically an intractable problem. Therefore, one usually confines the scope of diagnostic measurements to major reactive species that would be expected to be of special significance in processing plasmas. Indeed, these species will have greater interest for processing technology since they are the components of the plasma that would most likely find their way into the boundary layers near surfaces and mainly determine the characteristics of the plasma–surface interactions. We expect that the trend of future experimental work will be to make measurements that more fully describe the plasma–surface boundary layer. Hence, it is our intention to discuss the detection of those major gas-phase reactive species that would be of interest precisely because of their influence on surfaces exposed to the discharge.

We first discuss the various reactive species expected in processing plasmas and briefly review the associated production mechanisms. Next the requirements needed to make accurate profiling measurements in both space and time are reviewed along with the inherent difficulties in making these sorts of measurements especially near surfaces. In addition, we discuss the necessity and requirements for electric field profile measurements. The aim here is to critically examine profiling measurements for simple geometric situations.

In Section 3 (experimental techniques) we describe the basic experimental arrangements, advantages, and disadvantages of a variety of laser-based probe diagnostics. Measurements of reactive species and plasma parameters by these techniques are compared with those obtained by other familiar diagnostics. It is our bias that the refinement of laser-based probes for species detection provides the best means of making reliable and precise *in situ*, nonintrusive measurements on reactive species and plasma parameters. However, it is important to recognize the complementary nature of other diagnostic techniques.

Having discussed the basis of the experimental techniques, Section 4 reviews the experimental findings by researchers utilizing these techniques and the species and parameters thus far studied by laser-based probes.

Finally, we conclude with a discussion of future trends for reactive plasma research and point to some of the more difficult problems still to be solved.

11.2. An Overview of Reactive Plasmas

11.2.1. Species Identity

Reactive plasmas can be characterized by four qualities: fairly low total pressure (0.1–10 Torr), moderate electric fields (1–5000 V/cm), moderate electron energies (1–100 eV), and near-thermal translational velocity distributions for neutral and ionic heavy-particle species. Taken together these qualities ensure an environment that is complicated in the number of competing production mechanisms.[(2)]

The production of heavy-particle species can be grouped via two main pathways: those species resulting from chemical reactions between neutral species, either in the gas phase or through surface-assisted reactions, and those produced by electric field-driven charged species (usually electron) collisions. Certainly, these two pathways are never clearly separated in space and time in a discharge, but each may dominate under certain conditions or regions in the plasma discharge. From the point of view of plasma diagnostics, one would like to be able to account for the role of the each production path by analysis of careful population measurements for each species present as a function of spatial position and temporal behavior. Thus, one might ask: what are the important species to be expected in a reactive plasma and which ones could hopefully be detected and their population profiles measured?

11.2.1.1. Atoms

Atomic species production via electron impact dissociation of parent molecules ranks as one of the most prevalent processes in the bulk of a reactive plasma along with internal molecular electronic excitation and ionization caused by electron impact. As an example, for hydrogen, the electron collision cross sections are all comparable (10^{-17} cm^2 at 30 eV) and thus one would expect to measure a high degree of molecular dissociation in the H_2 discharge.[(3)] Atoms produced in this manner radiate rapidly to their ground electronic state or to excited long-lived metastable states. Most atoms of processing interest have in their ground state considerable chemical reactivity while metastable states can have, in addition, considerable energy content; both properties can have great importance in surface and gas-phase interactions.

The atoms produced by electron impact dissociation usually result from the breaking of a single molecular bond and as such are readily produced in the plasma. At typical operating pressures, the neutral atom recombination lifetimes are long (milliseconds) and thus the atoms can diffuse over the entire volume of the reactor system with large concentration gradients in sheath regions and near surfaces. The dissociation fraction of atoms to molecules in simple diatomic gas discharges can be very high, on the order of 50%. Given the high dissociation fraction and long recombination lifetimes, atomic production is not expected to be temporally modulated except in very low audio frequency or pulsed discharges.

11.2.1.2. Radicals

Highly reactive radicals can be readily produced in plasmas and have significant concentrations. Radical production is sustained by electron impact dissociation, charge transfer collisions between neutrals and ions, and photodissociation of neutral molecules. The spatial and temporal behavior are similar to those of atoms. Radicals readily diffuse throughout the discharge and can strongly influence surface deposition and etching processes.

For discharges in molecular gases such as fluoro- or hydrocarbons, many intermediate radical forms can be produced. The interpretation of radical production in discharges is complicated by uncertainties in the breakdown pathways for the parent molecule. One thing that is certain is that application discharges with high concentrations of radical species (comparable to atomic species) are finding more and more uses, so that measurements on radical production and distributions will be vitally important.

11.2.1.3. Ions

Molecular positive ions and negative ions obviously play a major role in sustaining reactive species production and determining plasma parameters. Positive ion production is maintained by electron impact ionization, charge transfer collisions, photoionization, and fast heavy particle or metastable collisions. Negative ions are mainly produced by slow electron impact attachment or by electron impact dissociative attachment. The average ionization fraction (over the entire discharge) of ions to neutral particle number densities is typically 10^{-4} or 10^{-3} for these low-temperature nonequilibrium reactive plasmas; however, the ionization fraction may be strongly modulated in sheath regions.

Positive ions are important to ion etching and implantation plasmas and can be detrimental to certain plasma deposition processes. Unfortunately, measurements of positive ion concentration profiles are sometimes difficult for laser-based techniques. Closed-shell positive ions usually can be thought of as arising from proton addition to a closed-shell neutral. Thus, for small-molecular-weight ions typically found in plasma environments, their electronic transitions typically lie in the VUV and the upper states are often dissociative making laser-induced fluorescence (LIF) detection difficult. Indeed, only IR laser absorption techniques using ground electronic state vibrational transitions seem widely applicable. Yet as we shall see later, such line-of-sight techniques suffer in terms of both spatial resolution and sensitivity. It is, however, important to point out that open-shell ions, e.g., N_2^+, often have bound–bound electronic transitions in regions easily accessible to dye lasers and can conveniently be detected by LIF.

Negative ion densities are known to be very high (even exceeding electron densities) in discharges in electronegative gases. Negative ions can be easily detected by laser photodetachment of the extra electron and subsequent measurement of the laser-induced transient current in the discharge. The identity of the negative ion is more difficult to ascertain, but can sometimes be inferred by the wavelength dependence of the photodetachment or by LIF of the photodetached neutral. Measurements on negative ions in reactive plasmas are important in order to analyze their influence on the local electric field as well as their interactions with surfaces exposed to electronegative plasmas.

11.2.1.4 Excited Molecular States

The relative populations of excited molecular electronic, rotational, and vibrational states are primarily determined by electron impact excitation and recombinative cross-sections. Most excited electronic states rapidly radiate because of short radiative lifetimes. Metastable electronic states which are not effectively coupled to lower states can have lifetimes as long as several milliseconds even in the plasma. Hence, as with atomic metastables, sustained molecular metastable concentrations develop, accounting for an appreciable percentage of the discharge energy content.

The high electron temperature distribution, along with photoemission from the discharge, can lead to nonthermal rotational and vibrational level populations in ground and metastable electronic states. These result from a variety of processes, including electron impact excitation, v'–v'' pumping, and even translational to vibrational excitation through neutral collisions.

Laser-based techniques such as LIF and resonantly enhanced multiphoton ionization (REMPI) can be used to probe excited molecular states with high spectral resolution. This allows the assigning of approximate rotational or vibrational temperatures to a molecule as well as elucidating excitation kinetics.

11.2.2. Sampling Considerations for Concentration Measurements

The ability to make laser-based concentration measurements in reactive plasmas depends on the investigator's interpretation of instrumental factors that critically affect photoluminescence or optogalvanic detection. While relative populations or concentrations may be based on the relative response of the detection apparatus, absolute measurements are much more difficult to achieve. Reliable absolute concentration determinations without a standard require the following: (1) accurate information regarding spectroscopic parameters such as excitation cross sections, (2) instrumental factors affecting the excitation and collection of fluorescence or optogalvanic signal, (3) analysis of systematic perturbations caused by nonlinear effects that may be convoluted into the collected data, and (4) careful error analysis and error propagation.

It should be clear that the total observed voltage change, ΔV at the input of a data collection system for a LIF experiment can be written

$$\Delta V = K n_2 y \tag{11-1}$$

where n_2 = number of atoms or molecules excited to the upper state (2) during the laser pulse, y = fraction of atoms or molecules that emit photons, i.e., quantum yield, and K = instrument and apparatus constant (V/photon). If we assume that only a small fraction of the laser's photons are absorbed and that only a small fraction of the atoms or molecules in the lower state (1) are excited, n_2 can be written

$$n_2 = \sigma_{12} \ell c_1 N_\ell \tag{11-2}$$

where σ = cross section for excitation of n_1 to n_2 (cm^2/photon), ℓ = effective absorption length (cm), c_1 = concentration of atoms or molecules in state 1 (species/cm^3), and N_ℓ = number of photons in the laser pulse into the sample's volume. Generally, the experimental apparatus will record the number of fluorescence photons as a charge pulse Q from a photomultiplier tube (PMT). Typically, this charge is translated to a voltage using a boxcar or integrator. In this case K in Eq. (11–1) becomes

$$K = kQ = k\eta_c \eta_g e G \tag{11-3}$$

where k = boxcar or integrator conversion factor (V/charge pulse per photon), η_c = fraction of emitted photons striking the active surface of the PMT, η_g = fraction of photons striking the active surface to yield a photoelectron, e = electron charge, G = gain of tube (electron multiplication factor). k, η_g, and G can be calibrated in the lab using a known photon source or less reliably gleaned from manufacturer's specifications. The collection efficiency η_c *is* apparatus dependent and can be written

$$\eta_c = \eta_T(\Omega_F/4\pi)$$

where η_T = wavelength-dependent transmission efficiency of the collection optics and $\Omega_F/4\pi$ = the fractional solid angle of collection for fluorescence. η_T can also be calibrated in the lab for a given set of lenses, filters, etc.

Perhaps the most difficult quantity to determine is the collection solid angle, $\Omega_F/4\pi$, which is in general a function of the relative positions of the laser probe interaction volume in the discharge, the focal point of the collection optics, and the reactor surfaces; mainly the electrodes. For most experimental apparatus, the geometry is arranged such that the laser probe beam path and the focal axis of the collection optics intersect either at right angles or are parallel. Typically, both axes are parallel to a planar electrode surface and either the collection optics and laser probe are translated simultaneously across the discharge or the reactor vessel is moved relative to the excitation point. The geometry for a plane parallel electrode cylindrical discharge reactor is shown in Fig. 11-1.

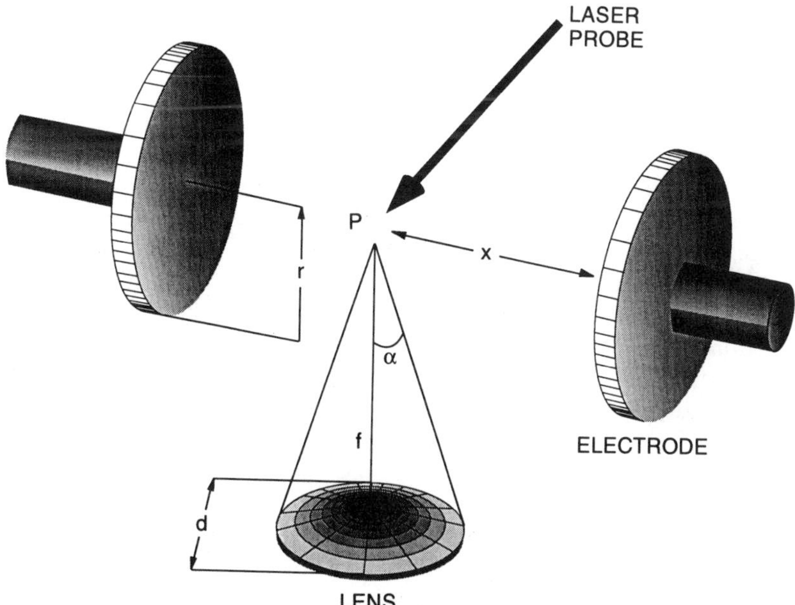

Figure 11-1. Geometry in a planar electrode cylindrically symmetric discharge reactor (symbols defined in text).

For most positions between the electrodes the fractional collection solid angle for an assumed point source is

$$\frac{\Omega_F}{4\pi} = \frac{d^2}{16f^2} \approx \sin^2\left(\frac{\alpha}{2}\right) \tag{11-4}$$

where d is the lens or aperture diameter at the entrance of the optical collection system and f is the focal length of the first lens and the last approximate-equality assumes $f \gg d$. However, near an electrode surface the collection solid angle is occulted. The critical distance at which this begins to happen is

$$x_c = \frac{rd}{2f} \tag{11-5}$$

where r is the difference in the radial position from the edge of the electrode to the radical position of the excitation point, P (see Fig. 11-1). Thus, for lateral positions, x, closer than x_c to an electrode the collection solid angle becomes

$$\frac{\Omega_F^*}{4\pi} = \frac{d^2}{16f^2}\left[1 - \frac{1}{\pi}\cos^{-1}\left(\frac{x}{x_c}\right) + \frac{1}{\pi}\left(\frac{x}{x_c}\right)\sqrt{1-\left(\frac{x}{x_c}\right)^2}\,\right] \quad 0 \leq x \leq x_c \tag{11-6}$$

For values of x more negative than the origin, the collected signal should drop discontinuously to zero as the probe beam then strikes the electrode surface. Clearly, this simple geometric analysis does not take into account reflections and fluorescent scattering from the electrode which could be sampled by the collection optics and assumes ideal optical properties for the collection system.

An alternative way of determining the collection solid angle variation is to fill the reactor system with a test gas that can be excited and will fluoresce at wavelengths similar to those used in the actual diagnostic experiments. By using the same excitation/collection optics and geometry, the relative position of the reactor and the excitation point can be translated in order to collect a signal that represents the true variation of any actual diagnostic data as a function of the collection solid angle across the discharge system. The advantage here is in being able to automatically account for many systematic optical effects that can influence the LIF signal.

In spatial and temporal profiling of species concentration, the resolution of the measurements is typically determined by the probe laser interaction volume and the experimental timing. For LIF experiments, the laser interaction volume or focal spot size can be made as small as 100 μm in diameter. For most practical applications, this is ideal as spatial concentration profiles would not be expected to vary discontinuously over such a small dimension. Near electrode surfaces, where concentration gradients are expected to be the most significant, the variation in collection solid angle as shown in Eq. (11-6) contributes to larger uncertainties in spatial resolution and concentration assessment. Careful consideration of probe beam geometry and instrumental factors can help to minimize these larger systematic uncertainties near surfaces, and allow refined measurements of concentration gradients.

Most LIF experiments in plasma discharges utilize pulsed lasers because of their high peak output power, which helps in the discrimination of LIF signal against ambient background contributions. Timing electronics today can accurately synchronize the firing trigger of a pulsed laser system with a particular phase of the current cycle of a discharge (rf or pulsed) with very low relative jitter. Thus, the limiting factor that determines temporal resolution is the temporal width of the laser probe light pulse. For most systems, this pulse width can be 10 nsec or less. This implies that if one wanted to make temporally resolved concentration measurements of say 10 points per rf current cycle, the discharge frequency could be as high as 10 MHz. Other factors that influence temporal resolution are the radiative lifetime of the LIF excited state and the species production kinetics. For many situations, the steady-state populations of important reactive species (as mentioned above) can be maintained throughout the rf current cycle and temporal variations are observed only for low-frequency or pulsed discharges.

Electric field profile measurements can also be influenced by the same factors as for species concentration measurements. Large electric field variations are observed near electrode surfaces in the sheath regions. Thus, excitation geometry will mostly determine the spatial resolution in these regions. For electric field measurements made by optogalvanic spectroscopy (OGS), the temporal resolution can be limited by the laser-induced photocurrent transient time which can couple strongly with the electron energy distribution function variation across the discharge. However, in most instances, the temporal resolution is again limited (for pulsed laser excitation) by the laser pulse temporal width and appropriate choice of impedance matching to measuring electronics can ensure a temporal resolution approaching this characteristic time.

11.3. Experimental Techniques

In this section we discuss the general experimental practices and techniques used in laser-based diagnostics for plasma species and characteristics.

A comparison to other known diagnostics is meant to support the basic utility of using state-selective laser probe techniques for making *in situ* non-perturbative measurements. However, in dealing with laser spectroscopic diagnostics, care must be taken to have a comprehensive understanding of the spectroscopy of any particular set of transitions used as a diagnostic or else one's results may be misleading. Furthermore, every effort should be made to delineate and minimize spurious systematic effects caused by the experimental apparatus in order to make a qualified interpretation of the data. For example, because of the high photon flux levels available from dye lasers, one should always verify assumptions concerning power-broadening and saturation effects. It is also important when using fluorescence techniques to avoid saturation of detector elements such as photomultipliers, which could lead to a false analysis of species profile data.

11.3.1. One-Photon LIF

One-photon LIF techniques make use of three quantum states of the species in question (see Fig. 11-2a), with the possibility of the initial and

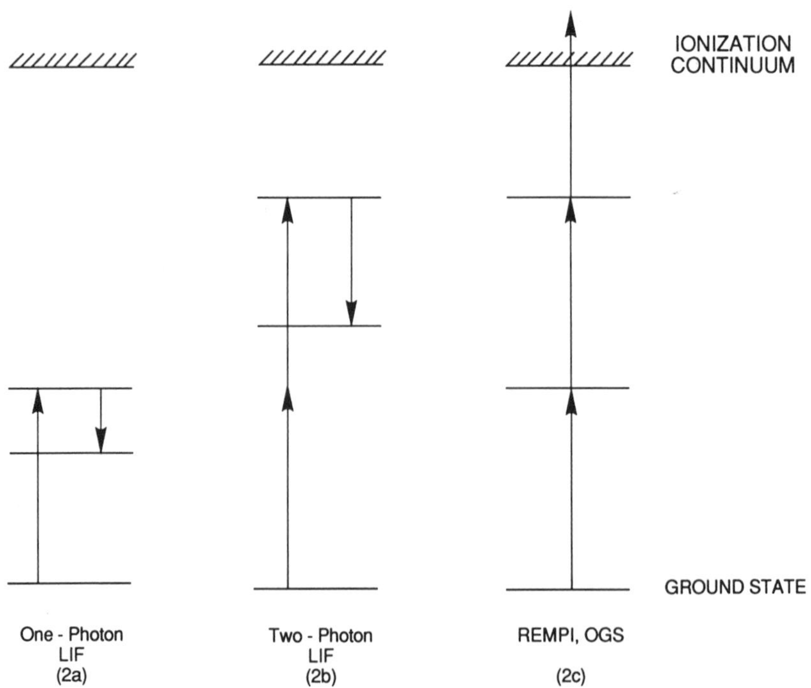

Figure 11-2. Laser-based diagnostic excitation and detection schemes.

final states being the same. The initial excitation transition from the lower (typically ground) state is pumped by the probe laser to an excited upper state with a fluorescence photon detected as this state radiates to an intermediate level. One-photon LIF is a well-developed technique and can typically take advantage of repetitive pulsed visible dye laser systems yielding intense probe beams with good spatial and temporal resolution.[4,5] Standard excitation wavelengths range from the visible to the near UV. The detected fluorescence typically appears to the longer wavelength side of the excitation wavelength, usually in the visible region to the near infrared, though some fluorescence transitions are in the near UV.

One-photon LIF as typically used for plasma measurements involves transitions in free radicals, molecular ions, and metastable states of atoms, and molecules. By analyzing the line shapes of transitions, information can be obtained concerning the velocity of the species from the Doppler broadening and the electric field strength by analyzing the Stark splitting components, in addition to species concentrations. For radicals and molecular ions, measurements over successive rovibronic transitions can yield information about relative populations between levels as well as determining rotational and vibrational temperatures. In discharges this information is useful since discharge kinetics do not necessarily preclude local thermal equilibrium population distributions. Diagnostic LIF transitions can be chosen to have a large transition moment between the lower (ground) and excited upper states in order to enhance the probe discrimination against the background plasma-induced emission. The excited upper state is usually weakly populated by the discharge. This minimizes spontaneous emission from nonlaser excited molecules and stimulated emission by the laser thereby providing a true measure of the ground state population. However, one-photon LIF laser probes are absorbed along the length of the probe path. If the probe laser is either a collimated or softly focused beam, the collected fluorescence corresponds to an integration of the lower state population along the beam path. By strongly focusing the laser probe and limiting the photon flux in the probe, it is possible to restrict the contribution of the probe interaction with the discharge to a specific small volume which can be translated around the discharge region. However, saturation and power-broadening effects are enhanced by this technique.

In a recent extension to typical concentration profiling experiments, Hargis and Greenberg[6] have demonstrated a one-photon LIF diagnostic scheme that uses a thin uniform sheet of laser probe light and an orthogonal CCD array detector to image the LIF in thin strips across the entire discharge gap as shown in Fig. 11-3. The advantage here is in being able to quickly measure the relative LIF spatial profile from electrode to electrode provided one is using strongly allowed transitions and has sufficient laser power available.

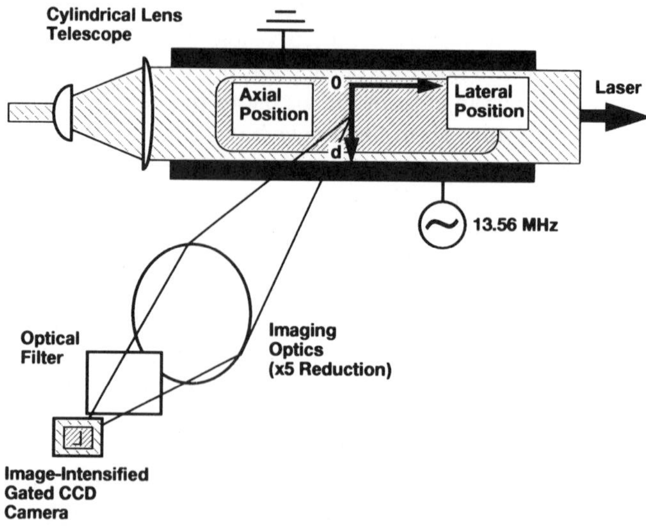

Figure 11-3. Strip LIF laser diagnostic detection. The experimental arrangement for detecting LIF across a reactor gap in thin strips is shown. The laser probe is formed into a uniform collimated thin sheet that bridges the gap. The collection optics can image a thin strip orthogonal to the laser probe onto the entrance slit of a spectrometer which provides wavelength selection. The exit slit output is measured along a diode array, recording the variation in LIF signal across the discharge.

11.3.2. Two-Photon and Multiphoton LIF

Measurements using transitions from ground states to higher electronic states in atoms or molecules usually require more than one photon, since the energy differences between such states would require the use of a single VUV photon which is very difficult to generate and propagate in appreciable quantities. An extension of the basic one-photon LIF technique (see Fig. 11-2b) makes use of nonlinear multiphoton and/or multicolor excitations to the excited upper state. Indeed, the two excitation arrangements are so closely related that often it is important to examine the role of multiphoton processes even for one-photon LIF; particularly since high photon densities caused by focusing of the probe beam can result in nonlinear processes such as multiphoton ionization thereby depleting the upper state population.

Multiphoton processes usually involve much smaller absorption cross sections for excitation than those in one-photon LIF. This fact implies it is much more difficult to ensure a sufficient transition rate to the upper level and hence to achieve a significant fluorescence photon yield that can be detected against the plasma-induced emission. However, from another perspective multiphoton techniques can have a decided advantage over their one-photon counterpart since the typically small absorption cross section means that the laser probe must be focused to a very small spot size in order to reach the high photon densities required for a transition. This means that multiphoton LIF can in practice achieve high spatial resolution (100 μm or less) without the adverse effects of saturation mentioned above. Furthermore, the requirement that the excitation transitions utilize two or more photons implies the possibility, by using particular excitation geometries, of making Doppler-free measurements or of measuring a particular directed velocity component for a species from the Doppler profile.

Currently, the most practical form of diagnostic multiphoton LIF is an excitation involving two photons of a single frequency as shown in Fig. 11-2b. Because the excitation absorption of the photons must be simultaneous, one of the photons is absorbed either to a specific allowed state or to an intermediate virtual state. The atomic selection rule for excitation conserves parity,[(7)] which allows the technique to probe transitions that are normally electric dipole forbidden. The detected fluorescence is emitted from a highly localized volume which gives the technique its excellent spatial resolution.

Two-photon LIF can be applied to measurements initiating from ground states of the lighter atoms such as hydrogen or oxygen or between the ground electronic state and higher electronic states in molecules. As mentioned above, this technique is usually required since a single photon excitation would require a coherent VUV source and increased experimental complications. For multiphoton LIF, photons with tunable wavelengths down to 200 nm can now be easily generated by using nonlinear mixing crystals such as beta-barium borate (BBO)[(8)] or Raman shifting in a hydrogen gas cell.

An alternative technique is two-photon stimulated emission spectroscopy,[(9)] which has been applied to the detection of carbon,[(10)] chlorine,[(11)] and hydrogen[(12)] in flames and cold gases. Here the advantage is in the directionality of the stimulated emission along the probe beam direction. This allows for a high collection efficiency of the stimulated fluorescence either in the forward or backward direction of the line of sight of the probe beam path. However, the relationship between emission intensity and species population is very nonlinear, making it quite difficult to infer species concentrations from these measurements.

11.3.3. Multiphoton Ionization

Absorption of additional photons to very high lying states in an atom or molecule and into the ionization continuum leads to the process of

REMPI (see Fig. 11-2c).[13] In this arrangement the ejected photoelectron is detected upon avalanche multiplication in the discharge as a transient change in the measured discharge impedance correlated in time with the firing of the pulsed excitation laser probe. The data obtained with REMPI can be complicated by a convolution with the position and energy dependence of the avalanche multiplication process making the interpretation of profile measurements essentially intractable.

11.3.4. Optogalvanic Detection

A technique related to REMPI is OGS in which a laser-induced impedance change is observed across the discharge.[14] Typically, upper electronic states of an atom or molecule are excited from either the ground or metastable states with ionization induced by absorption of an additional photon from the probe beam, photons from the discharge, and energetic collisions of the excited state molecule with other plasma species.[15]

The most reliable application of OGS has been for the measure of electric fields by observing Stark effect perturbations and mixing of energy levels in Rydberg atoms and molecules. The influence of Stark effect perturbations can generally be calculated in a straightforward manner,[16,17] so that a comparison of the degree of Stark splitting in a resonance manifold can yield a value of the magnitude of the local electric field strength as well as the electric field vector direction.[18] Since the wavelength location of the components of the Stark split resonance manifold are influenced only by the local electric field at the measurement point, this technique does not suffer the same drawback present with REMPI concentration determinations, because of the convolution of the abundance with the complicated avalanche process. OGS for measurement of electric fields has been demonstrated to have excellent spatial and temporal resolution. Indeed, the temporal response of OGS has been shown to be excellent, approaching the minimum time resolution set by the laser temporal pulse width.[19]

11.3.5. Absorption Techniques

Reliable measurements of species concentrations in reactive plasma systems using laser absorption techniques have been very difficult to achieve in the past. A most notable success of this technique has been the recent detection and measurement of rovibronic population of X $^1\Sigma_g^+$ H_2 in a plasma by VUV laser absorption.[20] The success of this experiment depended on the generation via harmonic techniques of coherent radiation at approximately 112 nm.

Nonetheless, for the most part, the absorption technique has thus far depended on the use of tunable IR lasers for absorption within a ground state energy level manifold of the atom or molecule. The difficulties that exist with this technique are varied. Since reactive plasmas at low pressures and species concentrations are optically thin, the observed fractional absorption tends to be exceedingly small and requires long absorption path lengths, which limits both the dynamic range and minimum detectable concentration for this approach. As the change in absorption is detected in the forward direction along the line of sight of the probe beam, this causes an integration of the species concentration behavior across the discharge necessitating the use of Abel inversions and assumptions that limit the spatial resolution. An additional complication results from the necessity of observing IR transitions which can coincide with a large IR background because of the thermal signature of the discharge and the surrounding environment. Similarly, the previous assumption of a weakly populated upper state may no longer always be valid.

Despite these difficulties, reactive halogen atoms have been detected in plasmas by this technique. Concentrations and translational temperatures for atomic chlorine have been measured in a chlorine glow discharge using diode laser absorption at 882.4 cm^{-1} by Wormhoudt and Stanton.[21] A successful detection measurement for one of the more elusive reactive plasma species, atomic fluorine, was demonstrated by Stanton and Kolb[22] using a tunable lead-salt diode laser to probe a ground state fine structure transition at 404 cm^{-1} (25 μm).

11.3.6. Comparison with Other Diagnostic Methods

We can compare the basic laser-based techniques mentioned above with other standard diagnostic methods such as plasma emission detection and actinometry for species concentration measurements, *ex situ* titration detection of transient species, nonlaser absorption methods, electric probe measurements of charged species, and mass spectrometric analysis of the reactive plasma constituents. Some or all of the following general problems can arise in utilizing these latter techniques. First, there can be a reduction in the precision or resolution of the data; the collected data may not truly reflect the presence or profile distribution of a species or parameter because of interference by other plasma processes. Second, the diagnostic technique itself could be significantly perturbative of the discharge environment. Finally, some methods make measurements *ex situ* to the plasma environment adding further assumptions in order to extrapolate data back into the plasma.

11.3.6.1. Emission Spectroscopy

Direct emission measurements from discharges are extremely useful particularly in conjunction with LIF techniques.[23] Indeed, emission measurements are still the only manner in which to detect some reactive species. Recent research efforts have been directed toward analyzing emission from chlorocarbon,[24] fluorocarbon,[25] and hydrocarbon[26] deposition and etching discharges. Carbon-bearing plasmas have extremely complicated emission spectra, as a large variety of hot intermediate radical and ionic species are formed, many of whose spectra have not been fully interpreted even in less complicated environments.

11.3.6.2. Actinometry

Actinometric concentration measurements are basically an emission-based technique where the concentration of a reactive species $[r]$ is given by

$$[r] = \left(\frac{I_r}{I_i}\right)\left(\frac{\alpha_r}{\alpha_i}\right)[i] \qquad (11\text{-}7)$$

where $[i]$ is the known controlled concentration of an inert trace gas species, such as argon, added to the discharge. I_r and I_i denote emission intensities for specific transitions in the reactive and inert species (normally in the same wavelength region), and α are constants of proportionality between concentration and emission intensity for each species.[27]

Actinometry is primarily used to measure concentrations of light reactive atoms and some radicals. (See Table 11-1.) An important criterion for the successful use of actinometry is that the ratio α_r/α_i must be constant regardless of position within the discharge or discharge conditions so long as the concentration ratio $[r]/[i]$ does not change. Typically, it is assumed that the addition of an inert gas in moderate trace amounts will not perturb the discharge environment. However, this assumption may not always be justified. Heavy inert gas ions will impact on electrode surfaces changing secondary electron and ion emission rates as well as possibly participating

Table 11-1. Various Reactive species r and Insert Gases i That Have Been Used for Actinometric Diagnostics

r	i
F	Ar
O	N_2
Br	
CO	
CO^+	
CF	
CF_2	
Cl	
CCl	

in three-body reactions and thus changing recombination rates or act as excited state quenchers in the bulk of the plasma. Moreover, in practice this technique often has reduced spatial and temporal resolution compared with LIF results.

11.3.6.3. Ex Situ Titration

Downstream chemiluminescent titration is used to detect transient atoms and radicals generated in flowing gas reactive plasmas. The method works by introducing a controlled amount of a chemical reactant downstream of a discharge or in the flowing afterglow. Through a known chemiluminescent reaction, the transient species combines to form an excited state product which may decay with a characteristic emission proportional to concentration or an alternative chemiluminescent reaction initiates at a known concentration after the transient species is nearly or completely scavenged from the flow discharge products.[28]

Titration has been used to detect atomic aluminum,[29] chlorine,[30] fluorine,[31] nitrogen,[28] and oxygen,[32] as well as radicals such as CN.[29] The obvious drawbacks of chemiluminescent titration are the complete loss of spatial concentration profile-dependent information and the possibility of further reactions and interactions with surfaces downstream which can alter the relative populations of species from those in the reactive plasma.

11.3.6.4. Absorption Spectroscopy

Absorption measurements using nonlaser sources, such as discharge lamps, can probe directly for the rotational and vibrational temperature of reactive species. High-resolution Fourier transform infrared (FTIR) techniques are able to measure simultaneously the complicated cold and hot band structures of radicals and molecules present in discharges using broadband sources. FTIR absorption and emission has been applied to silane (SiH_4) discharges to measure the rotational and vibrational temperature of the SiH radical as well as attempting to detect SiH_2 and SiH_3 radicals.[33]

VUV absorption spectroscopy has been used to measure concentrations of ozone, ground state, and metastable molecular oxygen in a dc glow discharge.[34] Absorption techniques generally require sophisticated instrumentation to handle the large amount of convoluted spectral information present in a reactive plasma and, as mentioned with laser-based absorption probes, lack good spatial resolution because of the lengthy absorption paths.

11.3.6.5. Charged Particle Methods

Two related measurements are *in situ* electric or Langmuir probes and quasi-*ex situ* mass spectrometry. Electric probes are used to measure positive ion density, electron temperature, and dc localized space potential in rf discharges.[35,36] In principle, electric probes could provide a wealth of well-resolved spatial and temporal information. However, there are systematic perturbations introduced by the very presence of the electric probe in the plasma environment. These effects make the accurate interpretation of Langmuir probe data difficult.

Mass spectrometry can be performed in very close proximity to a reactive plasma as a very general means of measuring ionic species and with the addition of electron- or photo-ionization neutral molecules can also be detected, often with excellent sensitivity.[37] The primary disadvantage of mass spectrometry lies in its *ex situ* nature, meaning that one is never quite sure that the species in the mass spectrometer are the same as in the plasma itself, with a corresponding loss of spatial and temporally resolved information. Additionally, fragmentation patterns that are not well known, upon ionization, particularly for radical species, lead to ambiguities in species identification.

11.4. Species Detected and Studied

Having reviewed the basic experimental techniques and practices for laser-based plasma diagnostics we now turn to a survey of reactive species and plasma characteristics that have been measured by these techniques. The material presented, while fairly inclusive, is probably not exhaustive but should serve as an excellent introductory survey for familiarizing the researcher with demonstrated measurement capabilities.

11.4.1. One-Photon LIF Detection

A number of reactive plasma species that have been measured with one-photon LIF techniques are listed in Table 11-2, which presents the particular species, the plasma source, the excitation wavelength and transition, the observed fluorescence wavelength and transition, and a recent reference.

Note from Table 11-2 that the majority of the species are radicals which have particular relevance to reactive plasma deposition or etching. As indicated in Section 11-3, absolute concentration measurements for radicals in discharges are both demanding and important for a complete description of these processes. However, current methods of determining instrumental factors can have large uncertainties (and vary from experiment to experiment), making the interpretation of data for absolute measurements difficult. Along with measurements for radicals, several measurements are presented for species that find use in processing discharges for ion impact etching and actinometry.

Table 11-2. Reactive Plasma Species Detected by One-Photon LIF

Species	Source discharge	Excitation wavelength (nm)	Excitation transition[a]	Fluorescence wavelength (nm)	Fluorescence transition	Ref.
CF_2	Pulsed, CF_4	248,266	$\tilde{A}^1A_1$–$\tilde{X}^1A_1$	257–271	A–X	38
CF	Pulsed, CF_4	193	$B^2\Delta$–$X^2\Pi$	194–220	B–X	38
BCl	rf, BCl_3	272	$A^1\Pi$–$X^1\Sigma^+$	272	A–X	39
OH	Afterglow	281–284	$A^2\Sigma^+$–$X^2\Pi$	312	A–X	40
N_2^*	Pulsed, N_2	337	$C^3\Pi_u$–$B^3\Pi g$	380–467	C–B	41
N_2^+	Pulsed, N_2	428	$B^2\Sigma_u^+$–$X^2\Sigma_g^+$	391	B–X	41
F*	$CF_4/O_2/Ar$	690	$^4P_{3/2}$–$^4D_{5/2}$	677	$^4D_{5/2}$–$^4P_{5/2}$	42
Ar*	$CF_4/O_2/Ar$	696	$1s_5$–$2p_2$	727,772	$2p_2$–$1s_3, 1s_4$	42
Cl_2^+	rf, Cl_2	386	$A^2\Pi$–$X^2\Pi$	396	A–X	43
CCl	rf, CCl_4	278	$A^2\Delta$–$X^2\Pi$	278	A–X	44, 45

[a]Following standard convention, for molecules the upper state is listed first; for atoms the convention is reversed, the lower state is listed first.

It should also be noted that many other important species have been detected either downstream from reactive plasmas or in chemical vapor deposition (CVD) systems. CVD can produce reactive species with concentrations comparable to those generated directly in reactive plasmas. The important point is that given a similar concentration in a reactive plasma and the ability to discriminate against the ambient background plasma-induced emission, the LIF detection schemes for these reactive species should be equally as straightforward for direct detection in the plasmas as from other sources.

In addition to the species listed in Table 11-2, LIF detection of rare gas metastable states has been demonstrated for He, Ne, Ar, Kr, and Xe downstream from a dc discharge.(46) Metastable helium has also been detected using optogalvanic techniques in dc discharges.(47) Rare gas atomic metastable detection and profiling in discharges is important because of the considerable energy content such metastables can supply during collisional processes in the bulk plasma and with surfaces. Detection of SiF_2 and CF_2 radicals has been performed during downstream silicon etching from a CF_4/O_2 microwave discharge.(48) The important observation of OH and SiO by LIF from CVD deposition of SiO_2 provides rotational and vibrational temperatures present during the reaction processes.(49)

11.4.2. Two-Photon LIF Detection

Species detected by two-photon allowed LIF (TALIF) results are shown in Table 11-3. The data shown here are presented in the same manner as for Table 11-2, with the addition of a column for the total excitation energy for the two-photon excitation transition. As can be seen, TALIF is well suited

Table 11-3. Reactive Plasma Species Detected by TALIF

Species	Source discharge	Excitation wavelength (nm)	Excitation transition	E_{total} 2-photon (cm^{-1})	Fluorescence wavelength (nm)	Fluorescence transition	Ref.
Cl	rf, $CClF_3$	233	$3p^2P^0$–$4p^4S^0$	85726	725–775	$4p^4S^0$–$4s^4P$	50
H	rf, H_2	205	$1s^2S$–$3d^2D$	97492	656	$3d^2D$–$2p^2P$	51
N	Afterglow	211	$2p^{34}S^0$–$2p^23p^23p^4D^0$	94787	869	$2p^23p^4D^0$–$2p^23s^4P$	52
O	rf, CF_4O_2	226	$2p^{43}P$–$3p^3P$	88495	845	$3p^3P$–$3s^3S$	19

to concentration measurements for light reactive atoms. Indeed, light atoms are perhaps the dominant reactive species to interact with surfaces on the boundary of the discharge. Thus, their detection and profiling in plasma processing systems is critical to a detailed interpretation of the plasma–surface interactions. The TALIF measurements for these atoms require the use of two UV photons typically near the short-wavelength generation limit of most tunable dye laser systems and as such are more demanding experimental techniques than those used for one-photon LIF.

As with the one-photon LIF measurements discussed above, several atomic species important to reactive plasmas have been detected by TALIF, though not as yet directly in plasmas. Multiphoton dissociation of molecules and the subsequent detection of an atomic fragment by TALIF has been exhibited for carbon,[53] silicon,[54] and sulfur.[55] Given the caveats mentioned for the *ex situ* species detected by one-photon LIF, the TALIF detection and profiling of these atoms directly in reactive plasmas should be possible.

11.4.3. OGS and REMPI Detection

In a plasma, it is sometimes difficult to distinguish between OGS and REMPI processes. Let us refer to these processes in general as laser-ion detection, since they both involve excitation of a bound–bound transition in an atom or molecule followed by detection of charged particles, as a change in the discharge current.

Perhaps the most direct use of laser-ion techniques was the pioneering work on N_2^+ and CO^+ in an rf plasma by Walkup *et al.*[56] These experiments were made possible by the fact that the charge-exchange cross section was a strong function of the electronic state of the ion. Thus, by laser pumping the ions from one state to another, a significant change in the discharge current or impedance could be created and detected. The technique was particularly sensitive in the cathode dark space.

In the case of molecular hydrogen, recent work has shown the possibility of detecting highly vibrationally excited, ground electronic state H_2 by 2 + 1

REMPI.[57,58] Hydrogen-bearing plasmas have very nonthermal vibrational level population distributions for H_2. Analysis of these expected high-lying vibrational population distributions has shown that such distributions can be an appreciable conversion source of vibrational to translational energy during collisions,[59] as well as perhaps the predominant source of the negative ion, H^-, from dissociative attachment.[60] Detection of excited electronic states by REMPI has been demonstrated for H_2.[61]

11.4.4. Plasma Characteristic Measurements by Laser-Based Techniques

Laser-based diagnostic probes are well suited for measuring plasma characteristics other than atomic and molecular species concentrations. Some of these characteristics are localized electric fields, negative ion densities, atomic or molecular velocity vectors via Doppler profiles, and microscopic particulate contamination.

11.4.4.1. Electric Field Strength

Laser-based diagnostics of electric fields have been performed using both LIF and optogalvanic detection. Both techniques rely on measuring Stark effect perturbations to molecular energy levels due to the local electric field.[62] The Stark effect can perturb the coulombic interactions in either an atom or molecule thereby shifting energy levels and causing mixing of electronic wave functions. The alterations to the wave functions lead to changes in transition probabilities between states. Thus, the Stark effect both induces energy shifts in allowed transitions and permits normally forbidden transitions to occur. The utility of detecting these Stark effect perturbations is that one can generally calculate these effects as a function of electric field. The predictions of the calculated model can then be fitted to the real data in an iterative fashion as a function of electric field. Both LIF and OGS techniques generate data showing wavelength shifts and intensity modulations as a function of the local electric field at the probe point. Analysis of influences of the excitation source polarization on the perturbed transitions is important and can yield further confirmation of the electric field magnitude and even vector direction.

LIF-based detection of electric fields has been very successfully demonstrated in electronegative BCl_3 rf etching discharges.[63] Here, Stark effect mixing of rotational wave functions with opposite parity in diatomic BCl permits forbidden transitions between two electronic states which are detected by LIF. The method can measure electric fields as small as 40 V/cm to 1–2 kV/cm with a spatial resolution of 100 μm along an axis orthogonal to the planar electrode surfaces. Since the excitation laser is repetitively pulsed, the diagnostic can be phase synchronized with the rf driving voltage across the discharge with a temporal resolution of 20 ns, thus allowing time-resolved measurements of the electric field temporal profile.[(64)]

A comparable technique to LIF measurements of electric fields is the method of Rydberg state Stark OGS.[(65)] High-lying Rydberg states of atoms or molecules exhibit a pseudohydrogenic behavior under the influence of the Stark effect; in other words, Rydberg molecules can be modeled as a singly charged positive ion core with a single valence electron in an extended orbit that is perturbed by the electric field in the same manner as hydrogen.[(62)] Usually excited intermediate or metastable states of atoms or molecules which can be populated by discharge kinetics are used as a starting point in the excitation scheme. A pulsed probe laser beam excites the metastable species to a perturbed Rydberg level. Again the excited Rydberg states are mixed as a function of the electric field so that line positions and intensities relative to the unperturbed lower state can be compared with calculated values. Further excitation of this level to the ionization continuum is provided by additional photons, collisions with other species, or field ionization. At this point the transient change in the discharge impedance is measured as an optogalvanic signal incorporated in the discharge current waveform.

Rydberg state OGS has been shown to be capable of measuring very small electric fields from 5 V/cm to fields over 5 kV/cm. As with LIF measurements, the technique has excellent spatial resolution (100 μm). In addition, analysis of forbidden transition intensity patterns as a function of excitation probe laser polarization can determine the electric field vector direction.[(65)] Furthermore, these measurements have been performed in a time-resolved fashion as a function of rf driving voltage phase.[(66)]

11.4.4.2. Photodetachment and Negative Ions

Negative ion densities have been measured by laser photodetachment techniques. Determination of negative ion production and kinetics in reactive plasmas has considerable current interest.[(67)] Negative ions in electronegative discharges used for etching of materials can reach high densities comparable to those for electrons. The role of negative ions in many reactive plasmas is not well understood, particularly since calculations of production rates often fall short of measured values.[(68,69)]

Photodetachment of electrons from negative ions usually require only a single relatively low-energy photon absorption. Once liberated in the discharge, the laser-induced electron current is detected either by optogalvanic detection across the discharge[(70)] or by measurements made from a metal electrode placed very near the laser probe interaction region.[(71)] Negative ion densities down to 10^8 cm^{-3} have been measured as well as electron affinities for various molecules and radicals[(72)] in reactive plasmas and negative ion kinetic energy.[(71)]

11.4.4.3. Energy Content by Doppler Profiles

The translational energies of reactive species in discharges can be determined by careful analysis of line-broadening effects cause by relative Doppler motion along a probe direction. Photodissociation of molecules by lasers and subsequent detection of molecular fragments by LIF or mass spectrometry has been used to determine bond dissociation energies.[73] Likewise, similar analysis of LIF measurements on molecular or atomic fragments in discharges yields translational energies for the species along a particular direction.[74] A full analysis of line shapes must take into account possible power-broadening effects and the contribution of the laser line profile to the data.

11.4.4.4. Particulates

Finally, recent advances have been made in laser scattering measurements for detection of microscopic scale particulate contamination in discharges. Significant amounts of particulate contaminates are formed from etching or sputtered products from the bounding surfaces or from gas-phase clustering. Their presence can adversely affect the production and performance of electronic devices and novel materials. Laser scattering can be used to determine the density profiles and size of the particulates.[75] Measurements using two-photon LIF on neutral atomic species, which have been formed from photodetached electrons from the particulates, indicate the particulates are negatively charged and can be suspended near sheath boundaries.[76]

11.5. Conclusions and Future Trends

The overall conclusion concerning laser-based diagnostics is very positive, portending increased use of this technique in the future. This prediction is based on several facts. The number of plasma species for which laser-based detection and characterization methods now exist is substantial as evidence by Tables 11-2 and 11-3. In every likelihood, it will continue to grow. Furthermore, the perception that laser-based diagnostics are too complicated and expensive is eroding. The large number of successful laser-based diagnostics that have been reported, some extending to the production line, belie most of the negative arguments concerning their complexity. Similarly, while the cost of the lasers is not low, this cost must be weighted against the much greater cost, both capital and operating, of plasma processing in general. Viewed in this manner, perhaps the better question is not who can afford laser-based diagnostics, but who can afford *not* to have laser-based diagnostics.

Predicting future technical advances for laser-based diagnostics is like all prognostications—uncertain at best. Clearly additional specific spectral applications will be found to monitor important intermediates. Probably the strongest driving force in this respect will not be spectroscopic advances, but new processes themselves, which require the identification and monitoring of different species.

Another growth area for laser-based diagnostics is their extension beyond the traditional analytical applications, qualitative and quantitative analysis of plasmas in terms of species and concentrations. It has recently been recognized that laser-based diagnostics offer nonperturbative probes for other characteristics of plasma. Examples of quantifiable entities include electric field measurements from Stark effects, kinetic energy or local temperature measurements from Doppler effects, and correlation of microscopic particulate formation with negative ion detection. One would expect to see further developments involving measurements of other plasma-related properties via their effect on the spectral transitions involved in laser-based diagnostics.

Acknowledgment

The authors gratefully acknowledge the support of this work by USAF Contract No. F33615-89-C-2921.

References

1. S. Veprek, *Plasma Chem. Plasma Process.* 9 Suppl. (1989) 31S.
2. B. Chapman, *Glow Discharge Processes.* Wiley, New York, 1980.
3. R. K. Janev, W. D. Langer, K. Evans, and D. E. Post, *Elementary Processes in Hydrogen–Helium Plasmas.* Springer-Verlag, Berlin, 1987.
4. A. Corney, *Atomic and Laser Spectroscopy*, Clarendon Press, Oxford, 1988.
5. W. Demtroder, *Laser Spectroscopy*, Springer-Verlag, Berlin, 1982.
6. P. J. Hargis and K. E. Greenberg, 42nd Annual Gaseous Electronics Conference, Palo Alto, 1989.
7. K. D. Bonin and T. J. McIlrath, *J. Opt. Soc. Am. B* 1 (1984) 52.
8. C. Chen, *Laser Focus World* 129 (Nov. 1989).
9. M. Alden, U. Westblom, and J. E. M. Goldsmith, *Opt. Lett.* 14 (1989) 305.
10. M. Alden, P.-E. Bengtsson, and U. Westblom, *Opt. Commun.* 71 (1989) 263.
11. A. D. Sappey and J. B. Jeffries, *Appl. Phys. Lett.* 55 (1989) 1182.
12. J. E. M. Goldsmith, *J. Opt. Soc. Am. B* 6 (1989) 1979.
13. E. W. Rothe, G. S. Ondrey, and P. Andersen, *Opt. Commun.* 58 (1986) 113.
14. J. E. M. Goldsmith and J. E. Lawler, *Contemp. Phys.* 22 (1981) 235.

15. M. Maeda, Y. Nomiyama, and Y. Miyazoe, *Opt. Commun.* 39 (1981) 64.
16. P. H. Heckmann and E. Trabert, *Introduction to the Spectroscopy of Atoms*, Chapter 11, North-Holland, Amsterdam, 1989.
17. C. Delsart, L. Cabaret, C. Blondel, and R.-J. Champeau, *J. Phys. B* 20 (1987) 4699.
18. B. N. Ganguly and A. Garscadden, *Phys. Rev. A* 32 (1985) 2544.
19. C. E. Gaebe, T. R. Hayes, and R. A. Gottscho, *Phys. Rev. A* 35 (1987) 2993.
20. G. C. Stutzin, A. T. Yang, A. S. Schlacter, K. N. Leung, and W. B. Kunkel, *Chem. Phys. Lett.* 155 (1989) 475.
21. J. Wormhoudt and A. C. Stanton, *J. Appl. Phys.* 61 (1987) 142.
22. A.C. Stanton and C. E. Kolb, *J. Chem. Phys.* 72 (1980) 6637.
23. R. E. Walkup, K. L. Saenger, and G. S. Selwyn, *J. Chem. Phys.* 84 (1986) 2668.
24. P. E. Clarke, D. Field, and D. F. Klemperer, *J. Appl. Phys.* 67 (1990) 1525.
25. H. Kawata, Y. Takao, K. Murata, and K. Nagami, *Plasma Chem. Plasma Process* 8 (1988) 189.
26. T. Kokubo, F. Tochikubo, and T. Makabe, *J. Phys. D* **22** (1989) 1281.
27. J. W. Coburn and M. Chen, *J. Appl. Phys.* 51 (1980) 3134.
28. S. Meikle and Y. Hatanaka, *Appl. Phys. Lett.* 54 (1989) 1648.
29. S. Meikle, H. Nomura, Y. Nakanishi, and Y. Hatanaka, *J. Appl. Phys.* 67 (1989) 483.
30. D. A. Daner and D. W. Hess, *J. Appl. Phys.* 59 940 (1985)
31. K. Ninomiya, K. Suzuki, S. Nishimatsu, and O. Okada, *J. Appl. Phys.* 58 (1985) 1177.
32. C. Vinckier, P. Coeckelberghs, G. Stevens, M. Heyns, and S. De Jaegere, *J. Appl. Phys.* 62 (1987) 1450.
33. J. C. Knights, J. P. M. Schmitt, J. Perrin, and G. Guelachvili, *J. Chem. Phys.* 76 (1982) 3414.
34. G. Gousset, P. Panafieu, M. Touzeau, and M. Vialle, *Plasma Chem. Plasma Process* 7 (1987) 409.
35. A. von Engel, *Electric Plasmas: Their Nature and Uses*, Taylor & Francis, London, 1983.
36. H. Sabaldi, S. Klagge, and M. Kammeyer, *Plasma Chem. Plasma Process* 8 (1988) 425.
37. J. E. Cantle, E. F. Hall, C. J. Shaw, and P. J. Turner, *Int. J. Mass Spectrom. Ion Phys.* 46 (1983) 11; J. W. Coburn, *Thin Solid Films* 171 (1989) 65.
38. S. G. Hanson, G. Luckman, and S. D. Colson, *Appl. Phys. Lett.* 53 (1988) 1588.
39. C. A. Moore, G. P. Davis, and R. A. Gottscho, *Phys. Rev. Lett.* 52 (1984) 538.
40. N. G. Adams, C. R. Herd, and D. Smith, *J. Chem. Phys.* 91 (1989) 963.
41. R. F. Wuerker, L. Schmitz, T. Fukuchi, and P. Straus, *Chem. Phys. Lett.* 150 (1988) 443.
42. S. G. Hansen, G. Luckman, G. C. Nieman, and S. D. Colson, *Appl. Phys. Lett.* 56 (1990) 719.
43. V. M. Donnelly, D. L. Flamm, and G. Collins, *J. Vac. Sci. Technol.* 21 (1982) 817.
44. R. A. Gottscho, R. H. Burton, and G. P. Davis, *J. Chem. Phys.* 77 (1982) 5298.
45. R. A. Gottscho, G. P. Davis, and R. H. Burton, *J. Vac. Sci. Technol. A* 1 (1983) 622.
46. V. E. Bondybey and T. A. Miller, *J. Chem. Phys.* 66 (1977) 3337.
47. B. N. Ganguly, *J. Appl. Phys.* **60** (1986) 571.
48. Y. Matasumi, S. Toyoda, T. Hayashi, and M. Miyamura, *J. Appl. Phys.* **60** (1986) 4102.
49. P. Van der Weijer and B. H. Zwerver, *Chem. Phys. Lett.* 163 (1989) 48.
50. G. S. Selwyn, L. D. Baston, and H. H. Sawin, *Appl. Phys. Lett.* 51 (1987) 898.
51. B. L. Preppernau, D. A. Dolson, R. A. Gottscho, and T. A. Miller, *Plasma Chem. Plasma Process.* 9 (1989) 157. See also B. L. Preppernau and T. A. Miller, *J. Vac. Sci. Technol. A.* 8 (1990) 1673.
52. W. K. Bischel, B. E. Perry, and D. R. Crosley, *Chem. Phys. Lett.* 82 (1981) 85.
53. P. Das, G. Ondrey, N. van Veen, and R. Bersohn, *J. Chem. Phys.* 79 (1983) 724.
54. P. D. Brewer, *Chem. Phys. Lett.* 136 (1987) 557.
55. P. Brewer, N. van Veen, and R. Bersohn, *Chem. Phys. Lett.* 91 (1982) 126.

56. R. Walkup, R. W. Dreyfus, and P. Avouris, *Phys. Rev. Lett.* 50 (1983) 1846.
57. D. C. Robie, L. E. Jusinski, and W. K. Bischel, *Appl. Phys. Lett.* 56 (1990) 722.
58. W. Meier, H. Zacharias, and K. H. Welge, *Chem. Phys. Lett.* 163 (1989) 88.
59. M. Cacciatore, M. Capitelli, and G. D. Billing, *Chem. Phys. Lett.* 157 (1989) 305.
60. K. N. Leung and W. B. Kunkel, *Phys. Rev. Lett.* 59 (1987) 787.
61. E. E. Marinero, R. Vasudev, and R. N. Zare, *J. Chem. Phys.* 78 (1983) 692.
62. H. A. Bethe and E. E. Salpeter, *Quantum Mechanics of One and Two Electron Atoms*, Plenum Press, New York, 1977.
63. C. A. Moore, G. P. Davis, and R. A. Gottscho, *Phys. Rev. Lett.* 52 (1984) 538.
64. M. L. Mandich, C. E. Gaebe, and R. A. Gottscho, *J. Chem. Phys.* 83 (1985) 3349.
65. B. N. Ganguly, J. R. Shoemaker, B. L. Preppernau, and A. Garscadden, *J. Appl. Phys.* 61 (1987) 2778.
66. B. L. Preppernau and B. N. Ganguly, Paper KB-5, 39th Annual Gaseous Electronics Conference, 1986.
67. R. A. Gottscho and C. E. Gaebe, *IEEE Trans. Plasma Sci.* PS-14 (1986) 92.
68. M. Bacal and G. W. Hamilton, *Phys. Rev. Lett.* 42 (1979) 1538.
69. A. Garscadden and W. F. Bailey, Paper No. 188, 12th Int. Symp. Rarefied Gas Dynamics, 1980.
70. C. E. Gaebe, T. R. Hayes, and R. A. Gottscho, *Phys. Rev. A* 35 (1987) 2993.
71. D. Devynck, A. M. Bacal, P. Berelmont, J. Bruneteau, R. Leroy, and R. A. Stern, *Rev. Sci. Instrum.* 60 (1989) 2873.
72. K. E. Greenberg, G. A. Hebner, and J. T. Verdeyen, *Appl. Phys. Lett.* 44 (1984) 299.
73. B. Koplitz, Z. Xu, and C. Wittig, *Chem. Phys. Lett.* 137 (1987) 505.
74. J. Dunlop, A. Tserepi, B. L. Preppernau, T. Cerny, and T. A. Miller, *Plasma Chem. Plasma Process.* 12 (1992) 1.
75. G. S. Selwyn, J. Singh, and R. S. Bennett, *J. Vac. Sci. Technol. A.* 7 (1989) 2758.
76. G. S. Selwyn, J. S. McKillop, and K. L. Haller, *Proc. SPIE Integ. Process.* (J. Bondur and A. Reinburg, eds.), Santa Clara, Calif., 1989.

Index